High-Pressure and
Low-Temperature Physics

High-Pressure and Low-Temperature Physics

Edited by

C. W. Chu
University of Houston
Houston, Texas

and

J. A. Woollam
NASA
Lewis Research Center
Cleveland, Ohio

PLENUM PRESS · NEW YORK AND LONDON

Library of Congress Cataloging in Publication Data

International Conference on High-Pressure and Low-Temperature Physics, Cleveland
    State University, 1977.
    High-pressure and low-temperature physics.

    Includes index.
    1. Superconductivity—Congresses. 2. Solids, Effect of radiation on—Congresses.
3. Materials at high pressures—Congresses. 4. Materials at low temperatures—Con-
gresses. I. Chu, Ching-wu, 1941-   II. Woollam, J. III. Title. IV. Title: High-
Pressure and Low-Temperature Physics.
QC612.S8I49 1977   537.6′23   78-7290
ISBN-13:978-1-4684-3353-1   e-ISBN-13:978-1-4684-3351-7
DOI:10.1007/978-1-4684-3351-7

Proceedings of the High-Pressure and Low-Temperature Physics
International Conference held at Cleveland State University,
Cleveland, Ohio, July 20—22, 1977

© 1978 Plenum Press, New York
Softcover reprint of the hardcover 1st edition 1978

A Division of Plenum Publishing Corporation
227 West 17th Street, New York, N.Y. 10011

Preface

High pressure science is a rapidly growing diverse field. The high pressure technique has become a powerful tool for both the study and preparation of materials. In spite of the many high pressure conferences held in recent years, I felt that there was a need for scientists within a well-defined area (not bound merely by the common experimental technique) to meet in an atmosphere conducive to frank exchange and close interaction. In this spirit, the Cleveland State University hosted such a conference from July 20 to 22, 1977, in which the physics of solids under high pressures and at low temperatures was specifically examined. Both the original and review papers presented at the conference and the candid discussions following their presentations appear in this volume. They clearly cover a rather complete spectrum of current research in the physics of solids at high pressures and low temperatures.

I wish to thank the National Aeronautics and Space Administration, the Office of Naval Research and the National Science Foundation for their financial support of the conference. In addition, I wish especially to thank Steinar Huang for his unceasing assistance in arranging this conference. I also wish to thank him and Francis Stephenson for their assistance in preparing this book.

C.W. Chu,
Chairman,
International Conference on
High Pressure and Low Temperature Physics

PREFACE

High pressure research, in a rapidly growing... the high pressure... physics...

...Antwerp...

Antwerp

Contents

CONFERENCE REVIEW
(Chairman: C.W. Chu)

PROSPECTS FOR METALLIC HYDROGEN

Arthur L. Ruoff

Department of Materials Science and Engineering
Cornell University
Ithaca, N.Y. 14853

ABSTRACT

The theoretical predictions of the transition of hydrogen from
an insulating molecular crystal to an electrical conducting alkali
metal are reviewed in the light of Hugoniot experiments carried
out to 0.9 Mbars and isentropic experiments carried out to a
claimed 8 Mbars. This isentropic experiment provides no evidence
for the existence of a first order transition contrary to earlier
claims.

If a pair potential model is accurate and if available isentropic and Hugoniot experiments are accurate, the transition to an alkali metal phase is quite high (several megabars). However, there
are theoretical predictions (which appear to have a sounder basis
than the pair potential model) which predict that molecular hydrogen itself becomes metallic at modest pressures; this result by
itself suggests that the pair potential model is inadequate. New
results of Ashcroft based on the behavior of metallic molecular hydrogen suggest that a first order transition may be possible at
modest pressures (near one megabar).

The state of static experimental research is reviewed. In general the pressure in such experiments has been overestimated. The
evidence of production of metallic hydrogen is, at this time, very
weak, and methods are suggested to test these claims.

Finally, the strength of perfect diamond crystals is considered
and shown to be very high. Experiments by Ruoff and Wanagel, in

which static contact pressures of 1.4 Mbars have been reached in
a contact circle having a radius of one micrometer, validate these
predictions. It is concluded that pressures as high as 3 - 5 Mbars
may be achieved in extremely small contact regions without yielding
and that, moreover, scientific observations will be possible in
those tiny regions.

INTRODUCTION

Two useful reviews of molecular and metallic hydrogen are the
thesis of Sensenig[1] and the comprehensive report of Ross and
Shishkevish.[2]   Calculated pressures for the assumed first order
transition to an alkali metal fall in the range 0.71 to 20 Mbars.

To compute a transition pressure, we need have only the
Helmholtz free energy, F, versus volume, V, for the two phases.
Then, as shown in Figure 1, we can obtain the common tangent
(if one exists) and the negative slope of this line is the transition pressure. If in fact we have the free energies at the minima at a given temperature and we have sufficient P - V data at
that temperature for each phase, the F - V curves can be constructed.
[Alternately the Gibbs free energy versus pressure curves could be
constructed for the two phases and the intersection (if it exists)
provides the transition pressure.] Our knowledge of the Helmholtz
free energy curves is limited. We accurately know the Helmholtz
free energy at the minimum (and up to 25 kbar) for the molecular
phase. The remainder of the molecular phase is not so well defined and the entirety of the metallic phase is hypothetical. We
note before proceeding that even a small change in an equation of
state of either phase can result in a large change in the transition pressure. The equations of state for both phases must be
reasonably accurate over the _entire_ range P = 0 to P_t, if an accurate transition pressure is to be obtained. Moreover, the minima in the free energy curves of the two phases much be known accurately. How well these free energies can be drawn and how accurately the phase transition pressure can be predicted is reviewed
in the section on theory.

On the experimental side, there have been:

1. Claims of pressures of over 500 kbars in uniaxial supported
    opposed cobalt cemented tungsten carbide anvil devices;
2. Claims of pressures over one Mbar in uniaxial opposed single
     crystal diamond devices;
3. Claims of 1 - 4 Mbars with cone-sphere carbonado indentors on
    carbonado anvils;
4. Claims of pressures of over one Mbar in a split sphere device.
5. Claims of producing metallic hydrogen.
    The accuracy and legitimacy of these claims will be reviewed.

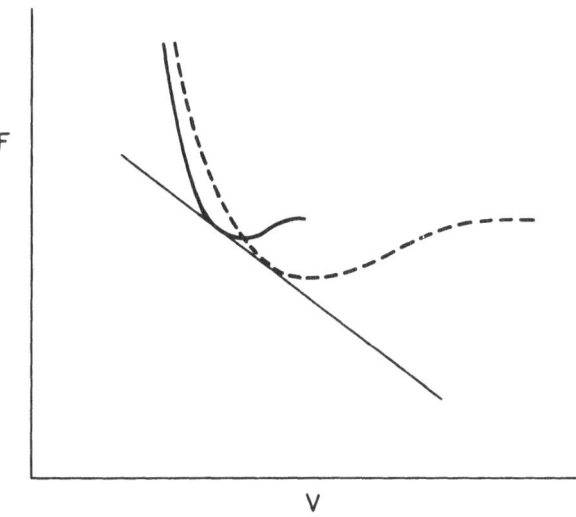

F

V

Figure 1.   Schematic of free energy versus volume curves.   Dashed
line show form which is stable at high pressures.

THEORETICAL REVIEW

In the case of molecular hydrogen we do know the free energy at
the minimum; we have an accurate experimental equation of state for
the solid to 25 kbars[3] and an approximate experimental Hugoniot equation of state for the liquid to 900 kbars[4] and a still less accurate
experimental isentropic equation of state for the liquid which may
in fact extend to 8 Mbars.[5] In the latter two cases the temperature
is very high. These data, of course, form the basis for judging the
development of the theoretical aspects of the molecular phase. An
accurate or rigorous solution is not available at the present time.
Various modeling approaches are used of necessity. In particular
the pair interaction method is used. There are several difficulties;
these are:

1.   Obtaining the appropriate pair potential;
2.   Accounting for the orientation dependence;
3.   Proving that the presence of a third molecule (or many others)
     does not affect the pair potential, i.e., proving that the pair
     potential concept has validity, or at least modifying it to in-
     clude many body effects.
This paper includes only the minimum discussion of these matters
which the author deems necessary. The discussions in the two review articles given earlier[1,2] provide much fuller details.

The pair potential at small separation corresponding to the
very high pressure region (megabars) can be calculated by the con-

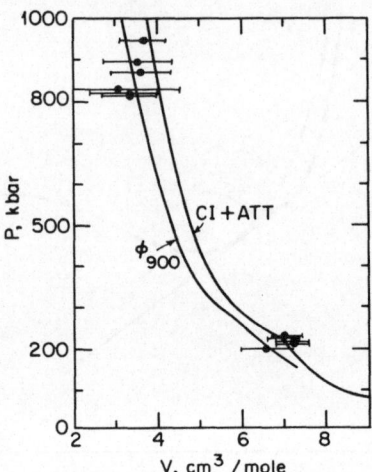

Figure 2.   40 kbar Hugoniot data of Van Thiel and Alder.[9] Calculated curves based on hard sphere fluid perturbation
theory after Ross and Shishkevish.[2]

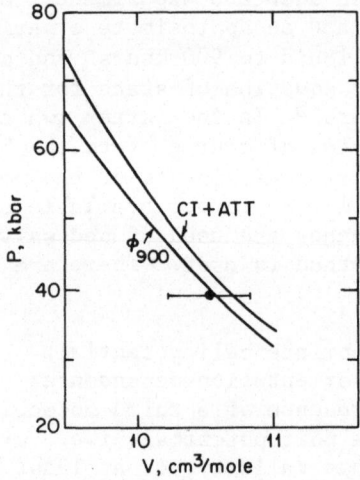

Figure 3.   900 kbar Hugoniot data of Van Thiel et al.[4] Calculated
curves based on hard sphere fluid perturbation theory
after Ross and Shishkevish.[2]

figuration interaction method.[6] McMahan and colleagues[7] believe that such calculations are correct to within better than 10%. One procedure is to add to this potential an attractive multipole potential (which is accurate at quite large separation) multiplied by a negative exponential scaling function as suggested by Trubitsyn[8] to give a resultant potential which, following the convention of Ross and Shishkevish, we call CI + ATT (for configurational interaction plus two attractive terms). This potential is

$$\phi(R) = 7.5e^{-1.69R} - (13/R^6 + 116/R^8)e^{-400/R^6} \qquad \text{(CI + ATT)} \qquad (1)$$

Strictly speaking the computed CI potential should include all attractive terms; however, there is some evidence that adding the attractive terms is approximately equivalent to adding many-body effects to the CI pair potential (see Figure 8 of Reference 2). We will also discuss a potential which we call the ϕ_{900} potential (because, as we shall see, it best fits the shock data at 900 kbars)

$$\phi_{900} = 1.555 \, e^{-1.495 \, R} \qquad (2)$$

Comparison with shock data will now be made. The theoretical shock Hugoniots for the liquid were computed by Ross using hard-sphere fluid perturbation theory.[2] The experimental result in Figure 2 is by Van Thiel and Alder[9] and in Figure 3 by Van Thiel et al.[4]

We note that the ϕ_{900}-potential gives the better fit to the 40 kbar point in Figure 2, and a considerably better fit to the 900 kbar points in Figure 3 while the CI + ATT-potential gives a better fit to the 200 kbar points.

Figure 4 shows the $0^{\circ}K$ isotherms for the two potentials used here. We note that at pressure above 1 Mbar, the volume difference between the two potentials at a fixed pressure is about 0.5 cm^3/mole. We will return to this point later.

Figure 5 shows the data points of Grigor'ev et al. There was a time when this data was interpreted as proof that metallic hydrogen had been made.[10] As is obvious the data is fitted by a single smooth curve. Detailed statistical analysis reveals that within 97% confidence limits, this data (taken by itself) does not reveal the presence of a first order transition. Stated in another way, there is no acceptable evidence here for a first order transition. Ross has in fact noted earlier that a straight line fit would also be acceptable.[2] [The data do not exclude the possibility of a transition; rather, they do not prove that one exists.]

Because of the absence of a complete description of the experimental procedure and the computer code used for attaining pressure,[5]

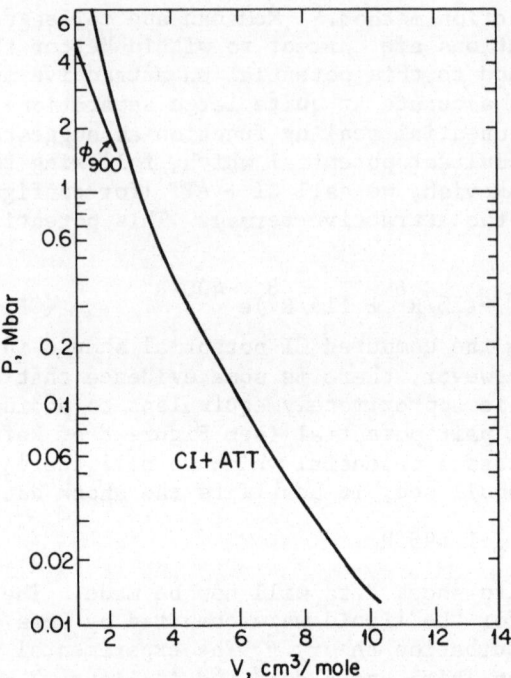

Figure 4.   0°K isotherms for two potentials.

it is impossible to form a judgment on the accuracy of this data.
The authors themselves[4] put no error limits on the pressures; it
is likely that these are at least as large as the error limits on
the density. It should be noted that the general form of the curve
drawn by the present author is that expected for an isentropic
equation of state curve in general. We note that

$$dP = -K \frac{dV}{V} = K \frac{d\rho}{\rho} \tag{3}$$

where

$$K = K_o + K_o' P + K_o'' \frac{P^2}{2} + \cdots . \tag{4}$$

For hydrogen at high pressure the K_o term is negligible. In general
$K_o K_o'' \overset{\sim}{K_o'} -10,$[11] and also K^t is a quantity which decreases from K_o'
to $K_\infty'.$[11] Hence we may write

$$K = \overline{K}^t P \tag{5}$$

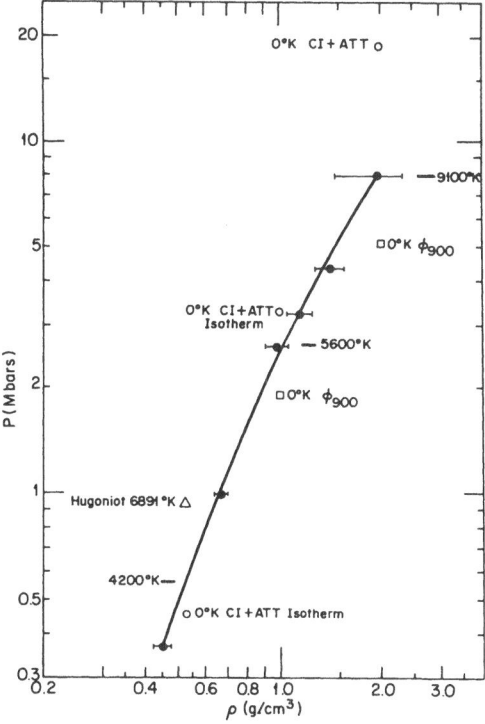

Figure 5. Isentropic data points of Grigor'ev et al.[5] Solid curve by present author.

where

$$\overline{K}' = K_o^\bullet + \frac{K_o'' P}{2} + \cdots .$$ (6)

Then (3) becomes

$$\frac{dP}{P} = \overline{K}' \frac{d\rho}{\rho}$$ (7)

or

$$\frac{d \ln P}{d \ln \rho} = \overrightarrow{K}'$$ (8)

Thus the slope of the curve in Figure 5 decreases as pressure increases as is expected.

The data of Grigor'ev et al.[5] is considered further in Figure 6. Also shown are other theoretical and experimental points of interest. We note the Hugoniot point lies considerably above the isentrope as expected because of the difference in temperature. Also shown are the $0^\circ K$ pressure points at $\rho = 1$ and 2 g/cm^3 for the

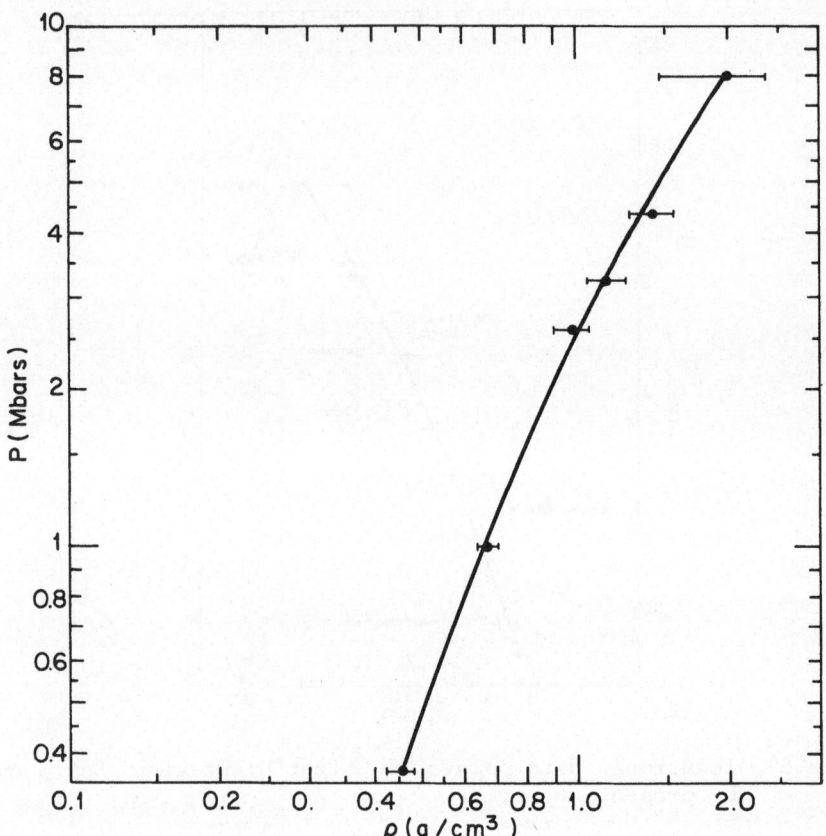

Figure 6.   Isentropic data points of Grigor'ev et al.[5] including
            various temperatures given by them at specific densities.
            Also shown at the same densities are a Hugoniot point[4]
            and $0°K$ points for the CI+ATT potential and for the
            ϕ_{900} potential.

two pairwise potential curves under consideration here.   Note that
these points would be moved to substantially higher pressures as a
result of thermal contributions.   It is clear that the CI+ATT pair
potential does not agree with the data.   If:

a.   the data are assumed correct,
b.   and the hard-sphere fluid perturbation method is accurate,
c.   and a pair potential model gives an accurate description,
then we would be forced to conclude that:

1.   the CI + ATT pair potential is too hard;

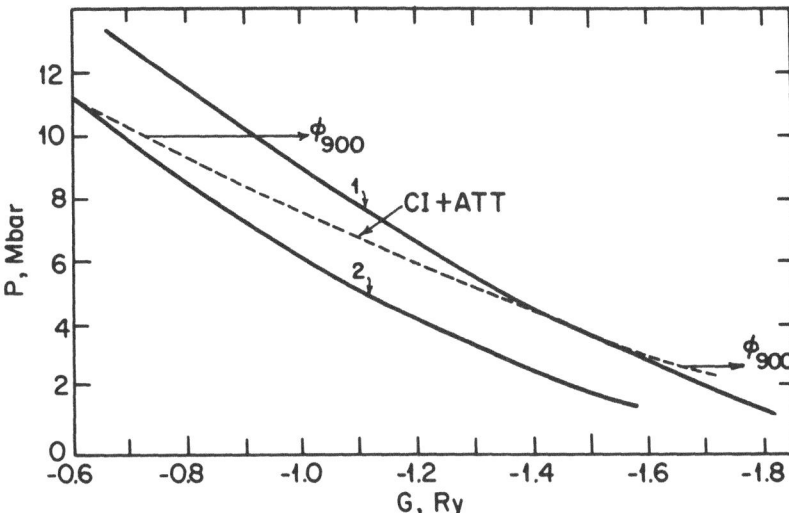

Figure 7.   Gibbs free energy plot versus pressure. Curve is based
            on the third order free electron perturbation theory
            of Hammerburg and Ashcroft with free electron correla-
            tion energy. Curve 2 is based on the same theory with
            the correlation energy set to zero. The dashed line
            corresponds to the molecular phase based on the CI+ATT
            potential. Arrows show (schematically) how points on
            this curve move if ϕ_{900} potential is used instead.

2. the ϕ_{900} potential is approximately correct;
3. no transition exists over the pressure range shown in the
    isentropic experiment;
4. the transition may not even occur in this pressure range in an
    isothermal experiment.
The latter point will be expanded on shortly.

    Figure 7 shows the Gibbs free energy versus pressure computed
for the CI + ATT potential for the molecular phase and for the metallic phase computed by the free electron third order perturbation
method.[12] The curves intersect at 3.8 Mbar. In this regard we
note that the detailed method used for the metallic phase (be it
the third order free electron perturbation method, augmented plane
wave method, linear combination of atomic orbitals method or the
Wigner-Seitz method) lead to nearly the same result. Recently,
theorists have suggested that the correlation energy computed on
a free electron basis is too large.[13] In the extreme case if we
assume the correlation energy is absent altogether, the second
metallic curve in Figure 7 is obtained. The curves now intersect

at 11 Mbars. The actual P vs G curve for the metallic phase probably lies closer to the free electron case.

We have already noted earlier that the ϕ_{900} potential gives a
better fit to the experimental data (we again caution that the
data and the analysis of it and the use of a pair potential description may be incorrect). We noted earlier that at pressures
above about 1 Mbar, the volume difference at a given pressure was
about 0.5 cm^3/mole. If the Gibbs free energy is computed using
the ϕ_{900} potential the points move as shown by the arrows in Figure 7. Consequently with the ϕ_{900} potential the molecular phase
is stable relative to the metal (the alkali metal form of hydrogen) at all the pressures shown. What this means (in the extreme
case and if the theoretical and experimental input is correct)
is that the U − V curve for the molecular phase may envelope the
U − V curve for the metallic phase and that no common tangent exists and that there will be no transition to the metallic phase.
In the less extreme case the U − V curve for the molecular phase
is simply very soft so the transition pressure is very high. Such
soft U − V curves appear to be a necessary consequence of assuming
a pair potential. It is therefore extremely important that we examine this assumption.

The pair potential model assumes that solid molecular hydrogen
is a collection of hydrogen molecules [each with a pair of two protons and two localized (in the molecule) electrons] and that pairs
of these molecules interact with each other in exactly the same
manner regardless of whether other molecules are nearby or not.
Such a solid is naturally an electrical insulator. Two important
papers have recently been published[14,15] which show that at high
pressures the molecular phase itself becomes conducting. Ramaker
et al.[14] note that at a volume of 5 cm^3/mole, electrons from the
fully occupied first Brillouin zone will begin to occupy states
in the second zone so that the molecular crystal becomes conducting.
Friedli has also made calculations which determined that the band
gap of the molecular crystal vanishes at about 2.5 cm^3/mole.[15] If
indeed such a metallic molecular phase exists at high pressures then
this would be direct evidence that a pair potential model is an incorrect description of the molecular phase at high pressures, and
the conclusions which one draws based on a pair potential model
are meaningless. An important new theoretical calculation by
Ashcroft[16] (not available at conference time) throws additional
light on this matter. Ashcroft has examined molecular hydrogen
at high densities (by high densities we mean densities similar
to those at the calculated minimum of the alkali metal phase of
hydrogen). In his calculations Ashcroft assumed that solid molecular hydrogen has the Pa3 crystal structure. For a given lattice
parameter (and hence density), the value of the bond distance which
gave the minimum energy for the system was obtained. When this

energy was plotted in turn against lattice parameter (or density),
the curve showed a minimum. Moreover this curve for the metallic
molecular phase closely approximates the curve for the alkali metal
phase of hydrogen in the neighborhood of its minimum. Thus the
molecular phase is characterized by two minima.

If this result in fact accurately describes the situation for
hydrogen, then:

1. there is no question that a first order transition exists;
2. this transition will not be exceptionally high but rather
   will be in the neighborhood of 1 Mbar;
3. there is a substantial maximum in the U - V curve between
   these minimum so the possibility of a metastable high pressure phase exists.

We are therefore presented with the following dilemma. If the
pair potential description of molecular hydrogen is correct, then
the transition pressure is probably very high (several megabars).
However, theoretical calculations suggest that the molecular phase
becomes an electrical conductor at high pressures so that (if these
theoretical calculations are correct) a pair potential model is incorrect. Furthermore, recent theoretical calculations suggest that
the molecular phase has two minima, one at zero pressure where
it is an electrical insulator and one at high pressure where it is
metallic. If accurate, these calculations predict a phase transition at modest pressures (around one megabar).

The only firm conclusion that the present author makes is this:
At the present time we do not approximately know the transition
pressure from molecular to alkali metal type metallic hydrogen.

EXPERIMENTAL

Uniaxial 3% Cobalt Cemented Tungsten Carbide

The yield stress of 3% cobalt cemented carbide is about 52 kbars.[17]
There are two ways to obtain the maximum attainable pressures in a
supported uniaxial opposed anvil device such as used by Drickamer[18],
ourselves[19] and others. Both lead, as is shown elsewhere, to a maximum or limiting pressure which is less than 200 kbars.[17] We might
note that it was pointed out many years ago by Lees[20] that such devices could only reach about 170 kbar; had he added the superposed
pressure in the support his result would have been 194 kbars. More
recently Dunn[21] has carried out an analysis of opposed anvil devices
and if the new and accurate value of the yield stress[17] is used in
his theory, it leads to a limiting pressure of 195 kbar. These results are summarized in Table 1.

X-ray Marker Method

The x-ray marker method proceeds as follows: The Hugoniot equation of state of a cubic metal is obtained. This is converted to an isotherm in the usual way (rightly or wrongly). This gives a P – V curve or if you wish a P – a curve where "a" is the lattice parameter. Then the cubic metal is placed in a static apparatus and from the measured lattice parameter the pressure is obtained.

Very hard materials such as the carbides having the NaCl type crystal structure (B1) have been studied in this way,[22] and from the data at pressures claimed to be as high as 350 kbar, values of the bulk modulus at zero pressure K_o and its pressure derivative K_o' were obtained. The expected value of K_o' is four; the values found are around fifteen. The explanation is simple: the apparent pressures are much too high. By assuming the maximum pressure is near 200 kbars one obtains not only agreement with the predictions of the previous subsection but also a value of K_o' consistent with physical expectations. Thus, in spite of the fact that the x-ray measurements are made with the greatest of care, the apparent pressures obtained by the marker method in highly nonhydrostatic environments are much too high. The subject is discussed further elsewhere.[17]

Uniaxial Diamond Devices

Here we only summarize the results described in detail elsewhere.[23] Two modes of failure can occur: fracture and yielding. If fracture were prevented (to date this had been the dominant failure mechanism), then the pressures shown in Table II would be obtained. Since macroscopic yielding has clearly not been observed, we are forced to conclude that pressures of 65 – 70 GPa have not been attained, contrary to claims which have been made based on the ruby scale and by use of the marker method described in the previous subsection.[24] Furthermore, until the beginning of yielding is clearly observed in such macroscopic devices (tip diameter of 600 μm), e.g. by the microprofilometer technique introduced by Ruoff,[17] claims of pressures as high as 50 GPa should be viewed with caution.

Sphere-core indentors

Vereshchagen[25] and his colleagues have used conical indentors (included angle of 168 – 172°) with spherical tips with R as small as 100 μm. The indentor and anvils were of carbonado, a cemented diamond composite which is an electrical conductor. Ruoff and Chan[26] analyzed these geometries in detail and concluded that at high pressures an effective saturation pressure was reached (even in the assumed absence of yielding or fracture). The saturation pressure found was 800 kbars for 172° angles and 1 Mbar for 168°

TABLE I

Limiting pressure in supported uniaxial opposed tungsten carbide
anvil system

Reference	P^{LIM} (kbars)
17	193
20	170[a]
21	195[b]

a. Computed without adding the support pressure of 24 kbars. Addition of this value would give agreement.
b. Computed from theory using the new value of yield stress.

TABLE II

Uniaxial Opposed Diamond Anvil Devices

Yielding	P (kbars)
Begins	500
Becomes large and macroscopically evident	650 - 700
Becomes extensive and limiting pressure is reached	800 - 850

angles. Of course as noted in the previous section yielding will occur (unless fracture precedes it). Fracturing will now be discussed briefly. (For more details see the paper by Ruoff.[23])

The fracture stress of carbonado anvils with R = 2 cm is 18 kbars;[24] unless there is a size effect on the fracture stress the pressure for fracturing is $P_o{}^f = 109$ kbars. If the fracture stress observed by them in anvils with R = 20,000 μm is constant down to

R = 2000 μm and increases thereafter as the radius decreases according to Auerbach's law

$$\sigma^f = k/R^{1/3} \tag{9}$$

or more generally

$$\sigma^f = c/a^{1/2} \tag{10}$$

where a is the radius of contact, then it can be shown [using the methods described by Ruoff and Chan[26]] that for the carbonado sphere-cone indentor with R = 100 μm used by Vereshchagen et. al.

$$P_o^f \lesssim 30 \text{ GPa}$$

[In general, workers who have studied Auerbach's law do not consider it likely that it would hold for R this large. In fact Tabor[27] suggests that the fracture stress is constant above R = 200 μm and increases according to Auerbach's law only for smaller radius. In this case P_o^f would be less than computed above.]

A summary of pressures attainable with the sphere-cone is shown in Table III. These are substantially less than the pressures of 1 - 4 Mbars claimed. It should be noted that the analysis given here is for diamond against diamond. If the sample is very thin, the analysis remains the same.[28] However if the sample is thick, two changes will occur: (1) the pressure to achieve a given force will be less and (2) the pressure distribution will be affected by the sample and it is conceivable that the sample will add some support to the anvil and will help delay fracture somewhat.

Metallic Hydrogen

Using sphere-cone carbonado tips (β = 4° to 6° and R = 100 μm to 1000 μm) Vereshchagen et al.[29] noted large resistance drops when the indentor was pressed against powdered samples of materials such as α-SiO$_2$, Al$_2$O$_3$, MgO, cubic BN, etc., as well as thin films (10 to 100 μm) of hydrogen. These were interpreted as phase transitions. It is noted by them that if after the resistance drop occurs and the force is lowered somewhat no reversal is seen suggesting metastable metal is now present; under these conditions slight subsequent heating causes a jump in resistance to the original value. However, the original drop in resistance could be explained by shorting, and the subsequent effect of heating in the low resistance phase may be to expand the material in the non-shorted region surrounding the middle and to lift the indentor the several angstroms needed to open the circuit. A useful experiment suggested by Keeler[30] would be to make identical experiments on neon (as made

TABLE III

Carbonado cone-sphere indentors

	P (kbars)	References
Saturation[a]	800 - 1000	26
Yielding[b]	500; 650; 800	23
Fracture[c]	300	23

a. Actual value depends on cone angle. See reference 26.
b. Different levels of yielding are found in Table II.
c. Based on assumption that carbonado gets stronger as tip radius
    decreases according to Auerbach's law with strengthening begin-
    ning at R = 2000 μm. Even if the extreme and unlikely condition
    prevails that Auerbach's law holds to R = 20,000 μm, fracture
    would occur by 700 kbars even in the absence of yielding.

on hydrogen) because neon is expected to have a very high transition
to the metallic state which surely could not be reached with the carbonado anvils.

At the Sixth Airapt International High Pressure Conference,
Yakovlev,[31] commenting on the paper by Mao and Bell[24] on ruby,
noted that according to his present pressure scale diamond became
metallic at about 1 Mbar and Al_2O_3 at about 0.4 Mbar. The behavior of the latter will be examined in the next year in the author's laboratory by the new nonshort technique using insulating
diamonds that has recently been developed by Ruoff and Chan[28] combined with the very high pressures recently produced and measured
by Ruoff and Wanagel[32]. There is dynamic work by Hawke[33] which
suggests that Al_2O_3 remains insulating under dynamic conditions
at 4+Mbar. There is, however, always the possibility that a
phase passes into a metastable region during a shock experiment
and the transition does not occur in the very short time available;
consequently it is vital that Al_2O_3 be tested carefully under static
conditions. In a sense the reliability of the entire set of ultra-
pressure experiments at the Moscow High Pressure Institute rests
on the results of this experiment.

To summarize, the evidence at present in the experiments of
Vereshchagen et al. that the resistance drop across a thin hydrogen film is not a short is weak. Hopefully further experiments in

the next year will clarify the issue on a number of reported insulator to metal transitions.

Kawai[34] has also claimed production of metallic hydrogen, but
in a recent paper coauthored by him[35], no mention is made of this
work although a long section is devoted to the subject of metallic
hydrogen. In Kawai's experiment, eight cubes, two opposite of
tungsten carbide and six of alumina are separated by porous MgO and
cardboard spacers. The eight-cube assembly was introduced into a
split-sphere device. Then hydrogen gas was introduced into the
assembly at 100 bars. Force was then applied to the entire system,
the assumption being that hydrogen gas was trapped in the center
region. Eventually a resistance drop was observed. There are two
reasons why a short may have been observed. First, the sample was
highly compliant so that extensive inward motion would be expected.
Second, the two tungsten carbide cubes are elastically much stiffer
than the six alumina cubes, so that, if the exterior surfaces of
the eight cubes remain parallel, the interior edges of the carbide
cubes must move inward much further than the alumina edges of the
alumina. Because of these two situations, it is extremely difficult for the author (who has worked considerably with the split-
sphere assembly[36]) to see how a short could be avoided. When the
sample region was filled with MgO instead of hydrogen gas no resistance drop was observed; of course MgO is much stiffer than
hydrogen gas or even solid hydrogen so that it would resist the
motion of the carbide pistons and forestall a short until a
larger external force were applied. It would be of interest to
see what would happen if neon gas (or for that matter a vacuum) were
used instead of hydrogen.

The present author does not think Kawai's claim of producing
metallic hydrogen is credible.

EXPERIMENTAL PROSPECTS

Perfect crystals of diamond can be very strong.[23,32,37] Compressive yield strengths can be very high: σ_o = 1.3 - 3.5 Mbars.[23,37]
Attainable pressures before yielding begins in perfect crystals
are predicted to be 1.8 Mbar for [100] loading and 5.3 Mbars for
[111] loading assuming no fracture occurs.[23]

To get very high pressures using diamonds it is necessary to
do two things:
1. Postpone fracture
2. Postpone yielding.
To postpone fracture, we should work with spherical diamond tips
of extremely small radii. We have found that the maximum stress
at fracture, P_o^f, computed from the Hertz equation,[23] varies with
the tip radius R according to

$$P_o^f \text{ (Mbars) } = 2.4/[R(\mu m)]^{1/3} \tag{11}$$

for a specific [100] face. Previous experience has suggested that the coefficient would be only 2/3 as large for a [111] face. In any case, if R = 0.1 μm the pressure at fracture may reach 5 Mbars (if no yielding occurs).

To postpone yielding, we should work with perfect crystals. Real crystals contain defects, e.g., about 10^6 cm/cm^3 of dislocation line. Hence on the average the distance between dislocations is about 10 μm. The space between them is dislocation free. Consider the following example:

Indentor tip radius is 2.5 μm.
P_o^f predicted is 1.8 Mbars from Equation (11).
Contact radius for P_o = 1.4 Mbars is about one micrometer.
Probability that highly stressed region in anvil or indentor is free of dislocation is very high, so that the pressure at yielding is expected to be at the limit for a perfect crystal, or at 1.8 Mbars on a [100] face.

Recently Ruoff and Wanagel[32] have used such small indentors to obtain pressures of 1.4 Mbars (computed from Hertz theory on the basis of applied force and tip radius, see reference 23). The author believes that the method of Ruoff and Chan[28] can be extended so that good scientific experiments can be made under these extremely tiny tips. In addition he believes that using synchotron radiation, even x-ray diffraction experiments will be made successfully. Finally, if the calculation of [111] compressive strengths are correct, pressures (in extremely tiny volumes) as high as 3 to 5 Mbars may eventually be obtained in the elastic region of diamond. Furthermore by utilizing the plastic range of the pressure vessel material, even higher pressures may be attained.[36,38]

Consequently, should the transition pressure be below 3 Mbars as Ashcroft predicts, and should perfect crystalline diamond be as strong as Ruoff predicts, the production of metallic hydrogen should be possible. Moreover, it should be possible, using the diamond scale developed by Ruoff and Wanagel[32] to know the pressure at which the transition occurs. Finally, if Ashcroft's calculations are correct, there is a reasonably good chance that the metastable phase can persist at low or zero pressure at liquid helium temperatures.

ACKNOWLEDGMENT

The author wishes to thank the National Aeronautics and Space Administration for support of this work and the National Science

Foundation for their support of the Cornell Materials Science Center.

REFERENCES

1. R.B. Sensenig, Survey of the Properties of Metallic Hydrogen and the Application to Superconductivity and Astrophysics, M.S. Thesis, Penn. State U., March (1975).
2. M. Ross and C. Shishkevish, Molecular and Metallic Hydrogen, ARPA Report R-2056, May (1977).
3. M.S. Anderson and C.A. Swenson, Phys. Rev. B, 10, 5184 (1974).
4. M. Van Thiel, L.B. Hord, W.H. Gust, A.C. Mitchell, M.A'Addario, K. Boutwell, E. Wilbarger and B. Barrett, Physics of the Earth and Planetary Interiors, 9, 57 (1974).
5. F.V. Grigor'ev, S.B. Kormer, O.L. Mikhaylova, A.P. Tolochko, V.D. Urlin, Zhurnal eksperimental'noy i teoreticheskoy fiziki, 69, No. 2(8), 743 (1975).
6. F.H. Ree and C.F. Bender, Phys. Rev. Letters, 32, 85 (1974).
7. A.K. McMahan, H. Beck, and J.A. Krumhansl, Physical Review A, 9, 1852 (1974).
8. V.P. Trubitsyn, Fizika tverdoga tela, 8, 862 (1966).
9. M. Van Thiel and B.J. Alder, Molecular Physics, 19, 427 (1966).
10. M. Van Thiel and B.J. Alder, Molecular Physics, 16, 286 (1972).
11. A.L. Ruoff and L.C. Chhabildas, Proceeding of the Sixth Airapt High Pressure Conference, Boulder, July 1977. Also Cornell Materials Science Center Report #2853, Cornell U., Ithaca, N.Y. 14853.
12. J. Hammerburg and N.W. Ashcroft, Physical Review B, 9, 409 (1974).
13. F.E. Harris, L. Kumar and H.J. Monkhorst, Phys. Rev. B, 7, 2850 (1973).
14. D.E. Ramaker, L. Kumar and F.E. Harris, Phys. Rev. Letters, 34, 812 (1975).
15. C. Friedli, "Band Structure of Highly Compressed Hydrogen", Ph.D. Thesis, Cornell U., Ithaca, NY 14853 (1975).
16. N.W. Ashcroft, private communication.
17. A.L. Ruoff, Sixth Airapt International High Pressure Conference, Session A-5. See also Cornell Materials Science Center Report #2851, Cornell U., Ithaca, NY 14853 (1977).
18. H.G. Drickamer, Rev. Sci. Instrum., 41, 1667 (1971).
19. J. Wanagel, V. Arnold and A.L. Ruoff, J. Appl. Phys., 47, 2821 (1976).
20. J. Lees, Ch. 1 in Advances in High Pressure Research, ed. by R.S. Bradley, Academic Press, New York, Vol. I (1966).
21. K.J. Dunn, J. Appl. Phys., 48, 1829 (1977).
22. A.R. Champion and H.G. Drickamer, J. Phys. Chem. Solids, 26, 1973 (1965).
23. A.L. Ruoff, Sixth Airapt International High Pressure Conference,

Session D-4, See also Cornell Materials Center Report #2855 (1977).

24. D. Mao, P.M. Bell, J. Shaner and D. Steinberg, Sixth Airapt International High Pressure Conference, Session A-5 (1977).

25. L.F. Vereshchagin, E.N. Yakovlev, B.V. Vinogradov, G.N. Stepanov, K. KhBibaev, T.I. Alaeva and V.P. Sakun, High Temperature-High Pressures, 6, 499 (1974).

26. A.L. Ruoff and K.S. Chan, J. Appl. Phys., 47, 5077 (1976).

27. F.B. Bowden and D. Tabor, Ch. 7 in Physical Properties of Diamond, ed. by R. Berman, Oxford U. Press, New York (1965).

28. A.L. Ruoff and K.S. Chan, Sixth Airapt International High Pressure Conference, Boulder (1977). Also Cornell U. Materials Science Center Report #2854, Cornell U.

29. L.F. Vereshchagen, E.N. Yakovlev and Yu A. Timofeev, JETP Letters, 21, 85 (1975).

30. K.N. Keeler, private communication.

31. E.N. Yakovlev, private communication.

32. A.L. Ruoff and J. Wanagel, Cornell Materials Science Center Report #2892, Cornell U., Ithaca, NY 14853 (1977). To be published in Science.

33. R. Hawke, submitted for publication.

34. N. Kawai, M. Togaya and O. Mishima, Proceedings of the Japan Academy, 51, 630 (1975).

35. B. LeNeindre, K. Suito and N. Kawai, High Temperature-High Pressures, 8, 1 (1976).

36. J. Wanagel and A.L. Ruoff, Sixth Airapt International High Pressure Conference Boulder (1977). Also Cornell Materials Science Center Report #2856, Cornell U., Ithaca, NY 14853.

37. A.L. Ruoff, Cornell Materials Science Center Report #2818, accepted for publication in J. Appl. Phys.

38. A.L. Ruoff, Advances in Cryogenic Engineering, 18, 435 (1973).

QUESTIONS AND COMMENTS

C.W. Chu: How do you compare your diamond scale with the ruby scale?

A.L. Ruoff: No direct comparison has been made. I will discuss the linear ruby scale and other things related to it at the Boulder Conference. I don't believe in the linear ruby scale, I think it drastically overestimates the pressure at one megabar.

C.W. Chu: Could you give an estimate of how much it is overestimated at one megabar?

A.L. Ruoff: In as much as they have not observed yielding of diamond, the maximum pressure that is present is probably less than 0.5 megabar.

I.L. Spain: I would object to that point. I think this is a highly
              controversial statement. There is a lot of evidence
              which will also be presented at the Boulder meeting to
              show that the equations of states of certain metals,
              MgO, NaCl and the ruby scale are in reasonable agree-
              ment. In fact, Mao and Bell at Carnegie feel that if
              the ruby scale is non-linear, it's actually in the op-
              posite sense to what Ruoff says. So I think there is
              going to be a very interesting discussion on this at
              Boulder.

J. Wittig: I didn't understand your argument concerning the dis-
              locations. Wouldn't you think that the diamond crystal
              is stronger the more dislocations there are?

A.L. Ruoff: No. But strength goes like this: If you have a perfect
              crystal, say a whisker, the strength is tremendous. If
              you have a few dislocations in it, like 10^4 to 10^8, the
              strength is quite low. If you plastically deform the
              materials such as face-centered cubic metals, and get
              the dislocation density back up to 10^{11} to 10^{12}, in that
              ballpark, then you have strong dislocation interactions
              and you start getting strengthening from that mechan-
              ism, but you're still way below the perfect crystal
              strength.

J. Wittig: Can you anneal the diamond crystal?

A.L. Ruoff: It's already annealed. The diamond crystal has been
              formed by a process at high temperatures that took
              millenia to cool it slowly. It still has 10^4 to 10^6
              dislocations and there is no way to get them out.

METALLIC HYDROGEN: RECENT THEORETICAL PROGRESS*

A.K. McMahan

University of California, Lawrence Livermore Laboratory

Livermore, California 94550

ABSTRACT

A summary is given of recent theoretical progress in determining the insulator-to-metal transition pressure in condensed hydrogen, and in understanding the nature of each of these phases. Some of the problems involved in retaining the metallic phase metastability at low pressures are reviewed.

INTRODUCTION

The behavior of condensed hydrogen under pressure has been a subject of some interest ever since Wigner and Huntington first discussed the possibility of a high pressure metallic phase in. 1935.[1] Since then much effort has been expended in theoretical calculations of both the molecular and metallic phase equations of state in order to determine the transition pressure,[2-15] as well as in experimental attempts to actually produce metallic hydrogen in the laboratory.[16-19] In recent years this effort has received impetus from attempts to understand the planetary interiors of Jupiter and Saturn[20] and from the prospect that metallic hydrogen might be a room temperature superconductor,[21] which could have dramatic consequences if it could also be maintained in metastable form at low pressures. The purpose of the present paper is to sum-

---
*Work performed under the auspices of the U.S. Energy Research and Development Administration, contract No. W-7405-Eng-48.

marize the considerable theoretical progress which has been made
recently in both determining the transition pressure as well as in
understanding the nature of both the metallic and insulating phases.
Some of the problems involved in retaining the metallic phase in
metastable form at low pressures are also discussed. No attempt
at a complete review is made here, for which the reader is referred
to the recent exhaustive work of Ross and Shishkevish.[22]

It has long been assumed that the insulator-to-metal transition
in hydrogen coincides with dissociation of the H_2 molecules, that
is when the diatomic molecular phase (presumed to be insulating)
transforms into the monatomic metallic phase. Theoretical calculations of "the transition" pressure have invariably been for this
dissociation process. As will be discussed, the best of such calculations now put this dissociation pressure in the range 300-700
GPa (3-7 Mbar).[22] However, recent theoretical evidence also suggests that the molecular insulating phase may itself transform into a metallic yet still molecular form prior to the dissociation of
the molecules[14,23] Thus there appear to be two distinct transitions,
dissociation and metalization. Tentative theoretical estimates put
metalization of the molecular phase at a pressure below 200 GPa.
This only partially reconciles theory with experimental claims to
have seen metallic conductivity in the vicinity of 100 GPa.[18,19]
Some of the experimental results[19] indicate a first order transition
to a conducting phase while these new theoretical developments suggest the transition from molecular insulator to molecular conductor
is of higher order.

The existence of metallic conductivity in the high pressure
molecular phase does not alter the fact that the only potentially
low pressure conducting phase is still the monatomic structure,
presuming it can indeed be maintained metastably with some non-
negligible lifetime. While the chance that the lifetime is of
practical consequence must still be considered remote, there have
been several recent promising results. At least for a significant
range of pressures below dissociation and at zero temperature, there
is now evidence that the monatomic phase is solid,[24] and furthermore will most likely assume some cubic structure[25] rather than
the semi-liquid-like filamentary structure which had been suggested
in the past.[26] Either liquid or liquid-like character for the phase
would effectively eliminate any possibility of metastability. The
major question in the problem of metastability is the mechanism by
which the conversion to the molecular phase is nucleated. In the
bulk, vibrational motion will eventually bring two or more protons
close enough together to permit formation of an electron bound
state. However, screening effects of the electron gas will impede
this process, and proper account of this fact has not yet been taken
in the lifetime calculations. Even should this lifetime be relatively long, one of course must still be concerned with nucleation at
interior defects and at the exterior surfaces.

DISSOCIATION

The compression of diatomic molecular hydrogen is expected to lead at some pressure to a first order transition to a monatomic metallic phase, i.e. the H_2 molecules will dissociate. The transition pressure is determined by comparison of the equations of state of each of the two phases, either by the common tangent construction, or equivalently, by the crossing of the two Gibbs free energy curves. The theoretical problem requires very accurate calculations for both phases. Two recent developments in this dual problem are responsible for pushing theoretical estimates to the transition pressure from the region around 200 GPa to perhaps as high as 700 GPa. They are: (1) the significant presence of many-body interactions in the high pressure diatomic phase, and (2) possible overestimation of the electron-electron correlation energy for the monatomic phase. The correction for each of these effects increases the transition pressure.

Many-body Effects in the Diatomic Phase

Equation of state calculations for the diatomic phase are based on an assumed pair potential describing the interaction between two H_2 molecules. Once this function is specified, the techniques of quantum crystal theory and of liquid hard-sphere perturbation theory can be used to calculate pressure-volume isotherms and Hugoniots for comparison with experiment. These theoretical techniques are believed to be reliable, and thus a comparison with experiment serves as a direct test of the assumed pair potential. Figure 1 compares four theoretical pair potentials with the phenomenological potential (cross-hatched area) needed to account for both the high pressure, high temperature P-V shock wave data, and the low pressure, and low temperature P-V static data. These results are taken from work by Ross.[10,11] All potentials are spherically averaged over the molecular orientations, an approximation which is believed to affect the transition pressure by only 20% or so.[10,13] The bare potential (between two isolated H_2 molecules, labelled CI) can be rigorously calculated using the configuration interaction technique, and is known to within about 10% over the range shown in the figure.[27] As can be seen it is a factor of two larger than the phenomenological potential! Ross[10] added an attractive term (CI + ATT) to generate an empirical potential in satisfactory agreement with all of the P-V data. Etters et al.[13] (EERD) did the same, but arrived at a potential which was still too large at small intermolecular separations to give results in agreement with the shock data. However, their potential was basically constructed to fit low pressure data (intermolecular separations R > 5 bohr), and in fact yields the best agreement with the static P-V data as well as giving accurate virial coefficients.

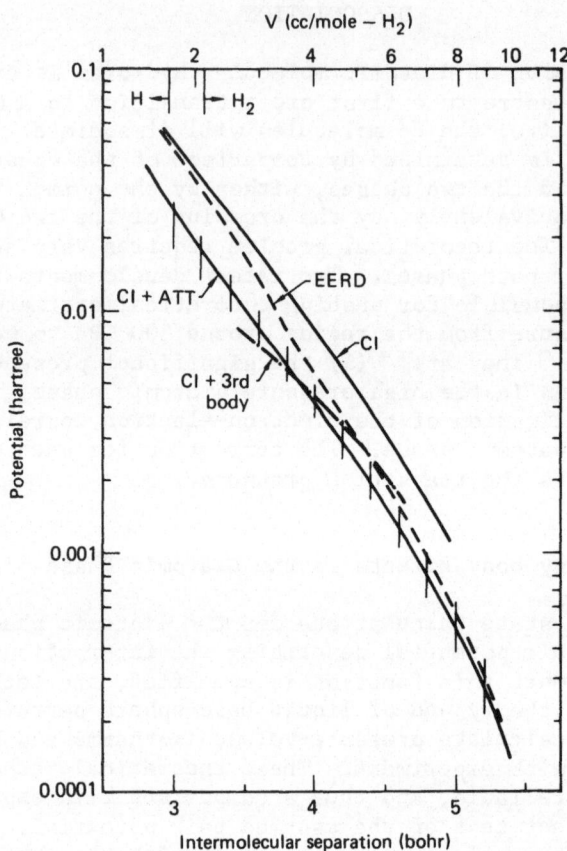

Figure 1. H_2-H_2 pair potentials as a function of intermolecular
            separation. Four theoretical potentials (labelled) are
            compared to the effective potential (cross-hatched area)
            which is consistent with high pressure shock wave data
            and low pressure static data. At the top left corner
            of the figure the volume change is shown which would
            occur if the dissociation pressure were 300 GPa.
            (Taken from Ref. 11)

     An explanation for the discrepancy between the best effective
pair potentials and the rigorous theoretical calculations was offered by Ree and Bender,[28] in terms of many-body (3-body and higher)
interactions. In dense molecular hydrogen one expects according
to the Pauli principle that the charge clouds of a given pair of
molecules should be distorted by the presence of other neighboring
molecules. Thus the bare pair potential should be replaced by some
effective pair potential which incorporates these many-body effects.

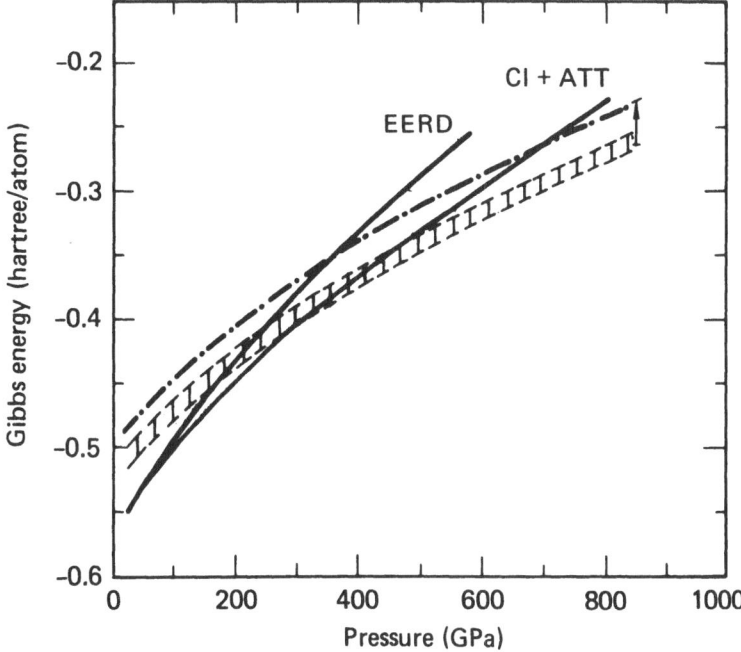

Figure 2. The T=0 Gibbs free energy curves for the diatomic molecular
and monatomic metal phases. The molecular phase curves
(solid lines) are labelled according to the pair potential
used. The cross-hatched area shows the spread amongst
results for the monatomic phase obtained using the APW,
LCAO, PERT and WS methods. The dash-dot curve shows the
amount by which these results would be shifted if the
free electron correlation energy were reduced by a
factor of two.

Using configuration interaction calculations for a triangular system of three molecules, Ree and Bender estimated an effective potential (CI + 3rd body), which can be seen to account for much of
the discrepancy. Their method breaks down for R < 3.5 bohr, as
higher than 3-body contributions must be included. Additional
evidence for the presence of many-body forces comes from the band
structure calculations of Liberman.[5] These calculations (which
use a spherical approximation for the molecules) rigorously include
the many-body effects and yield a T=0 P-V isotherm in excellent
agreement with that obtained from the CI + ATT potential over a
range which probes the pair potential down to about 3.4 bohr.[11]

To investigate the effect of these potentials on the dissociation pressure, the T=0 Gibbs free energy for the molecular phase
as calculated using the EERD and CI + ATT potentials is shown in

Figure 2. The transition pressure is given by the intersection of
these curves with corresponding Gibbs free energy calculations for
the monatomic metal. The shaded area represents the spread amongst
band structure calculations of the fcc monatomic metal as found
using the augmented-plane-wave (APW), linear-combination-of-atomic-
orbitals (LCAO), third order free electron perturbation theory
(PERT), and the Wigner-Seitz (WS) methods.[15] The EERD potential,
which may be viewed as being essentially the bare interaction at
small intermolecular separations, yields a Gibbs free energy curve
which crosses through the shaded area at 150-250 GPa. When many-
body effects are included as is the case for CI + ATT potential,
the resultant curve intersects the shaded region at 300-450 GPa.
The proper inclusion of many-body effects for the molecular phase
is thus seen to increase the stability of this phase and to roughly
double the transition pressure.

Correlation Energy in the Monatomic Phase

The relatively close agreement seen in Figure 2 amongst the
monatomic metal calculations by four quite different methods suggests
that the equation of state of this phase is well determined. Although an fcc structure has been assumed in these calculations, only
slight energy differences are expected for other possible structures.
However, Ross and McMahan[15] have observed that the free electron approximation for the electron-electron correlation energy, generally
used by each of the methods, may be seriously in error for the case
of condensed hydrogen. While this approximation is accurate for
slowly varying charge density away from the nuclei, there is some
doubt as to its reliability near the nuclei. Test calculations for
isolated atoms have shown the free electron approximation to the
correlation energy to be in error (too large in magnitude) by a
factor of two or three.[29] Since this is a negative quantity, the
total energy, and Gibbs free energy, will be corrected to more positive values, making the metal less stable with respect to the molecular phase. Supposing the correlation energy to be in error by
a factor of two, Figure 2 (dash-dot curve) shows that the predicted
dissociation pressure is 700 GPa, or again roughly doubled. For
most metals the correlation energy is generally considered a relatively small correction to be added to the dominant Hartree-Fock
energy. This is not the case for hydrogen which has a comparatively
small total energy. Thus the correlation energy now appears to be
the outstanding uncertainty facing determination of the monatomic
phase equation of state.

METALIZATION

Monatomic hydrogen is characterized by a half-filled band and
is thus a metal.[30] Dissociation of the H_2 molecules must therefore
lead to metallic conductivity. However, diatomic molecular hydrogen

may itself become a metal while still retaining its molecular
structure if compression results in the gap between the full valence band and the empty conduction band going to zero. The LCAO
calculations of Ramaker et. al.[14] and the mixed basis calculations
of Friedli and Ashcroft[23] both indicate that this does occur at
modest compression.[31] Figure 3 shows the electron band structure
for the $\alpha-N_2$ structure of molecular hydrogen shortly after the gap
has closed, as reported in the later work. It is evident that the
conduction band has dipped below the Fermi level at the R point in
the Brillouin zone. Such band overlap metalization results in a
continuous decrease in resistivity from that characteristic of an
insulator to that of a metal.[32] The transition is customarily
referred to as second order, in that there is no volume change.
Using a free electron model to analyze the effect of the abrupt
appearance of new electron pockets (such as at R), Lifshitz[33]
has shown that both the pressure and its first volume derivative
are continuous (the second derivative is discontinuous) at the
critical volume where the pockets first appear, i.e. in this case
where the gap becomes zero.

The important point which is suggested by both of the papers
by Ramaker et al. and by Friedli and Ashcroft is that the gap
closes at a pressure which is less than that predicted for dissociation. One may thus be concerned with two transitions in the behavior of T=0 condensed hydrogen under pressure. As shown in
Figure 4, insulating molecular hydrogen may first undergo the second
order transition under compression to a conducting molecular phase.
Then with continued compression, this molecular metal will undergo
the expected first order transition to the monatomic phase. The
location of both transitions are shown with error bars in the figure. In the case of the dissociation pressure, the large error
bar is principally due to the monatomic phase correlation energy
as discussed earlier. The preferred CI + ATT potential of Ross was
used to get the high pressure molecular isotherm. In the case of
the molecular insulator to molecular metal transition, Friedli and
Ashcroft find band overlap at 2.4 cc/mole-H_2 in a calculation using
a fixed 1.4 bohr bond length. They show that the overlap will occur
sooner (at larger volumes) if the bond length is larger. Ramaker
et al. do indeed find that when the bond length is optimized at each
volume, that the molecules do expand with compression. They find
some occupancy of the second Brillouin zone already at the largest
volume they considered. Since they worked with a simple cubic array
of molecules, the near neighbor distance at this volume was used to
get an equivalent fcc (or $\alpha-N_2$) volume, shown as the lower bound
at 3.5 cc/mole-H_2.

The point of view espoused in Figure 4 is not without question.
While there are experimental claims to have seen metallic conductivity in the vicinity of 100 GPa,[18,19] as was noted earlier the static
work of Vereshchagen et al.[19] suggests in addition that the tran-

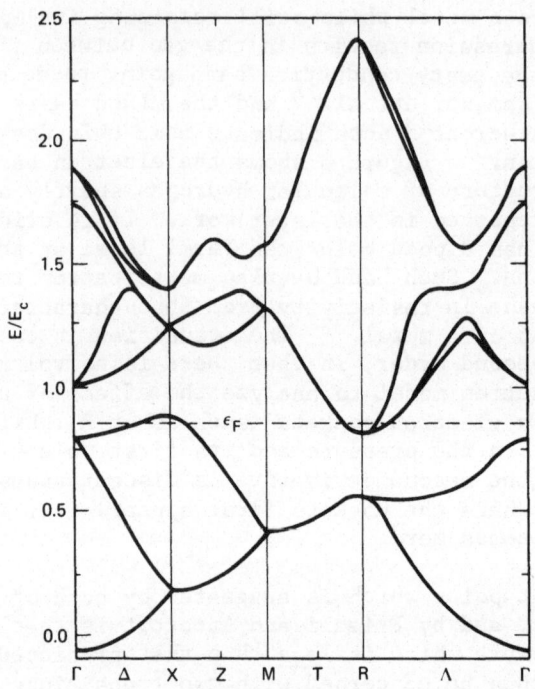

Figure 3.   Band structure of the molecular phase somewhat after
             the gap has closed.   The lattice structure is α-N$_2$,
             the volume is V = 2.033 cc/mole-H$_2$, the near neighbor
             distance R_{nn} = 3.18 bohr, and E_c = 1.9496 Ry.   (Taken
             from Ref. 23)

sition is first order, and thus apparently inconsistent with the
band overlap mechanism.   Given the extreme complexity of these very
high pressure measurements, however, it is advisable to also consider the analogous insulator-to-metal transition in diatomic molecular iodine which occurs at more manageable pressures.   Here the
transition has been observed in the range 4-17 GPa under static
compression, and is known to be second order with a continuous
decrease in resistivity[34]—a seemingly perfect example of band overlap metalization.   The question in the case of iodine is then where
does dissociation occur.   Shock wave data exists up to compressions
and temperatures sufficient to insure dissociation, and yet no evidence is seen for such dissociation.[35]   It may simply be that the
volume change is too small (3% or less) to be detected within the
scatter of the data.   However, the monatomic iodine equation of
state also serves to accurately reproduce all of the shock data

above metalization.[36] The intriguing possibility remains for the
case of iodine that the molecular structure may gradually and continuously evolve under compression towards a monatomic arrangement
in such a manner that band overlap and dissociation are either one
and the same or closely related phenomena. The same possibility
can be raised for the case of hydrogen. However, at the present
time the calculations of Ramaker et al.[14] are the best indication
that the two transitions in hydrogen are indeed distinct.

NATURE OF THE MONATOMIC METAL

Because of the interest in producing monatomic hydrogen metal
in a metastable form below the dissociation pressure, the precise
details of the structure of this phase are important. There is
now evidence, at least over a considerable range of pressures extending below dissociation, that the monatomic solid is likely
to assume an isotropic structure, and not one of the highly
anisotropic forms which have been suggested. Furthermore, it
also seems likely that monatomic hydrogen will indeed be a solid
and not a liquid at zero temperature. For reference in the discussion below it should be noted that at dissociation, monatomic
hydrogen is characterized by $r_s = 1.2 - 1.3$, according to the uncertainties shown in Figure 4; while at zero pressure,[37] $r_s = 1.7$.
The electron gas parameter r_s is given as usual by

$$\frac{4\pi}{3} (r_s a_0)^3 = n^{-1},$$

where n is the electron density.

Anisotropy

Brovman et al.[26] first predicted anisotropic structures for
solid monatomic hydrogen based on static lattice calculations using
third order free electron perturbation theory. Similar results
are shown in Figure 5 for the family of face-centered-tetragonal
(fct) structures at $r_s = 1.36$, taken from the recent (second order)
work of Straus and Ashcroft (which was carried out only at this
density).[25] As the ratio c/a is adjusted while keeping density
fixed, it is seen that the static energy is lowest for c/a = 0.45
and not for the cubic fcc structure at c/a = 1. Brovman et al.
systematically investigated all of the Bravais lattice types, and
some non-Bravais lattices. They found the lowest static energy
for closely spaced planes of triangular lattices, for which there
was almost no restoring force for the motion of filaments of atoms
in a direction perpendicular to the planes.

The correct test of stability at zero temperature must come of

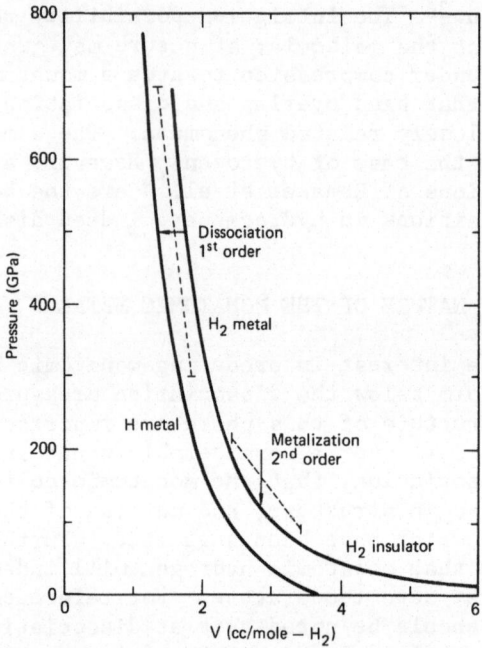

Figure 4.   The T=0 pressure-volume isotherms for both molecular
            and monatomic phases. These were obtained using the
            CI + APT potential and the APW method respectively.
            The two transitions (arrows) are shown with error bars
            (dashed lines) indicating uncertainties in their
            locations.

course from the total energy and not the static energy. Brovman
et al. in fact calculated the zero-point energy in test cases, and
found the situation unchanged. In sharp contrast the dashed lines
in Figure 5 show the total energy results of Straus and Ashcroft,
and it is seen that in their calculations the zero-point energy
has <u>reversed</u> the situation. According to the total energy it is
now the isotropic fcc structure which is most favored. The difference between the two calculations is that Brovman et al. used
the harmonic approximation to determine the zero-point energy,
while Straus and Ashcroft used the self-consistent harmonic approximation. In the former, the force constants are computed according
to infinitesimal displacements of the atoms from equilibrium,
while in the latter they are determined in an average and self-
consistent manner over the full range of the atomic zero-point
excursions. Since these excursions are large in monatomic hydrogen,
and the pair potential between atoms highly anharmonic, the latter
approach should give a more accurate representation of the zero-
point motion.

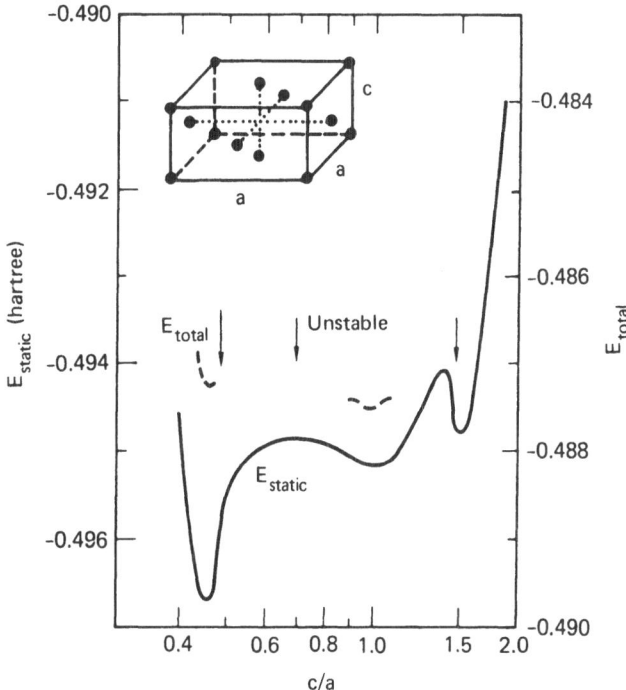

Figure 5.   Energy of T=0 face-centered-tetragonal monatomic hydrogen
            as a function of c/a ratio, at a density corresponding
            to r_s = 1.36.   Results for the static (solid curve) and
            total (dashed curve) energy are shown.   Arrows indicate
            dynamic instabilities.   (Taken from Ref. 25)

    It is not difficult to see that the zero-point motion will favor
isotropic structures.   Figure 6 shows a representative pair potential
between two hydrogen atoms in the solid, taken from the work of Beck
and Straus.[38]   The interaction between two immobile atoms separated
by a distance R is given by the solid line, while the dashed line
shows the effective interaction obtained by averaging this potential
over the zero-point motion of the two atoms.   The source of the tendency towards anisotropy in the static energy is related to the
Friedel oscillations at large R.   For isotropic structures, the
spacing between the most densely populated planes of atoms is too
small to mesh with the troughs in the Friedel oscillations in the
potential.   Thus the solid can lower its energy at fixed density
by enlarging the spacing between some sets of these planes, even
though atoms within the planes must be packed more closely.   As
can be seen in the dashed curve, the zero-point motion serves to
smear out and reduce these oscillations, and thus weaken this
mechanism for anisotropy.   In addition, the hard-core of the averaged

potential is seen to have moved out to larger values of R. This
simply means that when they are undergoing zero-point motion, the
atoms must be kept farther apart in order to avoid overlap of their
hard cores. Since anisotropic structures generally have smaller
near-neighbor separations than is the case for isotropic structures,
this too serves to favor isotropy.

At the present time it is not possible to answer the question
as to whether the monatomic solid will assume an isotropic structure at all pressures below dissociation. The tendency in the
static energy favoring anisotropy increases with decreasing density.[26] However, for $r_s > 0.7$, so too does the size of the zero-
point motion (which favors isotropy), at least as measured by the
rms displacement of the atoms relative to their near neighbor
spacing.[9] From this point of view it would seem plausible that
the zero-point motion might continue to stabilize isotropic structures all the way down to zero pressure, $r_s = 1.7$. However, there
is another consideration. As Beck and Straus have noted, the same

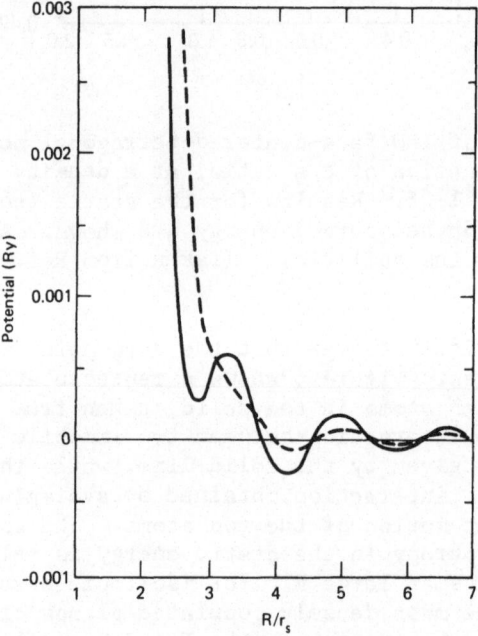

Figure 6.   Representative H-H pair potential in fcc monatomic hy-
            drogen. The solid curve shows the potential between
            two immobile atoms separated by a distance R. The
            dashed curve is the average potential between two atoms
            undergoing zero-point motion whose equilibrium separ-
            ation is R. (Taken from Ref. 38)

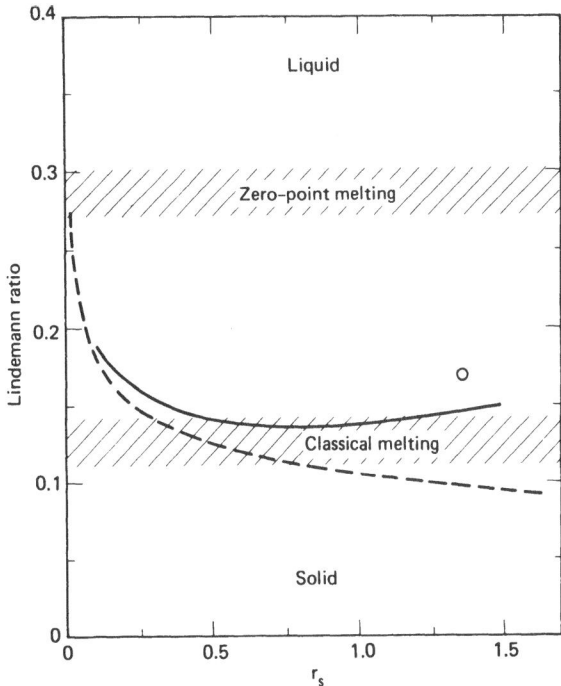

Figure 7.   The Lindemann ratio for T=0 fcc monatomic hydrogen as
a function of r_s. The dashed curve is the result obtained in the harmonic approximation and neglecting
screening. The solid curve is obtained from the
Einstein approximation (Ref. 9), while the data point
is obtained from the self-consistent harmonic approximation (Ref. 25). The liquid is presumed to be more
stable than the solid if this ratio lies above the
critical value (shaded region) which should be in this
case for zero-point melting and not for the more familiar case of classical melting.

mechanism that favors anisotropy in the static energy also intro-
duces instabilities into the lattice dynamics, in the form of Kohn
anomalies which depress certain of the transverse vibrational frequencies. In self-consistent harmonic calculations for the fcc
lattice, Caron[9] has shown that this effect is sufficient for
$r_s > 1.5$ to drive some of the transverse vibration frequencies to
imaginary values. That is, the fcc crystal becomes dynamically
unstable. A tentative assessment at the present time, then, is
that isotropic structures are favored for $r_s \lesssim 1.5$. At lower densities, and especially at zero pressure, the question of structure
appears to still be open. It should be kept in mind, however, that
the differences in Gibbs free energy between any of these possible

structures is small, and will have little bearing on the pressure at which the molecules dissociate.

Liquid or Solid?

Because of the small mass of the hydrogen atom and the consequent large zero-point motion, the possibility has been raised that the monatomic hydrogen might be a liquid even at zero temperature.[20,22] The recent Monte Carlo variational calculations of Chester[24] suggest that this is not the case, except perhaps at the very lowest densities. This work assumes the same atom-atom pair potential as used by Straus and Ashcroft, and investigates the relative stability of the solid and liquid phases over the range r_s = 0.8-1.6. Throughout most of this density interval, the solid is found to have the lower energy. The energy difference is small, however, as for example about 1% at r_s = 1.36. Near r_s = 1.6 the numbers are too close to determine the more stable phase given the current uncertainties in the calculations.

Additional evidence that the solid is the more stable phase at zero temperature can be seen from the Lindemann melting criterion. This empirical condition asserts that melting will occur when the rms deviation of the atoms from their lattice sites $<u^2>^{1/2}$ becomes larger than some prescribed fraction of the near neighbor distance R_{nn}. It is essential to note that the critical value of the fraction

$$f = <u_x^2 + u_y^2 + u_z^2>^{1/2}/R_{nn}$$

at melting f_m is different in the case of interest here—T=0 melting due to zero-point motion—than it is for the more familiar case of melting in classical crystals due to thermal motion. For classical melting, experimental data suggests f_m = 0.11-0.14.[39] Similar values are obtained from classical Monte Carlo calculations for both soft and hard sphere potentials.[40] However, for T=0 melting due to zero-point motion, quantum Monte Carlo variational calculations suggest $f_m \sim 0.3$ for both soft sphere[41] and hard sphere[42] systems. Solid helium (under modest pressure) is an actual example of a system which can support zero-point motion corresponding to $f \lesssim 0.3$ and yet still not melt (although helium is bcc, see Ref. 39).

The distinction between the classical and quantum values for the Lindemann melting ratio is of crucial importance to the question of whether T=0 monatomic hydrogen is liquid or solid, as can be seen in Figure 7. This figure shows various calculations of f for fcc T=0 hydrogen as a function of r_s. Hubbard and Smoluchowski[20] apparently used the classical value of f_m, in conjunction with the harmonic approximation to the atom-atom pair potential obtained in the absence of screening (dashed curve), in suggesting that the solid phase of hydrogen would cease to be stable for $r_s \lesssim 0.5$ or

P $\gtrsim$ 100 TPa. They noted that proper inclusion of screening might result in the liquid being the stable phase at all densities. Indeed, the Einstein calculations of Caron[9] (solid line) and the self-consistent harmonic calculations of Straus and Ashcroft[25] (data point)--both of which include screening-- lie generally above the classical melting ratio. As has been emphasized above, however, it is the quantum value of f_m ($\sim$0.3) which is the correct choice in the present case. According to Fig. 7, it then seems quite clear that T=0 hydrogen should be solid at essentially all densities.

The structure of the Einstein curve in Fig. 7 is typical of a quantum crystal such as solid helium. For both hydrogen and helium, the initial effect of compression is to <u>reduce</u> the relative amplitude of the zero-point motion, presumably because the atoms begin to encounter more of each others hard-core repulsive forces. Since hydrogen is only a mild quantum crystal ($f\sim$0.15), this effect has no dramatic consequence. For helium ($f\sim$0.3), however, it is well known that such compression causes the stable T=0, P=0 liquid to solidify. At extremely high densities the situation reverses, and the relative amplitude of the zero-point motion increases with compression. In this free particle limit, the repulsive forces are simply no longer able to compete with the much larger kinetic energy of the atoms and the crystal melts.

METASTABILITY OF THE MONATOMIC METAL

One source of continuing interest in monatomic solid hydrogen is its potential as a room temperature superconductor. To be practically useful, though, the solid would have to be maintained in a metastable state at some modest pressure. The difficulties facing this achievement are considerable, to say the least, and may well be impossible to surmount. One must first establish the monatomic phase at low pressure, presumably by rapid depressurization from the region where it is stable. Then one is confronted with the roughly 1 eV energy difference favoring the molecular phase. Given the apparently simple geometric rearrangement necessary for restoration of the molecular structure, this suggests an immediate and explosive recombination of the molecules. However, this intuitive assessment is not sufficiently precise. The major question is the mechanism by which the first few recombinations take place, and the probable time which will elapse before this occurs. Once the process has started, of course, the conversion to the molecular phase will be extremely rapid.

Crystal surfaces and interior defects are obvious regions for nucleation of this conversion. Salpeter[43] has examined the evaporative recombination from surfaces, and suggested the lifetime of

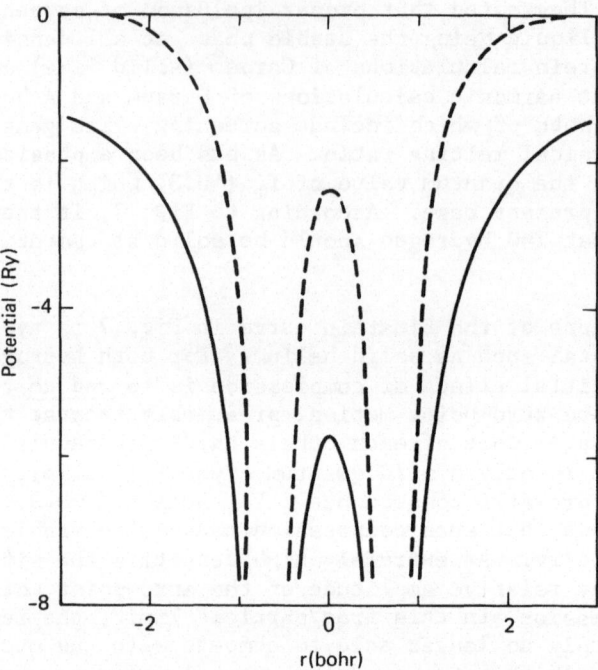

Figure 8.   Coulomb potential for two protons separated by a distance
            of 1.4 bohr. The solid curve shows the bare potential.
            The dashed curve shows the extent to which this potential
            is reduced due to dielectric screening by an electron gas
            corresponding to r_s = 1.7.

a macroscopic sample may be only 10^{-3} sec. It has been suggested
that this mechanism could be impeded by coating the solid with an
appropriate material.[26] A metal with a comparable electron density
to hydrogen would be needed to minimize rapid spatial variations
of the electron density at the interface.[44] As to interior faults
such as dislocations, it is a general property of quantum crystals
that they tend to self-anneal. Possibly a defect free crystal
could be obtained in the high pressure stable region, and maintained
as such throughout depressurization. There has been some concern
that vacancies and interstitials might be energetically favorable
even in the ground state of the monatomic solid. However, calculations by Caron[9] and by Dobson and Ashcroft[45] suggest that this
is not the case.

     Zero-point or thermal vibration of the atoms in the bulk
solid must also be considered a source of nucleation.[46] Should
two protons approach one another to within a distance comparable
to the 1.4 bohr equilibrium bond length of an H_2 molecule, it

might be energetically favorable for two electrons to become bound
in the resulting potential well, thus creating a molecule. In this
regard the suggested isotropic and solid structure for monatomic
hydrogen discussed in the previous section is helpful. The protons
will approach each other less closely in the solid than in the
liquid, and isotropic solids have generally the largest near neighbor
separations. Even so, Chapline[46] has suggested that zero-point
motion alone will lead to recombination in the zero pressure solid
within the order of 10^{-3} sec. This estimate may not be entirely
reliable, however, as the effects of dielective screening were not
properly accounted for. Figure 8 illustrates such effects in regard
to the potential well of two closely spaced protons. It is seen
that screening significantly reduces the size of the potential
well (dashed curve) over that which would obtain in the absence
of the electron gas (solid curve). It may well be that this more
shallow potential well _cannot_ support an electron bound state.[44]
Should this be the case, then the simultaneous close approach of
three or more protons would be required in order to obtain a combined
well deep enough to bind an electron. This would constitute
a more rare fluctuation in the lattice vibrational motion, and so
the lifetime involved in such a process could be considerably longer
than the 10^{-3} sec estimated by Chapline for the close approach
of just two protons. That such a process would be required is only
speculation at this point. However, the problem is of crucial importance
and warrants further investigation. One can at least talk
of coating the surface to inhibit evaporation, but if there is a
short lifetime for nucleation due to vibrational motion in the
bulk, then the prospect of practical low pressure metastability
must be irrevocably dismissed.

ACKNOWLEDGMENTS

In the course of preparing this review I have benefited from
many discussions with my colleagues Drs. M. Ross, W. Hoover,
R. Grover, and D. Young. I am also grateful for conversations
with Profs. N.W. Ashcroft and G.V. Chester. Prof. Chester kindly
permitted reference to his work on the relative stability of the
solid and liquid phases, which is yet to be published.

REFERENCES

1. E. Wigner and H.B. Huntington, J. Chem. Phys. _3_, 764 (1935).
2. A.A. Abrikosov, Astron. Zh. _31_, 112 (1954).
3. V.P. Trubitsyn, Fiz. Tverd. Tela. _8_, 862 (1966) [Sov. Phys.
   Solid State _8_, 688 (1966)].
4. G.A. Neece, F.J. Rogers, and W.G. Hoover, J. Compt. Phys. _7_,
   621 (1971).

5. D.A. Liberman, "Equation of State of Molecular Hydrogen at
    High Pressure", LA-4727-MS, Los Alamos Scientific Laboratory
    (1971); Int. J. Quan. Chem. Symp. No. 10, 297 (1976).
6. E.A. Dynin, Fiz. Tverd. Tela. 13, 2488 (1971) [Sov. Phys. Solid
    State 13, 2089 (1972)].
7. G.F. Chapline, Jr., Phys. Rev. B6, 2067 (1972).
8. E. Ostgaard, Phys. Lett. 45A, 371 (1973).
9. L.G. Caron, Phys. Rev. B9, 5025 (1974). Caron's Fig. 4 has
    been used to obtain the rms atomic vibrational amplitude for
    fcc monatomic hydrogen shown in Fib. 7 (solid curve) of the
    present paper.
10. M. Ross, J. Chem. Phys. 60, 3634 (1974).
11. M. Ross, R.H. Ree, and R.N. Keeler, Proc. of the Fourth Int.
    Conf. on High Pressure (Kyoto, 1974), pp. 852.
12. J. Hammerberg and N.W. Ashcroft, Phys. Rev. B9, 409 (1974).
13. R. Etters, R. Danilowicz and W. England, Phys. Rev. A12, 2199
    (1975).
14. D.E. Ramaker, L. Kumar and F.E. Harris, Phys. Rev. Lett. 34,
    812 (1975).
15. M. Ross and A.K. McMahan, Phys. Rev. B13, 5154 (1976).
16. F.V. Grigor'ev, S.B. Kormer, O.L. Mikhailova, A.P. Tolochke,
    and V.D. Urlin, ZhETF Pis. Red. 16, 286 (1972) [JETP Lett.
    16, 201 (1972)]; Zh. Eksp. Teor. Fiz. 69, 743 (1975) [Sov. Phys.
    JETP 42, 378 (1975)].
17. R.S. Hawke, D.E. Duerre, J.G. Huebel, R.N. Keeler, and
    H. Klapper, Phys. Earth Planet. Interiors 6, 44 (1972).
18. N. Kawai, M. Togaya and O. Mishima, Proc. Japan Acad. 51, 630
    (1975).
19. L.F. Vereshchagen, E.N. Yakovlev and Yu. A. Timofeev, ZhETF
    Pis. Red. 21, 190 (1975) [JETP Lett. 21, 85 (1975)];
    E.N. Yakovlev, Priroda, No. 4, 66-69 (1976).
20. W.G. Hubbard and R. Smoluchowski, Space Sci. Rev. 14, 599 (1973).
21. N.W. Ashcroft, Phys. Rev. Lett. 21, 1748 (1968).
22. M. Ross and C. Shishkevish, "Molecular and Metallic Hydrogen",
    Rept. R-2056-ARPA, May 1977, Rand Corporation, Santa Monica,
    Ca 90406.
23. C. Friedli and N.W. Ashcroft, Phys. Rev. B, to be published.
24. G.V. Chester, to be published.
25. D.M. Straus and N.W. Ashcroft, Phys. Rev. Lett. 38, 415 (1977).
26. E.G. Brovman, Yu. Kagan and A. Kholas, Zh. Eksp. Teor. Fiz. 61,
    2429 (1971); 62, 1492 (1974) [Sov. Phys. JETP 34, 1300 (1972);
    35, 783 (1972)].
27. A.K. McMahan, H. Beck and J.A. Krumhansl, Phys. Rev. A9, 1852
    (1974).
28. F.H. Ree and C.F. Bender, Phys. Rev. Lett. 32, 85 (1974).
29. B.K. Tong and L.J. Sham, Phys. Rev. A1, 139 (1965); Y.S. Kim
    and R.G. Gordon, J. Chem. Phys. 60, 1842 (1974).
30. At low densitits where the correlation energy becomes very
    important, conventional electron band theory breaks down. The
    monatomic phase of hydrogen may in fact undergo a Mott transi-

tion to an insulating state. However, the calculations of
Ambladh et al. [C.O. Almbladh, U. von Barth, Z.D. Popovic,
and M.J. Stott, Phys. Rev. B14, 2250 (1976)] indicate that
this transition will occur at such a low density (r_s = 1.9) as
to correspond to a negative pressure. Throughout this paper,
then, it is assumed that the monatomic phase is indeed every-
where metallic. Similar conclusions have been reached by
Aviram, et.al. [I. Aviram, S. Goshen, Y. Rosenfeld, and
R. Thieberger, J. Chem. Phys. 65, 846 (1976).]

31. Liberman's Korringa-Kohn-Rostocker calculations for the molecular phase (Ref. 5) do not show band overlap. It is possible that
the spherical approximation used for the molecules in this work
is the source of this difference. Liberman finds that the optimized diameter of these spherical molecules (analogous to
the bond length) steadily decreases with compression, whereas
using a rigorous treatment of the molecular configuration
Ramaker et al. (Ref. 14) find the bond length to first stay
the same and then increase with compression. It is consistent
with the work of Ref. 23 that a decreasing bond length will delay and perhaps eliminate altogether closing of the gap.

32. N.F. Mott, Metal-Insulator Transitions (Taylor & Francis, London, 1974), pp. 19.

33. I.M. Lifshitz, Z. Expl. Theor. Phys. 38, 1569 (1960) [Sov. Phys.-
JETP 11, 1130 (1960)].

34. A.S. Balchan and H.G. Drickamer, J. Chem. Phys. 34, 1948 (1961);
L.F. Vereshchagin, A.A. Semerchan, N.N. Kuzin and Yu A. Sadkov,
Dokl. Akad. Nauk SSSR, 209, 1311 (1973) [Sov. Phys. Dokl. 18, 249
(1973)].

35. Early work by Alder and Christian [B.J. Alder and R.H. Christian,
Phys. Rev. Lett. 4, 450 (1960)] claimed to see such a transition,
but is now believed to be in error. See Ref. 36.

36. A.K. McMahan, B.L. Hord and M. Ross, Phys. Rev. B15, 726 (1977).

37. This zero pressure density comes from the augmented-plane-wave
calculations reported in Ref. 15. Free electron perturbation
theory gives a zero pressure value of r_s closer to 1.6.

38. H. Beck and D. Straus, Helvetica Physica Acta 48, 655 (1975).

39. This is the range typically given for the Lindemann ratio in
the case of melting due to thermal motion. The ratio is generally presumed to be independent of structure. However,
Shapiro has recently suggested that f_m is in fact structure
dependent with values of about 0.12 and 0.20 for fcc and bcc
lattices respectively (again for classical melting).
(J.N. Shapiro, Phys. Rev. B1, 3982 (1970). Note that his
δ_f is smaller than our f_m by a factor of 1/$\sqrt{3}$]. The structure dependence is not an issue in the present discussion of
classical vs. zero-point melting as the calculations cited
in Refs. 9, 25, 40-42 are all for the fcc phase.

40. W.G. Hoover, S.G. Gray and K.W. Johnson, J. Chem. Phys. 55,
1128 (1971).

41. D.Ceperley, G.V. Chester and M.H. Kalos, "A Monte Carlo Study

of the Ground State of Bosons Interacting with Yukawa Potentials", "Monte Carlo Simulation of a Many Fermion System",
(preprints, May 1977).

42. J.-P. Hansen, D. Levesque and D. Schiff, Phys. Rev. A3, 776
     (1971).

43. E.E. Salpeter, Phys. Rev. Lett. 28, 560 (1972).

44. N.W. Ashcroft (private communication).

45. J.F. Dobson and N.W. Ashcroft, "Einstein-Kanzaki Model of
     Static and Dynamic Lattice Relaxation: Application to Vacan-
     cies in Metallic Hydrogen" (preprint, December 1976).

46. G.F. Chapline, Jr., Phys. Rev. B6, 2067 (1972).

QUESTIONS AND COMMENTS

E.F. Skelton: You mentioned the fact that iodine may be a good
               material to look at as a model of hydrogen. One
               should note that in collaboration with Ian Spain's
               lab of the University of Maryland, we have just last
               week observed a structural transition in iodine. The
               pressure hasn't been measured yet, but there is clear
               x-ray evidence for a structural transformation.

A.K. McMahan: Would you like to make a guess at the pressure?

E.F. Skelton: It is somewhere between 150 and 200 kbar. There's
               a ruby crystal in there to determine the pressure in
               the near future. (note added in proof: Based on the
               ruby-R line shift, the transition pressure is estimated
               to be 162±3 kbar.)

A.K. McMahan: Is it definitely involved with the onset of metal-
               lization? Did you measure the conductivity?

E.F. Skelton: No. Not yet.

T.M. Rice: My question concerns the distinction between the semi-
               metallic state caused by band overlap in the molecular
               phase of hydrogen and the metallic monatomic phase of
               hydrogen. The energy gap in the molecular phase you
               estimated is collapsing at a rate of 5 eV per megabar
               so that the band overlap rapidly becomes large and it
               is not clear to me how to distinguish between the
               molecular phase with a fairly large band overlap and
               the monatomic metal.

A.K. McMahan: According to the calculations of Friedli and Ashcroft,
               the width of the energy gap between the full valence
               band and the empty conduction band in the molecular
               phase depends linearly on some fractional power of

the volume. Thus it is only because molecular hydrogen is so very compressible at pressures below a Mbar
that an applied pressure of a few Mbar can reduce
the initial 9 eV gap to zero. A subsequent increase
in pressure to, say, 5 Mbar will actually result in
only a small overlap (negative gap), equivalent to
about 10 percent of the valence band width at that
density. The change in electronic structure brought
about by such small overlap may not be large enough
to disrupt the covalent bonds in the H_2 molecules,
so that a molecular phase may still exist. The distinction between such a molecular but metallic phase
and the more familiar monatomic metal will lie in the
ground state energy having two local minima as a function of lattice structure. The calculations of
Ramaker et al. are the best indication to date that
at pressures somewhat below dissociation there are
indeed two such local minima, corresponding to two
metallic phases, one stable and one metastable, and
characterized by different crystalline symmetries.

A.L. Ruoff: If the molecular hydrogen as a result of band overlap becomes metallic molecular hydrogen, would not
the eventual transition pressure to the monatomic
metal be even higher than calculated on the assumption
that the molecular hydrogen remains an insulator?

A.K. McMahan: The decreasing band gap and eventual overlap of the
bands in the molecular phase is really a symptom of
the increasing importance of many-body interactions
amongst the H_2 molecules as the solid is compressed.
Since the Ross CI + ATT potential has been empirically
deduced from high pressure shock data, I would guess
that these effects may already be included in some
rough way in the current molecular equation of state,
and consequent estimates of the transition pressure
to the monatomic phase. As I mentioned in the talk,
the effect of many-body interactions in the molecular
phase is such as to raise the pressure at which the
H_2 molecules dissociate.

I.L. Spain: I'd like to make a comment. You said that it hadn't
been realized that the Lindemann criterion was different for zero-point motion and for thermal motion. In
fact, this point was thoroughly discussed by Dugdale
and Domb with respect to helium, some 20 years ago.

A.K. McMahan: I was referring only to the discussions of the liquid
or solid nature of monatomic hydrogen.

I.L. Spain: It seems to me that we're talking about metastability
                of something which we may not be able to make. On
                the other hand, it is well known, for instance, that
                the metallic form of InSb can be brought back to
                Earth. Also, it is well known that if silicon and
                germanium are compressed and pressure is then lowered,
                a metastable structure results--not the high pressure
                β tin-like structure, but of some other structure.
                I'd like to ask the question, is it now possible to
                apply some of these theoretical ideas to those sub-
                stances for which we have actually done the experi-
                ments to see how the theory checks out?

A.K. McMahan: Yes it would be possible. However, iodine is a better
                candidate than these materials if one is looking for
                insight into the hydrogen problem. Like hydrogen,
                iodine has a diatomic molecular structure in the in-
                sulating phase, whereas these other materials are,
                of course, tetrahedrally coordinated.

HIGH HYDROGEN PRESSURES IN SUPERCONDUCTIVITY

B. Baranowski, T. Skoskiewicz

Institute of Physical Chemistry, Polish Academy of
Sciences
Warsaw, Poland

The experimental conditions for the preparation of metallic hydrogen are not easily available. Therefore it is of interest to study the eventual metallic behaviour of hydrogen in the form of its alloys[1] as these can be probably obtained in much less extremal conditions than the pure component. This means —first of all— a much smaller equilibrium pressure of hydrogen above such alloys than is the case for metallic hydrogen. Let us stress here the analogy to metallic mercury whose equilibrium pressure above amalgams can be by orders of magnitude smaller than above the pure component. It is obvious that the higher the thermodynamic activity of hydrogen the closer the approach to its eventual metallic behaviour.

In connection with these considerations of special interest are hydrides of transition metals which are characterized by a high number density of hydrogen particles and typical metallic properties.

For several metallic hydrides an unavoidable condition for their preparation is a high thermodynamic activity of hydrogen. As already known examples the preparation of hydrides of nickel,[2] chromium[3] and manganese[4] can be mentioned. A similar situation exists in a couple of alloys based on these metals. Further non-stoichiometric hydrides prepared at low hydrogen activities —like palladium hydride — can still absorb hydrogen when kept in contact with a high thermodynamic activity of this component. Higher concentration of hydrogen inside the metallic matrix means an in-

creasing interaction between these particles and an approach to a
metallic behaviour of this component.

Low temperature investigations of metal hydrides have been
promoted by the results of theoretical speculations of Ashcroft,
who estimated that the metallic hydrogen should be a high temperature superconductor.[5]

Superconductivity of higher hydride of thorium Th_4H_{15} discovered in 1970 by Satterthwaite[6] and particularly superconductivity of
palladium hydride discovered by one of us in 1972[7] stimulated many
experimental and theoretical investigations of superconductivity
of metal hydrogen system.

In recent years several methods of hydrogenation of metals have
been applied in preparation of superconducting metal hydrides.

-- most commonly used was the electrolytical cathodic charging
   at room temperature as well as at somewhat lower temperature
   (about - 70°C) (Satterthwaite,[8] Harper,[9] Burger[10] and others).
-- ion implantation at helium temperature applied to hydride
   samples preparation by Stritzker, Buckel[11] led to the highest critical temperature ($T_c \simeq 17K$ for $Pd_{0.6}Cu_{0.4}$ +H)[12]
   among metal hydrogen systems.
-- evaporation of metal into a layer of solid hydrogen previously deposited on a cold substrate (L.E. Sansores,[13]
   Silverman[14]).
-- high hydrogen pressures.
Besides the last one, in all of these methods hydrogenation
is performed in non-equilibrium conditions. Hydrogen atoms are produced here, being easily recombined to hydrogen molecules, especially in contact with a metallic surface.   This thermodynamically
preferred process proceeds simultaneously with the adsorption and
absorption inside the metallic sample.   In fact a competition of
many relaxation processes takes place which can simply lead to
poor reproducibility.   Further, a nonhomogenious hydrogen distribution in the samples is obtained.   In some cases even an estimation of the mean hydrogen concentration is not available.

Superconductivity of palladium hydride was found on samples
hydrogenized electrochemically at room temperature.

At H/Pd atomic ratios higher than 0.77 strong dependence of
critical temperature on H/Pd ratio was observed.   The maximum
critical temperature T_c = 6.6K at H/Pd = 0.94 was found.   Very
broad transitions were typical for the samples prepared electrochemically.[15]

Further increase of critical temperature of superconducting
palladium hydride was achieved when high hydrogen pressure technique was applied for sample preparation.[16]

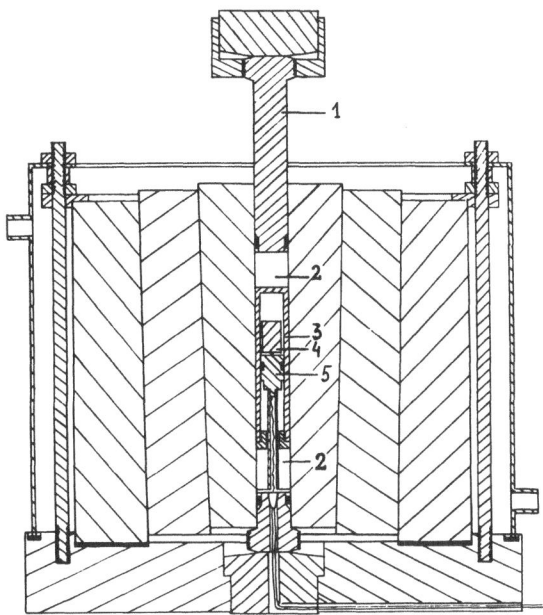

Figure 1. High hydrogen pressure device: 1-piston, 2-pressure
transmitting liquid, 3-hydrogen chamber, 4-sample holder,
5-sealing system separating hydrogen and liquid.

The high pressure apparatus is shown in Fig. 1. The metallic
samples were hydrogenized at room temperature under hydrogen pressures up to 25 kbar. After a time necessary for achievement of
homogeneous distribution of hydrogen in the samples the whole apparatus was cooled down to about - 50°C and then samples were moved
into the cryostat. However, even at this temperature (-50°C) a
partial desorption of hydrogen after pressure reduction was detected.
This reduced the homogeneity of hydrogen distribution and the average hydrogen concentration. High diffusion coefficient of hydrogen
in palladium and small thickness of samples worked in favour of this
process. To avoid these disadvantages another device was developed.[17]
It is shown in Fig. 2. It can stand pressures only up to 13 kbar
but on the other hand it can be cooled down even to helium temperatures without releasing the pressure.

In Fig. 3 superconducting transitions observed on the samples
prepared under high hydrogen pressures are shown. Critical temperatures exceeding 9K were observed for H/Pd approaching unity.[17] The
transitions observed were much narrower than in the case of samples
prepared electrochemically. This device was also suitable for the
estimation of influence of hydrostatic pressure on critical point of
samples.

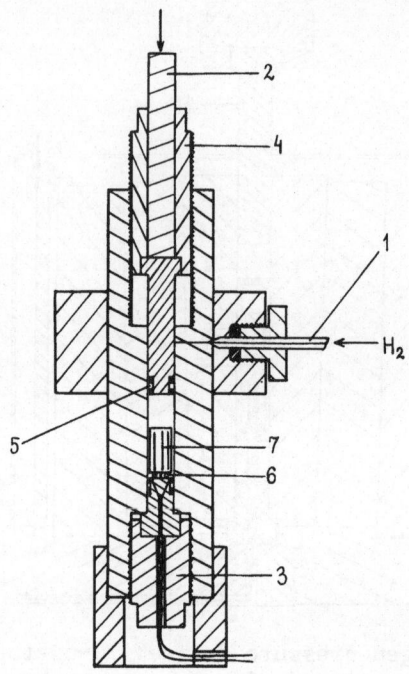

Figure 2.   High hydrogen pressure device working at low temperatures
            (17):   1-gas inlet, 2-piston, 3- and 4-screws, 5-sealing,
            6-manganin coil, 7-samples.

In Fig. 4 a negative influence of pressure on critical temperature of palladium hydride and deuteride is presented. Lower
critical temperature appeared when measurement was performed without deflation of hydrogen from the high pressure vessel. The pressure in solidified hydrogen was estimated in the following way:
knowing the hydrogen pressure in the vessel at room temperature we
were able to estimate the corresponding molar volume of hydrogen
from De Graaff (1960) data.

Compressibility of solid hydrogen at 4.2K reported by Stewart,[25]
served for the calculation of the pressure at 4.2 corresponding to
the estimated molar volume. These pressures are given in the figure. After deflation of hydrogen at liquid nitrogen temperature
the critical temperature increased for about 0.5 deg.

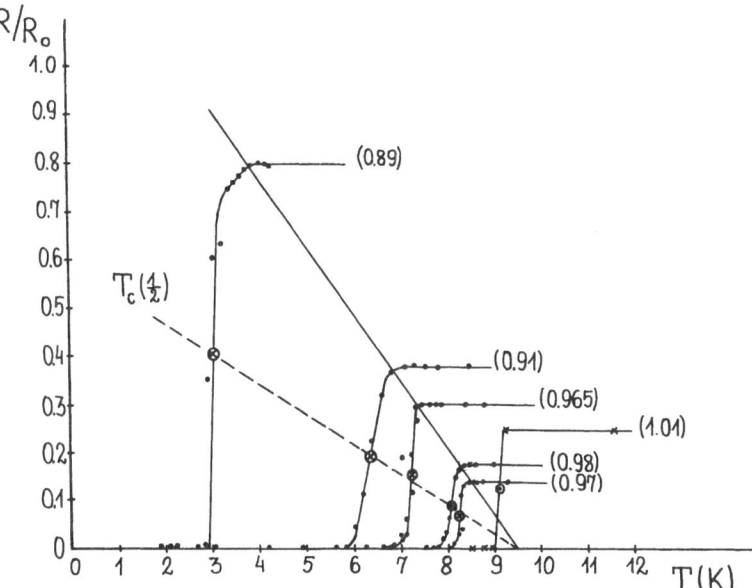

Figure 3. Relative resistance R/R_o of palladium hydride samples of different hydrogen content vs. temperature. R_o is the resistance of the hydrogen free sample at 25°C.

Using the above high pressure device we could investigate the isotope effect.[17] Results are shown in Fig. 5.

High hydrogen pressures in the study of superconductivity were also used by:
-- Schirber, who measured the dependence of T_c in Pd-H and $Pd_{.95}Rh_{.05}H$ on hydrogenation pressure[18] and independently measured the isotope effect.[19]
-- Igalson[20] and Eichler[21] who measured the energy gap of Pd-H in tunneling experiments.
-- Horobiowski[22] who made magnetization measurements and critical magnetic field measurements.
The advantages of high hydrogen pressure technique can be illustrated by several examples:
-- Satterthwaite was able to prepare massive and homogeneous samples of superconducting thorium hydride using a simple quartz device for the reaction of thorium and hydrogen at elevated temperature and hydrogen pressure increased up to 750 bar.[23] The synthesis of normal hydrogen pressure gave

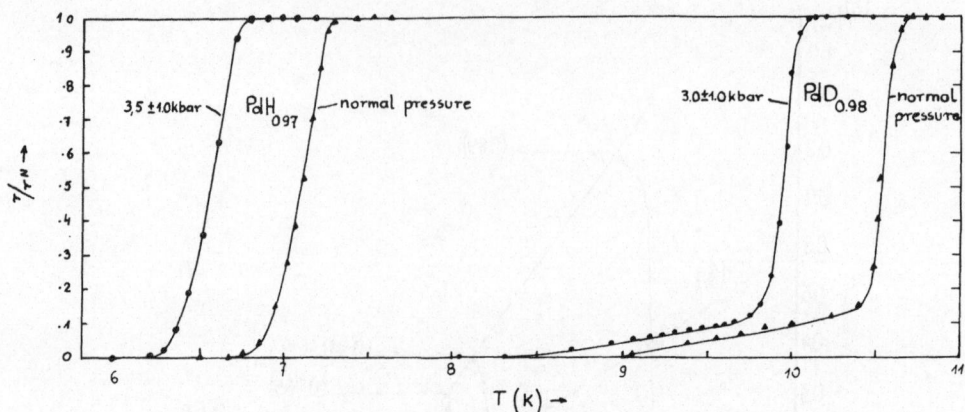

Figure 4. Effect of hydrostatic pressure on superconductive transitions of palladium hydride and deuteride samples (17).

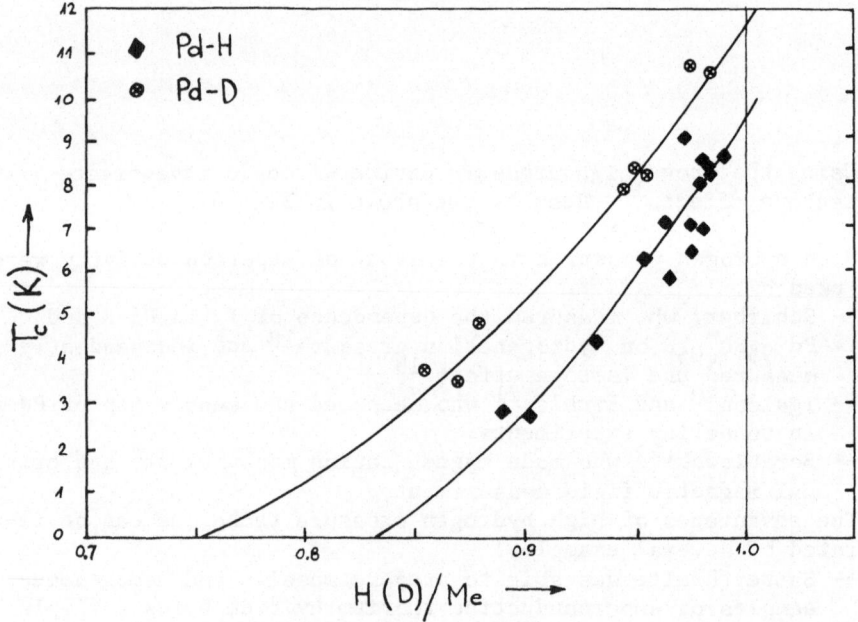

Figure 5. Critical temperature in Pd-H and Pd-D systems against
hydrogen or deuterium content (17).

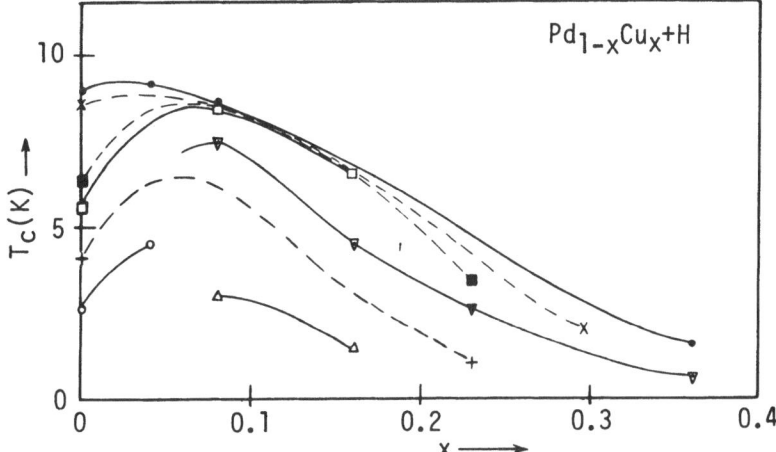

Figure 7. Superconducting transition temperatures of Pd–Ni alloy series saturated with hydrogen or deuterium at different pressures, as a function of nickel concentration.

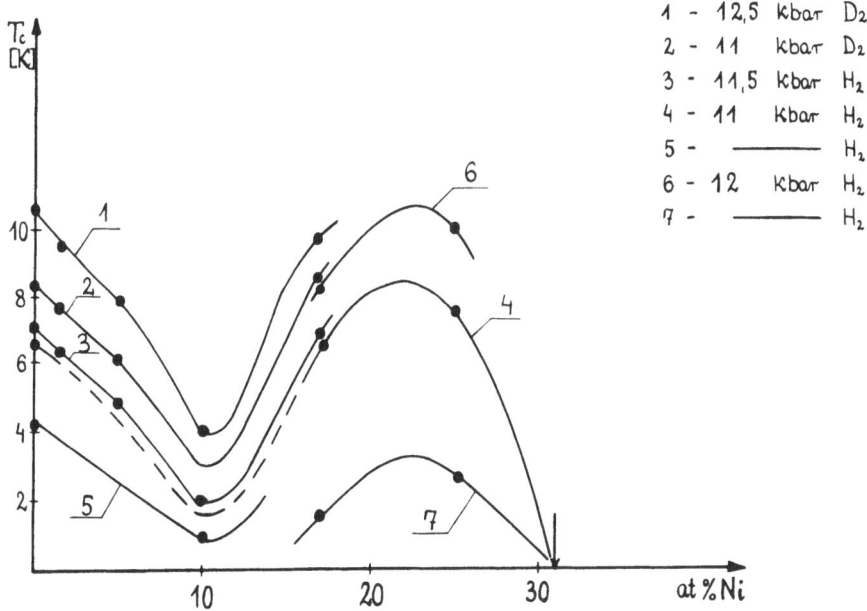

Figure 6. Superconducting transition temperatures of Pd–Cu alloy series saturated with hydrogen or deuterium (filled marks) at various pressures, as a function of copper concentration.

  only powdered samples.
-- preparation of palladium hydride samples for tunneling experiments is only possible by high pressure method or
  by metal condensation on solid hydrogen.
-- Horobiowski[22] was able to state type I superconductivity
  of Pd-H, synthesized under high hydrogen pressure in contradiction to McLachlan et al.[10] who stated type II superconductivity probably due to inhomogeneity of their samples
  prepared electrochemically.

These findings on superconductivity of palladium – hydride have
stimulated investigations of palladium based alloy hydrides. Particularly interesting results were achieved for palladium – noble
metal and palladium – nickel alloys. In the case of noble metal
additions, solubility of hydrogen decreases. This is the reason
why ion implantation technique provides probably higher hydrogen
concentrations than other methods of sample preparation. In Fig. 6
critical temperatures for Pd-Cu-H samples under high pressure are
shown.

  The samples hydrogenized by ion implantation exhibit higher
critical temperatures.

  However, in the case of Pd-Ni-H system the high pressure
hydrogenation has advantages over other methods. In Fig. 7 critical temperature of Pd-Ni alloys saturated with hydrogen or deuterium at various pressures as a function of Ni concentration is
shown.[24]

  It is seen that after hydrogenation, superconductivity was observed at nickel concentrations up to 30 at% while pure alloys exhibit ferromagnetic ordering at nickel concentrations higher than
2.5 at %. Therefore the magnetic ordering had to be totally reduced by homogeneous hydrogenation before superconductivity could
appear.

  Let us stress that high pressures of gaseous hydrogen provide
a very large thermodynamic activity of this component. Due to a
strong deviation from ideality the effective thermodynamic potential is several times larger than that corresponding to an ideal
gas: for example at 12 kbar and 25°C the fugacity approaches the
value of 10^7 bar - that is about three orders of magnitude higher
than the hydrostatic pressure. This feature is an additional positive aspect of the application of high pressure technique in the
investigations of metal – hydrogen systems.

REFERENCES

1. B. Baranowski, T. Skoskiewicz, A.W. Szafranski, Fiz. Niz. Temp., 1, (1975), 616.
2. B. Baranowski, K. Bochenska, S. Majchrzak, Roczn. Chem., 41, (1967), 2071.
3. B. Baranowski, K. Bojarski, Roczn. Chem. 46, (1972), 525.
4. M. Krukowski, B. Baranowski, Roczn. Chem., 49, (1975), 1183.
5. N.W. Ashcroft, Phys. Rev. Lett., 21, (1968), 1748.
6. C.B. Satterthwaite, I.L. Toepke, Phys. Rev. Letters 25, (1970), 741.
7. T. Skoskiewicz, Phys. Status Solidi (a) 11, (1972), K123.
8. R.J. Miller, C.B. Satterthwaite, Phys. Rev. Lett., 34, (1975), 144.
9. J.M. Harper, Phys. Letters, 47A, (1974), 69.
10. D.S. Mac-Lachlan, T.B. Doyle, J.P. Burger, Proc. XIV-th Int. Conf. Low Temp. Phys. Otaniemi 1975, Vol. 2, p. 44.
11. B. Stritzker, W. Buckel, Z. Phys. 257, (1972), 1.
12. B. Stritzker, Z. Phys. 268, (1974), 261.
13. L. Sansores, R.E. Glover, Bull. Am. Phys. Soc., 19, (1974), 437.
14. P.I. Silverman, C.V. Briscoe, Phys. Letters, 53A, (1975), 221.
15. T. Skoskiewicz, Phys. Stat. Sol. (b), 59, (1973), 329.
16. B. Baranowski, W. Bujnowski, Roczn. Chem., 44, (1970), 2271.
17. T. Skoskiewicz, A.W. Szafranski, W. Bujnowski, B. Baranowski, J. Phys. C: Solid State Phys., 7, (1974), 2670.
18. J.E. Schirber, Phys. Letters 45A, (1973), 141.
19. J.E. Schirber, Phys. Rev. B 10, (1974), 3818.
20. J. Igalson, et al, Solid State Comm., 17, (1975), 309.
21. A. Eichler, H. Wuhl, B. Stritzker, Solid State Comm., 17, (1975), 213.
22. M. Horobiowski, T. Skoskiewicz, E. Trojnar, Phys. Stat. Solidi (b) 79, (1977), K147.
23. C.B. Satterthwaite, D.T. Peterson, J. Less Common Metals, 26, (1972), 361.
24. T. Skoskiewicz, to be published.
25. J.W. Stewart, J. Phys. Chem. Solids 1, (1956), 146.

QUESTIONS AND COMMENTS

J. Wittig: What is the initial hydrogen concentration at the loading pressure of 25kbar?

B. Baranowski: I didn't speak about it. We measured last time the absorption isotherms of hydrogen and palladium and

it seems that at room temperature it reaches already a one-to-one ratio at a pressure of 12 kbar.
This is still an unanswered question because upon
increasing the pressure, one observes the relaxation
process by resistivity measurements, which could be
explained by a further uptake of hydrogen. This is
not to be expected from the absorption isotherm.
So it's hardly understood what's going on between
12 and 25 kbar.

J. Wittig: So what would you guess of it?

B. Baranowski: Well, we are speaking only about direct evidence.
                     The direct evidence indicates that we have never ex-
                     ceeded one-to-one. The calculations give you a con-
                     centration of 1.2, or even higher, but we have shown
                     that these calculations break down at high pressures.
                     It's meaningless for the range of 10 kbar.

D. Bloch: Have you an idea about the interpretation of these
                     complex temperature dependences, especially those
                     which exist above 10 kbar?

T. Skoskiewicz: What I can say for sure, is that it's not a minimum
                     due to the hydrogen concentration. That means the
                     behavior of the critical temperature doesn't reflect
                     the variation of the hydrogen concentration. In this
                     pressure region, we've always had about one-to-one
                     H/metal ratio.

B. Baranowski: The hydrogen-to-metal ratio, for both nickel and
                     palladium, is more or less the same in the whole
                     region.

C.G. Homan: Is there any neutron scattering data on the phonon
                     spectrum?

B. Baranowski: I know of none at such high concentrations.

E.F. Skelton: Did you do your compositional analyses before pres-
                     surization, and after pressurization, and is there
                     any possibility of squeezing some of the hydrogen
                     out?

B. Baranowski: Well, we expected there was no change at all because
                     a reduction of hydrogen pressure was performed at
                     liquid nitrogen temperature.

E.F. Skelton: So, do you know that the structure is invariant to
                     your pressure?

B. Baranowski: Well, we expect there was no change of the composition at all, at this low temperature. On the other hand, it was performed rather quickly.

T. Skoskiewicz: And there was no change in the electrical resistance.

B. Baranowski: True.

J.E. Schirber: I have a comment regarding that. I made the same measurements of the pressure dependence of T_c in palladium hydrides with one-to-one ratio, or near one-to-one ratio. Again making the samples the same way, by high pressure hydrogen, so that you get a very isotropic specimen. And I've done the pressure measurements with both hydrogen as a pressure transmitter and with helium and so I can corroborate with what you're saying. It's a real pressure effect, not a loss of H.

B.T. Matthias: If it's the hydrogen that does it, how come the isotope effect has the wrong sign?

B. Baranowski: For metallic hydrogen, you are right, one would expect it to be just the opposite isotope effect. It doesn't work in the right direction.

B.T. Matthias: Why?

B. Baranowski: I didn't say that metallic hydrogen is responsible for superconductivity. It's not my opinion, so I have no defense for opinions of other people.

THE ROLE OF PRESSURE IN THE STUDY OF THE FERMI SURFACE*

J.E. Schirber

Sandia Laboratories

Alburquerque, New Mexico 87115

ABSTRACT

Pressure studies of the Fermi surface are providing critical tests for the physical significance of model descriptions for the electron structure of metals. Work since the most recent reviews is cited. The solid and fluid He pressure techniques used in ongoing studies on Gd, U and Th are described briefly and results on these materials are discussed.

I. INTRODUCTION

Pressure studies of the Fermi surfaces of metals have proven to be critical in evaluating various theoretical models and descriptions of the metal. The philosophy involved is to make whatever approximations and parameterizations necessary to achieve the best fit to the available normal volume data and then redo the model calculation at a reduced volume. A physically significant model should account for the observed variations with pressure of Fermi surface cross sections and should reflect properties related to corresponding changes in the density of states. Most of the studies to date have been on elemental metals but an increasing effort is being mounted in the area of intermetallic compounds.

It is probably fair to say that most of the easily obtainable single crystal elemental metals have been subjected to some level

*This work was supported by the United States Energy Research and Development Administration, ERDA, under Contract AT(29-1)789.

of investigation as a function of pressure. The bulk of these
studies has been reviewed by Schirber[1] and by Brandt, Itskevich
and Minina.[2] The effects of pressure on the electronic structure
of d-band transition metals was reviewed by Svecharev and Panfilov.[3]
Therefore, only the most recent work will be cited here. We will
also confine our comments to pressure studies and not discuss in
the interest of brevity the similarly useful information obtained
from uniaxial stress[4] and magnetostriction techniques.[5]

Summarizing first recent work on the elements, Cs has been
studied by dHvA measurements in both fluid He and in solid He to
nearly 10 kbar.[6] Templeton[7] has substantially reduced the uncertainty in the pressure derivatives of Fermi surface cross
sections along symmetry directions for the noble metals Cu, Ag
and Au. Bryant and Vuillemin[8] have reported fluid He phase shift
measurements in Cd which extend and complement the earlier work
of Schirber and O'Sullivan.[9] Beardsley, et al.[10] have completed
a rather complete experimental study of the Fermi surface of Mg
which points out dramatically the importance of convergence in
using local pseudopotential descriptions. There has been recent
work[11] on In which extends and appears to be in quantitative disagreement with the earlier studies.[12] Brandt et al.[13] reported
studies of the Fermi surface of Sn to 10 kbar by measuring the
oscillatory surface impedance. Anderson, et al.[14] report comparison of Nb Fermi surface derivatives with a self-consistent
APW band structure at normal and reduced lattice spacings. Fermi
surface cross sections of both the hole and electron sheets of
As were measured as a function of pressure and analyzed in terms
of the Lin and Falicov pseudopotential model by Schirber and
Van Dyke.[15]

The first direct Fermi surface studies under pressure on a
rare earth material was the work of Slavins and Datars[16] on hcp Yb.
They measured the pressure derivatives of three dHvA frequencies
for H || [0001] and could see no oscillations in the fcc phase.
Ribault,[17] on the other hand, reports a single frequency for the
fcc phase which extrapolates to a magnitude of zero near 12 kbar.

Studies of the ferromagnetic d-band metal Fermi surfaces
as a function of pressure have been initiated[18,19] in Co and Ni.
These data are as yet fragmentary but promise to be sensitive to
the details of the model description chosen to represent a ferromagnetic metal.

An increasing number of compound metals are being studied
using high-field magnetoresistance and dHvA techniques. It is
therefore expected that pressure studies will ultimately be con-
ducted in these materials. Several sheets of the Fermi surface of
ReO_3 have been studied as a function of pressure by Schirber,
et al.[20] Reasonable agreement with these results was obtained by

Myron, Gupta and Liu[21] who performed a Korringa-Kohn-Rostoker band
calculation as a function of interatomic spacing. To date the series $AuGa_2$, $AuIn_2$, $AuAl_2$ has received the most attention primarily
because of anomalies in the magnetic and superconducting properties
of $AuGa_2$. These anomalies can be understood on the basis of a band
structure-electron transition model involving pushing the second band
at Γ through the Fermi level at 7 kbar and liquid He temperatures.[22]
Anomalies in elastic properties and nmr were explained in InBi by a
similar band structure based argument which could account for the
pressure dependence of the Fermi surface measured by Schirber and
Van Dyke.[23]

Section II will describe very briefly the experimental techniques which have been used in recent studies of Gd, U and Th. The
results of these studies will be given and discussed in Section III.
Section IV will summarize where we are in this field today and give
an opinion as to where the emphasis should be in future studies.

II. EXPERIMENTAL

High-quality long electron mean-free-path single crystal specimens are typically required for detailed measurements of the Fermi
surface. This means that measurements are usually performed in high
magnetic fields and at temperatures near 1 K. Under these conditions
all pressure media are of necessity solid above a few bar making it
difficult to maintain strain-free conditions particularly on anisotropic materials. Although isolated examples of pressure studies of
the Fermi surface were reported over two decades ago, it was not until the development of the field modulation deHaas-vanAlphen (dHvA)
technique[24] that comprehensive studies could be performed. The
field modulation technique allowed highly accurate Fermi surface
cross section determinations without the necessity of introducing
leads into the high-pressure environment. A second impetus to this
type of study was the demonstration[25] that extremely hydrostatic conditions to 10 kbar could be achieved by careful isobaric freezing of
He even with the most fragile and anisotropic single crystal specimens.

The dHvA measurements further lend themselves to pressure
studies in that the oscillatory property, for example the magnetization which is given by $M = M_0 \sin 2\pi F/B$, typically has a very
high phase F/B. (Here F is the dHvA frequency and B is the magnetic induction.) Thus very small changes in the frequency or
cross-sectional area of the Fermi surface are related to shifts
in position ΔH of a given feature in the pattern by

$$d\ln F/dP = \frac{1}{B}\frac{\Delta H}{\Delta P}$$

In some situations [26,27] the sensitivity is such that the pressure
derivative of the frequency can be measured accurately in fluid
He, i.e., at pressures <25 bar at 1 K. These phase shift tech-

niques have also been applied to situations where F/B is not large and/or the pressure derivative is small using solid He to effect measurable shifts in the oscillation position.[28] Typically, it is possible to obtain useful information about the volume dependence of most cross sections of the Fermi surface in a given material by either one or a combination of i) direct measurement of the frequency in solid He to 10 kbar, ii) phase shift determination of the pressure derivative in fluid He, iii) phase shift determination of the derivative in solid He to 10 kbar.

In the studies of 4f and 5f materials discussed below, the work was all done in a 55 kOe superconducting solenoid at temperatures in the 1 - 4 K range. All three pressure techniques described above were used as appropriate. The Gd sample was a 1/8 in. diameter sphere cut by spark erosion from a rod of residual resistance ratio ∿260 prepared by an electrotransport process.[29]

The Th was also purified by electrotransport and then heated 144 hours at 4 x 10^{-12} Torr. Crystals were formed by a temperature cycling technique[30] with residual resistance ratios of ∿2000. Samples were cut by spark erosion into 1 mm diam. x 3 mm long cylinders oriented along principal crystallographic directions.

The U samples were cut by spark erosion into ∿1 mm diam. x ∿3 mm long cylinders from single crystal material grown by recrystallization within the α phase.[31] The residual resistance ratio as measured was about 30 but it is not clear what the significance of this ratio is in view of the numerous phase changes U undergoes below 50 K. It was found that pressures >7.5 kbar quenched these transitions allowing retention of the α phase to 1 K.

III. RESULTS

Gadolinium

Our results[32] for Gd are shown in Table I. These are the most complete studies thus far of the pressure dependence of the Fermi surface of a rare earth metal. Several of the cross sections reported to date were measured by more than one of the three techniques outlined in the previous section for determination of pressure derivatives of dHvA frequencies. The agreement between the various techniques is excellent. These results show a surprising variation in both sign and magnitude. This is particularly striking in view of the near isotropy of the linear compressibilities of Gd. In such cases, the dominant effect (except for small cross sections) is typically given by the scaling of the Fermi surface with the Brillouin zone at the rate of 2/3 K_T where K_T is the bulk compressibility. For Gd this would corre-

Table I: de Haas-van Alphen frequencies and their pressure deriv-
          atives $d\ln F/dP \equiv [F(P) - F(0)]/F(0)\Delta P$ for Gd, in units
          of kbar^{-1}. The frequencies were determined using 19 kG
          for $8\pi M_s/3$ and are in units of 10^6 G. For spherical
          samples, $B = H + 8\pi M_s/3$ where M_s is the saturation mag-
          netization. The uncertainties given are estimates cor-
          responding to the most accurate determination when more
          than one method was employed.

Field Direction	Frequency	$d\ln F/dP$
[0001]	0.28	0.025 (± 0.005)[b]
	1.96	-0.021 (± 0.001)[a,b,c]
	16.9	0.002 (± 0.001)[a]
	13.9	0.005 (± 0.002)[a]
	40.7	0.012 (± 0.003)[a]
[10$\bar{1}$0]	1.5	-
[11$\bar{2}$0]	0.52	-0.12 (± 0.01)[c]
	1.7	-0.04 (± 0.02)[a,c]
	2.8	-
	46.6	0.007 (± 0.001)[a,c]

[a] fluid He phase shift
[b] solid He phase shift
[c] solid He frequency measurement

spond to a value of 0.0017 kbar^{-1}. Of particular interest is the
large negative derivative observed for the 0.52×10^6 G frequency
for H $||$ [11$\bar{2}$0]. This result obtained in solid He to 4 kbar in-
dicates that an electron transition will occur near 8 kbar at
liquid He temperatures. To date there are not data in this pressure-temperature range to corroborate this prediction.

It has proven very difficult to obtain a satisfactory model
description for the Fermi surface of Gd. The usual practice of

Table II: Frequency and mass data obtained at 9.3 kbar for
α-uranium. The pressure derivative $d\ell nF/dP$ is defined as $[F(P_1) - F(P_2)]/[(P_1 - P_2)F]$ and was determined between 8.5 and 9.3 kbar.

Direction	Frequency	Effective Mass	$d\ell nF/dP$
[010]	1.269×10^7G	0.91	3.7×10^{-3}kbar^{-1}
[100]	1.39×10^7G	∿0.9	–
[001]	–	–	–

rigidly splitting a paramagnetic band calculation cannot simultaneously account for the topology of the Fermi surface observed
and correctly give the proper value for the magnetization.[32] It
appears that fully self-consistent relativistic ferromagnetic band
calculations will be required.

Uranium

Results for pressure studies of the Fermi surface of α -U are
summarized[33] in Table II. Previous attempts to obtain information
pertaining to the Fermi surface of α -U have been thwarted by the
occurrence of several crystallographic phase changes as α -U is
cooled below 43 K. We found that cooling the specimen at pressures
above ∿7.5 kbar permitted retention of the α -U phase allowing
dHvA oscillations to be observed. The data as yet are very limited.
One cross section was measured as a function of pressure using the
solid He phase shift technique.[28]

This work has inspired an attempt[34] to make a band calculation
for α -U. Heretofore only high-temperature cubic structures had
been treated. This is a formidable challenge because α -U crystallizes in an orthorhombic lattice with 4 atoms/cell. In addition a
relativistic calculation is necessary.

Thorium

The results[35] for pressure derivatives of all three sheets of
the Fermi surface of Th are given in Table III. Again many of the
cross section derivatives were obtained by more than one technique
and found to be in good agreement. The striking result is that all

Table III: Pressure Dependence of Fermi Surface Cross Sections of Th

Cross Section	Orientation	Frequency (in 10^6 G)	$d\ell nF/dP$ (in 10^{-4}kbar^{-1})	
hole superegg (N)	[100]	22.1	−38(±4)	a
electron lung (L)	[100]	10	−40(±1.0) −4.4(±0.4)	b c
hole superegg (F)	[110]	24.7	−38(±4)	a
electron lung (C)	[110]	9.6	−3.4(±0.3)	c
electron lung (A)	[110]	2.014	−50(±10)	a
			−60(±10)	b
hole dumbbell (E)	[110]	19.8	−12(±7)	a
			−11(±2)	c
			−13(±5)	b
hole dumbbell (H)	[111]	10.9 x 10^6	−39(±5)	a
electron lung (I)	[111]	11.6 x 10^6	−14(±6)	a

a fluid He phase shift
b direct frequency measurement in solid He
c solid He phase shift
Orbit designation that of Boyle and Gold, Phys. Rev. Lett. 22, 461
1969.

of the cross-sectional areas <u>decrease</u> as the interatomic spacing is
reduced. This behavior is completely unprecedented for a cubic
material.

Th begins the 5f series. Although it is tempting to view the
5f series by analogy with the 4f rare earth metals, considerable
experimental evidence points to itinerant character for the 5f
electrons for the early part of the series. Koelling and Freeman[36]
showed that treatment of the 5f electrons as itinerant substantially improved the agreement between their relativistic band
structure description and experiment over earlier treatments in

which the 5f electrons were considered to be very localized. They
also made several predictions as to the way the Fermi surface would
change with pressure. In particular they predicted that the super-
egg would decrease in size compensated by an increase in size of
the dumbbell. The electron sheet (the lung) was expected to more
or less scale with the Brillouin zone. The results in Table III
do not agree with these predictions in that all the sheets are observed to decrease.

Further attempts to account for these results using RAPW
band descriptions as a function of interatomic spacing with and
without the presence of the f resonance indicate that the unexpected behavior of the Fermi surface of Th with volume is probably due to changes in s-d hybridization.[35] It is likely that
the actinides will also require self-consistent calculations in
order to account in detail for the observed volume dependence.

IV. CONCLUSIONS

This update of recent studies of the effect of pressure on
the Fermi surface of metals and a brief look at ongoing work on
Gd, U and Th should indicate that this field is still very active.
The experimental studies continue to prompt new theoretical en-
deavors and may in the case of the 4f and 5f metals lead to new
insight into the way these materials must be treated from a band
theoretical point of view.

Future work will probably center on the increasingly more
available rare earth materials and on intermetallic compounds.
The latter area is very active at normal volumes with high field
magnetoresistance, dHvA, Schubnikov-deHaas, magnetoacoustic and
magnetothermal oscillations and cyclotron resonance techniques
all being employed to map out experimentally the electronic
structure. Pressure studies should, as they have in the elemental
metals, provide critical tests for the model description. The
compounds have the advantage that an enormous diversity exists
permitting detailed examination of properties through an isostructural or isoelectronic series, a flexibility just not usually
available with elements.

ACKNOWLEDGMENTS

I would like to thank my collaborators A.J. Arko, D.D.
Koelling, B.N. Harmon, E. Fisher and F.A. Schmidt without whom
the ongoing studies on the 4f and 5f materials would not be
possible. I also thank R.L. White and D.L. Overmyer for excellent
technical assistance.

REFERENCES

1. J. E. Schirber, Materials Under Pressure, p. 141 (Tokyo, 1974).
2. N.B. Brandt, E.S. Itskevich and N. Ya. Minina, Usp. Fiz. Nauk 104, 459 (1971) Soviet Phys. -- Uspekhi 14, 438 (1972).
3. I.V. Svechkarev and A.S. Panfilov, Phys. Stat. Sol. 63, 11 (1974).
4. D. Gamble and B.R. Watts, J. Phys. F: Metal Physics 3, 98 (1973).
5. R. Griessen and R.S. Sorbello, Phys. Rev. B6, 2198 (1972).
6. G.M. Beardsley and J.E. Schirber, Bull. Am. Phys. Soc. Series II 17, 693 (1972).
7. I.M. Templeton, Can. J. Phys. 52, 1628 (1974).
8. H.J. Bryand and J.J. Vuillemin, Phys. Rev. 9, 3193 (1974).
9. J.E. Schirber and W.J. O'Sullivan, Proc. 11th Int. Conf. on Low Temp. Phys., edited by J.F. Allen, D.M. Finlayson and D.M. McCall (St. Andrews, 1963) p. 1141.
10. G.M. Beardsley, J.E. Schirber and J.P. Van Dyke, Phys. Rev. B6, 3569 (1972).
11. C.J.P.M. Harmans, "The Pressure Dependence of the Fermi Surfaces of Indium and Aluminum," (Doctoral dissertation, Vrije Universiteit te Amsterdam), 1975.
12. W.J. O'Sullivan, J.E. Schirber and J.R. Anderson, Solid State Commun. 5, 525 (1967).
13. N.B. Brandt, S.V. Kuvshinnikov and Ya. G. Ponomarev, Zk. ETF Pis. Red. 19, 201 (1974).
14. J.R. Anderson, D.A. Papaconstantopoulos, J.W. McCaffrey and J.E. Schirber, Phys. Rev. B7, 5115 (1973).
15. J.E. Schirber and J.P. Van Dyke, Phys. Rev. Lett. 26, 246 (1971).
16. A.J. Slavin and W.R. Datars, Can. J. Phys. 52, 1622 (1974).
17. M. Ribault, Ann. Phys. 2, 53 (1977).
18. J.R. Anderson, J.J. Hudak, D.R. Stone and J.E. Schirber, Proc. International Conf. Magnetism (Moscow, 1973).
19. J.R. Anderson, P.Heimann, D.R. Stone and J.E. Schirber, Proc. 21st Conf. on Magnetism and Magnetic Materials (Philadelphia, 1975).
20. J.E. Schirber, B. Morosin, J.R. Anderson and D.R. Stone, Phys. Rev. B5, 752 (1972).
21. H.W. Myron, R.P. Gupta and S.H. Liu, Phys. Rev. B8, 1292 (1973).
22. H.T. Weaver, J.E. Schirber and A. Narath, Phys. Rev. B8, 5443 (1973).
23. J.E. Schirber and J.P. Van Dyke, Phys. Rev. B15, 890 (1977).
24. D. Shoenberg and P.J. Stiles, Proc. Roy. Soc. (London) 281, 62 (1964).
25. J.E. Schirber, Cryogenics 10, 418 (1970).
26. I.M. Templeton, Proc. Roy. Soc. (London), A292, 413 (1966).

27. J.E. Schirber and W.J. O'Sullivan, Phys. Rev. 184, 628 (1969).
28. J.E. Schirber and R.L. White, J. Low Temp. Phys. 23, 445 (1976).
29. D.T. Peterson and F.A. Schmidt, J. Less- Common Met. 29, 321
     (1972).
30. F.A. Schmidt, B.K. Lunde, and D.E. Williams, USERDA Report
     ID-4125, 1977, p. 72.
31. E.S. Fisher and D. Dever, Phys. Rev. 170, 607 (1968).
32. J.E. Schirber, F.A. Schmidt, B.W. Harmon and D.D. Koelling,
     Phys. Rev. Lett. 36, 448 (1976) and to be published.
33. J.E. Schirber, A.J. Arko and E.S. Fisher, Solid State Commun.
     17, 553 (1975).
34. T.J. Watson-Yang, D.D. Koelling and A.J. Freeman, Bull. Am.
     Phys. Soc. Series II, 22, 445 (1977).
35. J.E. Schirber, F.A. Schmidt and D.D. Koelling, to be published.
36. D.D. Koelling and A.J. Freeman, Phys. Rev. B12, 5622 (1975).

QUESTIONS AND COMMENTS

T.F. Smith: I would just like to comment about the transitions
of uranium which Jim was talking about. As far as
we can tell, there is no evidence to suggest that
there is any actual crystallographic phase change.
The picture that Ed Fisher and I have of this is
that these are essentially electronic transitions
and this is consistent with all the other information available. One aspect of this behavior that
we are looking at, at the present time, is the Hall
coefficient. Berlincourt published some work on
the Hall coefficient of uranium about 20 years ago
in which he recorded a change in sign associated
with that first alpha transition: the so-called
α_1 phase. We are currently redoing this work and
eventually hope to be doing it under pressure. So
far we've confirmed the reversal in sign of the
Hall coefficient at the α-α_1 transition. So I
suggest that Jim is seeing the unmodified phase that
has a positive Hall coefficient rather than the low-
temperature negative Hall coefficient phase we obtain when we cool down the sample at zero pressure.

B.T. Matthias: I didn't get it quite straight. Are there any
f-electrons in thorium or not?

J.E. Schirber: The model can't tell. The band structure guys can't
distinguish. They can account for the data with
either model at this point. I don't think they
think there are any though. I think they think the
f-bands are above the Fermi level and empty.

B.T. Matthias: Then, for the second question, why would you call them actinides?

J.E. Schirber: I refuse to try to defend that. I agree.

D.K. Finnemore: Is the basic shape of the Fermi surface for thorium consistent with your data?

J.E. Schirber: Yes, Boyle and Gold, and we, agree to a fraction of a percent with everything they're saying. And both calculations reproduce that really quite well.

H.R. Ott: I want to comment on the sign of the Hall effect. That can also be obtained by an anisotropic relaxation time. It doesn't have to be basically related directly to an electronic transition.

J. Wittig: Did you conclude from the disappearance of the de Haas-van Alphen signal that there is a Lifshitz transition in uranium at 8 kbar?

J.E. Schirber: We actually have to extrapolate. I've only taken the data up to 4 kbar, and the frequency is heading toward 0. But I would expect there to be, and I would greatly appreciate someone making some measurements in that temperature/pressure regime and see if they can pick up any anomalies. To my knowledge, there is no information.

W.P. Crummett: I'm just curious. Have you made any comparisons of previous calculations with your data?

J.E. Schirber: For which material?

W.P. Crummett: α-uranium.

J.E. Schirber: No. We have the thorium results. For α-uranium, Freeman might have the calculations but they are not talking yet. At least I haven't gotten their results yet.

ELECTRONIC STRUCTURES OF FERROMAGNETIC 3d-TRANSITION METALS AT

NORMAL AND HIGH PRESSURES

J. R. Anderson

Department of Physics and Astronomy
University of Maryland
College Park, Maryland

ABSTRACT

The de Haas-van Alphen effect has been measured at normal and higher pressures for both nickel and hcp cobalt. For nickel the changes with pressure have been compared with similar studies in copper and used to estimate the pressure dependence of the s-d band separation. The pressure measurements have also been used to suggest corrections to present models of the cobalt band structure.

I. INTRODUCTION

Over the last several years there have been a number of experimental and theoretical studies of the 3d ferromagnetic metals in an attempt to relate their magnetic behavior to their electronic structures. Conventional Fermi surface techniques, especially measurements of the deHaas van Alphen effect, dHvA, have been successfully applied to iron, nickel, and cobalt[1] and the results of these experiments have been related with partial success to band structure calculations. Many of these band structures have been calculated ignoring the ferromagnetic character of the material; subsequently the spin-degenerate one-electron bands have been split in an ad hoc manner by a ferromagnetic exchange interaction in order to account for the non-integral magneton numbers.[2,3] Recently, however, there have been a few calculations in which separate Slater-type exchange potentials for the spin-up and spin-down bands have been obtained (spin-polarized approach).[4-8] In addition there have been attempts to actually calculate this exchange splitting explicitly.[9,10]

Large sheets of the Fermi surface have been studied in iron[11,12] and possibly nickel,[7] but no large portions of Fermi surface have been found so far in cobalt.[13,14,15] The features of the large pieces that have been studied agree at least qualitatively with the bands obtained by rigidly splitting the spin-degenerate bands. On the other hand, the band calculations frequently have given quite different predictions for smaller portions of the Fermi surfaces. In fact, it is often not possible to say whether a particular sur- face obtained from dHvA measurements corresponds to holes or elec- trons. In addition the degree of itineracy of the electrons that produce the net magnetic moment in ferromagnetic metals has not been established. Accurate and complete de Haas-van Alphen data plus a good band structure calculation would be extremely useful in sorting out this problem.

To this end, we are carrying out studies of electronic structures and their changes due to changes in lattice spacing. Such experi- ments provide checks on band calculations and allow one to estimate the relative shifts of s and d bands and spin-up and spin-down bands with lattice spacing.

Both the fluid-He phase shift technique[16,17] and the high pres- sure solid-He phase shift technique,[18] developed by Schirber and White, have been applied to study the effect of pressure on dHvA frequencies in cobalt and nickel. The solid-He phase shift experi- ments have been carried out by Dr. J.E. Schirber. Preliminary dHvA measurements have been made on nickel for the applied magnetic field along the easy direction of magnetization, [111], and along [100]. Measurements in cobalt have been made on one set of oscillations for H// [0001]. Some experimental results have been reported pre- viously.[14,19]

In Section II we describe our experimental techniques; most of the details have been published before. In Section III we briefly describe the accepted band structures and Fermi surfaces of cobalt and nickel for convenience in presenting the experimental results in Section IV. Finally in Section V we compare our results with existing calculations.

II. EXPERIMENTAL TECHNIQUE

The dHvA measurements were made using the conventional field- modulation technique with a 55 kOe Nb-Zr superconducting solenoid. The modulation frequency was typically 44 Hz and the signal was detected at a harmonic of this frequency, usually the second. In some cases it was necessary to use higher harmonics in order to selectively study one dHvA frequency.

Measurements were made on both nickel and cobalt spheres, spark- cut from single crystal rods.[20] The samples were lightly etched to remove material damaged during the spark-cutting process. It was necessary to anneal the nickel spheres at temperatures within about 30°C of the melting point for about 5 hours in order to obtain acceptable dHvA signals. (Annealing at about 150°C below the melting point for 48 hours did not significantly improve the samples.) Two nickel samples, one with diameter about 4.5 mm and the other with diameter about 1.5 mm were used in these studies. Several cobalt spheres, ranging in diameter from about 1.7 mm to 3.3 mm were investigated. Measurements were made by means of both the fluid-He phase shift and the solid-He phase shift techniques. In both approaches one takes advantage of the fact that the oscillating dHvA magnetization has a large phase $2\pi F/B$, where F is the dHvA frequency and B is the magnetic induction. The pressure derivative of this frequency is given by

$$\frac{d\ell nF}{dP} = \frac{1}{B}\left(1 + q\,\frac{\partial M}{\partial H}\right)\frac{\Delta H}{\Delta P} + \frac{qM}{B}\frac{d\ell nM}{dP} \quad . \tag{1}$$

Here P is the pressure, M is the saturation magnetization, H is the applied field, $q = 4\pi\,(1-D)$ where the demagnetizing factor D is 1/3 for a sphere, and ΔH is the shift in magnetic field position of one dHvA oscillation at fixed magnetic induction B for a pressure change ΔP.

The larger ΔP is the larger the shift ΔH, but the fluid-He technique is limited to a pressure of about 125 bar at 4K and about 25 bar at 1 K. In these measurements the important quantity is not ΔH but the ratio of ΔH to the field change for one dHvA cycle δH, where $\delta H = \frac{B^2}{F}$. The quantity $\Delta H/\delta H$ can be written

$$\frac{\Delta H}{\delta H} = \frac{F}{B}\frac{d\ell nF}{dP}\,\Delta P \quad , \tag{2}$$

if one omits the terms containing M in Eq. 1 for simplicity.

From Eq. 2 it can be seen that the largest relative shift is obtained at the smallest B and the largest ΔP. However, since the signal decreases with decreasing B and with the higher temperatures required to increase ΔP without solidifying the helium, there are limits to what can be obtained by increasing ΔP and decreasing B. Consider, for example, the nickel neck oscillations, for which the frequency is fairly small ($F \simeq 2.65$ MG), and a value of B at which the signal-to-noise ratio is acceptable for phase shift studies (~ 20 kG). At a temperature of about 1 K, if one assumes that $d\ell nF/dP$ is equal to the compressibility, then $\Delta H/\delta H \approx 0.002$, a shift which is extremely difficult to measure. At 4.2 K, since the value of ΔP can be increased by a factor of 5, the resulting shift is about 1% which is also difficult to measure especially in view of the reduced signal to noise ratio at the higher temperature. In addition such small shifts could be masked easily by shifts result-

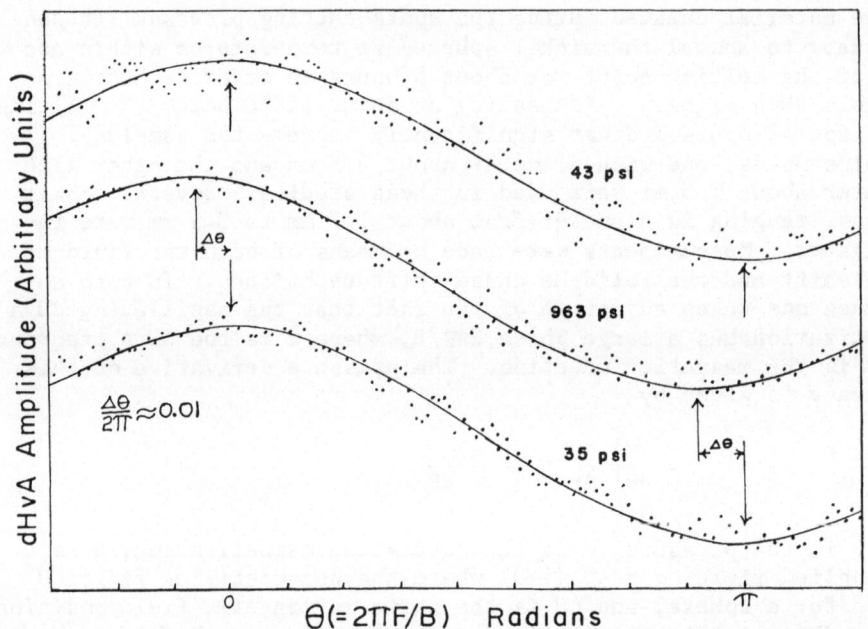

Figure 1.   Example of field sweeps at 35 psi, 963 psi, and 43 psi.
            Each solid line is the average of sixteen sweeps.   The
            oscillations at a frequency of 2.65×10^6 Gauss originate
            from the necks in nickel for H// [111].   The magnetic
            induction $B_o = H + 8\pi M_s/3 = 17.22$ kG.

ing from variations in sample orientation due to application of
pressure.

In Fig. 1 we show an example of a fluid-He phase shift experiment for the necks in nickel for H// [111].   The large field from
the superconducting solenoid was held constant to about 2 or 3 parts
in 10^6 and a sweep over one cycle was obtained by passing a small
direct current (in addition to the AC modulation) through the superconducting modulation coil.   Signal averaging techniques were employed by repeating the sweep over one cycle 16 times and storing
the results on magnetic tape.   Finally a least-squares fit of the
signal averaged sweeps at each pressure was carried out in order
to determine the shift in phase.   (Reasonable agreement with the
solid-He results, described below, was obtained in this case.)

Although the experiments were carried out at symmetry directions
for which the dHvA frequency is least sensitive to changes in orientation, problems still arose with the fluid-He technique.   A tilt
of about 0.6° could produce roughly the same shift on the nickel
neck oscillations as a change of pressure of 25 bar.   This appears

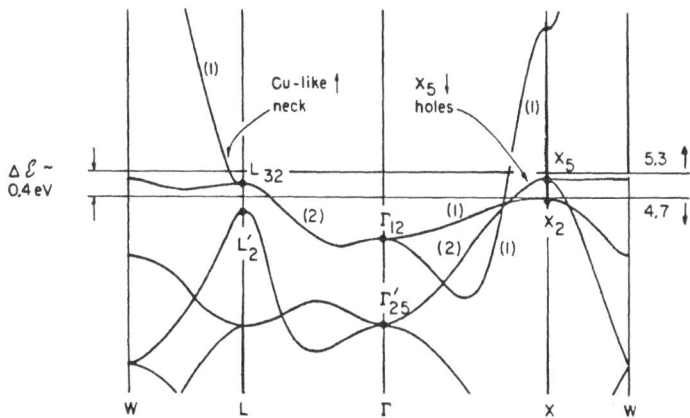

Figure 2.   Schematic sketch of the energy bands of nickel as summarized by Gold.[1] Spin-orbit splittings are not shown.

to have been the problem in our later fluid-He studies of the nickel oscillations for which on-line computing with a PDP-9 computer
was used. Shifts of the order of 0.001 of a cycle could be detected,
but these were not reproducible, probably as a result of sample
tilting.

In order to measure these small shifts a different approach
using solid-He phase shifts was employed. Because the pressure
changes could be as much as 400 times the fluid-He pressures at
1 K, a shift of a few tenths of a cycle could be obtained which is
easy to measure. This technique is similar to the fluid-He technique except that the phase shifts are determined after the sample
and He are warmed up to change the pressure and then cooled, solidifying the He once more. There are two disadvantages to this
process; one, it is tedious and two, the sample may shift orientation as the pressure is varied. However, studies at several pressures and magnetic fields showed that sample tilting was not significant.

III. ELECTRONIC STRUCTURES OF Ni AND hcp Co

In order to discuss the experimental results it is helpful to
have a model. Therefore results from recent band calculations in
Ni and Co will be summarized.

There have been many studies of the band structure of nickel.
The main checks on these calculations have been comparisons with
small pieces of Fermi surface obtained from dHvA studies. In Fig.
2 we show a schematic sketch of the energy bands for nickel. The

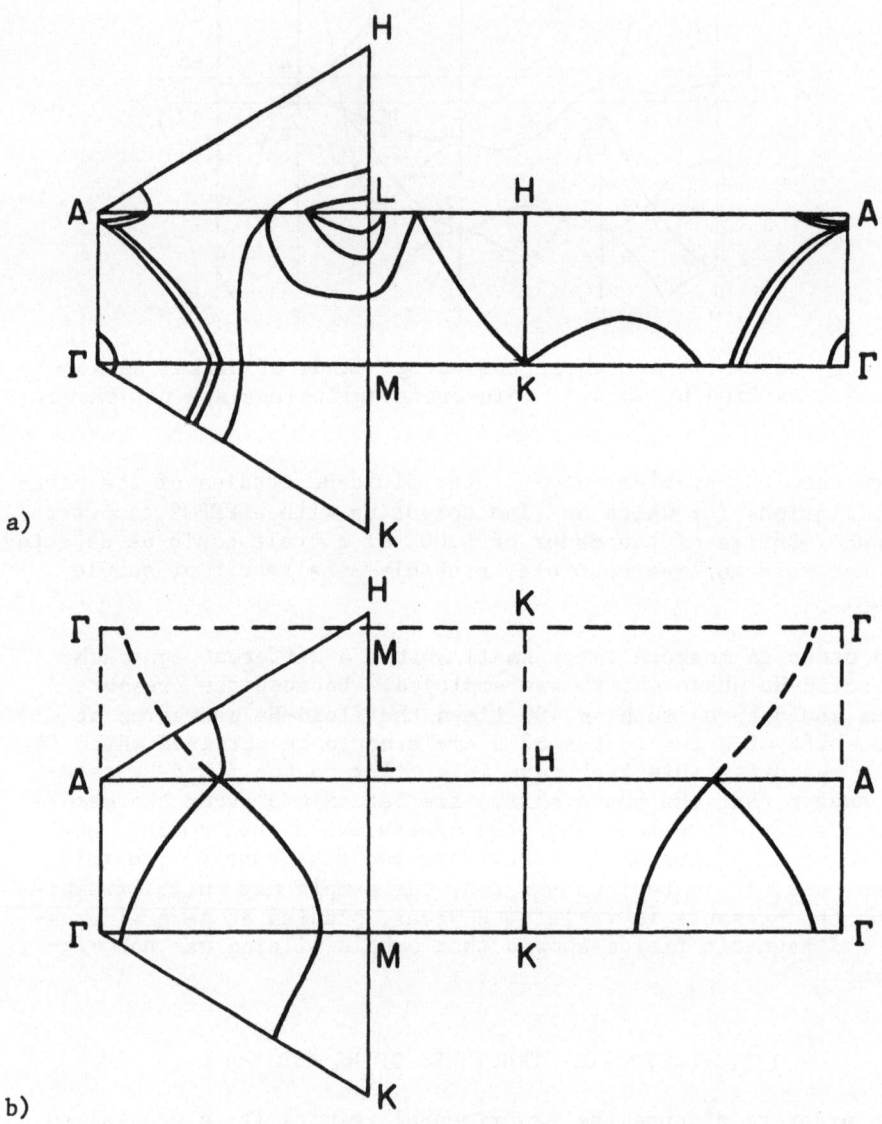

Figure 3.   Cross-sections of the cobalt Fermi surface in selected
            symmetry planes as given by Singal and Das.[10] a-majority
            spin Fermi surface; b- minority spin Fermi surface;
            -- refers to extended lines for double-zone scheme.

parts of the structure relevant to our experiment are the [111]
Cu-like necks at L produced by the majority spin bands and the hole
pockets at X resulting from the minority spin bands. The X_2 level
has been placed below the Fermi level E_F because a second set of
hole ellipsoids centered at X, which would result if the X_2 level
were greater than E_F, has not been observed experimentally. The
necks are mainly of s-p character while the hole pockets are d-like.
Thus a study of the pressure dependence of these two portions of the
Fermi surface should lead to an estimate of the relative shifts of
the majority s and minority d bands due to changes in atomic spacing.

Calculations of the band structure of hcp Co have been performed by Connolly,[5] Hodges and Ehrenreich,[21] Wakoh and Yamashita,[3]
Ishida,[22] and Singal and Das.[10] Because the Singal and Das hy-
bridized-tight-binding-orthorgonalized-plane-wave treatment is most
recent and therefore has been compared with recent experimental
data, we will assume their result represents the best calculation
to date. As a starting configuration they used $3d^{5\uparrow}$ $3d^{3\downarrow}$ $4s^{1/2\uparrow}$
$4s^{1/2\downarrow}$, where $\uparrow$ and $\downarrow$ represent majority and minority bands, respectively, and they made an explicit calculation of the exchange at a
number of points in $\vec{k}$-space. They found an overall 3d bandwidth of
about 0.34 Ry for the majority and minority spins. The minority 3d
bands were shifted upward from the majority 3d bands by about 0.29
Ry. A partial check on the success of their calculation is the
fact that they obtained very nearly the experimental spin magnetization of 1.56 μ_B per atom, where μ_B is a Bohr magneton.

In Fig. 3 we show cross-sections of the cobalt Fermi surface
in various symmetry planes for both majority and minority spins as
given by Singal and Das. They have correlated their results with
the small pieces of Fermi surface observed in the dHvA effect. At
this point we only wish to point out that the [0001]β oscillations,
which we have studied as a function of pressure, have been attributed
by Singal and Das to majority-spin electrons in a small piece of
Fermi surface in the ΓMK plane.

IV. EXPERIMENTAL RESULTS

A. Nickel

We have measured the shift with pressure $\frac{1}{B}\frac{\Delta H}{\Delta P}$ of the [111] oscillations from the majority-carrier necks at L and [100] oscilla-
tions from the minority-carrier d band hole pockets at X. The values are presented in column 4 of Table 1. The solid-He measurements were made at 1,2, and 3 kbar and are more reliable than the
fluid-He results as was discussed earlier. Since the samples were
spherical, we assumed that $\vec{B} = \vec{H} + \frac{8\pi}{3}\vec{M}$. According to Eq. 1 meas-
urements at different values of $|\vec{B}|$ will permit not only determina-
tion of $\frac{d\ell nF}{dP}$ but also $\frac{d\ell nM}{dP}$. However, within the accuracy of our

Table I   Pressure Dependence of dHvA Frequencies

Element	Orbit	Frequency MG	$\frac{1}{F}\frac{dF}{dP} \simeq \frac{1}{B}\frac{\Delta H}{\Delta P}$ $(10^{-3}\ \text{kbar}^{-1})$	$\frac{1}{M}\frac{dM}{dP}^{a}$ $(10^{-3}\ \text{kbar}^{-1})$	K $(10^{-3}\ \text{kbar}^{-1})$
Co	β[001]	3.53	−2.2 ± 0.3[b]	0.31[c]	0.513[d]
Ni	neck[111]	2.64	0.4 ± 0.3[b] 0.6 ± 0.1[e]	0.26[f]	0.55[g]
	pockets at X[100]	10.0	0.12± 0.03[e]		

a $4\pi M$ was assumed to be 17.97 kOe and 6.40 kOe for cobalt and nickel, respectively.
b   Fluid-He phase shift.
c   Kouvel and Hartelius. (R.T)[24]
d   Fisher and Dever.[23]
e   Solid-He phase shift.
f   Kondorskii and Sedov (4.2K).[35]
g   G.A. Alers et al.[36]

experiment, $\frac{1}{B}\frac{\Delta H}{\Delta P}$ did not vary with field over a range from 16 kG to
40 kG. Consequently we were unable to determine $\frac{d\ell nM}{dP}$ in these experiments. However, our errors are consistent with the small values
given by other experiments as presented in column 5. The compressibility K is given in column 6 of Table 1 and one can see that our values of $d\ell nF/dP$, although small, are of the same order of magnitude as
K.

B. Cobalt

We have measured the shift with pressure of the β-oscillations
for H//[0001] and the result is also given in Table 1. The result
is somewhat surprising since the shift is negative and about four
times the compressibility. Distortion of the lattice is not a likely explanation since the change in c/a with pressure is very small,
$\frac{d\ell n~c/a}{dP} \approx 6.17 \times 10^{-6}$ kbar $^{-1}$. Again we were able to give only

an upper limit for $\frac{d\ell nM}{dP}$, $\frac{d\ell nM}{dP} < 10^{-3}$ kbar^{-1}, which is consistent
with the value 0.31×10^{-3} kbar^{-1} measured by Kouvel and Hartelius.[24]

V. DISCUSSION

A. Nickel

The shifts with pressure of the dHvA frequencies for nickel and
cobalt are given in Table 1. The result for nickel is small compared
with copper, which has a similar piece of Fermi surface with [111]
necks at L. (See Table II.) (We should also note that the neck
cross-sectional area of copper is nearly an order of magnitude larger
than the neck area of nickel.) The variation with pressure of the
[100] cross-sectional area for the hole pockets is even smaller, about
1/5 the compressibility.

In order to estimate the relative shifts of s and d bands in
nickel we will follow the approach of Svechkarev and Pluzhnikov
(SV),[26] who compared pressure effects in several transition metals.
Some experimental parameters, which are needed for consideration of
the pressure results, are presented in Table II. In particular, we
can estimate the shift of the minority carrier d-bands relative to
the majority carrier s-bands by comparing the pressure shifts of
the hole pockets and necks. We assume that the d bands are split
into d↑ and d↓ with different densities of states at the Fermi surface and that the s↑ and s↓ bands are not split. For simplicity we
will also neglect any change in the magnetization with pressure.†

Table II Parameters for Nickel and Copper

Orbit	F MG	m*/m$_o$	ζ ev (Eq. 4)	dℓnA/dℓn V
Ni [111] necks (s-p)	2.65	0.22	−0.140	−1.09
Ni [100] pockets (d)	10.0	0.77	−0.141	−0.22
Cu [111] necks (s-p)	21.77	0.45	−0.560	−2.55[25]

SV assumed that the relevant small pieces of Fermi surface are centered at symmetry points and that the energy difference from these points to the Fermi energy ζ_i can be expressed as

$$\zeta_i = \frac{\hbar^2}{2\pi m_i^*} A_i \quad , \tag{3}$$

where $2\pi\hbar$ is Planck's constant and A_i and m_i^* are the cross-sectional area and cyclotron mass for that piece of Fermi surface, respectively. They also assumed that the relevant levels are not hybridized but either entirely s-p or entirely d. Finally they assumed for the width of the d-bands Δd as well as the energy separation between any non-hybridized d-levels,[27]

$$\frac{d\ell n\, \Delta_d}{d\ell n\, V} = -5/3 \quad , \tag{4}$$

where V is the volume. For s-p bands if the free electron approximation is assumed,

$$\frac{d\ell n \Delta_s}{d\ell n\, V} = -2/3 \quad .$$

As a result of applied pressure there is a relative shift of s-p and d bands and a redistribution of the electrons. Following SV, we have obtained for the shift in the s-bands $E_s - E_d\!\downarrow$ the following relations,

$$\frac{dE_{s-d\downarrow}}{d\ell n\, V} = -\zeta_d \left[1 + \frac{2N_{d\downarrow}}{N_s} \right] \left[\frac{d\ell n\, A_{d\downarrow}}{d\ell n\, V} + \frac{2}{3} \right] \quad , \tag{5}$$

and

$$\frac{dE_{s-d\downarrow}}{d\ell n \ V} = \zeta_s \ \left(1 + \frac{N_s}{2N_{d\downarrow}}\right) \left(\frac{d\ell n \ A_s}{d\ell n \ V} + \frac{2}{3}\right) \tag{6}$$

Here N_s and $N_{d\downarrow}$ are the densities of states at the Fermi level for the s-states and the minority d-states, respectively. Values for these densities of states have been estimated from the band calculations of Callaway and Wang[6] as follows: $N_{d\downarrow} \simeq 18.7$ electron/atom-Ry and $N_s \simeq 2N_{s\uparrow} \simeq 4.8 \ \frac{electrons}{atom-Ry}$. (Since the value from the measured

electronic contribution to the specific heat is approximately 42 electrons/atom-Ry, an enhancement of 1.79 is implied.)

From Eq. 5 we obtain $\frac{dE_{s-d\downarrow}}{d\ell n \ V} = -0.6$ ev, which disagrees with the result obtained from Eq. 6,

$\frac{dE_{s-d\downarrow}}{d\ell n \ V} = +0.07$ ev. Clearly this approximation of SV does not work for nickel. However, the numbers used in Eq. 6 are more reliable and therefore we assume Eq. 6 gives a more reasonable value. In Table III we show this value for nickel as well as some of the results of SV. There does not appear to be any systematic variation among these values but the value for nickel is small compared to the results for the other metals.

Connolly[4] has shown that the position of the d-bands relative to the s-p-bands, charaterized by $E(\Gamma_{25'}) - E(\Gamma_1)$, and the d-band width, characterized by $E(X_5) - E(X_1)$, are approximately linearly related. In Fig. 4 results for both copper and nickel, taken in part from Connolly's paper, are given. The dashed and dot-dashed lines correspond to calculations for copper for different lattice spacings by Davis et al.[29] and O'Sullivan et al.[30] respectively. The solid line is drawn to indicate the trend for nickel.

If we assume that $\frac{dE_{s-d\downarrow}}{d\ell n \ V} = 0.07$ ev corresponds to the change in the s-d band separation, and we use the linear relation from Fig. 4, then $\frac{dD}{d\ell n \ V} \sim 0.09$ eV where D is the d-band width.

The calculation for copper[29] also shows that

$$\frac{d\ell n \ (E_F - E(L_2'))}{d\ell n \ V} \simeq \frac{d\ell n \ F_{neck}^{Cu}}{d\ell n \ V} , \tag{7}$$

where F_{neck}^{Cu} is the Cu neck dHvA frequency. If we assumed that Eq. 7 is valid for nickel as well, we find

Table III

Element	$\dfrac{dE_{s-d}}{d\ell n\ V}$ (ev)
Mo	-2.1 ± 0.4*
W	-3.1 ± 0.5*
Re	-1.1 ± 0.2*
Ru	-1.15*
Os	-0.9*
Ir	$+0.4$*
Ni	$+0.07$

* Results presented by SV.

$$\left. \frac{d\ell n\ (E_F - E(L_2'))}{d\ell n\ V} \right\}_{Ni} \simeq -1.09$$

If one makes a similar assumption for the hole pockets, we obtain

$$\frac{d\ell n(E(X_5) - E_F)}{d\ell n\ V} \simeq -0.22 \quad .$$

(For copper $\dfrac{d\ell n\ (E(X_5) - E_F)}{d\ell n\ V} \simeq -.435$.)

We conclude that the model of SV does not work for nickel. The changes with pressure are significantly smaller for nickel than for copper. Since the estimates from our pressure studies of shifts in energy levels are based upon unreliable models, we have begun APW calculations as a function of lattice spacing in order to check our predictions of shifts of energy levels with change in volume and to calculate changes in the Fermi surface with pressure.

B. Cobalt

The pressure dependence of the β-oscillations in cobalt, shown in Table 1, is interesting not only because it is much larger than the compressibility but also because the Fermi surface cross-section decreases with pressure. The frequency of the β-oscillations has been shown to have a minimum at [0001] and to increase rapidly with magnetic field direction until the oscillations seem to break up

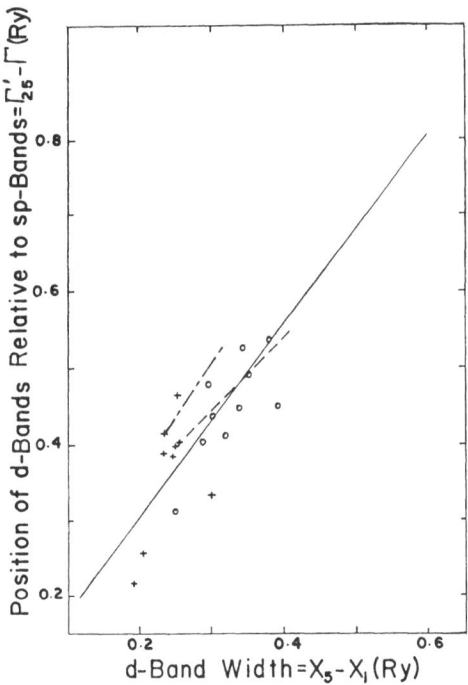

Figure 4.   Comparison of separation between d and s-p bands
($\Gamma_{25}'-\Gamma_1$) vs. d-band width (X_5-X_1) following Connolly[4].
(+)-copper; (0) nickel. Calculations at different
lattice spacing for copper: (---)29, (— — — —)[30].

into several components and then disappear abruptly at about 49°.[15]
This behavior is indicative of a hyperboloidal neck along [0001].
Rosenman and Batallan[13] have suggested that these β-oscillations
arise from a neck at Γ in the majority spin electron surface. Such
a neck was found in the calculations of Connolly[5] and Singal and
Das,[10] but Wakoh and Yamashita[3] and Ishida[22] did not obtain such
a neck in their majority-spin bands. We would like to present a
plausibility argument to suggest that the negative pressure shift
we have observed implies that the β-oscillations are produced by
a hole surface instead of an electron surface.[8]

The sign of $\dfrac{\partial A_\beta}{\partial \epsilon_F}$, where A_β is the cross-sectional area cor-

responding to the β-oscillations and ϵ_F is the Fermi energy, tells
whether these carriers are electrons or holes. In order to determine this sign we consider

$$\frac{dA}{dP} = \frac{\partial A}{\partial \varepsilon_F} \frac{d\varepsilon_F}{dP} + \frac{\partial A}{\partial P}\Bigg)_{\varepsilon_F} \tag{8}$$

We assume that the main effect of pressure is to change the Fermi energy and that the change in A_β is reflected only through the change in ε_F. Thus we ignore distortions in the bands due to pressure and obtain

$$\frac{\partial A_\beta}{\partial \varepsilon} \sim \frac{dA_\beta/dP}{d\varepsilon_F/dP} \quad . \tag{9}$$

In order to determine $d\varepsilon_F/dP$ we consider the electronic Gruneisen constant γ_e for cobalt given by White,[31]

$$\gamma_e = 1 + \frac{d\ln N(\varepsilon_F)}{d\ln V} = \frac{\beta V K}{C_V e} \quad . \tag{10}$$

Here $N(\varepsilon_F)$ is the density of states at the Fermi surface, V is the volume, β is the coefficient of thermal expansion, $C_V e$ is the electronic contribution to the specific heat, K is the compressibility, and the temperature is held constant in taking the derivative. If we assume the density of states is exchange enhanced, we may write $N(\varepsilon_F) = N_0(\varepsilon_F)Q$, where Q is an exchange enhancement factor. It also seems reasonable to assume that Q is proportional to the Curie temperature, T_c, so that $\frac{\partial\ln Q}{\partial\ln V} = \frac{d\ln T_c}{\partial\ln V}$. Patrick[17] has found that the

Curie temperature (1128°C at 1 atm.) is extremely insensitive to pressure. Therefore we assume

$$\frac{d\ln N(\varepsilon_F)}{d\ln V} \approx \frac{d\ln N_0(\varepsilon_F)}{d\ln V} \quad , \tag{11}$$

and the derivative can be written

$$\frac{d\ln N(\varepsilon_F)}{d\ln V} = \frac{d\ln N(\varepsilon_F)}{\partial\ln V}\Bigg)_{\varepsilon_F} + \frac{\partial\ln N(\varepsilon_F)}{\partial\varepsilon_F}\Bigg)_V \frac{d\varepsilon_F}{d\ln V} \quad . \tag{12}$$

From Eqs. 10 and 12 we obtain

$$\frac{d\varepsilon_F}{dP} = \frac{K\left(\frac{\partial\ln N(\varepsilon_F)}{\partial\ln V}\Bigg)_{\varepsilon_F} - (\gamma_e - 1)\right)}{\frac{\partial\ln N(\varepsilon_F)}{\partial\varepsilon_F}\Bigg)_V} \quad . \tag{13}$$

From the photemission data of Eastman[18] we see that

$$\frac{\partial \ln N(\varepsilon_F)}{\partial \varepsilon_F}\bigg)_V < 0 \ . \tag{14}$$

Unfortunately we have no information about $\dfrac{\partial \ln N(\varepsilon_F)}{\partial \ln V}\bigg)_{\varepsilon_F}$, but

if we ignore this term in the same spirit that we ignored the second term in Eq. 8, we see that $\dfrac{d\varepsilon_F}{dP} > 0$. Since $dA_\beta/dP < 0$, we see from

Eq. 9 that $\dfrac{\partial A_\beta}{\partial \varepsilon_F} < 0$, which implies that the carriers producing the

β-oscillations are holes. Consequently we now suggest that the β-oscillations result from a hole surface instead of a majority electron surface intersecting the ΓMK plane to form necks. (See Fig. 3a.)

Coleman et al.[34] have studied magnetoresistance in cobalt and have proposed a modification of the minority-spin Fermi surface which gives a hole surface with necks at U as shown in Fig. 5. This surface could account for the β-oscillations. In addition, breakdown could occur between this hyperboloidal surface and the smaller surface enclosed by it, a result which is suggested by the breakup of the β-oscillations at angles greater than 45° from the c axis. If the β-oscillations can be obtained in this manner, then there is no need to modify the majority-spin bands to produce necks in the ΓMK plane. However, if such necks in the majority-spin Fermi surface do exist, there will be open orbits along the c axis; it is not clear from the work of Coleman et al. whether such open orbits are present or not. Also, if this second hyperboloidal surface exists, the extremal areas should be observable in the de Haas-van Alphen effect. From the de Haas-van Alphen experiments there is no definite evidence for such a surface.

Therefore, we have suggested another possible modification of the minority-carrier bands in order to obtain a neck-like hole surface. The minority carrier bands have been shifted by about 0.3 ev with respect to the Fermi energy and slightly modified as shown in Fig. 6, thus producing a hole surface intersecting the AHL plane. The intersection of this minority-spin Fermi surface with symmetry planes is shown in Fig. 7. The neck-like surface consists of cylinders with axes along the K-H lines. The small closed pieces near L and along the line ML could explain most of the dHvA frequencies that have been observed. In Table IV we compare the experimental extremal areas with those obtained from

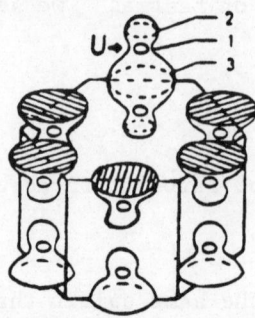

Figure 5.   Minority spin band around U point for cobalt as pro-
            posed by Coleman et al.[34]

this crude model; it can be seen that qualitative agreement is obtained.

    Band structure calculations at different lattice spacings are
necessary for interpretation of the pressure dependence of the
β-oscillations for comparison with proposed Fermi surface models.
In addition pressure studies of other dHvA frequencies in cobalt
would aid in developing a suitable model for the Fermi surface.

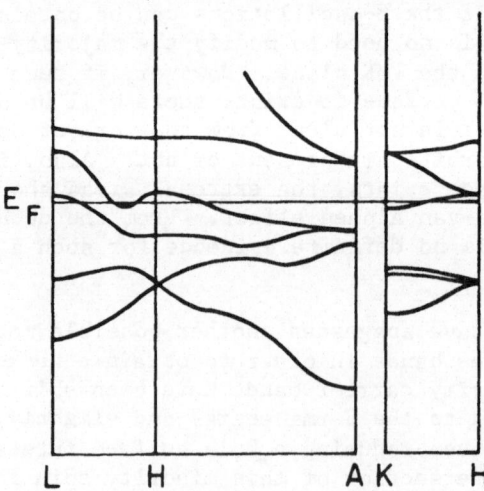

Figure 6.   Schematic representation of the minority carrier bands
            in cobalt shifted upward by about 0.3 ev compared with
            the bands of Wakoh and Yamashita[3] and slightly modified
            to produce a hyperboloidal-like hole surface along K to H.

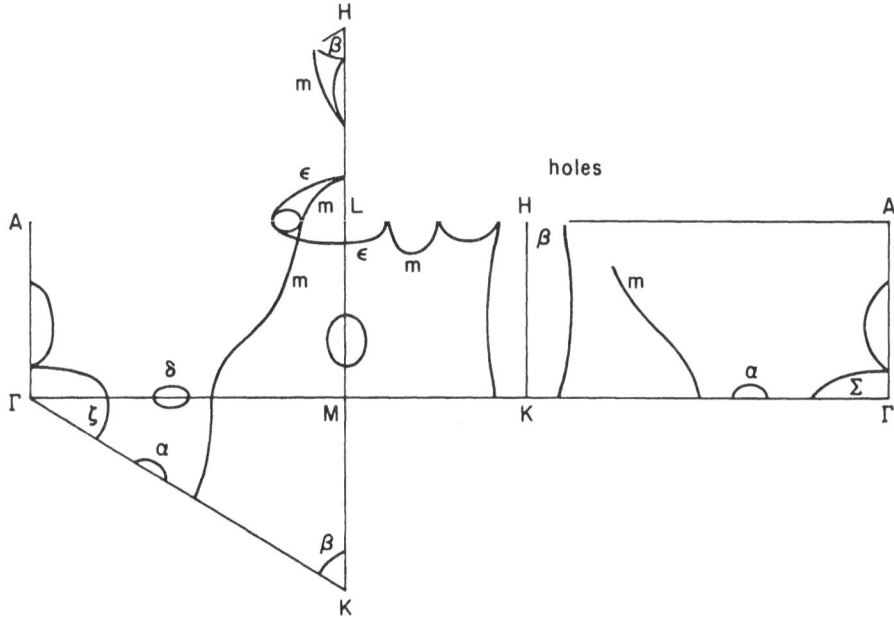

Figure 7. Schematic of the intersection of the minority spin Fermi
surface of cobalt with selected symmetry planes. The
hole surface proposed to explain the β−oscillations is
shown along K to H.

Table IV dHvA Frequencies and Areas for hcp Co

Nomenclature	Orientation	F (MG)	A_{exp} (units of $\left(\frac{2\pi}{a}\right)^2$)	A_{Calc} (units of $\left(\frac{2\pi}{a}\right)$)
α	[0001]	1.04_6	0.00159	0.001
β	[0001]	3.53	0.00536	0.009
ε	[0001]	11.75	0.0178	0.03
ζ	[0001]	13.50	0.0205	0.05
δ	[10$\bar{1}$0]	1.31	0.00199	0.001
ε	[10$\bar{1}$0]	9.39	0.0143	0.008
ε	[11$\bar{2}$0]	10.27	0.0156	0.014

ACKNOWLEDGEMENTS

This work has been carried out in collaboration with Peter Heimann, J.E. Schirber, and D.R. Stone and many useful discussions with them are acknowledged. We also wish to thank T.L. Einstein for his helpful suggestions concerning this work and G. Lonzarich for his useful comments concerning the redistribution of electrons between spin-up and spin-down bands.

† Dr. G. Lonzarich has pointed out that it may not be possible to neglect the effect due to the change in magnetization with pressure. For nickel $d\ell n\ \sigma_s/dP = -2.9 \times 10^{-4}$ kbar^{-1},[35] where σ_s is the magnetization in emu/gm. If one neglects changes in the shapes of the bands, then the change in σ_s is due only to a shift of carriers from spin-up to spin-down bands. This effect would produce a decrease in area with pressure for both the majority carrier electron necks at L and the minority carrier hole pockets at X. The change in area with pressure due to this effect can be estimated by means of the expression,

$$\left.\frac{d\ell n\ A_i}{dP}\right)_m \simeq \frac{Z\ \sigma_s}{2\ \zeta_i N_i(\varepsilon_F)\mu_B N}\ \frac{d\ell n\ \sigma_s}{dP}\ .$$

Here Z is the gram molecular weight, N is Avogadro's number, and ζ_i is given by Eq. 3. The density of states $N(\varepsilon_F)$ must be obtained from band calculations for both the majority and minority carriers. For nickel $N_\uparrow(\varepsilon_F) \simeq 2.4$ elec/atom-Ry and $N_\downarrow(\varepsilon_F) \simeq 21.1$ elec/atom-Ry[6].

For the necks we calculate $d\ell n\ A_{s\uparrow}/dP)_m \simeq -3.6 \times 10^{-3}$ kbar^{-1}, a very large change. If one assumes that there is another contribution due to a change in the energy bands, which is the same as that for the necks in copper, $d\ell n\ A_{cu}/dP \simeq 1.8 \times 10^{-3}$ kbar^{-1}, then the total change would be approximately $d\ell n\ A_{s\uparrow}/dP)_T \simeq -1.8 \times 10^{-3}$ kbar^{-1}. The calculated result is too large in magnitude but shows that the redistribution of electrons accompanying the change in magnetization in nickel may be an important contribution to the Fermi surface changes with pressure.

For the hole pockets we obtain $d\ell n\ A_{\downarrow d}/dP)_m \simeq -4.1 \times 10^{-4}$ kbar^{-1}. This result combined with 2/3 K (compressibility scaling) gives $d\ell n\ A_{d\downarrow}/dP)_T \simeq -0.5 \times 10^{-4}$ kbar^{-1}, which is close to the small experimental value (Table III).

§ The variation of magnetization with pressure may be used to estimate the transfer of electrons from the majority to minority spin bands as the pressure is increased following the approach discussed for nickel. For cobalt the densities of states for spin-up and spin-down bands are 1.92 electrons/atom-Ry and 15.31 electrons/atom-Ry, respectively.[10] For the β-oscillations $\zeta_{co} = -0.178$ and $\sigma_s \simeq 161$ emu/gm.[15] The calculation gives

$$\left. d\ell n \ A/dP \right\}_m = \begin{array}{l} \overline{+}7.5 \times 10^{-3} \ kbar^{-1} \ (majority \ bands) \\ \\ \pm 9.4 \times 10^{-4} \ kbar^{-1} \ (minority \ bands). \end{array}$$

The upper signs correspond to electrons and the lower to holes. If this result is combined with 2/3 K (scaling), an area that decreases with pressure would result from either majority electrons or minority holes. Although neither estimate agrees well with the experimental value, this result shows that the redistribution of electrons between spin-up and spin-down bands is probably important for cobalt as well as nickel.

REFERENCES

1. See A.V. Gold, Jour. of Low Temp. Phys. 16, 3 (1974) for a recent review of the experimental situation.
2. J. Yamashita, M. Fukuchi, and S. Wakoh, J. Phys. Soc. Japan 18, 999 (1963).
3. S. Wakoh and J. Yamashita, J. Phys. Soc. Japan 28, 1151 (1970).
4. J.W.D. Connolly, Phys. Rev. 159, 415 (1967).
5. J.W.D. Connolly, International Journal of Quantum Chemistry IIs, 257 (1968).
6. J. Callaway and C.S. Wang, Phys. Rev. B7, 1096 (1973).
7. C.S. Wang and J. Callaway, Phys. Rev. B9, 4897 (1974).
8. S. Wakoh, J. Phys. Soc. Japan 20, 1894 (1965).
9. K.J. Duff and T.P. Das, Phys. Rev. B3, 192 & 2294 (1971).
10. C.M. Singal and T.P. Das - to be published.
11. D.R. Baratt, Phys. Rev. B8, 3439 (1973).
12. A.V. Gold, L. Hodges, P.T. Panousis, and D.R. Stone, Intern. J. Magnetism 2, 357 (1971).
13. I. Rosenman and F. Batallan, Phys. Rev. B5, 1340 (1972).
14. J.R. Anderson, J.J. Hudak, D.R. Stone, and J.E. Schirber, Proceedings of the International Conference of Magnetism ICM-73, Moscow, Volume III, 344 (1974).
15. J.R. Anderson, J.J. Hudak, and D.R. Stone, in Magnetism and Magnetic Materials: AIP Conf. Proc. 10, C.D. Graham, Jr. and J.J. Rhyne, eds., (American Institute of Physics, 1973), p. 46.
16. I.M. Templeton, Proc. Roy. Soc. (London) A292, 413 (1966).
17. J.F. Schirber and W.J. O'Sullivan, Phys. Rev. 184, 628 (1969).
18. J.E. Schirber and R.L. White, Jour. of Low Temp. Phys. 23, 445 (1976).
19. J.R. Anderson, Peter Heimann, J.E. Schirber, and D.R. Stone, in Magnetism and Magnetic Materials: AIP Conf. Proc. 29, J.J. Becker, G.H. Lander, and J.J. Rhyne eds. (American Institute of Physics, 1975), p. 529.
20. The single crystals used in this study were obtained from

Materials Research Corporation, Orangeburg, New York.

21. L. Hodges and H. Ehrenreich, J. Appl. Phys. 39, 1280 (1968).
22. S. Ishida, J. Phys. Soc. Japan 33, 369 (1972).
23. E.S. Fisher and D. Dever, Trans. of Metallurgical Society of AIME 239, 48 1967).
24. J.S. Kouvel and C.C. Hartelius, J. Appl. Phys. 35, 940 (1964).
25. J.E. Schirber and W.J. O'Sullivan, Colloque International Du C. N.R.S., Sur Les Proprietes Physiques Des Solides Sous Pression, Grenoble 188, 113 (1970).
26. I.V. Svechkarev and V.B. Pluzhnikov, Phys. Stat. Sol. (b) 55, 315 (1973).
27. V. Heine, Phys. Rev. 153, 673 (1967).
28. W.H. Keeson and C.W. Clark, Physica 2, 513 (1935).
29. H.L. Davis, J.S. Faulkner, and H.W. Joy, Phys. Rev. 167, 601 (1968).
30. W.J. O'Sullivan, A.C. Switendick, and J.E. Schirber, Phys. Stat. Sol. (b) 68, K29 (1975).
31. G.K. White, Proc. Phys. Soc. 86, 159 (1965).
32. L. Patrick, Phys. Rev. 93, 384 (1954).
33. D.E. Eastman, J. Appl. Phys. 40, 1387 (1959).
34. R.V. Coleman, R.C. Morris, and D.J. Sellmyer, Phys. Rev. B8, 317 (1973).
35. E.I. Kondorskii and V.L. Sedov, Zh. Eksperimi Teor. Fiz. 38, 773 (1960) [English Transl.: Soviet Physics-JETP 11, 561 (1960)].
36. G.A. Alers, J.R. Neighbors, and H.Sata, J. Phys. Chem. Sol. 13, 40 (1960).

NEUTRON SCATTERING AT HIGH PRESSURE IN SINGLET-TRIPLET SYSTEMS

D.B. McWhan

Bell Laboratories

Murray Hill, New Jersey 07974

ABSTRACT

Recent measurements on the $\Gamma_1 - \Gamma_4$ exciton at 1 atm and at high pressure in PrSb are reviewed. Measurements on the J = 0 $\rightarrow$ 1 excitation and the magnetic form factor of SmS above and below the transition to the mixed valence phase are summarized.

In many rare earth compounds there is a partial lifting of the degeneracy of the 4f multiplet by spin orbit coupling and by the crystalline electric field, and in the case of Pr^{3+} ($^3H_4$) and Sm^{2+} ($^7F_0$) the ground state in the cubic environment of the NaCl structure is a singlet. Interest in these two ions has focussed on two areas, the competition between exchange and crystal field[1] in Pr^{3+} and the mixed valence nature of a number of Sm compounds.

The theory of crystal field effects in insulators is well understood, but the situation is far more complicated in metals as a result of screening by the conduction electrons. The technique of inelastic neutron scattering spectroscopy has been used to map out the crystal field splittings in a number of rare earth compounds, in particular the pnictides.[2] The spectra have been successfully analyzed using a Hamiltonian of the form:

$$H = H_{ex} + W[xO_4 + (1-x)O_6]$$

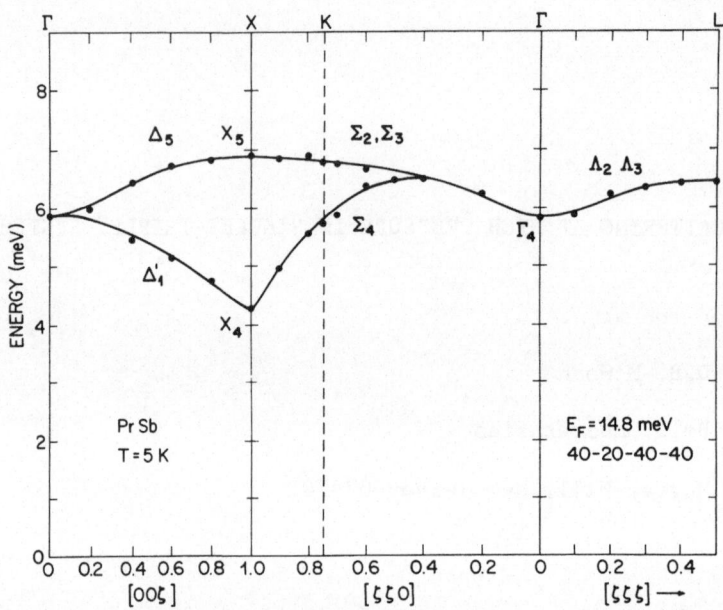

Figure 1. The lifting of the triple degeneracy of the Γ_1-Γ_4 exciton
in the $4f^2$ configuration of Pr^{3+} in PrSb by the aniso-
tropy in the exchange interaction. The lines are guides
to the eye.

where the exchange is taken to be isotropic and the crystal field
is given in terms of an overall splitting, W, and the ratio of the
4th to 6th order terms, x. Calculated values for these parameters
based on an ionic point charge model gave relatively good agreement across the whole rare earth series.[2] In this model the splitting varies directly as the effective charge on the ligands and the
average radius of the 4f wavefunctions, $<r^4>$, and it varies inverse-
ly as the interatomic spacing to the fifth power (a_o^{-5}). In metallic systems the effective charge should be screened by the conduction electrons, but it is not clear how to calculate this decrease in effective charge. The volume dependence of W and x in
PrSb is discussed below.

Mean field-RPA theories predict that in singlet-singlet or
singlet-triplet systems of this type a unique type of magnetic transition should occur with decreasing temperature if the ratio of the
exchange interaction to that of the crystal field interaction exceeds
a critical value. The ferromagnetic transition in Pr_3Tl is believed
to result from this type of spontaneous polarization of the ground
state wavefunction.[3] This balance of exchange to crystal field is

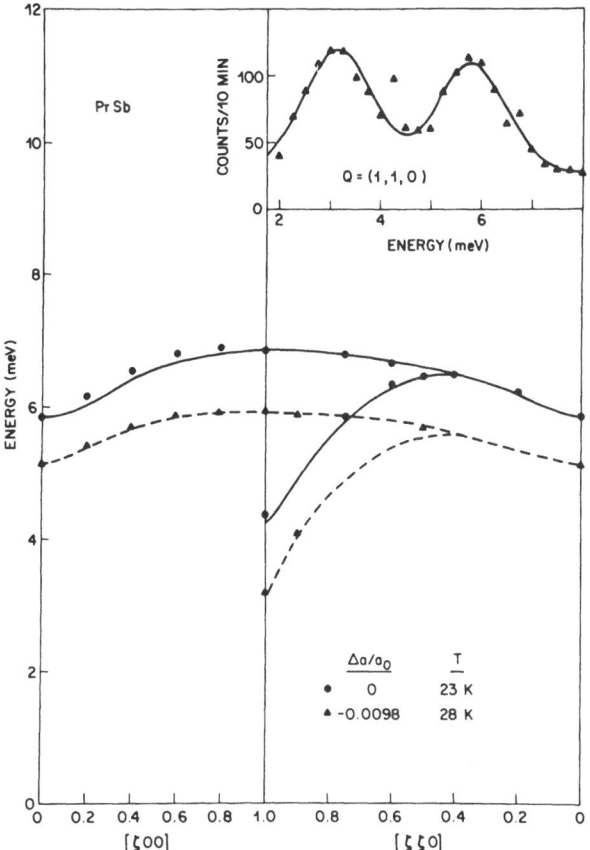

Figure 2.   The effect of pressure on the crystal field levels.
The inset shows typical data obtained at P = 1.6 GPa
and T = 28 K.

very sensitive to pressure, and, in fact, the Curie temperature of
Pr$_3$Tl is suppressed to 0 K at a pressure of 12 kbar.[4]   The theory
has been quantitatively tested in the case of the semiconducting
phase of SmS at 1 atm where the ratio of exchange to crystal field
is below the critical value.[5]   The temperature dependence of the
dispersion relation for the triplet exciton is well represented
assuming isotropic exchange and including terms up to third nearest
neighbors.

We have recently made inelastic neutron scattering measurements on single crystals of PrSb and find that the exciton dispersion is far more complicated than expected.[6]   The degeneracy of the

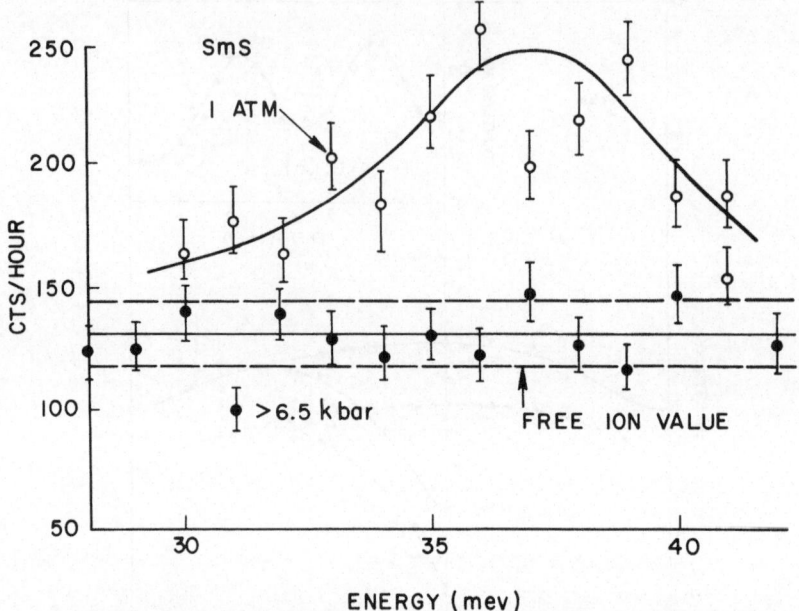

Figure 3.   The effect of pressure on the J = 0 → 1 transition of
            Sm^{2+} at 1 atm (open circles) and in the mixed valence
            phase above 0.65 GPa (closed circles).   The energy
            expected for the free ion is indicated.   The top line
            is a guide to the eye.   The lower solid line and the
            dashed lines are a least squares fit to a straight line
            and plus and minus σ respectively.

$\Gamma_1-\Gamma_4$ triplet exciton is lifted along the [100] direction but not
along [111] as shown in Fig. 1.   This splitting results neither
from a coupling of the excitons to the phonons because they have
opposite parity at X where the splitting is a maximum, nor from
an anisotropic exchange of the form used in Pr metal (axially symmetric about the line connecting sites).[7]   Comparison with the
band structure,[8] which has pockets of electrons at X and pockets
of holes at Γ suggests that the softening of the longitudinal exciton along [100] results from processes which involve the electrostatic couplings between f-shell and outer electrons ($J\vec{f}l$ or $J\vec{f}s$).
These are processes by which electrons are scattered from X to Γ
while an exciton in an f-shell is deexcited and then scattered
back as the exciton is created again at another site.

The energy of the $\Gamma_1-\Gamma_4$ triplet exciton was observed to decrease rapidly with decreasing lattice spacing as shown in Fig. 2.
This decrease had been inferred both from measurements of the

Knight shift[9] and magnetic susceptibility[10] as a function of pressure. In addition to the Γ_1-Γ_4 transition, the energy of the Γ_4-Γ_5 transition was observed to decrease with pressure. These two measurements completely determine the crystal field levels and show that x, the ratio of the 4th to 6th order terms in the crystal field Hamiltonian, is not a strong function of pressure and that it is the overall splitting, W, which is decreasing. None of the presently considered models for crystal field interactions in metals are compatible with these pressure measurements and the measurements at 1 atm across the rare earth series.

The decrease of the longitudinal exciton frequency at X with increasing pressure suggests that a soft mode magnetic transition will occur at higher pressures. As a function of pressure PrSb will undergo a transition from a paramagnetic to an antiferromagnetic state at low temperatures. Preliminary experiments performed by C. Vettier in a clamp device at the Institute Laue Langevin suggests that the critical pressure for this transition is ~ 2.8 GPa.

Turning to the transition to the mixed valence state in SmS, neutron scattering provides two fingerprints of the electronic configurations of Sm^{2+} ($4f^6$) and Sm^{3+} ($4f^5$) ions. As is well known at 1 atm SmS is a semiconductor with a well-characterized Sm^{2+} ($^7F_0$) ground state.[11] At a pressure of ~ 0.6 GPa there is an electronic transition to a mixed valence phase.[12] The lattice parameter is consistent with a mixture of Sm^{2+} and Sm^{3+} ions existing together. This phase is characterized by a magnetic susceptibility which saturates at low temperature,[13] a large linear term in the specific heat[14] and a large increase in resistivity at low temperatures.[14] It has been suggested that interconfiguration fluctuations somehow suppress the expected magnetic ordering. The J = 0 → 1 excitation of the Sm^{3+} ion has been studied at high pressure to determine if a sharp excitation exists in the collapsed phase. Preliminary results obtained at Brookhaven National Laboratory on a polycrystalline sample in a beryllium-copper clamp device are shown in Fig. 3.[15] The signal-to-noise is poor and further experiments are planned using a larger sample. The results shown in Fig. 3 have been confirmed recently using time of flight techniques instead of a triple axis geometry.[16] It appears that a sharp excitation corresponding to a J = 0 → 1 transition of the Sm^{2+} ions does not exist in the collapsed phase. However, further work will be needed to establish whether the excitation has been broadened substantially and/or shifted to higher energy as has been observed in the Ce_xTh_{1-x} mixed valence system.[17]

The second fingerprint is the induced magnetic form factor of the Sm^{3+} ion. For an ionic compound this form factor is

$$f(q) = \langle j_0(q) \rangle + c_2 \langle j_2(q) \rangle$$

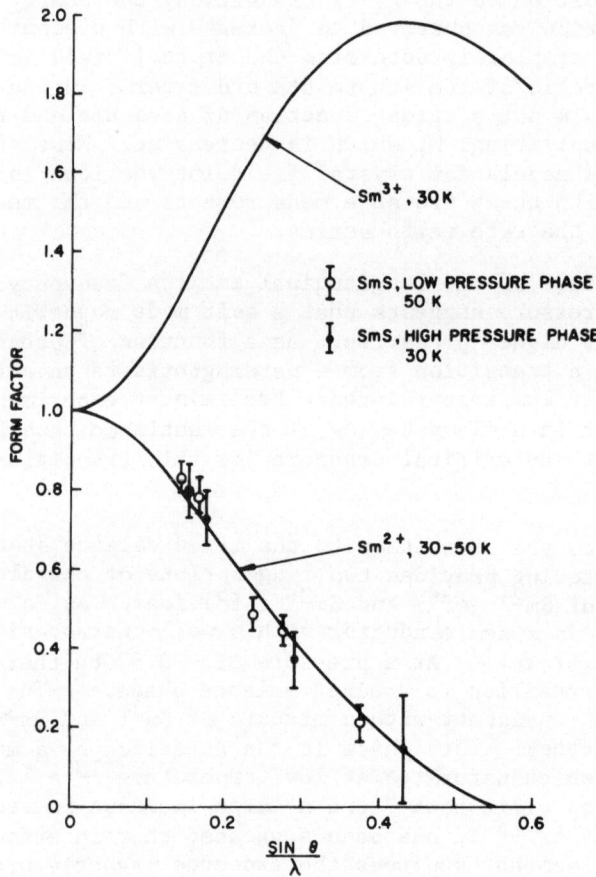

Figure 4. The induced magnetic form factor of SmS at 1 atm and
           at ~ 0.7 GPa. The calculated curves are for ions in
           a cubic environment with crystal field and exchange
           parameters appropriate for SmS.

where $\langle j_n \rangle$ is an integral of the product of the 4f radial electron
density and a spherical Bessel function of order n and

$$c_2 \approx \frac{\overline{L_z}}{\overline{L_z + 2S_z}}$$

The dominant term in Sm^{2+} is the field induced admixture of the
J = 1 level into the J = 0 ground state leading to $c_2 \approx -1$. A similar result is expected for Sm^{3+} at high temperature, but at low
temperature the 1/T terms of the J = 5/2 ground state lead to
$c_2 \approx +5.4$ for an isolated Sm^{3+} ion. In the mixed valence compound

$Sm_{0.76}Y_{0.24}S$ the fraction of Sm^{3+} increases as the temperature is
raised. The form factor of this compound has been determined using
the polarized neutron technique to measure the elastic magnetic
structure factors in an applied field of 42.5 kOe.[18] No evidence
for a normal Sm^{3+} form factor was found. Recently the form factor
of Sm in pure SmS was measured at pressures below and above the
transition to the mixed valence phase using a polycrystalline sample and a beryllium-copper clamp device designed to fit into the
superconducting magnet.[19] The results were normalized to the
measured susceptibility to give the reduced form factors shown in
Fig. 4. The open circles are the results at 1 atm and the closed
circles those at ~ 0.7 GPa. The curves are calculated for Sm^{3+}
and Sm^{2+} at T = 30° K. The low pressure results are in good agreement with the calculated curve. The high pressure results indicate
that only the Van Vleck contribution arising from the field-induced
admixture of J = 7/2 into the J = 5/2 ground state is observed and
that the 1/T terms have been quenched. This is consistent with the
fact that the magnetic susceptibility saturates at low temperatures
and shows that whatever mechanism causes this saturation at $\vec{q}$ = 0
is independent of $\vec{q}$.

In conclusion, the neutron scattering studies at high pressure
in SmS give further evidence that the fluctuations in the mixed
valence phase drastically affect the local 4f electronic states
of both the Sm^{2+} and Sm^{3+} ions. The studies on PrSb suggest a
coupling to the conduction electrons and indicate that the theory
of crystal fields in metals is far from complete.

All the high pressure studies reviewed here were done in collaboration with a number of people and I thank R.J. Birgeneau,
E.I. Blount, W.C. Koehler, R.M. Moon, S.M. Shapiro, G. Shirane,
and most of all C. Vettier for helpful discussions.

REFERENCES

1. For a review see R.J. Birgeneau, AIP Conf. Proc. 10, 1664
    (1972).
2. K.·C. Turberfield, L. Passell, R.J. Birgeneau, and E. Bucher,
    J. Appl. Phys. 42, 1746 (1971).
3. R.J. Birgeneau, J. Als-Nielsen and E. Bucher, Phys. Rev. Lett.
    27, 1530 (1971).
4. S. Huang and P.W. Chu, American Physical Society 19, 174 (1974).
5. S.M. Shapiro, R.J. Birgeneau, and E. Bucher, Phys. Rev. Lett.
    34, 470 (1975).
6. C. Vettier, D.B. McWhan, E.I. Blount, and S. Shirane, to be
    published.
7. P. Bak, Phys. Rev. B 12, 5203 (1975).
8. H.L. Davis in The Actinides: Electronic Structure and Related
    Properties Vol. II, Academic Press, New York, 1974.

9. H.T. Weaver and J.E. Schirber, AIP Conf. Proc. 24, 49 (1974)
    and Phys. Rev. B 14, 951 (1976).
10. R.P. Guertin, J.E. Crow, L.D. Longinotti, E. Bucher,
    L. Kupferberg, and S. Foner, Phys. Rev. B 12, 1005 (1975).
11. For a review see C.M. Varma, Rev. Mod. Phys. 48, 219 (1976).
12. A. Jayaraman, V. Narayanamurti, E. Bucher, and R.G. Maines,
    Phys. Rev. Lett. 25, 1430 (1970).
13. M.B. Maple and D. Wohleben, Phys. Rev. Lett. 27, 511 (1971).
14. S.D. Bader, N.E. Phillips and D.B. McWhan, Phys. Rev. B 7,
    4686 (1973).
15. S.M. Shapiro, R.J. Birgeneau, D.B. McWhan and E. Bucher, Bull.
    Amer. Phys. Soc. 20, 383 (1975).
16. H. Mook, Private Communication.
17. R.J. Birgeneau and S.M. Shapiro in <u>Valence Instabilities and
    Related Narrow-Band Phenomena</u>, R.D. Parks, ed., Plenum Press,
    New York, 1977, p. 49.
18. R.M. Moon and W.C. Koehler, Bull. Amer. Phys. Soc. 22, 292
    (1977).
19. R.M. Moon, W.C. Koehler, D.B. McWhan and F. Holtzberg, to be
    published.

QUESTIONS AND COMMENTS

C. Cordero: The splitting between the J_0 and J_1 states of
               samarium shows an increase with temperature value
               towards its higher, free ion value. Could you say
               whether there is a discrepancy between that result,
               as reported by Nathan et al. (1975), and the pre-
               vious measurements of Bucher et al. (1971), who
               obtained a good fitting to the susceptibility versus
               temperature curve using only the low temperature
               value of the splitting?

D.B. McWhan: I wasn't aware of the discrepancy. The temperature
               dependence is well represented by mean field-RPA
               theory for the idealized paramagnetic singlet-triplet
               system at 1 atm according to the measurements of
               Shapiro, Birgeneau and Bucher (Phys. Rev. Letts. 34,
               470 (1975)).

C. Cordero: Do your measurements give any information that could
               help in distinguishing, for samarium sulfide in the
               collapsed state, between the Pauli enhanced and the
               Van Vleck low temperature susceptibilities suggested
               by different models?

D.B. McWhan: You see only the Van Vleck term.

J.E. Schirber: I have a comment to make on the praseodymium-nitride situation. We have, as Denis has mentioned, been assuming the answer to this strange behavior the wrong way for the point charge model, so the answer must lie in the conduction electron screening or what have you. Unfortunately, we weren't willing to let well enough be, and made one too many measurements. We made similarly accurate measurements as a function of pressure on praseodymium-nitride which supposedly is an insulator at low temperature, and there were therefore no conduction electrons, no screening, and we see the same effect: i.e., a decrease of the crystal field with pressure. So we're sort of fresh out of a model.

E.F. Skelton: I noticed some of the neutron dispersion curves from antimonide compounds seemed to show extrema in the acoustic modes. I remember several years ago how Harold Smith at Oak Ridge correlated these with high superconducting transition temperatures. I was wondering if these things have been found to be superconducting?

D.B. McWhan: Not to my knowledge. There's no evidence of them being superconducting. But the argument that you are making is in terms of FCC metals having softening at the zone boundary which you would not expect without having electron-phonon coupling or something like that.

E.F. Skelton: It's in the middle region of the zone where he saw these extrema in high T_c materials, whereas isomorphic structures with very low T_c's were "normal," e.g. NbC and HfC.

D.B. McWhan: What we see there in the phonons is nothing different than what you see in sodium chloride. There's nothing in that phonon spectrum that is in any way unexpected. You always have an interaction of the longitudinal acoustic and optical modes, where they would cross.

H.R. Ott: I just wanted to say that only the singlet ground state antimonides might be candidates for being superconductors. All the others order magnetically. In addition they all seem to be semimetals.

B.T. Matthias: Not every soft mode leads to superconductivity.

T.A. Kaplan: Are you saying that you have new evidence that forces one to talk about the non-equilibrium fluctuation, and that there's not some ground state of this system that denotes a mixture of $4f^6$ and $4f^5 5d$?

D.B. McWhan: The n-scattering results are consistent with the model of a homogeneous mixed valence phase in SmS at high pressure.

SOME MAGNETIZATION STUDIES AT HIGH HYDROSTATIC PRESSURES

R.P. Guertin*

Physics Department, Tufts University

Medford, MA, 02155

ABSTRACT

Magnetization measurements in hydrostatic pressures, P, to
$\simeq$ 10 kbar are presented for several magnetically ordered rare earth
intermetallic systems. The details of the experimental techniques
are discussed. For an induced ferromagnet, Pr_3Tl, the Curie temperature and the zero field saturation moment are both reduced to zero
for P $\sim$ 8 kbar. The results are compared with a molecular field
model which takes into account the tetragonal symmetry experienced
by the magnetic Pr ions. The pressure dependence is presented of
the Neel temperature, T_N, for several RSb antiferromagnets;
R = Nd, Gd, Tb, Dy, Ho and Er. Also discussed is the effect of
pressure on spin reorientation fields for R = Dy, Ho and Er. For
DySb the effect is large and consistent with the behavior of T_N
under pressure.

INTRODUCTION

In this paper we present the results of recent experiments which
probe the effect of hydrostatic pressure on the magnetic properties
of several magnetically ordered rare earth intermetallic compounds.
The experiments are carried out at low temperatures and in high
magnetic fields. For most of the materials investigated, crystalline electric field (CEF) effects play an important, sometimes

*Supported by the National Science Foundation. Grant #DMR 77-05959.

97

dominant, role in the description of the magnetic properties. The discussion will emphasize the balance between the exchange and the CEF interaction, and how this balance is affected by a uniform reduction of the lattice constant. A description of the experimental procedures is followed by a discussion of pressure induced loss of ferromagnetism in Pr_3Tl. Finally, we present the influence of pressure on the ordering temperatures and transition fields of several antiferromagnetic rare earth monoantimonides.

EXPERIMENTAL PROCEDURES

The method employed for making magnetization measurements under pressure[1] utilizes a commercially available[2] vibrating sample magnetometer[3] (VSM) adapted to high pressure clamp devices.[4] In Fig. 1 we show a somewhat simplified view of the pressure clamp body (CB) which is attached to the drive rod (D) of the VSM, and which is thermally isolated from the liquid helium bath (LHe) by a thin-walled stainless steel dewar.[5] Pressure is applied outside the cryostat at room temperature and held in by the locknut (L)-piston (P) assembly. The pressure at low temperatures is determined by measuring, with ac susceptibility, the superconducting transition temperature[6] of a piece of tin (Sn) located with the sample (S) in the pressure-transmitting fluid, which is 1:1 isoamyl alcohol:n-pentane. Magnetic fields to 60 kG are provided by a superconducting solenoid (M). The oscillating ($\sim$90 Hz) magnetic dipole moment of the clamp-plus-sample assembly is detected by counterwound pickup coils (PC). Although the clamp generally outweighs the sample by $10^2 - 10^3$:1, there are two reasons why the clamp contribution to the total moment is quite small ($\simeq 10^{-2}$ emu at 50 kG): First, the clamp is constructed entirely of cobalt-free beryllium copper[7]; secondly, much of the clamp moment is self-cancelling because it extends beyond the region of the pickup coils. As a result, only about 2% of the entire clamp moment is detected. A high homogeneity supercon-ducting solenoid (1:10^4 over a 3 cm diameter sphere) is used in order to reduce eddy currents in the moving clamp.

For operation of the system at T < 4.2 K, liquid helium is condensed in the inner dewar, then pumped and controlled in the usual manner. For T > 4.2 K, the dewar is filled with a low pressure of helium gas (He). The temperature is measured with a low magnetoresistance calibrated resistance thermometer (ST) and, if necessary, stabilized (to $\simeq$0.01 K) by a control resistor (CT)-heater (H) feedback system located just below the 4.2 K contact point (HC) of the thin walled dewar. The baffle (B) reduces convection in the region of the clamp.

Magnetization, σ, data is recorded continuously either at fixed

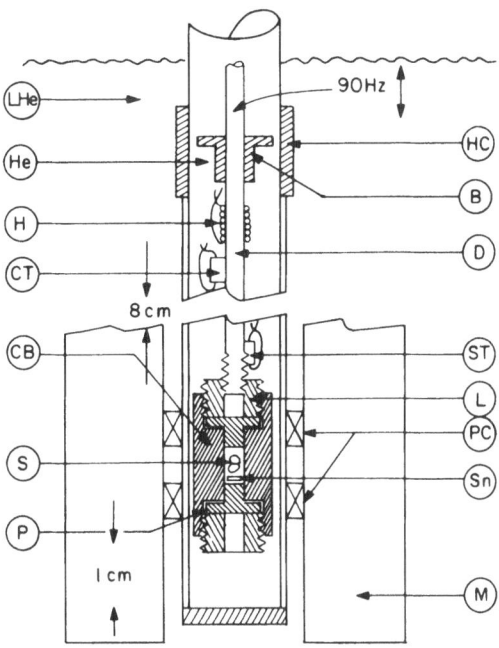

Figure 1. View of the hydrostatic pressure clamp assembly in the 60 kG superconducting solenoid (M). The sample (S) is enclosed in the clamp body (CB) and attached to the drive rod (D) of the vibrating sample magnetometer, which is operated at ≃ 90 Hz. The assembly is thermally isolated from the 4.2 K (LHe) bath by a thin-walled stainless steel dewar. The various clamp components and associated thermometry are explained in the text.

temperature-swept field, or at fixed field-swept temperature. In this way field or temperature induced transitions may be accurately determined, by way of the derivatives, $(d\sigma/dH)_T$ or $(d\sigma/dT)_H$, which are obtained with numerical differentiation of the recorded data. The sensitivity of the system is such as to detect changes in σ of $\simeq 10^{-5}$ emu in a single run; however, because of spurious background signals, relative changes in σ between different pressures is limited to $\simeq 10^{-3}$ emu. At present, the magnetic field, temperature and pressure limits are $0 \leq H \leq 60$ kG, $1.4 \leq T \leq 35$ K and $0 < P < 10$ kbar, respectively.

PRESSURE INDUCED LOSS OF FERROMAGNETISM IN Pr_3Tl

Among the rare earth metals and their metallic compounds the exchange interaction between localized magnetic moments, which is mediated by the conduction electrons, is generally strong enough

to produce magnetic order at finite temperature. The CEF interaction, whose origin is the electric fields from surrounding ions,
is often treated as a perturbation and is used to explain, for example, magnetic anisotropy in the ordered state.[8] However, for
some compounds, particularly for those of light rare earths such
as Pr, where exchange is small and the radial extent of the 4f shell
is large, the reverse situation may exist. In this case, exchange
is treated as a perturbation on the CEF levels. The competition
between the CEF and exchange interaction is manifested to some extent in nearly all rare earth systems, but perhaps the most striking
examples of this competition are found among "singlet ground state"
systems such as Pr_3Tl. Pr_3Tl crystallizes in the cubic Cu_3Au
structure and orders ferromagnetically at T_C = 12 K. The satura-
tion moment, $\sigma_o[\equiv\sigma(H=0)$ at $T \ll T_C]$ is 0.7 μ_B/Pr-atom, or only 22%
of the free ion Pr^{3+} value. The magnetic properties of Pr_3Tl depend on a very sensitive balance between the CEF and exchange interactions, and it is reasonable to suppose that a fundamental perturbation such as lattice constant reduction will strongly affect
this balance and hence the magnetic properties.

The CEF Interaction in Pr_3Tl

For rare earth ions with integral total angular momentum quantum
number, J, (non-Kramer ions), the CEF of surrounding ions, which partially lifts the 2J+1 degeneracy of the spin-orbit ground state, can
give rise to a (non-magnetic) singlet ground state sublevel. For
Pr^{3+} with cubic symmetry in metallic systems, the ground state is
normally the Γ_1 singlet with the Γ_4 (magnetic) triplet the first
excited state.[1] For $T\ll\Delta$, where Δ is the $\Gamma_1 - \Gamma_4$ splitting, the
CEF-only low field susceptibility, χ_c, is van Vleck-like and independent of temperature[9]:

$$\chi_c = \frac{2Ng^2\mu_B^2\alpha^2}{k\,\Delta} \tag{1}$$

In Eq. (1), g is the Lande g factor (=4/5 for Pr^{3+}), k is Boltzmann's
constant, and α is the matrix element, $<\Gamma_1|J_z|\Gamma_4>$. For Pr_3Tl, Δ
(=76 K) is determined from the magnetic exciton dispersion[3] measured
by neutron inelastic scattering[10]. This gives χ_c = 1.99 x 10^{-4}
emu/gm-Oe.

The influence of pressure on the susceptibility of a system where
Eq. (1) is applicable would seem to be straightforward: as the
neighboring ions are brought closer to the rare earth ion, the crystalline electric fields (and hence Δ) should increase, causing χ_c
to decrease. Details of this expected decrease were reported earlier[11] along with the results of pressure experiments on the susceptibility of several nonordering singlet systems. For some of the
systems, (the Pr- and Tm- monopnictides), the results contradicted

the expected decrease in χ_c. Subsequent experiments[12] showed that for some singlet systems the effect of pressure is to cause a <u>decrease</u> in Δ, (increase in χ_c) presumably due to band structure-charge screening effects. Nevertheless, we shall see that in Pr_3Tl, the results of pressure experiments are consistent with an increase of the CEF interaction with increasing pressure.

The cubic CEF level structure sequence for Pr_3Tl is Γ_1-Γ_4-Γ_3-Γ_5, with $\Delta = 76$ K, as mentioned. The energies are only weakly perturbed by the molecular field which produces the ferromagnetism (see next section). In terms of the Lea, Leask, and Wolf[13] cubic CEF parameters, we have $W = 3.36$ K and $x = -0.877$ (point charge value). However, a discussion of CEF- only effects in Pr_3Tl would be incomplete without discussion of the tetragonal symmetry experienced by a Pr ion. This comes about because four of the twelve nearest neighbors to a Pr ion in Pr_3Tl are Tl atoms, and these, in principle, can carry a different effective charge from the remaining eight nearest neighbors. The complete nearest neighbor CEF Hamiltonian for the Cu sites in the Cu_3Au structure is[14]:

$$\mathcal{H}_c = B_4[O_4^0 + 5O_4^4] + B_6[O_6^0 - 21O_6^4] + C_2O_2^0 + C_4[3O_4^0 - 35O_4^4]$$

$$+ C_6[5O_6^0 - 63O_6^4]. \qquad (2)$$

In Eq. (2) the symbols O are the appropriate Stevens operator equivalents and the coefficients B_n and C_n may be written

$$B_n = b_n \frac{Ze^2}{R^{n+1}} <r^n> \gamma_n$$

$$C_n = c_n \frac{\delta Ze^2}{R^{n+1}} <r^n> \gamma_n \quad . \qquad (3)$$

Here Z is the charge on the nearest neighbor Pr ions, R is the nearest neighbor distance, $<r^n>$ are the mean nth powers of the radii of the 4f electrons and the γ_n are Stevens multiplicative factors. In the expression for C_n, δZ is the Tl-Pr charge <u>difference</u>. The point charge model numerical values of the b_n and c_n coefficients appropriate to the 12-fold coordination are $b_4 = -7/32$, $b_6 = -39/256$, $c_2 = +1$, $c_4 = -1/16$ and $c_6 = +1/64$. For purely cubic symmetry $\delta Z = 0$ and only the B_n terms are non-zero. The effect of the C_n terms is to cause a splitting of the Γ_4 triplet into a doublet and a singlet. The two higher lying excited states are also split. This Hamiltonian was diagonalized in the J_z representation, and we find that the Γ_4 splitting is extremely sensitive to the value, and sign, of δZ. For example, for $0 \leq \delta Z \leq 0.01$, one obtains $76 \leq \Delta \leq 154$ K. (Δ is now the gap between Γ_1 and the singlet of the tetragonally split triplet, Γ_4). Neutron scattering experiments[10] show that the Γ_4 splitting must be $\lesssim 20$ K, however, so we are restricted to values of $|\delta Z|/Z \lesssim 0.002$. Nevertheless, the magnetic properties of Pr_3Tl, which are strongly dependent on Δ, are clearly sensitive to the tetragonality, which is parameterized by δZ.

Induced Ferromagnetism

The ferromagnetism of Pr_3Tl comes about because the singlet ground state is magnetically polarized by a molecular field (the exchange interaction). This phenomenon is known as "induced" ferromagnetism and has been discussed by several authors.[15-17] The exchange part of the total magnetic Hamiltonian is described within the molecular field approximation as [18]

$$H_{ex} = -2K <J_z> \sum_i \vec{J}_{zi} \tag{4}$$

where K is the exchange constant and $<J_z>$ the averaged magnetization. The terms $\vec{J}_{zi}$ are angular momentum operators which connect Γ_1 with the excited states in Pr_3Tl.

If we consider for the moment only the Γ_1 and Γ_4 CEF levels, it is convenient to describe the influence of the molecular field by a dimensionless parameter $\eta = 4K\alpha^2/\Delta$. For $\eta > 1$ the system will order magnetically, whereas for $\eta \lesssim 1$ the system will remain in an enhanced van Vleck paramagnetic state down to T=0 K. (For Pr_3Tl at P=0, η is 1.05, barely above criticality). For $\eta < 1$ (which will turn out to be the case for Pr_3Tl at P > 8 kbar) the enhanced low field susceptibility is given by[15]

$$\chi = \frac{\chi_c}{1 - \lambda \chi_c} \tag{5}$$

where χ_c is given in Eq. (1) and λ (which is proportional to K) is the molecular field constant.

The magnetization per Pr ion is determined self-consistently from:

$$<J_z> = \frac{\sum_{n=1}^{9} <n|J_z|n> e^{-E_n/kT}}{\sum_{n=1}^{9} e^{-E_n/kT}} \tag{6}$$

The molecular field alters the eigenvalues, E_n, and eigenstates, $|n>$, originally found from the diagonalization of Eq. (2). The results may then be compared to the experimental data. The molecular field model for the cubic ($\delta Z=0$) case including all 9 CEF levels of the J=4 multiplet does not simultaneously fit $T_C(=12K)$ and the measured saturation moment at T << T_C, $\sigma_o (=0.7 \mu_B/Pr$-atom). If the exchange constant K_1 is used which fixes $T_C = 12$ K, we obtain $\sigma_o = 0.32 \mu_B/Pr$-atom; if the exchange constant K_2 is used which fixes σ_0 = 0.7 μ_B/Pr-atom, we obtain T_C = 15.5 K. (The analogous singlet-triplet model[17] yields T_C = 18.8 K). Nevertheless, the molecular field model is a useful and convenient means to understand the gross features of the experimental data.[19]

Results of the Pressure Experiments

The Pr_3Tl sample used in the pressure studies was a long thin piece of carefully annealed polycrystalline material placed parallel to the external field. The transition temperatures as a function of pressure are shown in Fig. 2. The insert in Fig. 2 shows the criterion[20] by which the T_C's were determined. It can be seen in Fig. 2 that for P $\gtrsim$ 8 kbar the sample should be in the highly enhanced paramagnetic state (see Eq. (5)). As mentioned previously, the depression of T_C with increasing pressure is consistent with the increase in Δ expected with increasing pressure. Our calculations, using the cubic molecular field model with K = K_1, show that a lattice constant reduction of only 0.04 % (P $\simeq$ 0.5 kbar) is sufficient to suppress T_C to 0 K. However, if K = K_2, we obtain $dT_C/dP \simeq$ -1.8 K/kbar at P=0.

Magnetization data vs field (to 50 kG) were obtained for a wide variety of temperatures from 1.4 K to above T_C(P). Of particular importance are data at the lowest attainable temperature and for various pressures. In principle, these data provide rough values of σ_0 as a function of T_C. Some of the data are shown in Fig. 3, where we plot σ for several pressures vs the internal field in the sample; i.e., a correction is made for the demagnetizing factor of the sample. (The dramatic effect of pressure on the magnetic properties of Pr_3Tl is well illustrated by the data of Fig.3. For example, at H_{int} = 0.5 kG, σ decreases by more than a factor of three for a decrease in lattice constant of < 1%). Also shown in Fig. 3 are the predictions of the cubic molecular field (MF) model using K = K_2 (upper dashed curve) and of the CEF-only value of σ vs H_{int} (lower dashed curve), which is determined from Eq. (1). The data for P = 8.0 and 9.3 kbar represent magnetization curves for "borderline" ferromagnets.

Because of the large field dependence for all the data of Fig. 3 (even for P > 8 kbar), it is somewhat difficult to determine the H_{int}=0 intercept, σ_0, of the σ vs H data. Nevertheless, a reasonable extrapolation is given in Fig. 4 where we show values of σ_0 vs T_C plotted on a normalized scale. The circles in Fig. 4 are direct extrapolations from data such as that in Fig. 3. [These data are similar to analogous data of Andres et al.[18] for the $(La_{1-x}Pr_x)_3Tl$ system]. The squares in Fig. 4 are extrapolations relative to that for P = 8.0 kbar, which is approximately equal to the critical pressure for complete suppression of the ferromagnetism. Also shown in Fig. 4 are the predictions of the singlet-

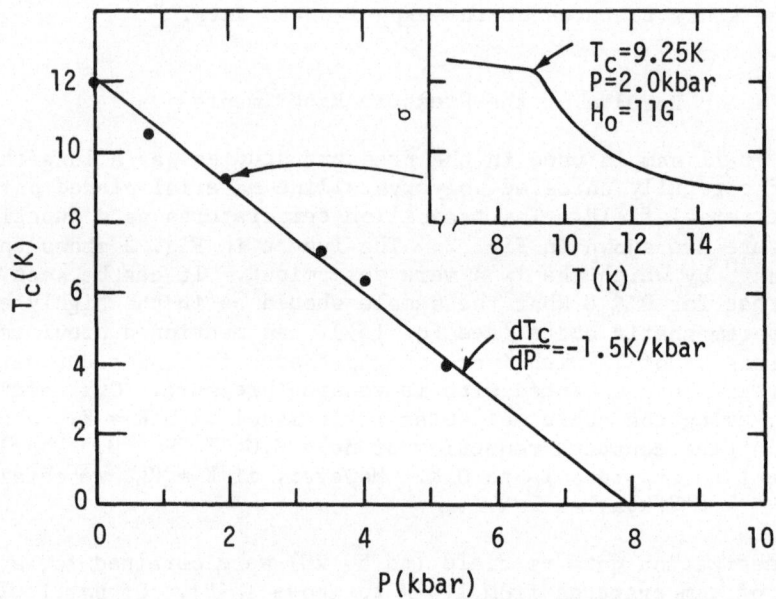

Figure 2.   Curie temperature, T_C, vs pressure, P, for a polycrys-
            talline sample of Pr_3Tl. The solid line is a linear
            least squares fit to the data. The insert shows the
            method by which the T_C's were determined.

triplet collective excitation (CE) model[18,19] and of our calculations for the molecular field (MF) model in the cubic approximation
using $K = K_1$. Fits of the MF model to the data of Figs. 2, 3 and
4 are further improved by including the tetragonality ($\delta Z \neq 0$) experienced by the Pr ions. These details will be presented in a
forthcoming publication.[21] It would be very interesting to have
calculations of σ_0 vs T_C for a collective excitation model which
incorporated the tetragonality and which took into account all 9 CEF
levels of the Pr J=4 multiplet.

ANTIFERROMAGNETIC RARE EARTH MONOANTIMONIDES (RSb)

Selected magnetic properties of several single crystal metallic
antiferromagnetic rare earth monoantimonides (RSb) were studied as
a function of hydrostatic pressure. The RSb materials examined were
for R = Nd, Gd, Tb, Dy, Ho and Er. The magnetic properties of the

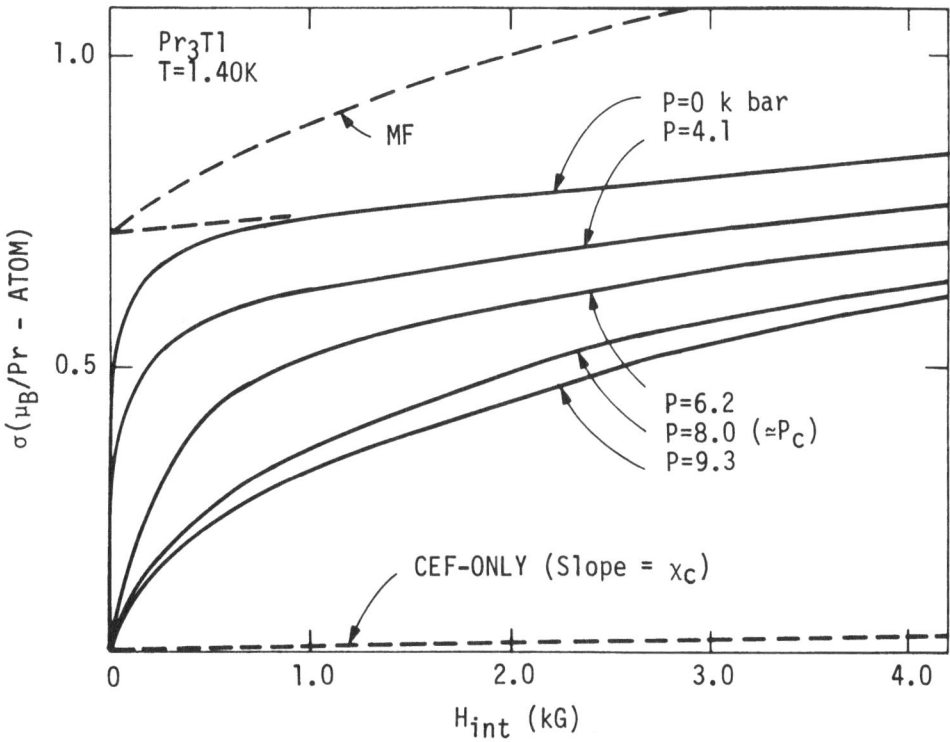

Figure 3.   Magnetization, σ, vs internal field, H_{int}, for low fields
at T = 1.40 K and for several pressures. For P $\gtrsim$ 8 kbar,
the sample is in the highly enhanced paramagnetic state.
The upper dashed line shows the prediction of the molecular field model for P=0; the lower dashed line shows the
behavior expected for zero exchange (CEF only).

ordered RSb[22] are not as sensitive to the balance between the CEF
and exchange interactions as was the case for Pr_3Tl above. Despite
similarities of the basic magnetic characteristics of these antiferromagnets, the pressure dependence of the Neel temperature, T_N,
shows no regular variation. Also measured was the pressure dependence of the critical fields for spin realignment in DySb, HoSb and
ErSb.[23]

Pressure Dependence of T_N

The pressure dependence of T_N for the ordered RSb is of interest
because a study of the systematics of T_N for all antiferromagnetic
rare earth monopnictides reveals that T_N decreases with decreasing
cation radius.[22] This trend would imply $dT_N/dP < 0$ for RSb. (How-

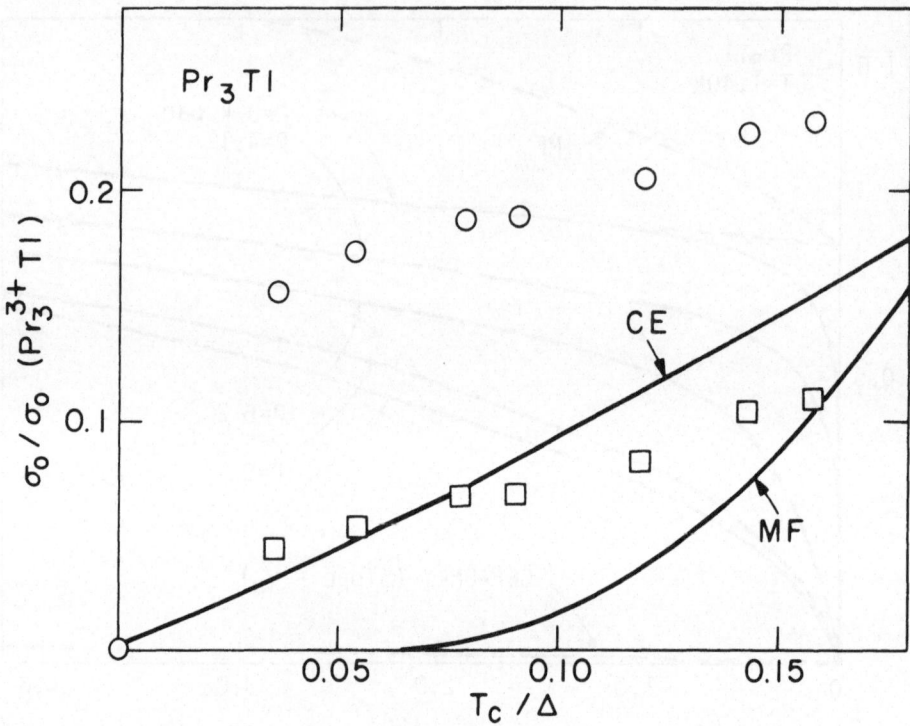

Figure 4.   Zero field extrapolated moment, σ_o, normalized to the
free ion Pr_3Tl value vs T_C/Δ. The circles are extrapolations of data such as that in Fig. 3; the squares are
extrapolations relative to that of P = 8.0 kbar, which
is close to the critical pressure for complete suppression
of ferromagnetism. Also shown are predictions of the
singlet-triplet collective excitation (CE) model and of
a cubic molecular field model (MF).

ever, this turns out to be the case for R = Dy only). In spite of
the fact that the basic magnetic characteristics of the RSb antiferromagnets are not nearly as sensitive as Pr_3Tl to the exchange-
CEF interaction competition, this competition may be used to make
plausible the conjecture that $dT_N/dP < 0$: As discussed above in
connection with Eq. (1), the CEF interaction may be expected to increase with decreasing lattice constant. T_N is thus expected to
decrease with pressure, because the CEF interaction, which lowers
T_N from the exchange-only value, will become relatively more important. (We assume here that the exchange interaction is relatively
independent of pressure since we find $dT_N/dP \simeq 0$ for GdSb; Gd^{3+}
being an S-state ion with negligible CEF interaction). On the other
hand, as discussed above, Δ for PrSb was found to decrease with de-

creasing lattice constant,[11,12] and to the extent that this effect is reflected in the other RSb, T_N may increase with pressure.

In Fig. 5 we show T_N vs P for all the RSb studied. Note the variable and broken temperature scale, which is necessary in order to display all the data. The transitions, determined as the maxima in $d\chi/dT$ by numerical differentiation of the data, were very sharp for R = Dy, Ho and Er, but were rather broad for the others. As can be seen in Fig. 5, there is no consistent trend of T_N vs P for the RSb, and only the data for DySb agrees with the original conjecture that $dT_N/dP < 0$. For NdSb, the large increase in T_N probably reflects $d\Delta/dP < 0$ found for neighboring PrSb.[12] No change in T_N with P could be detected in any of the other materials within experimental error except for ErSb, where $dT_N/dP = +1.5 \pm 0.1 \times 10^{-2}$ K/kbar for a single crystal sample. TbSb is interesting in light of the discussion related to Pr$_3$Tl: TbSb is an induced antiferromagnet, with a singlet-triplet, Γ_1-Γ_4, splitting $\Delta = 26$ K.[24] Unlike Pr$_3$Tl, however, the exchange is well above criticality $(\eta \gtrsim 3)$, so the magnetic properties are not nearly so sensitive to the CEF-exchange interaction balance, and this probably explains the lack of response to pressure of the ordering temperature in TbSb.

Pressure Dependence of Transition Fields

All the RSb materials studied are Type II antiferromagnets showing transitions back to the paramagnetic state at sufficiently high magnetic fields. The transition fields for GdSb, NdSb, and TbSb were above the range attainable with these experiments. It has been reported that DySb and HoSb show metamagnetic transitions at a critical field, H_L, to the HoP flopside spin structure[25] and then a higher field transition, H_U, to the paramagnetic state. However, in the case of HoSb, data of our own[23] and by others[26] on several samples show that the precise spin structure may be somewhat sample dependent. Nevertheless, no pressure dependence of H_L or H_U could be detected for any HoSb samples; thus this seems to be an intrinsic property, at least to ~ 7 kbar. The antiferromagnetic to paramagnetic transition field for ErSb increased at a relative rate of $+12 \times 10^{-3}$/kbar, or about three times the rate of the relative transition temperature increase. These data may reflect the pressure induced reduction in the CEF interaction noted above in connection with PrSb and NdSb.

The case of DySb is particularly interesting. In Fig. 6 we show the pressure dependence of the magnetization of DySb at T = 1.40 K for two pressures. As can be seen, H_U moves rapidly to higher fields ($dH_U/dP = +1.6$ kG/kbar). However, H_L moves to lower fields with increasing pressure, and for an extrapolated pressure of about 60 kbar, $H_L \rightarrow 0$, which means that for P $\gtrsim 60$ kbar, the

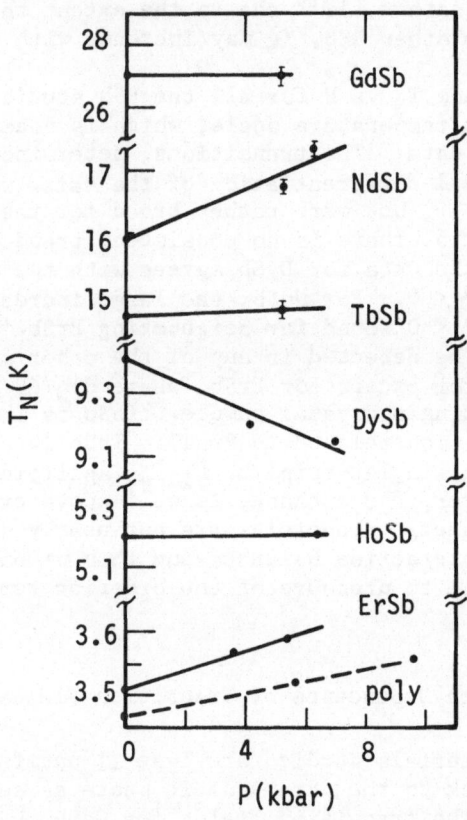

Figure 5.   Neel temperature vs pressure for several single crystal
            RSb antiferromagnets. The dashed line is for a poly-
            crystalline ErSb sample.

HoP ferromagnetic structure would occur in zero field for DySb.
This is consistent with the data for DyAs,[22] which has a smaller
ionic radius and which is ferrimagnetic, with the HoP structure
for T = 8.5 K.   Thus the behavior of both the ordering temperature
and the lower critical field in DySb under pressure are in agreement with the simple predictions based on the Neel temperature
systematics of the other rare earth monopnictides. As illustrated
in the lower portion of Fig. 6, a splitting develops in H_L at
lower temperatures with increasing pressure. The insert in Fig. 6
shows $(d\sigma/dH)_T$ at H_L, determined by numerical differentiation of
the data. Because the pressure was hydrostatic to $\lesssim$ 5%, established
by comparing the width of the Sn superconducting transition temperatures at various pressures, it is difficult to see how this

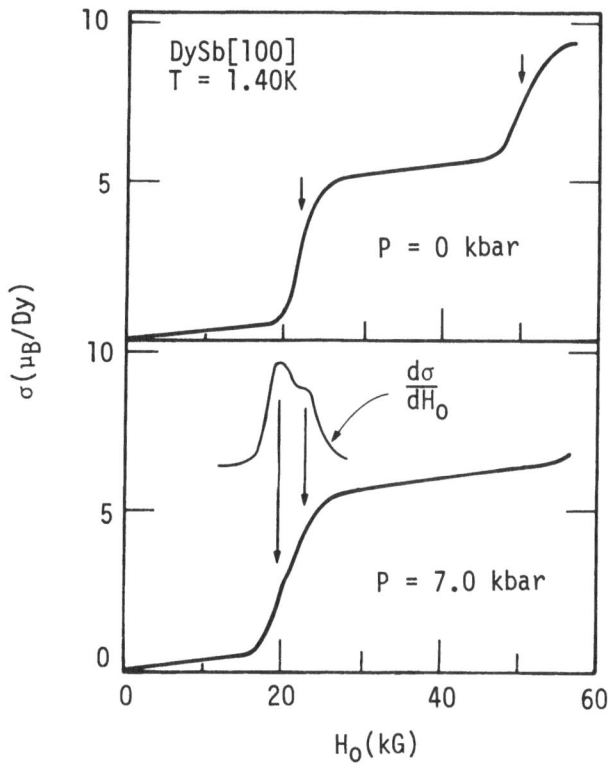

Figure 6. Pressure dependence of σ vs external field, H_o, for DySb
at T = 1.40 K. The arrows indicate the positions of
transition fields. The low field transition of the
P = 7.0 kbar data shows some structure (see curve of
$d\sigma/dH_o$).

splitting could arise. The effect suggests that a new magnetic
sublattice is induced by pressure, but this is not easy to understand with a simple uniform reduction of lattice constant.

ACKNOWLEDGMENTS

The author is grateful to F.P. Missell, S. Foner, and J. Crow,
with whom he has collaborated in the work described above. The
research was carried out at the Francis Bitter National Magnet
Laboratory[27], M.I.T., where the author is a Visiting Scientist.

REFERENCES

1. R.P. Guertin and S. Foner, Rev. Sci. Instrum. $\underline{45}$, 863 (1974).
2. Princeton Applied Research, Princeton, N.J. 08540.
3. S. Foner, Rev. Sci. Instrum. $\underline{30}$, 548 (1959)
4. Patterned after D. Wohlleben and M.B. Maple, Rev. Sci. Instrum. $\underline{42}$, 1573, (1971).
5. S. Foner and E.J. McNiff, Jr., Rev. Sci. Instrum. $\underline{47}$, 1304 (1976).
6. T.F. Smith, C.W. Chu and M.B. Maple, Cryogenics $\underline{9}$, 53 (1969).
7. Special order from Kawecki Berylco Corporation, Reading, Pa 19603.
8. "Magnetic Properties of Rare Earth Metals", (Ed. by R.J. Elliott), Plenum, New York (1972).
9. B.R. Cooper and O. Vogt, Phys. Rev. $\underline{B1}$,1211 (1970).
10. R.J. Birgeneau, J. Als-Neilsen and E. Bucher, Phys. Rev. $\underline{B6}$, 2724 (1972).
11. R.P. Guertin, J.E. Crow, L.D. Longinotti, E. Bucher, L. Kupferberg and S. Foner, Phys. Rev. $\underline{B12}$, 1005 (1975).
12. C. Vettier, D.B. McWhan and G. Shirane, Bull. Am. Phys. Soc. $\underline{22}$, 291 (1977). See also article by D.B. McWhan for this conference.
13. K.R. Lea, M.J.M. Leask and W.P. Wolf, J. Phys. Chem. Solids $\underline{23}$, 1381 (1962).
14. The calculation follows the method of M.T. Hutchings, in "Solid State Physics", (Ed. by F. Seitz and D. Turnbull). Academic Press, New York, 1964, Vol. XVI, p. 227.
15. B. Bleaney, Proc. Roy. Soc. (London) $\underline{267A}$, 19 (1963).
16. B. Grover, Phys. Rev. $\underline{140}$, A1944 (1965).
17. B.R. Cooper, Phys. Rev. $\underline{B6}$, 2730 (1972).
18. K. Andres, E. Bucher, S. Darack, and J.P. Maita, Phys. Rev. $\underline{B6}$, 2716 (1972).
19. The collective excitation model will not be detailed here. See Ref. 16, and T.M. Holden and W.J.L. Buyers, Phys. Rev. $\underline{B9}$, 3797 (1974).
20. P. Wojtowicz and M. Rayl, Phys. Rev. Letters $\underline{20}$, 1489 (1968).
21. J.E. Crow, R.P. Guertin, F.P. Missell and S. Foner. Manuscript in preparation.
22. G. Busch, J. Appl. Phys. $\underline{38}$, 1386 (1967).
23. F.P. Missell, R.P. Guertin and S. Foner, Solid State Commun. $\underline{23}$, 369 (1977).
24. T.M. Holden, E.C. Svensson, W.J.L. Buyers and O. Vogt, Phys. Rev. $\underline{B10}$, 3864 (1974).
25. P. Streit, G.E. Everett and A.W. Lawson, Phys. Lett. $\underline{50A}$, 199 (1974).
26. G. Busch, P. Schwob and O. Vogt, Phys. Lett. $\underline{23}$, 636 (1966). G. Busch and O. Vogt, Phys. Lett. $\underline{22}$, 388 (1966).
27. Supported by the National Science Foundation.

QUESTIONS AND COMMENTS

H.R. Ott: Has anybody measured elastic anomalies at the Curie temperature?

R.P. Guertin: Jack Crow (Temple Univ.) has looked at the thermal expansion of this material, which, unfortunately, is polycrystalline and cannot be made in single crystal form to the best of my knowledge, and they find no expansion anomalies at the transition temperature. Now this is not a terribly accurate measurement that's being made on polycrystalline material, but that's the only information I have. It would be very interesting to find that out.

H.R. Ott: Was it done under high pressure?

R.P. Guertin: No.

M.B. Maple: Have you tried fitting the data to Arrott plots?

R.P. Guertin: We have attempted to make Arrott plots and they don't look very good. Arrott plots work beautifully for $ZrZn_2$, as you well know, but not for this kind of material, where the origin of the magnetism is totally different.

A.R. Moodenbaugh: I was wondering how you determine the T_c of Pr_3Tl in the first case.

R.P. Guertin: We are looking at what amounts to very low field susceptibility. The susceptibility increases with decreasing temperature to a value which is limited by the demagnetizing factor of the sample. We do this at several different fields, until we find a field which is low enough so that the "kink" does not change with decreasing field. And that's what we use to determine T_c. Incidentally, as we squeeze on this material, it gets more and more difficult to define T_c because every T_c criterion you can think of is getting squeezed away. Hence, all of our data below 3.9 K is extrapolated as far as T_c is concerned.

B.T. Matthias: I just don't understand that at all, the comparison of Pr_3Tl with $ZrZn_2$. The two have absolutely nothing in common. They may be ferromagnetic. With $ZrZn_2$ one doesn't know exactly where the ferromagnetism comes from. Whereas in Pr_3Tl we know perfectly well where it does come from. And if you want to com-

pare them to where magnetism disappears, why not
compare to something like uranium-platinum in which
the Curie point also disappears. But, $ZrZn_2$ and
Pr_3Tl really have nothing in common.

R.P. Guertin: Yes, I know. I'm fully aware of that. The comparison was made only because $ZrZn_2$ is probably the
best known case for pressure induced loss of ferromagnetism. That's the only reason that the comparison was made.

H.R. Ott: I would like to make a comment in regard to erbium
antimonide. You have an increase in T_N under pressure and you expect a negative thermal expansion
coefficient above T_N. This is actually true. I
have measured it quite recently.

R.P. Guertin: That's very nice. That hangs together very well.
That puts ErSb in the same category possibly with
the case of NdSb.

T.F. Smith: You commented that in the thermal expansion measurements, you haven't seen any effects at the magnetic
transition.

R.P. Guertin: They're not my measurements, but as I understand,
there was no kind of anomaly occurring at all.

T.F. Smith: That really surprises me, because you would expect
an anomaly in the expansion just like you would
expect an anomaly in the heat capacity. Particularly since you've got a positive value for the
pressure dependence of that transition. So if
you've got a cusp in the heat capacity, then you'd
expect to get a cusp in the thermal expansion.

R.P. Guertin: There may be one, but they just haven't been able
to observe it. There is a heat capacity anomaly
but as you know it is quite small because the entropy
from the higher levels is extremely small.

H.R. Ott: If a lattice distortion is to be expected, x-ray
measurements are probably more conducive. The effect would have to be large enough though.

L.N. Mulay: What was the magnitude of maximum magnetization
or magnetic susceptibility in emu/gm for any of your
typical antiferromagnets? Could you measure the
changes with pressure with the kind of precision
you mentioned--in the 10^{-4} or 10^{-5} range?

R.P. Guertin: In the case of GdSb, by the time you get down to
                  low temperatures, 4K and below, the susceptibility
                  is so small that we couldn't see it. It drops off
                  very rapidly. But, for the cases I've observed,
                  near the region of the transition temperatures,
                  there was no problem whatsoever in seeing the sig-
                  nal from any of these samples.

L.N. Mulay: What was the magnitude or the range of magnitudes
                  of the exchange interaction for the antiferromagnets
                  you mentioned? How did that interaction change with
                  pressure?

R.P. Guertin: We just don't know because these measurements repre-
                  sent the bulk property which is a function of both
                  J and Δ, the crystal field interaction, and we just
                  have no handle on that. As far as what the numbers
                  of the exchange constants are, I just don't know.

Unknown: Do you have any particular ideas on why some of the
                  singlet systems like TbSb are independent of pres-
                  sure?

R.P. Guertin: In the case of TbSb, I think it's just a case of
                  being way over critical, and the exchange just
                  completely dominates the magnetic properties. The
                  exchange is probably not a very sensitive function
                  of pressure, because the GdSb data does not show any
                  change. We therefore just don't see any changes
                  with pressure. Those are rather wide transitions
                  so there are relatively large error bars in those
                  cases.

D. Bloch: Should the T_c vs. P curve for Pr_3Tl approach the
                  T_c = OK point as you have drawn it?

R.P. Guertin: Our calculations show that it comes down and then
                  angles into a vertical slope. The molecular field
                  predictions come down like this. That straight
                  line should have been dashed before it hit 8 kbar.

PRESSURE-INDUCED ELECTRONIC TRANSITION IN CeAl$_2$

C. Probst

Zentralinstitut für Tieftemperaturforschung der
Bayerischen Akademie der Wissenschaften
D-8046 Garching

and

J. Wittig

Institut für Festkörperforschung, Kernforschungsanlage
Jülich
D-5170 Jülich, West-Germany

ABSTRACT

The room-termperature resistance of CeAl$_2$ passes through a faint
maximum around 60 kBar and falls off at higher pressure. Strong
anomalies were previously reported for the temperature-dependence
of the resistivity under pressure up to 15 kBar by Nicolas-Francillon
et al. It has been found that these anomalies gradually disappear
with even higher pressures. They are entirely suppressed at pressures exceeding $\cong$120 kBar resulting in a R-T characteristic being
very similar to that of the nonmagnetic compound LaAl$_2$. There is
a striking similarity to the R vs. T behavior of pure cerium in the
β or γ phase on the one hand and in the α phase on the other hand.
We conclude that CeAl$_2$ undergoes a continuous electronic transition
with pressure just as Ce above the critical point.

The occurrence of superconductivity or magnetic order in
Ce-intermetallics shows an interesting correlation with the Ce-Ce
interatomic distance[1,2]. In Fig. 1 the superconducting transition
temperature T_c or the magnetic ordering temperature T_m is plotted

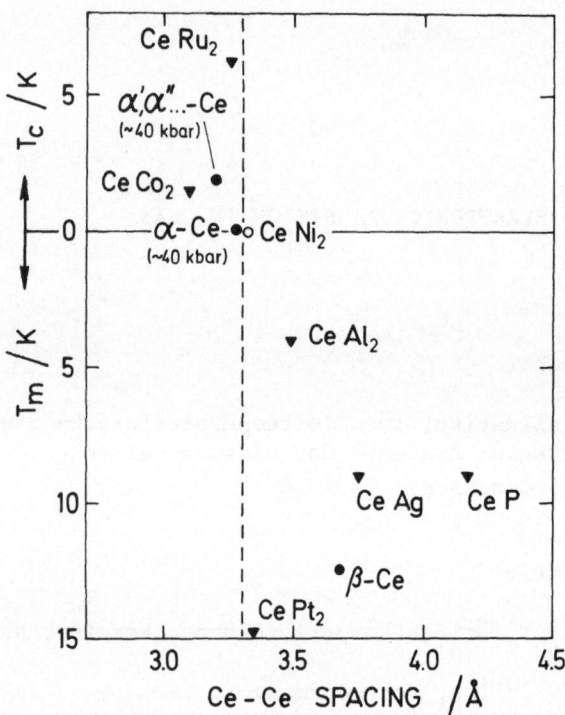

Figure 1.   Superconducting T_c or magnetic ordering temperature
T_m vs. interatomic Ce-Ce spacing for the β-, α-,
α', α''-phases of Ce-metal and several Ce-intermetallics, (After Hill[2], see also ref. 1).

against the Ce-Ce spacing for several Ce-compounds. It is seen, that
for the cubic Laves phase compounds CeX_2 (X = Pt,Al, Ni, Ru, Co)
the magnetism of the 4f-shell apparently breaks down at a critical
Ce-Ce distance of about 3.3 Å. The compound $CeNi_2$ with just the
critical spacing is neither superconducting down to 15 mK nor magnetic. The critical distance for the delocalization of the 4f
electron may vary for different lattice structures, since other
parameters will probably also be important, e.g. the metallic
radius of Ce in a particular compound, its coordination number or
the specific properties of the neighboring atoms.

For Ce-metal we see from Fig. 1 approximately the same critical
spacing as for the cubic Laves phase compounds. In the β-phase of
Ce a magnetic moment is observed, corresponding to the Hund's rule
moment of the first localized 4f electron. β-Ce orders antiferro-
magnetically at 13 K[3]. α-Ce, which has just the critical Ce-Ce
spacing, is called an intermediate valence phase. α-Ce is distin-

guished at low pressure by a rather high, but roughly temperature-
independent Pauli susceptibility and a high electronic specific
heat[4]. Recently, α-Ce has been found to be a superconductor with
a T_c increasing from 20 mK at 22 kBar to 50 mK at 40 kBar [5]. The
high pressure phases of Ce (α',α'',...) stable above $\simeq$40 kBar are
known to be fairly "good" superconductors with T_c's between 1 and
2 K [5,6,7]. From the lattice constant it was concluded that Ce is
tetravalent in the high pressure phase [8].

CeAl$_2$ has at normal pressure a Ce-Ce spacing close to the
critical value (Fig. 1). It was therefore regarded as a candidate
for a valence transition at high pressure and subsequent superconductivity [1], similar to the other cubic Laves phase compounds
CeRu$_2$ and CeCo$_2$, which have a small Ce-Ce distance and are good
superconductors. The magnetic moment of the Ce ion in CeAl$_2$ at
ambient pressure, deduced from the high temperature susceptibility,
corresponds to the 4f^1 configuration [9]. The possibility of a
Ce^{3+} $\rightarrow$ Ce^{4+} transition on cooling, similar to the electronic transition in pure Ce, was excluded from dilatation experiments[9]. At
about 4 K CeAl$_2$ undergoes a transition to an ordered low temperature phase with probably complex magnetic structure. The transition manifest's itself in strong anomalies in the specific heat,
the thermal expansion, the resistivity etc. [9,10,11].

In the Y$_{1-x}$Ce$_x$Al$_2$ system the appearance of tetravalent Ce was
inferred from magnetic and resistivity measurements for alloys with
low Ce content [9,10]. For the Y$_{1-x}$Ce$_x$Al$_2$ system it is presumed that
the mean lattice parameter predominantly determines the magnetic
behavior of the Ce ions rather than band structure and local environment effects[10]. Since the dilution of CeAl$_2$ with Y results in a
lattice-contraction of up to 2.5%, we conclude, that in pure CeAl$_2$
a magnetic$\rightarrow$nonmagnetic transition of the Ce ions may occur at
moderate pressure.

Here we report first results of a high pressure experiment on
CeAl$_2$. The resistivity has been measured up to about 180 kBar and
in the temperature range 1.2 .. 300 K, employing standard high pressure technique [12,13]. Single-crystal CeAl$_2$ was powdered in an agate
mortar. We succeeded in placing a powder-sample together with a
strip of lead foil in a high pressure cell, using steatite as a pressure transmitting medium. For increasing pressure the pressure
inhomogeneity $\Delta P/P$, read from the lead manometer, did not exceed
7 - 10%, while for the load releasing cycle the pressure gradient
across the sample was roughly 25 kBar.

In Fig. 2 the room-temperature resistance of such a powder-
sample is plotted vs. the load of the press. At small pressure
the resistance decreases strongly due to the compression of the
powder. With increasing pressure a rather flat region occurs near

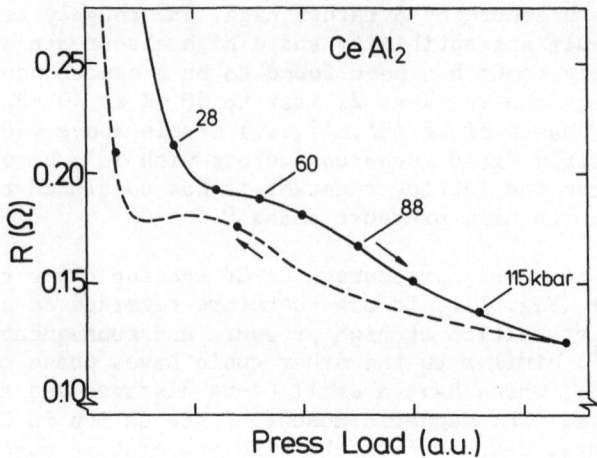

Figure 2.   Room-temperature resistance of a CeAl$_2$-sample vs.
press load, solid line for increasing pressure,
dashed line for decreasing pressure. The strong decrease of the resistance at small pressure is due to
the compression of the powdered sample.

60 kBar, followed by a steeper decrease. On the pressure releasing
cycle a shallow maximum was clearly resolved. We suppose, that
under increasing pressure the faint maximum is masked by the decrease of the contact resistance between the grains of the powder.
From the smooth variation of the resistance with pressure and the
absence of any excessive drift we have at present no indication
that we are crossing a crystallographic phase boundary in the whole
pressure range. However, due to the sluggish relaxation processes
in the powder, we cannot exclude that a small discontinuity in the
R-P characteristic of CeAl$_2$ escaped the detection.

The sample of Fig. 2 was isobarically cooled to liquid He temperature at 28, 44,..., 115, and 130 kBar (at the data points of
Fig. 2). Fig. 3 shows the corresponding R-T curves. These curves
are reproducibly observed during cooling and warming. On releasing
the load the pressure distribution across the sample was more in-
homogeneous (≈25 kBar), resulting in a slight smearing of the shape

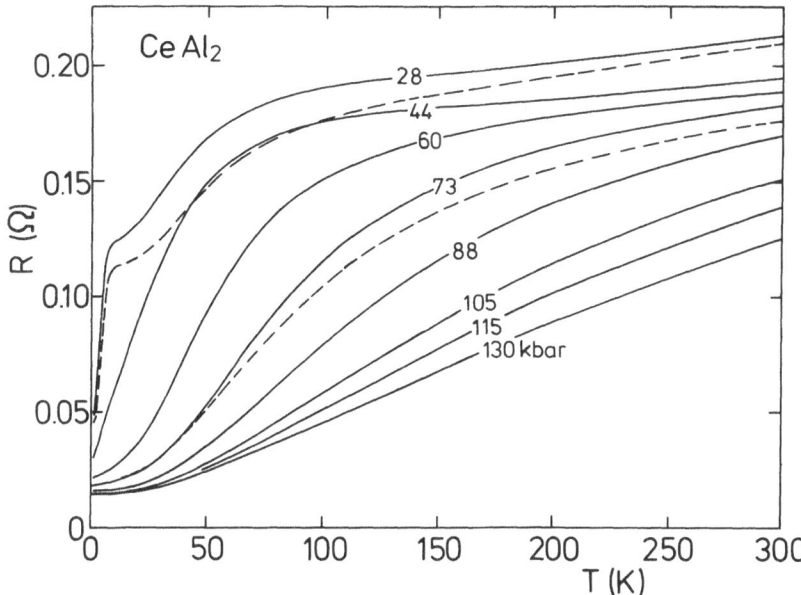

Figure 3. Temperature dependence of the resistance of the CeAl$_2$-
           sample of Fig. 2 at various pressures. The dashed
           curves were measured at decreasing pressure, where the
           pressure inhomogeneity was roughly 25 kBar. The mean
           pressure for the lower curve was 76 kBar, for the upper
           curve 28 kBar.

of the R-T curves (dashed curves) in comparison to the curves recorded at increasing pressure. However, the distinct anomalies
of the 28 kBar curve appeared reversibly with decreasing pressure.
From the R-T curves we had no indication for a 1st order crystallographic transformation below room-temperature.

The striking feature of the resistance curves of Fig. 3 is the
continuous disappearance of the distinct anomalies (cf. the 28 kBar
curve) with increasing pressure. The anomalies are entirely suppressed at pressures exceeding ≃120 kBar, resulting in a R-T characteristic being very similar to that of the nonmagnetic Laves phase
compound LaAl$_2$.

The strong anomalies of the 28 kBar isobar have been already
previously found by Nicolas-Francillon et al.[14], who carried out
resistivity measurements under hydrostatic conditions up to 15 kBar.

The authors subtracted the resistivity of the nonmagnetic compound
$LaAl_2$ from the CeAl$_2$-data to separate the magnetic part of the resistivity ρ_m for CeAl$_2$. The broad maximum of ρ_m at 50 - 100 K was
explained with a Kondo-type scattering of the conduction electrons
by the Hund's rule ground state of the Ce^{3+} ions split by the crystalline electric field. The small maximum at low temperature indicates the formation of the ordered low-temperature phase.

To explain our results we suggest that a continuous electronic
transition takes place in CeAl$_2$ in the pressure range of 50 - 100
kBar, similar to the continuous electronic transition in pure Ce
above the critical point of the γ/α- phase boundary. The experimental curves of Fig. 2 and Fig. 3 show a striking similarity to
the resistivity behavior of Ce in the pressure range around the
γ/α-transition, both with respect to their pressure dependence and
their temperature dependence: If we accept that the actual R-P
characteristic of CeAl$_2$ (cf. Fig. 2 and its discussion) is rather
flat for P $\leq$ 60 kBar or even shows a shallow maximum near 60 kBar,
then we find that the R-P curvature closely resembles the pressure-
resistance isotherms measured by Jayaraman[15] for fcc-Ce above the
critical point. Regarding the R-T behavior of CeAl$_2$ we notice a
great similarity of the 28 kBar isobar (Fig. 3) with the R-T curve
of the magnetic β-phase (and probably also of the magnetic γ-phase)
of Ce, while the 130 kBar isobar closely resembles the R-T behavior
of the nonmagnetic α-phase of Ce[16]. Recently Gschneider, Jr.
et al.[17] succeeded in measuring the resistivity of pure β-Ce. Similar to CeAl$_2$ the resistance anomaly of β-Ce was explained with a
Kondo-type scattering of the conduction electrons[18].

With reference to Fig. 1, it is of course interesting whether
CeAl$_2$ will eventually become superconducting at very high pressures.
In a preliminary experiment we have cooled a sample of CeAl$_2$ at a
pressure of about 180 kBar down to∿ 0.15 K, without any sign of
superconductivity. However, if the electronic structure of CeAl$_2$
at 180 kBar is similar to that of α-Ce, which has a maximum T_c
of 50 mK, superconductivity is expected for CeAl$_2$ at lower temperatures. Experiments to much lower temperatures are in progress.

REFERENCES

1. C. Probst and J. Wittig, in Handbook on the Physics and Chemistry of Rare Earths, edited by K.A. Gschneidner, Jr.
   and L. Eyring, (North-Holland Publ. Comp., Amsterdam),
   Chapter 10, in print.
2. H.H. Hill, in Plutonium 1970 and Other Actinides, editor W.N.
   Miner, (New York:AIME), p.2, 1970.
3. P. Burgardt, K.A. Gschneidner, Jr., D.C. Koskenmaki, D.K.
   Finnemore, J.O. Moorman, S. Legvold, C. Stassis, T.A. Vyrostek,
   Phys. Rev.B., 14, 2995, (1976).

4. D.C. Koskimaki and K.A. Gschneidner, Jr., Phys. Rev. B 11, 4463, (1975).
5. C. Probst and J. Wittig, Proc. 14th Internat. Conf. on Low Temp. Phys., Otaniemi 1975, editors M. Krusius and M. Vuorio, (North-Holland Publ. Comp., Amsterdam), Vol. 5, p.453, (1975).
6. J. Wittig, Phys. Rev. Lett. 21, 1250, (1968).
7. C. Probst and J. Wittig, to be published.
8. F.H. Ellinger and W.H. Zachariasen, Phys. Rev. Lett. 32, 773, (1974).
9. E. Walker, H.-G.Purwins, M. Landolt, F. Hullinger, Jour. Less-Common Metals 33, 203, (1973).
10. F. Steglich, Magnetic Moments of Rare Earth Ions in a Metallic Environment, in Festkörperprobleme – Advances in Solid State Physics, (Vieweg, Braunschweig), in print.
11. M. Croft, I. Zorić, J. Markovics and R. Parks, Proc. Int. Conf. on Valence Instabilities and Related Narrow Band Phenomena, Rochester, N.Y. (1976), in print.
12. A. Eichler and J. Wittig, Z. Angew, Physik 25, 319, (1968).
13. See also paper by J. Wittig and C. Probst, presented at this conference.
14. M. Nicolas-Francillon, A. Percheron, J.C. Achard, O. Gorochov, B. Cornut, D. Jerome and B. Coqblin, Solid State Commun. 11, 845, (1972).
15. A. Jayaraman, Phys. Rev. 137, A179, (1965).
16. M. Nicolas-Francillon and D. Jerome, Solid State Commun. 12, 523, (1973).
17. K.A. Gschneidner, Jr., P. Burgardt, S. Legvold, J.O. Moorman, T.A. Vyrostek and C. Stassis, J. Phys. F. 6, L49, (1976).
18. S.H. Liu, P. Burgardt, K.A. Gschneidner, Jr., S. Legvold, J. Phys. F. 6, L55, (1976).

QUESTIONS AND COMMENTS

H.R. Ott: Did you observe any hysteresis?

C. Probst: No. No hysteresis with respect to pressure or to temperature.

D. Jérome: I think you could present your data in a more convincing way, if you subtract the phonon contribution. What you are interested in is the spin contribution. You should have a much greater drop in resistivity when you consider only the spin contribution. This way you will have to measure the phonon contribution curve under pressure. But, if you do that you will get a far different effect.

H.R. Ott: You have no anomaly in the resistance. Did you check

by any other means for the onset of magnetic ordering?

C. Probst: No, the only indication of a magnetic ordering was
                    the small maximum at low temperatures.

H.R. Ott: That's not necessarily true, because I had data on
                    cerium arsenide which shows the anomaly in some
                    samples, and in others nothing at all, but the mag-
                    netic ordering occurred in all the samples.

C. Probst: Okay. It is not my observation that these small
                    maxima are due to magnetic ordering.

B.T. Matthias: Which one of these phases was the one which is
                    isomorphous to α-uranium?

C. Probst: Alpha prime. It's orthorhombic.

B. T. Matthias: Is it superconducting?

C. Probst: I would think so. From our data, we have obtained
                    transitions indicating that both phases are super-
                    conducting.

D. Jérome: Have you ever looked at $CeAl_3$? It may have a mag-
                    netic to non-magnetic transition at much lower
                    pressure--15-20 kbar.

C. Probst: No.

D. Bloch: Very often to get the magnetic contribution, one
                    tends to subtract the phonon contribution of the
                    nonmagnetic phase. But, I think that is something
                    very very dangerous, and that can result in many
                    mistakes because you know that the compressibility,
                    for instance, is changed at the magnetic to nonmag-
                    netic transition which means that the Debye tempera-
                    ture is simultaneously modified.

A.R. Moodenbaugh: Going back to the problem of the magnetic trans-
                    ition, do you have plans to measure the sample
                    magnetically under pressure?

C. Probst: No. The transition is too high in pressure to
                    measure the susceptibility.

VALENCE INSTABILITIES OF RARE EARTH IONS IN METALS*

M.B. Maple

Institute for Pure and Applied Physical Sciences
University of California, San Diego
La Jolla, California 92093

ABSTRACT

Rare earth ions are capable of producing "Kondo-like" anomalies in the physical properties of both concentrated and dilute rare earth metallic systems which are qualitatively similar. The anomalies can be viewed as originating from the hybridization localized 4f electron states and extended states of the conduction band as a result of the unstable valence of certain rare earth ions. The physical properties of a concentrated and a dilute rare earth metallic system which exemplify Kondo-like behavior, SmS in its collapsed metallic phase and $(\underline{LaSm})Sn_3$, respectively, are briefly reviewed.

I. INTRODUCTION

The purpose of this paper is to briefly review some of the experimental aspects concerning valence instabilities of rare earth (RE) ions in metals. Several more extensive reviews of this subject can be found elsewhere in the literature to which the reader is referred.[1,2,3,4]

---

*Research supported by the U.S. Energy Research and Development Administration under Contract No. ERDA E(04-3)-34PA227.

In most metallic RE compounds and alloys, the RE ion has an integral valence, or equivalently, an integral occupation of its
4f electron shell. Therefore, only one 4f electron configuration,
$4f^n$, is involved and there is negligible hybridization between the
localized 4f states and the extended states of the conduction band.
As a result, most RE compounds and alloys exhibit ionic magnetism.
At elevated temperatures, the magnetic susceptibility follows a
Curie-Weiss law

$$\chi(T) = \frac{N\mu_{eff}^2}{3k_B(T-\theta)} \tag{1}$$

where N is the number of RE ions, μ_{eff} is the effective magnetic
moment which can be derived from Hund's rules and θ represents a
Curie-Weiss temperature near which some type of ordering of the RE
magnetic moments generally occurs. In addition, both the magnetic
susceptibility and the specific heat, for example, exhibit Van Vleck
anomalies due to the small multiplet splittings of Sm and Eu, and
crystal field effects which can be as pronounced as in insulators.

In contrast, the metallic RE compounds and alloys considered
herein are those in which the RE ion has a nonintegral valence or
a nonintegral average occupation of its 4f electron shell. Therefore two 4f electron configurations are involved, $4f^n$ and $4f^{n-1}5d^1$,
where the d electron is assumed to be delocalized and to reside
in the conduction band. In this situation, there is appreciable
hybridization between the localized 4f states and the extended
states of the conduction band, which leads to nonmagnetism below
a characteristic temperature T_O. For temperatures much larger than
T_O, the susceptibility can be approximated by an expression of the
form

$$\chi(T) \sim N \frac{\epsilon(n)[\mu_{eff}(n)]^2 + [1-\epsilon(n)][\mu_{eff}(n-1)]^2}{3k_B(T+T_O)} \tag{2}$$

where N is the number of RE ions, $\epsilon(n)$ is the fraction of time the
configuration $4f^n$ is occupied, and μ_{eff} is the effective magnetic
moment associated with the configuration $4f^n$. For temperatures
much smaller than T_O, the susceptibility approaches a finite value
as $T \to 0$, i.e.,

$$\chi(T) \to \text{constant as } T \to 0 \tag{3}$$

rather than diverging as a Curie-law or undergoing a transition to
a magnetically ordered state. This nonmagnetic behavior below the
characteristic temperature T_O is accompanied by Kondo-like anomalies
in the physical properties near T_O. For example, there can be a
resistance minimum followed by a large increase in resistivity as
the temperature is lowered, a peak in the specific heat near T_O,
as well as a peak in the thermoelectric power near T_O. This striking behavior appears to originate from the unstable valence of

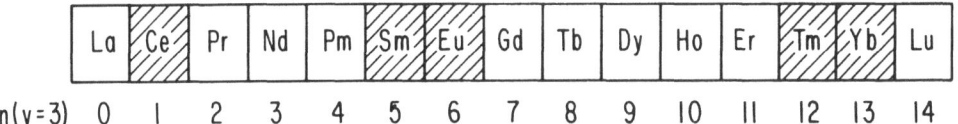

Figure 1.   The rare earth series. Trivalent (v = 3) 4f electron
            shell occupation numbers are indicated. Valence in-
            stabilities frequently occur for the rare earth ions
            Ce, Sm, Eu, Tm and Yb (shaded).

certain RE ions and, phenomenologically, is qualitatively similar
in both <u>concentrated</u> and <u>dilute</u> RE metallic systems.

In the concentrated RE metallic systems, the anomalies are often
attributed to valence or interconfiguration fluctuations (ICF)[1,5,6]
between the configurations $4f^n$ and $4f^{n-1}5d^1$ with a frequency τ_0^{-1}
where τ_0 is the ICF lifetime. The characteristic temperature T_0 is
related to τ_0 by $T_0 \equiv h/k_B\tau_0$, and for a large number of metallic
RE systems, T_0 has been found to be $\sim 10^2$K. This type of behavior
has been well documented in many <u>concentrated</u> RE metallic systems
for the RE ions Ce, Sm, Eu, Tm and Yb.[1,2,4] As indicated in Fig. 1,
valence instabilities for RE ions have been observed to occur at
the beginning (Ce), near the middle (Sm, Eu), and at the end (Tm,
Yb) of the RE series as the 4f electron shell is progressively
filled.

In <u>dilute</u> RE metallic systems, the anomalies in the physical properties are generally ascribed to the Kondo effect which originates
from the exchange interaction

$$\mathcal{H}_{int} = -2\,\mathcal{J}\underset{\sim}{S} \cdot \underset{\sim}{s} \qquad\qquad (4)$$

where $\underset{\sim}{S}$ is the spin of the RE ion, $\underset{\sim}{s}$ is the conduction electron
spin density at the site of the RE ion, and $\mathcal{J}$ is the exchange interaction parameter which characterizes both the strength and the sign
of the interaction. The appreciable hybridization between 4f and
conduction electron states for RE ions with unstable valence results
in a large <u>dominant</u> negative (or antiferromagnetic) contribution to
$\mathcal{J}$ which leads to Kondo-like behavior with the Kondo temperature
T_K given approximately by

$$T_K \sim T_F \exp\left(-1/N(E_F)|\mathcal{J}|\right) \qquad\qquad (5)$$

where T_F is the Fermi temperature and $N(E_F)$ is the density of states
at the Fermi level of the conduction band. Kondo-like behavior in
<u>dilute</u> RE metallic systems has been well established for the RE ions
Ce and Yb, but relatively little is known with regard to dilute

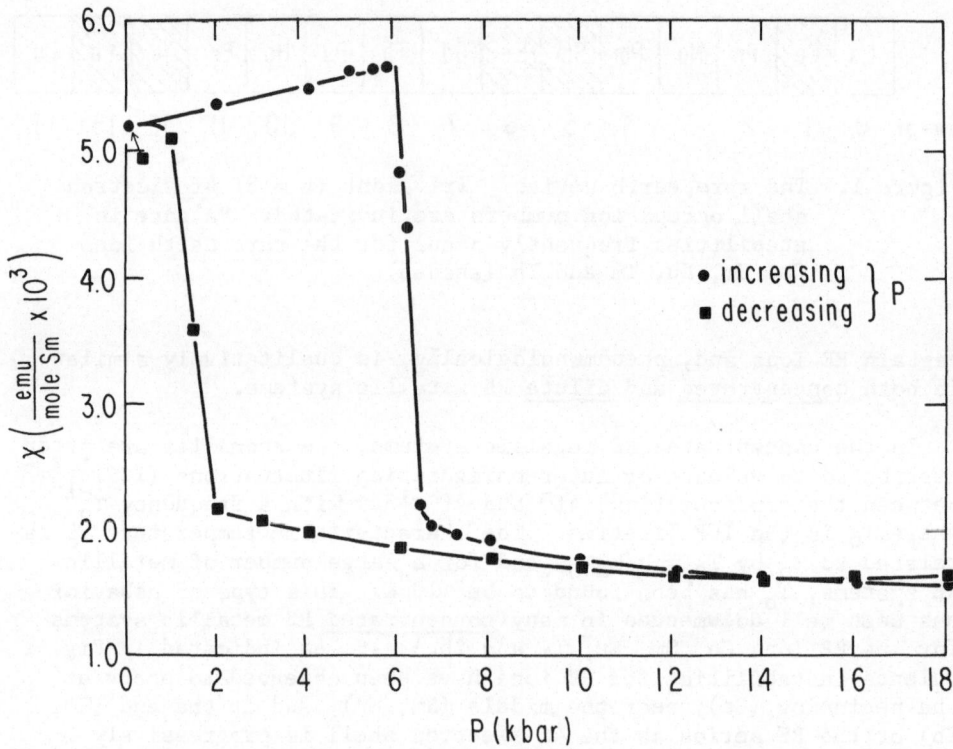

Figure 2. Magnetic susceptibility χ of SmS vs. pressure P at room
temperature (from Ref. 5).

RE Kondo-like behavior for the RE ions Sm, Eu and Tm. However,
definitive evidence for dilute RE Kondo-like behavior has recently
emerged from studies of the superconducting matrix-impurity system $(\underline{La}Sm)Sn_3$. In light of these recent results, we will re-
strict our considerations to the Kondo-like behavior of the RE
ion Sm in an exemplary concentrated RE metallic system, SmS in its
collapsed metallic phase above 6.5 kbar, and the recently documented
dilute RE metallic system, $(\underline{La}Sm)Sn_3$.

II. CONCENTRATED RARE EARTH METALLIC SYSTEMS –
SmS IN ITS HIGH PRESSURE COLLAPSED METALLIC PHASE

At zero pressure, the compound SmS is a semiconductor with the
cubic NaCl crystal structure and the Sm ions are divalent.[7] Electrical conductivity occurs via thermal activation of localized
electrons from the Sm 4f electron shells into the conduction band
with a small activation energy $\sim$0.2eV. The Sm 4f electron shell
contains 6 electrons and the compound exhibits ionic Van Vleck

paramagnetism with a nonmagnetic J = 0 ground state.[7]

Under pressure, SmS undergoes a discontinuous transition from the semiconducting state to a metallic state as first reported by Jayaraman et al.[8] The transition is accompanied by a large decrease in volume ($\Delta V/V \sim -8\%$) without a change in crystal structure. In the collapsed metallic phase, the lattice parameter[8] of SmS indicates that the Sm valence is ~ 2.7 or, equivalently, the average occupation number of the Sm 4f electron shell is ~ 5.3.[5] Thus, two 4f electron shell configurations, $4f^6$ and $4f^5$, are involved in the ground state of SmS in its collapsed metallic phase. Additional evidence for a nonintegral Sm valence in the collapsed phase has been provided by Mössbauer isomer shift measurements by Coey et al.[9] on SmS in the collapsed high pressure metallic phase and the "chemically collapsed" pseudobinary compound $Sm_{0.77}Y_{0.23}S$, and X-ray photoemission spectroscopy (XPS) studies by Campagna et al.[10], Freeouf et al.[11] and Pollak et al.[12] on various chemically collapsed pseudobinary compounds formed by alloying SmS with a third element. It is interesting to note that whereas the Mössbauer isomer shift measurements are unable to resolve the presence of the two Sm 4f electron shell configurations $4f^6$ and $4f^5$ in the high pressure and chemically collapsed metallic phases of SmS, the XPS measurements reveal their simultaneous presence in chemically collapsed metallic SmS. Within the context of the ICF model, this is consistent with the fact that the characteristic ICF lifetime is of the order of 5×10^{-13} sec., the measuring time is $\sim 10^{-9} - 10^{-7}$ sec. for the Mössbauer isomer shift measurements and $\sim 10^{-17}$ sec. for the XPS measurements. The 6.5 kbar electronic transition in SmS is the analogue of the well-known γ-α electronic transition in Ce metal.[13]

The magnetic susceptibility of SmS as a function of pressure at room temperature[5] is shown in Fig. 2. Upon application of pressure the susceptibility first increases slowly, then drops abruptly at the semiconductor-metal transition at 6.5 kbar, and thereafter decreases slowly. Upon release of pressure, the transition does not occur until ~ 2kbar. The pressure hysteresis in the susceptibility is similar to that found in the resistivity by Jayaraman et al.[8]

The magnetic susceptibility of SmS as a function of temperature in its high pressure collapsed metallic phase[5] is shown in Fig. 3. At higher temperatures, the susceptibility of SmS is intermediate between normal $4f^5$ and $4f^6$ behavior. It exhibits weak, but definite temperature dependence below ~ 200K and saturates to a constant value below ~ 40K with no indication of magnetic order. The constant low temperature susceptibility is in sharp contrast to that which would be expected if the transition had proceeded directly to the trivalent $4f^5$ configuration. In the $4f^5$ configuration, the ground state must be at least a doublet, which would result in a low temperature divergence of the magnetic susceptibility or some type of magnetic order.

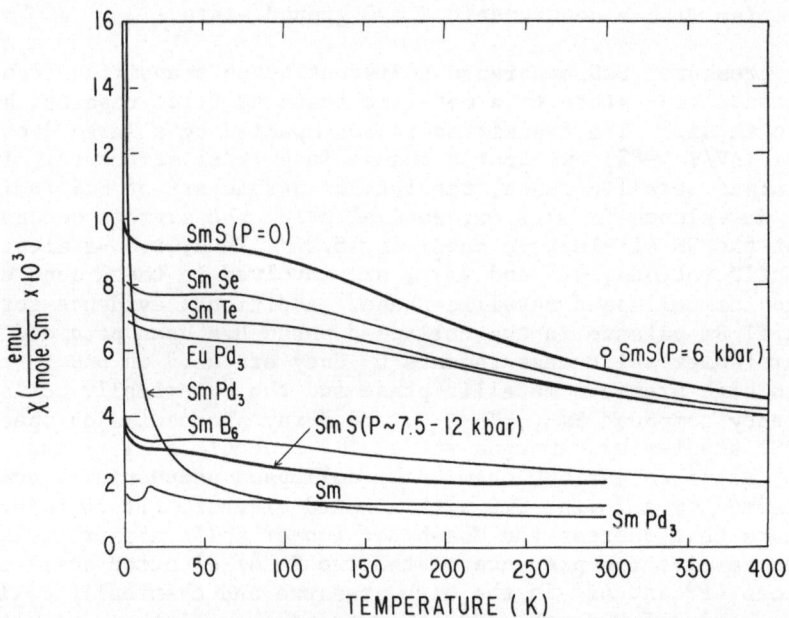

Figure 3.   Magnetic susceptibility χ of SmS at zero pressure and in
its high pressure collapsed metallic phase vs. temperature.
Also shown for comparison are $\chi(T)$ data for compounds
with the configuration $4f^6$[SmSe, SmTe and EuPd$_3$], the
configuration $4f^5$[Sm and SmPd$_3$], and the intermediate
valence ICF compound SmB$_6$ (from Ref. 5).

In its collapsed metallic phase, the weakly temperature dependent
magnetic susceptibility of SmS is accompanied by anomalies in other
physical properties such as the specific heat and electrical resistivity. Shown in Figs. 4 and 5, respectively, are specific heat[14]
and electrical resistivity[14] data for SmS in its high pressure collapsed metallic phase. The data reveal a large γT contribution to
the heat capacity (γ = 145 mJ/mole-K^2) and a striking low temperature resistivity anomaly. The large magnitude of the resistivity
anomaly is probably due to the fact that the electrons which are
responsible for the conductivity are the same electrons which are
emitted and absorbed by the 4f shell as it undergoes temporal fluctuations between the configurations $4f^6$ and $4f^5$.

III.   DILUTE RARE EARTH METALLIC SYSTEMS –
THE SUPERCONDUCTING MATRIX-IMPURITY SYSTEM (LaSm)Sn$_3$

Recent experiments by DeLong et al.[15] and Bakanowski et al.[16]
on the superconducting matrix-impurity system (LaSm)Sn$_3$ revealed
the first definitive evidence for a Kondo effect due to Sm impurity

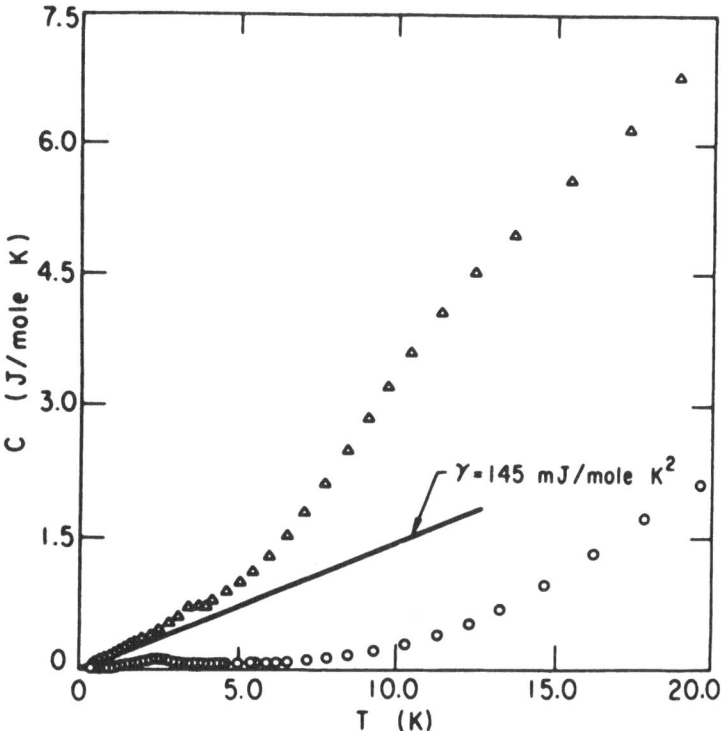

Figure 4. Heat capacity C vs. temperature of SmS at approximately
15 kbar (triangles) and at zero pressure (circles). A
plot of C/T vs. T^2 shows that the limiting coefficient
γ of the linear term in the heat capacity of the collapsed metallic phase is $\gamma = 145$ mJ/mole K^2 (from Ref.
14).

ions. Prior evidence for a Kondo effect due to Sm ions was reported
by Chouteau et al.[17] who attributed various anomalies in the physical properties of concentrated (Sm,La)S alloys to the spurious presence of small amounts of Sm^{3+} ions, the concentrations of which de-
pended upon lattice defects and vacancies. In the (LaSm)Sn_3 system,
the Kondo effect is manifested in both the superconducting and normal state physical properties.

Measurements of the initial depression of T_c, $(-dT_c/dn)_{n=0}$, of
$LaSn_3$ by various RE impurity additions by Schmid and Umlauf (18) and
Abou-Aly et al.[19] provided the first indication of a possible Kondo
effect in the (LaSm)Sn_3 system. From these measurements, the initial
depression of T_c, $(-dT_c/dn)_{n=0}$, for Sm impurity additions was found
to be anomalously large compared to other RE impurity additions
(Fig. 6). In analogy with the anomalously large depression of T_c

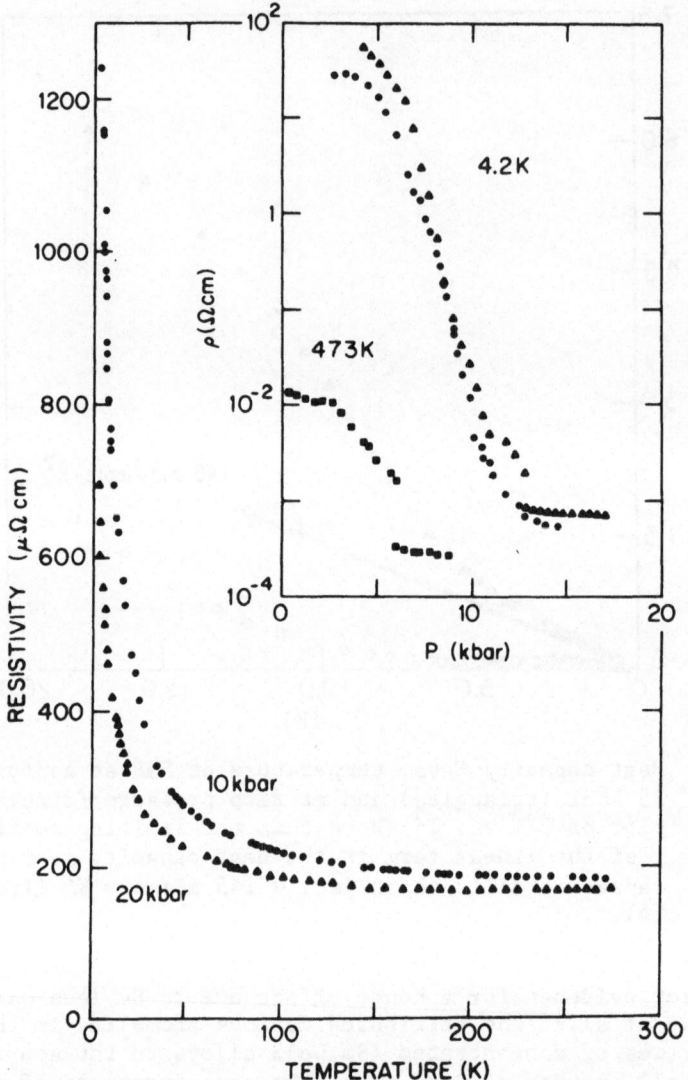

Figure 5. Electrical resistivity ρ vs. temperature of SmS at 10 and
           20 kbar. The inset compares the insulator-metal transi-
           tion as a function of pressure at 4.2 and 473 K.   The
           pressure transmitting medium was AgCℓ (circles), frozen
           (triangles) or liquid (squares) n-pentane isoamyl alcohol
           (from Ref. 14).

in superconducting matrices containing small concentrations of Ce
impurities which exhibit a Kondo effect, this result suggests that
the exchange interaction parameter $\mathcal{J}$ for Sm in LaSn$_3$ is large and
negative.[28,29]

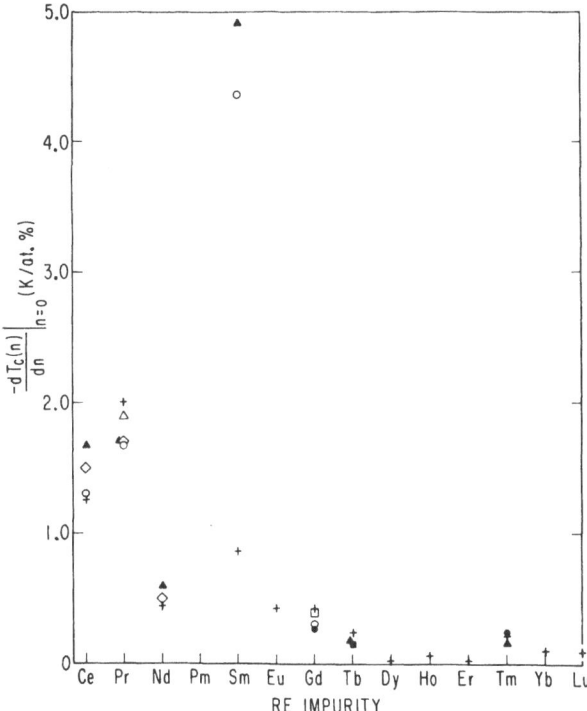

Figure 6. Initial rate of depression of the superconducting transi-
           tion temperature of (LaRE)Sn$_3$ alloys vs. RE impurity.
           The data are taken from: Ref. 18-crosses; Ref. 15,20 and
           21-open circles; Ref. 19 solid triangles; Ref. 22-open
           diamonds; Ref. 23-open triangles; Ref. 24-open squares;
           Ref. 25 and 26-solid circles; Ref. 27-solid squares.

In order to further explore the anticipated manifestation of the
Kondo effect in the superconducting state properties of the (LaSm)Sn$_3$
system, DeLong et al.[15] measured the detailed dependence of superconducting transition temperature T_c on Sm impurity concentration
and the reduction of the specific heat jump at T_c as a function of
T_c. The results are shown in the form of a plot of the reduced transition temperature T_c/T_{c_0} vs. Sm concentration n in Fig. 7 (a) and
a plot of the reduced specific heat jump $\Delta C/\Delta C_0$ vs. T_c/T_{c_0} in
Fig. 7(b). Fig. 7(a) shows that the T_c/T_{c_0} vs. n curve deviates
from the initial linear behavior below $T_c/T_{c_0} \simeq 0.5$ [for (LaSm)Sn$_3$,
T_c = 6.38 K and $(-dT_c/dn)_{n=0}$ = 4.32 K/at.%Sm]. Fig. 7(b) illustrates
the marked negative deviation of the $\Delta C/\Delta C_0$ vs. T_c/T_{c_0} curve from

the behavior predicted by the BCS law of corresponding states and
the Abrikosov-Gor'kov (AG) theory.[30]

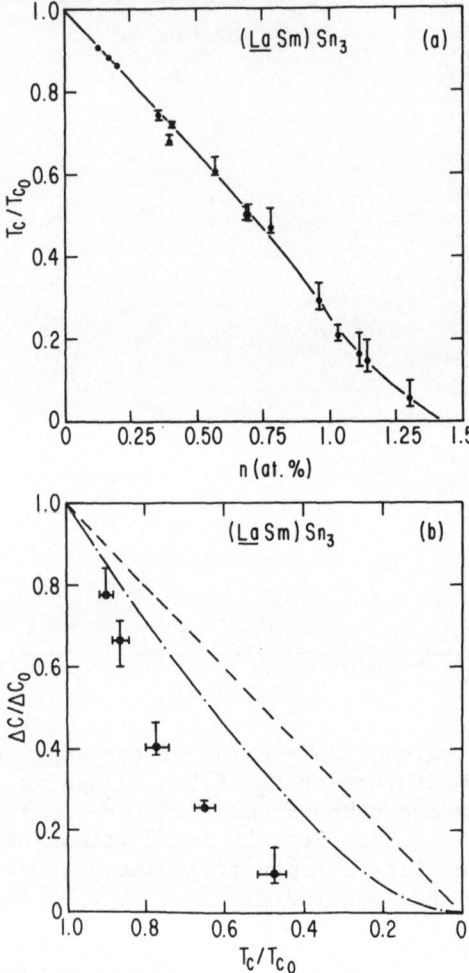

Figure 7. (a) Reduced superconducting transition temperature
           T_c/T_{c_0} vs. Sm impurity concentration n for the (LaSm)Sn$_3$
           system. The solid line is a smooth curve drawn through
           the data. The vertical bars denote 10% and 90% points
           of inductively measured transition amplitudes. (b) Redu-
           ced specific heat jump $\Delta C/\Delta C_0$ vs. reduced transition tem-
           perature T_c/T_{c0} for the (LaSm)Sn$_3$ system. The dashed

           line represents the BCS law of corresponding states
           and the dot-dashed line indicates the AG result. The
           vertical bars represent estimated experimental uncer-
           tainty in determining the magnitude of the specific
           heat jumps at T_c, and the horizontal bars represent es-
           timates of the total width of the transitions (from
           Ref. 15).

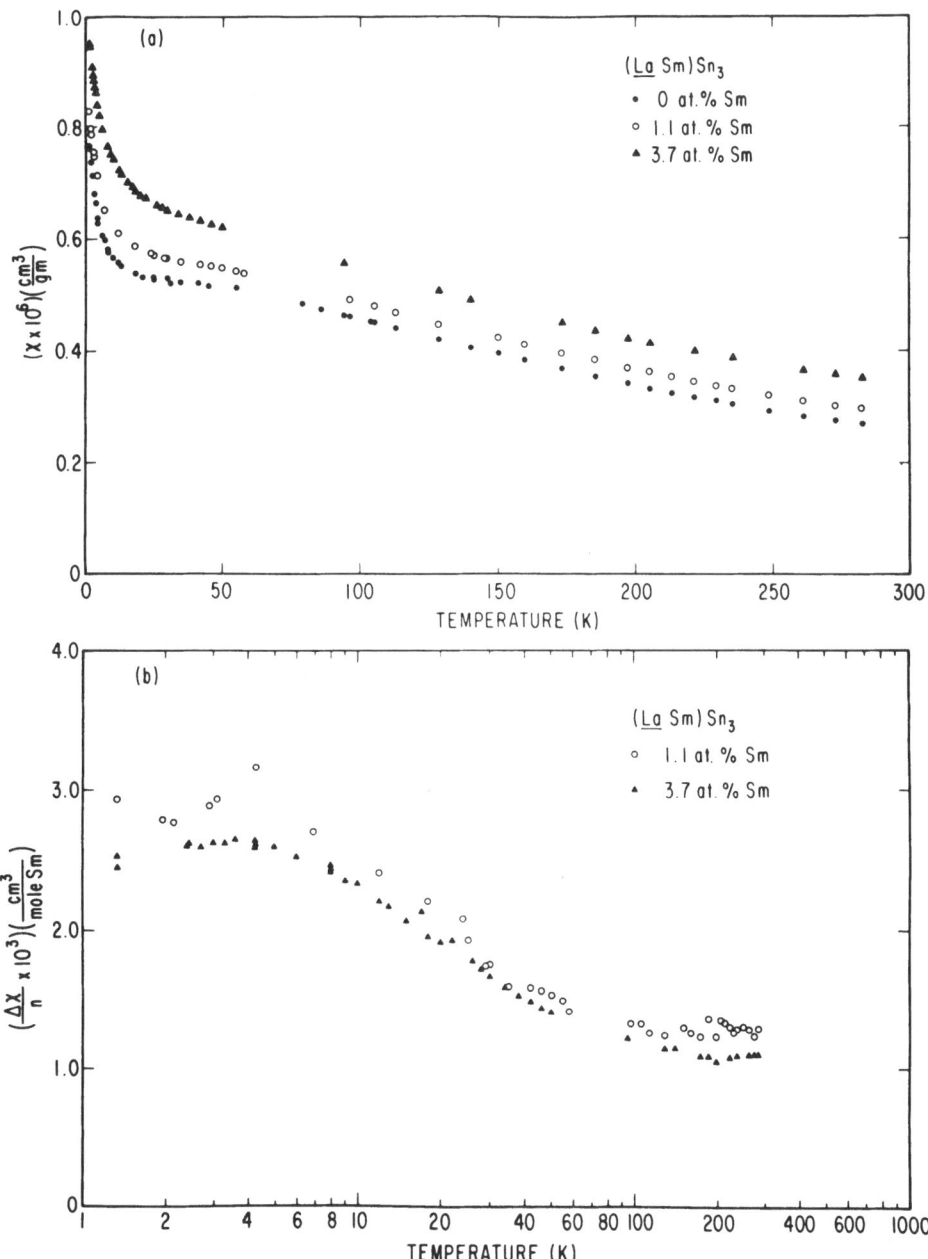

Figure 8.   (a) Magnetic susceptibility χ vs. temperature for $LaSn_3$
and two $(\underline{LaSm})Sn_3$ alloys with Sm concentrations of 1.1
and 3.7 at.%.   (b) Magnetic susceptibility (corrected
for the $LaSn_3$ background susceptibility) of $(\underline{LaSm})Sn_3$
alloys vs. temperature (note logarithmic scale) (from
Ref. 15).

The form of T_c/T_{c_0} vs. n and $\Delta C/\Delta C_0$ vs T_c/T_{c_0} are very similar
to the behavior exhibited by the (La,Th)Ce system near an equiatom-
ic La,Th matrix composition.[29,31,32] Thus a qualitative comparison
of the data of (LaSm)Sn$_3$ and (La,Th)Ce alloys can be made by assuming that the modification of the superconducting properties by
the Kondo effect scales with T_K/T_{c_0} in the same manner in both
systems. Since T_K is of the order of one to ten times T_{c_0} near the
equiatomic La,Th host composition, T_K for the (LaSm)Sn$_3$ system is
estimated to lie in the range $\sim$5-100 K.

Magnetic susceptibility χ vs. temperature data for a pure LaSn$_3$
sample and two (LaSm)Sn$_3$ samples with Sm concentrations of 1.1 and
3.7 at.% are displayed in Fig. 8(a).[15] The susceptibilities of
the dilute alloys were found to be slightly enhanced by weakly temperature dependent contributions over the host susceptibility. The
incremental susceptibility $\Delta\chi$, defined as $\Delta\chi = \chi[(LaSm)Sn_3]$ -
$\chi[LaSn_3]$, is shown in the form of a $\Delta\chi/n$ vs. log T plot for the
1.1 and 3.7 at.% Sm samples in Fig. 8(b). The weak dependence of
the Sm contribution to the magnetic susceptibility is reminiscent of
the behavior of the susceptibility of SmS in its collapsed metallic
phase described in Section II. At higher temperatures, the susceptibility is much smaller than the weakly temperature dependent
Van Vleck susceptibility of the configuration $4f^6$, while at low
temperatures, it does not diverge as a Curie law for the configuration $4f^5$. The susceptibility is nearly constant between room
temperature and $\sim$100 K, increases by a factor $\sim$2.5 below 100 K,
and approaches a finite value as $T \to 0$.

Normal state specific heat measurements on a series of (LaSm)Sn$_3$
alloys with concentrations between 0 and 0.8 at % Sm from 0.6 to
10 K[15] indicate that there is a large enhancement of the electronic
specific heat coefficient γ of 1.06 ± 0.16 J/mole-Sm K^2. This enhancement of γ is equivalent to an average enhancement of the LaSn$_3$
host density of states of 230 states/eV-atom per spin direction.
To our knowledge, this is the largest enhancement of γ by a dilute
impurity reported for any system to date. Like the behavior of the
magnetic susceptibility, this extraordinary enhancement of γ is similar to that observed for SmS in its collapsed metallic phase.
The weak dependence of the magnetic susceptibility and the large
value of the enhancement of γ in the low temperature specific heat
for (LaSm)Sn$_3$ are consistent with the relatively high value of T_K
inferred from the superconductivity studies. The specific heat
generally exhibits a maximum near T_K - the γT term may be the low
temperature tail of a Kondo specific heat anomaly with a maximum
well above the 10 K upper limit of the specific heat measurements.

Bakanowski et al.[16] recently reported the existence of a minimum

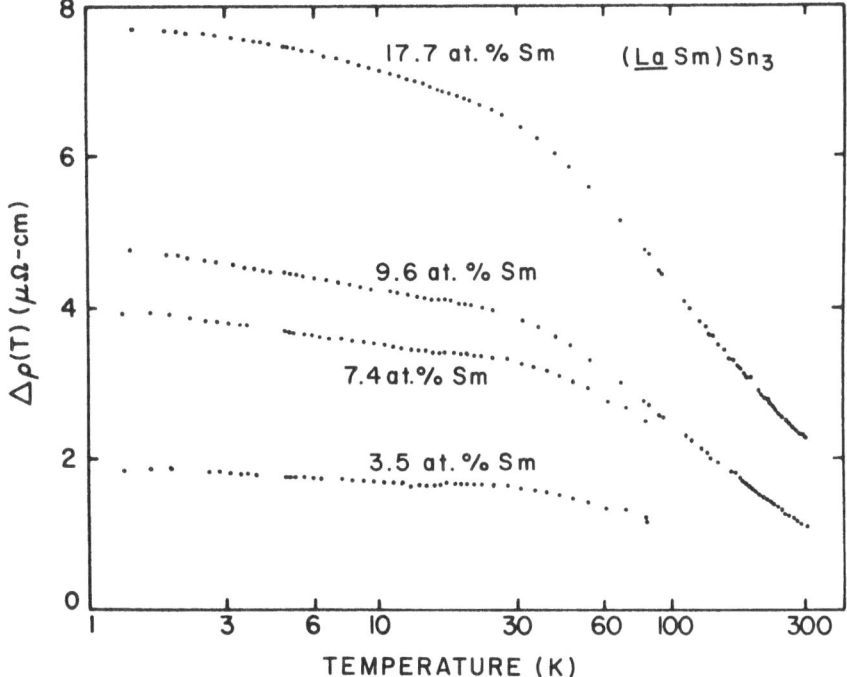

Figure 9. Magnetic contribution to the electrical resistivity $\Delta\rho$ vs. temperature for (LaSm)Sn$_3$ (from Ref. 16).

in the normal state electrical resistivity for several (LaSm)Sn$_3$ alloys (Figure 9). The incremental resistivity contributed by the Sm ions can be described well by a calculation based on the exchange model by Hamann[33] with an effective spin of 0.27 ± 0.05 and T_K = 110 ± 20 K. This value of T_K is of the order of the value inferred from the T_c/T_{c_0} vs. n and $\Delta C/\Delta C_0$ vs. T_c/T_{c_0} superconductivity data and the normal state magnetic susceptibility and low temperature specific heat data.

The experiments on the (LaSm)Sn$_3$ system reviewed in this section document the Kondo anomalies in the superconducting and normal state physical properties with a relatively large Kondo temperature $T_K \sim$ 100 K. The strong resemblance between the anomalies in the normal state physical properties of the (LaSm)Sn$_3$ system and the system SmS in its collapsed metallic phase show that the anomalies in dilute RE metallic systems which are ascribed to the Kondo effect are qualitatively similar to those in concentrated RE metallic systems which are attributed to ICF.

IV.   CONCLUDING REMARKS

In this paper, we have briefly reviewed the striking behavior
imparted to the physical properties of both <u>concentrated</u> and <u>dilute</u>
RE metallic systems by RE ions with unstable valence. We have concentrated on the RE ion Sm for two reasons. First, the physical
properties of the exemplary system SmS in its collapsed metallic
phase have been more extensively and thoroughly documented than any
other concentrated system with unstable RE valence. Second, the
(<u>La</u>Sm)Sn_3 system provides the first definitive evidence for Kondo-
like behavior due to Sm impurity ions in a dilute RE metallic system.

Both concentrated and dilute RE metallic systems appear to be
<u>nonmagnetic</u> below a characteristic temperature T_0 (or T_K) with tem-
perature dependent anomalies in their physical properties near
T_0 (or T_K). In the case of the concentrated RE metallic systems,
and in some dilute RE metallic systems, the <u>nonmagnetic</u> behavior
can be correlated with a <u>nonintegral</u> valence, or nonintegral
average occupation of the 4f electron shell. The anomalies can be
viewed as originating from the hybridization of the localized 4f
electron states and the extended states of the conduction band as
a result of the unstable valence of certain RE ions. In concentrated
RE metallic systems, the Kondo-like anomalies are often attributed
to ICF and have been well-documented for the RE ions Ce, Sm, Eu, Tm
and Yb. However, in dilute RE metallic systems, the Kondo-like
anomalies are generally ascribed to the "Kondo effect" but, until
recently, have only been well-established for Ce and its "4f hole"
counterpart Yb. In analogy with the concentrated RE metallic systems, it is anticipated that examples of Kondo-like behavior will be
found in dilute metallic systems containing Eu and Tm impurities as
well. In fact, recent experimental evidence indicates that Eu may
exhibit Kondo-like behavior in the Eu molybdenum chalcogenides
$Eu_xMo_6X_8$ where X = S or Se.[34]

We have also tried to show that superconductivity provides a
sensitive probe for detecting, identifying and studying the effects
of RE ions with unstable valence when the matrix is a superconductor.

Finally, a fundamental understanding of the origin and nature of
concentrated and dilute RE metallic systems which exhibit Kondo-like
anomalies constitutes a challenging problem for theoretical description. In fact, during the past several years, a tremendous amount
of effort has been expended by both experimentalists and theorists
alike towards developing an understanding of this intriguing class
of solid state materials.

REFERENCES

1. M.B. Maple and D. Wohlleben, AIP Conf. Proc. No. 18, Magnetism and Magnetic Materials - 1973, ed. by C.D. Graham, Jr. and J.J. Rhyne, Am. Inst. of Phys., 1974, p. 447.

2. D. Wohlleben and B.R. Coles, in "MAGNETISM: A Treatise on Modern Theory and Materials," Vol. V, ed. by H. Suhl, Academic Press, New York, 1973, Ch. 1.

3. C.M. Varma, Rev. Mod. Phys. 48, 219 (1976).

4. M.B. Maple, L.E. DeLong and B.C. Sales, to appear in the Handbook on the Physics and Chemistry of Rare Earths, ed. by K.A. Gschneidner, Jr., and L. Eyring, North-Holland Publishing Company, Amsterdam, The Netherlands, Chapter 11.

5. M.B. Maple and D. Wohlleben, Phys. Rev. Lett. 27, 511 (1971).

6. L.L. Hirst, Phys. Kondens. Mat. 11, 255 (1970).

7. E. Bucher, V. Narayanamurti and A. Jayaraman, J. Appl. Phys. 42, 1741 (1971).

8. A. Jayaraman, V. Narayanamurti, E. Bucher and R.G. Maines, Phys. Rev. Lett. 25, 1430 (1970).

9. J.M.D. Coey, S.K. Ghatak, M. Avignon and F. Holtzberg, Phys. Rev. B14, 3744 (1976).

10. M. Campagna, E. Bucher, G.K. Wertheim and L.D. Longinotti, Phys. Rev. Lett. 33, 165 (1974).

11. J.L. Freeouf, D.E. Eastman, W.D. Grobman, F. Holtzberg and J.B. Torrance, Phys. Rev. Lett. 33, 161 (1974).

12. R.A. Pollak, F. Holtzberg, J.L. Freeouf and D.E. Eastman, Phys. Rev. Lett. 33, 820 (1974).

13. A.W. Lawson and T.Y. Tang, Phys. Rev. 76, 301 (1949).

14. S.D. Bader, N.E. Phillips and D.B. McWhan, Phys. Rev. B7, 4686 (1973).

15. L.E. DeLong, R.W. McCallum, W.A. Fertig, M.B. Maple and J.G. Huber, Solid State Commun. 22, 245 (1977).

16. S. Bakanowski, J.E. Crow and T. Mihalisin, Solid State Commun. 22, 241 (1977).

17. G. Chouteau, F. Holtzberg, O. Peña, T. Penney, R. Tournier and S. von Molnar, Int. CNRS Colloquium on Metal-Nonmetal Transitions, Autrans, France, June 1976.

18. W. Schmid and E. Umlauf, Commun. Phys. 1, 67 (1976).

19. A.I. Abou-Aly, S. Bakanowski, N.F. Berk, J.E. Crow and T. Mihalisin, AIP Conf. Proc. No. 29, Magnetism and Magnetic Materials 1975, ed. by J.J. Becker, G.H. Lander and J.J. Rhyne, Am. Inst. of Phys., 1976, p. 358.

20. R.W. McCallum, W.A. Fertig, C.A. Luengo, M.B. Maple, E. Bucher, J.P. Maita, A.R. Sweedler, L. Mattix, P. Fulde and J. Keller, Phys. Rev. Lett. 34, 1620 (1975).

21. L.E. DeLong, R.W. McCallum and M.B. Maple, in Proceedings of the Fourteenth International Conference on Low Temperature Physics, Otaniemi, Finland, 1975, Vol. 2, ed. by M. Krusius and M. Vuorio, North Holland, Amsterdam, 1975, pp. 541-544.

22. M.H. van Maaren and M. van Haeringen, in Proceedings of the
     Fourteenth International Conference on Low Temperature Physics,
     Otaniemi, Finland, 1975, Vol. 2, ed. by M. Krusius and M. Vuorio
     North Holland, Amsterdam, 1975, pp. 533-536.
23. P. Lethullier, Phys. Rev. 12B, 4836 (1975).
24. A.M. Toxen, P.C. Kwok and R.J. Gambino, Phys. Rev. Lett. 21, 792
     (1968).
25. R.P. Guertin, J.E. Crow, A.R. Sweedler and S. Foner, Solid State
     Commun. 13, 25 (1973).
26. E. Bucher, K. Andres, J.P. Maita and G.W. Hull, Jr., Helv. Phys.
     Acta 41, 723 (1968).
27. H.E. Hoenig, H. Happel, H.K. Njoo and H. Seim, in Proceedings
     of the First Conference on Crystalline Electric Field Effects
     in Metals and Alloys; Montreal, 1974, unpublished, ed. by
     R.A.B. Devine, pp. 298-322.
28. M.B. Maple, in "MAGNETISM: A treatise on Modern Theory and
     Materials," Vol. V, ed. by H. Suhl, Academic Press, New York,
     1973, Ch. 10.
29. M.B. Maple, Appl. Phys. 9, 179 (1976).
30. A.A. Abrikosov and L.P. Gor'kov, Zh. Eksp. Teor. Fiz. 39, 1781
     (1960); Sov. Phys. JETP 12, 1243 (1961).
31. J.G. Huber, W.A. Fertig and M.B. Maple, Solid State Commun.
     15, 453 (1974).
32. C.A. Luengo, J.G. Huber, M.B. Maple and M. Roth, Phys. Rev.
     Lett. 32, 54 (1974) and J. Low Temp. Phys. 21, 129 (1975).
33. D.R. Hamann, Phys. Rev. 158, 570 (1970).
34. M.B. Maple, L.E. DeLong, W.A. Fertig, D.C. Johnston, R.W. Mc-
     Callum, and R.N. Shelton, in Valence Instabilities and Re-
     lated Narrow Band Phenomena, ed. by R.D. Parks, Plenum Press,
     1977, pp. 17-29.

QUESTIONS AND COMMENTS

D. Debray: You did talk about valence fluctuation analysis?
                Do you really have any evidence for the temporal
                fluctuation?

M.B. Maple: If I understand what you mean, I would say, as I
                said before, that the susceptibility itself is
                evidence for temporal fluctuations if there are any
                fluctuations at all, because if you had a spatial
                distribution of two configurations...

D. Debray: If it has spatial distribution, it doesn't have to
                be temporal. You've got to interpret the suscepti-
                bility of yours without assuming they are temporal.

M.B. Maple: If you want to talk about it in a quantum-mechanical sense, then there is no real evidence. Now let me just say that the Mössbauer measuring time is about 10^{-9} seconds and the x-ray photoelectron spectroscopy (XPS) measuring time is about 10^{-17} or 10^{-18} seconds, or something like this, whereas the valence fluctuation lifetime is expected to be about 10^{-13} seconds. What happens is that the Mössbauer measurements cannot resolve the two lines corresponding to the isomer shifts for both the divalent and trivalent configurations, whereas the XPS measurements actually do see both spectra simultaneously. Now, whether those have anything to do with intrinsic fluctuations, or whether the XPS measurements project out those two states, I really couldn't say. So, in that sense, there is no real evidence for fluctuations if you talk about them in a quantitative way.

D. Debray: Yes, I understand. I should add, however, that in the XPS method, what you study is the surface, not the bulk. Moreover, you have to forget those data because they are not reproducible.

M.B. Maple: What you say is true, but I think that the possible discrepancy between the results of those two groups (Bell Laboratories and IBM) are not significant within the context of answering your question. I mean that it is true that in samarium sulfide, for example, if you scratch the surface, you transform the surface, even though the rest of the material won't be transformed. But there still remains a lot of material which is not transformed. I think that the measurements that have been done by both groups are pretty reliable. Personally, I find them convincing.

C. Cordero: With regard to the measurements which you showed on the susceptibility of samarium compounds with decreasing temperature, the temperature for the onset of the constant susceptibility region was lower than in samarium sulfide itself. What would the reason be for the different values and large error limits of this temperature which are reported for intermediate valent samarium sulfide in the literature?

M.B. Maple: Well, I think this characteristic temperature is something that is an order of magnitude quantity. I don't think it can ever be possible to pin it down within better than half an order of magnitude. Because the transition itself is very smooth in nature. It's not abrupt.

C. Cordero: The temperature dependence of the susceptibility
                    shown by samarium in trivalent samarium compounds
                    is also different from that expected for trivalent
                    samarium sulfide.

M.B. Maple: That's how I see it. You mean in $(\underline{La}Sm)Sn_3$. Yes,
                    it is different.

THEORETICAL APPROACH TO THE CONFIGURATION FLUCTUATION IN Sm-CHALCOGENIDES*

T.A. Kaplan and S.D. Mahanti

Physics Department, Michigan State University

East Lansing, Michigan 48824

and

Mustansir Barma

Tata Institute of Fundamental Research

Colaba, Bombay-400 005 India

ABSTRACT

We present a theoretical approach to the problem of configuration mixing in Sm chalcogenides. Properties in the collapsed phase of SmS are discussed in terms of an essentially localized or excitonic picture of most of the Sm 5d-electrons, with a small number ($\approx .1$ electrons per Sm) occupying the free-electron-like states at the bottom of a broad 5d-band. We review how such an essentially localized model for the 5d-electrons can lead to the observed results on volume and degree of mixing vs. pressure, low-T specific heat, dc electrical conductivity, plasma frequency, XPS intensities and magnetism. We also present new calculations of phase boundaries in the T-p plane, obtaining for SmS both a first order and a second order boundary within a simplified model. The latter boundary, which occurs e.g. at high p and low T makes contact with a recent experimental result of Güntherodt et al. The unusual shape of the observed first-order boundary is shown to be in accord with our general model. A new evaluation of the low frequency dielectric con-

* Supported by NSF Research contract #DMR-76-16597.

stant ($\hbar\omega^S$.1ev) shows behavior very similar to the unusual experimental results obtained recently by Batlogg et al. and by Allen. We compare the above picture with the more common models where all the 5d-electrons (~.7 per Sm in SmS) occupy free-electron like states, and give a critique of the latter.

I. INTRODUCTION

Rare earth atoms involved in materials exhibiting fluctuations of the number of 4f electrons ("valence fluctuations") have recently attracted considerable interest. A great deal of work has been presented in the proceedings[1] of the recently held conference on the subject at Rochester, N.Y. That the experimental results pose a difficult, and perhaps very fundamental, challenge to present-day solid state theory can be seen by perusal of those proceedings. While a variety of rare earth materials show these effects, we have focused on the Sm chalcogenides SmS, SmSe, SmTe and related compounds-- these show the interesting properties of fluctuating or mixed valence, but maintain the simplicity of the cubic rocksalt structure. The present paper is quite similar to our contribution[2] to the Rochester Conference, differing in that (a) it amplifies various discussions e.g., of magnetic properties and Mössbauer experiments, (b) it extends aspects of our earlier calculations; included in particular, are calculations of the phase diagram in the temperature-pressure (T-p) plane, and a calculation of the optical conductivity down to the far infrared, and (c) it includes new arguments that bear on the controversy, band vs. essentially-localized models.

At room temperature, SmS undergoes[3-5] a first-order transition to a mixed-configuration (or mixed-valence or fluctuating-valence) phase as pressure increases through $p_c \tilde{=} 6$ kbar, with a decrease in volume of ~10%,but with no other apparent change in crystal structure. In the mixed phase, which is apparently nonmagnetic[5], the system is in a mixture of $4f^6$ (i.e. Sm^{2+}) and $4f^5 5d$ (i.e. Sm^{3+}) configurations. The amount of admixture has been determined from the lattice constant[3], isomer shift[4], XPS data[6]; a ratio of f^6 to $f^5 d$ of roughly 3:7 has been found. Furthermore, the Mössbauer measurements[4] lead to a spatially homogeneous picture, where every Sm ion is in the f^6-$f^5 d$ mixture. Actually these measurements are interpreted[4] as implying that the Sm ion fluctuates between f^6 and $f^5 d$ configurations faster than ~10^{-9} sec^{-1}; quantum mechanically this means in simplified terms that the splitting between the energy eigenstates which are bonding and antibonding linear combinations of f^6 and $f^5 d$ is $> \varepsilon_o$ where $\varepsilon_o \simeq 10^{-5}$ eV (assuming the system is in thermal equilibrium). Also, the magnetic susceptibility[5] supports this picture, probably being inconsistent with a static array of $f^5 d$ ions (since the $f^5 d$ configuration is probably paramagnetic). While the low-p phase is a semiconductor, the mixed phase is a peculiar

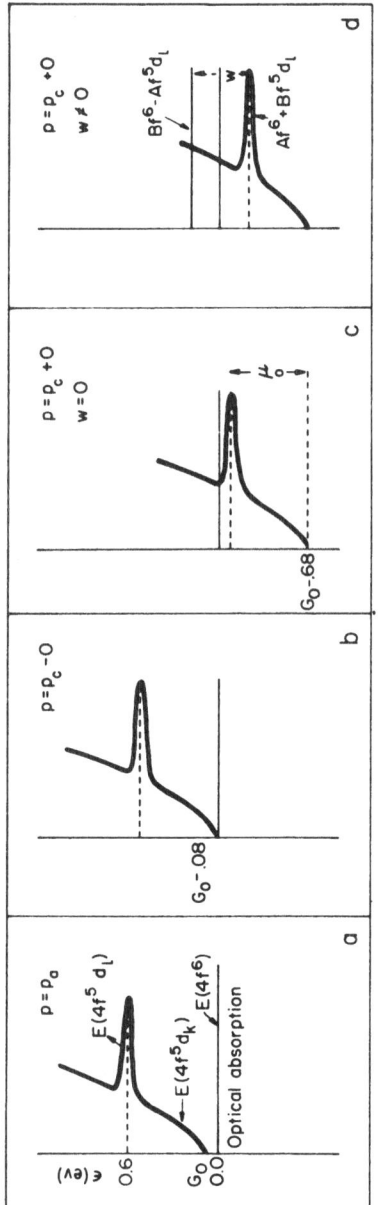

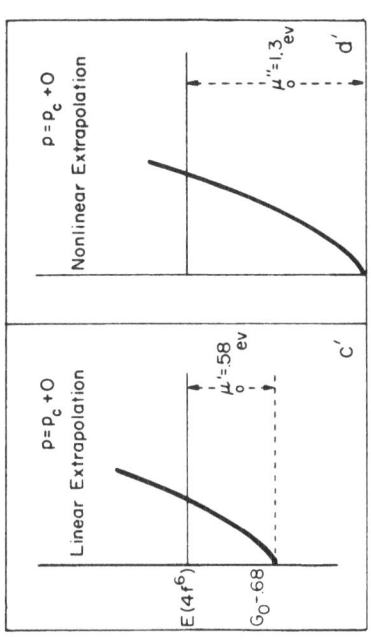

Figure 1.   Electron-hole excitations from semiconducting ground state ($4f^6$).   (a) and (b): Semi-conducting phase. (c) and (d): Extrapolation to collapsed phase (present picture). (c') and (d'): Extrapolation to collapsed phase (common picture).

metal, its resistivity ρ at room temperature being ~100 times larger than that of Cu, increasing with decreasing T, and the low-T
specific heat being $\propto$ T and anomalously high[7]. A plasma edge in
the optical absorption was found to occur at a bare frequency
(after correction for the high-frequency dielectric constant) corresponding to n $\simeq$ .9 electrons/Sm in the formula $\omega_p^2 = 4\pi n e^2/m^*$.[8]
Further, there is an anomaly in the optical response at low fre-
quency, the real part of the dielectric constant $\varepsilon_1(\omega)$ showing[9]
a precipitous rise as $\hbar\omega$ decreases below about 0.1 ev. The phase
boundary in the T-p plane[3,10] corresponding to this semiconductor-
metal transition is interesting: It implies[3] that the entropy dis-
continuity $\Delta S = S_{metal} - S_{semiconductor}$ is negative for $T > T_o \approx 100^\circ K$,
and probably[11] changes sign as T decreases through T_o (very similar
behavior occurs in the T vs. x diagram[3,11]--see footnote 6 for definition of x). Finally, SmSe and SmTe change smoothly but rather
rapidly from the semiconducting to the "metallic" phase[3] at much
higher p.

In most of the models of the Sm-chalcogenide mixed phase, all the
d-electrons (~0.7/Sm atom) are in a broad one-electron type of band[11,
12,13]. We[2,14,15] were led to a sharply contrasting picture, in which
only about .1 electrons/Sm atom are of this "band-like" character,
the remaining d-electrons being essentially localized[15a,16]. One
of our motivating considerations may be discussed with the help of
Fig. 1, where (a) and (b) show a schematized picture of the observed
optical absorption (abscissa) vs. energy (ordinate) in the semiconducting phase of SmS. In this phase the ground state has each
Sm^{2+} ion in $4f^6$, ^7F_o. The pressures in (a) and in (b) are atmospheric and a value just smaller than the critical pressure p_c, respectively. In the data[8,17], there are two sharp and intense low-
lying peaks at ~.8 and 1.5 eV, each with a width of ~.5 eV. These
were interpreted[17] as intraatomic transitions, $f^6 \to f^5 d_\ell$, where d_ℓ is
a localized t_{2g}-state, and the doublet originates in the $^6H_1$ terms
of $4f^5$; furthermore the .5 eV width is attributed to the fine structure of these terms[18]. Using this interpretation, but simplifying,
we have chosen to schematize the actual data by showing a single
peak, considerably narrower than .5 eV at .6 eV which would be
about the energy of the lowest state of $^6H_1$ given this interpretation. The broader background is also a schematized version of the
absorption thought to occur[8] in transition $f^6 \to f^5 d_k$ where d_k is a
conduction-band Bloch state. G_o is the band gap thought to be
about .1 to .2 eV.[8,19,11]

The nature of the sharp peaks deserves further comment. A band
calculation[20] for semiconducting SmS shows a band gap (4f-5d) at
about .1 eV with the lower part of the conduction band being broad,
with a structureless density of states starting nearly free-electron-
like at the bottom and extending to about 2.5 eV above the bottom
of the band. (Above this, more complicated 5d structure occurs).

In other words, band theory does not account for these prominent
features. The situation is quite similar to that in EuO (see Freiser
et al, footnote 18), where as noted by Lendi[21], the very prominent
peak in the optical absorption near the bottom of the conduction band
could not be explained, according to his band calculation, by excitation to the one-electron Bloch functions. He[21] therefore suggested
an excitonic mechanism, apparently unaware of the earlier, quantitatively rather successful, interpretation[18] on the basis of a Frenkel exciton. There is a notable difference between the "extra" peak
in EuO and those in SmS, namely, the latter apparently lie in a continuum of excitations. Nevertheless, we feel that the interpretation
of these peaks as virtual excitons (see ref. 22) is reasonable, and
indeed, strongly suggested (a) by comparison of experiment with the
band calculation on SmS and (b) by the success of the analogous
interpretation[18] in Eu compounds.

Returning to Fig. 1, as pressure increases from p_a, the excitation energies decrease[8,23] at a rate of ~12 meV/kbar. This decrease
would yield a small band gap (Fig. 1b) at $p=p_c-0$ of $\sim G_o-.08(=.02$ eV if
$G_o=.1$ eV). A slight increase in pressure then causes a volume collapse of about 12%. Making the plausible assumption that this still
rather small volume change continues to cause these excitation energies to decrease approximately linearly with volume, we are led to
Fig. 1c[24]. But this represents an instability: the state of the
system in which every Sm is in f^6 is no longer the ground state,
system states in which some of the f electrons go into d-states being
lower. If it were not for the broad background, overlapping the exciton, the system ground-state would simply have each Sm in the state
f^5d_ℓ (d_ℓ is an essentially localized, t_{2g}, state). However, the fact
that the conduction band extends below the localized state means that
the energy will be lowered further by dumping some of the d_ℓ-electrons
into the band-like states. The Fermi energy measured from the bottom
of the band, $\mu_o=.6-G_o\cong.5$ eV. To estimate the number of band electrons
we take a free electron density of states for energy up to a cutoff
e_c: $D(e)=3e^{1/2}/e_c^{3/2}$, giving the number of conduction electrons (per
Sm) of $b=2(\mu/e_c)^{3/2}$ (e is the energy measured from the bottom of the
conduction band.)[25] Thus with $e_c=2.5$ eV (the bandwidth used by
Varma and Heine[26]) and $\mu=.5$ eV we obtain $b=.18$. Hence a small number
of electrons per Sm (much less than .7) goes into the band. To appreciate the difficulty of getting b close to .7, we consider the conventional picture shown in (1c'): i.e. there is no excitonic peak
in the collapsed phase, but we assume still a linear descent of the
band with volume. Then μ would be .58 eV giving $b=.22$ - still about
a factor of 3 too small. In order to get $b=.7$, μ must be 1.3 eV
(depicted in Fig. 1d'), more than a factor of 2 larger than the value
in the linear extrapolation; it does not seem plausible to expect
such large nonlinearity effects as a function of volume for only a
10-15% volume change in connection with the variation of the rather
broad 5d band.[27]

As just noted, the excitonic level remains in our model (and is in fact appreciably occupied), whereas in the other model it has disappeared. Presumably this level could disappear because of screening of the 4f hole by the very mobile conduction electrons; however, given that b is a small number per Sm as discussed above, and that therefore there is almost one localized d-electron per Sm, such screening could not be a large effect (no matter how small the screening length). The overlap of the now occupied localized d's on neighboring sites could also tend to delocalize these d's; however, this overlap of d-states in the presence of f^5 might cause a band width considerably smaller than that of the calculated[20] d-states in the presence of an f^6 core (see section III), so that d-d Coulomb repulsion would have a strong tendency to keep these d-electrons localized.

Another major motivation for our looking in a direction essentially different from that suggested by the conventional model is that we find it difficult to believe that it can lead to the nonmagnetic behavior observed[5]. The conventional model may be described as follows: The zeroth order ground state is one in which most (70%)of the Sm's are in $4f^5$ states, the corresponding f-electrons having been dumped into the d-band; the remaining Sm's are in $4f^6$. The lowest $4f^5$ state in the crystal field is a Kramers doublet and is therefore magnetic. Qualitative arguments were given[5,28], based on the example considered in the Kondo effect, namely the extremely dilute limit, which led to a nonmagnetic ground state. The only attempt[29] at a calculation in the case of interest, namely where most of the Sm-sites are magnetic "impurities" (Sm,f^5 ions) was made quite recently. It was concluded that the nonmagnetic ground state still held for this case. But an approximation was made[29] that resembles very closely the one made by Hubbard in his treatment of the single-band Hubbard model.[30] This Hubbard decoupling leads[31] to nonmagnetic behavior in an example which is known to be magnetic (the half-filled narrow-single-band Hubbard model, which is antiferromagnetic). Therefore this calculation[29] does not bring us closer to the answer to the question as to the magnetism in the model studied. Furthermore, a comparison with nickel suggests strongly that the model considered[29] (a 2-band (f and d) Hubbard model with one-electron f-d mixing, large f-f intrasite Coulomb repulsion U, the bare f and d bands being narrow and broad, respectively; this is also the Anderson impurity model with an "impurity" in every unit cell) should lead to a magnetic ground state. As already pointed out[11], this model should be applicable to Ni with f replaced by 3d and d replaced by 4s; furthermore the requirement that the Fermi level lie in the narrow band, considered to be crucial,[11,29] is satisfied in nickel. But of course the parameters are appreciably different[11]. However, the change in parameter values in going from Ni to SmS, seems to us to be in the direction of <u>favoring</u> magnetism (the width of the narrow band <u>decreases</u>, giving a <u>larger</u> density of states at μ, and the density of broadband electrons decreases somewhat, preventing any additional screening of U).[32] It seems to us that one of the most fundamental ques-

tions is why the enormous difference between the two mixed-valence
systems Ni and collapsed SmS.

A different and common type of argument for nonmagnetism that
is unconvincing to us goes as follows. If there is fast fluctuation of an electron in and out of a localized atomic orbital, and
if when it is in the localized orbital the atom is in a nonmagnetic state, then the behavior of the system will be nonmagnetic
(if, according to some authors, the fluctuations are fast enough).
A simple counterexample is a 2-site 3-electron model of double exchange[33] where there is one spatial orbital localized at each
"magnetic" site. Taking the two orbitals, at sites a and b, as
degenerate [so an appropriate Hamiltonian in the absence of an ex-
ternal magnetic field is $H=\varepsilon_0(n_a+n_b)+t\sum_\sigma(c_{a\sigma}^\dagger c_{b\sigma}+h.c.)$ where
$n_\nu=\Sigma_\sigma c_{\nu\sigma}^\dagger c_{\nu\sigma}$, and $c_{\nu\sigma}^\dagger$ creates an electron at site ν with spin σ]
it follows that the fluctuation frequency $\omega=2|t|/\hbar$; that is, if
at time 0 the system is in the state $n_a=2$, $n_b=1$, it will fluctuate
between this and the state $n_a=1$, $n_b=2$ with frequency ω. But, as
seen immediately by Kramer's famous theorem, the ground energy
level is a Kramer's doublet and therefore no matter how large ω
is, this model will give a Curie-law magnetic susceptibility no
matter how low the temperature T. It is interesting to note that
this model is identical to the one discussed recently by Varma[34]
in connection with his physical argument for non-magnetism. Namely, his argument goes, "with one site being a singlet, either an up
spin or a down spin can equally well be transferred to the other
site and the gain in kinetic energy is indifferent to the spin at
the other site," or in other words "there is no preference for a net
moment." Although this conclusion is correct, it does not lead
to the nonmagnetism desired, a Curie law holding here as we have
seen [an isolated electron (or Sm^{3+} ion, as another example) also
has no preference for a net moment — but it certainly gives rise
to a Curie law]. Finally we note that our counterexample to Varma's
argument is by no means limited to models with a small (non-macroscop-
ic) odd number of electrons. For example, the linear chain Hubbard
model in the limit where the intrasite interaction $U\to\infty$ gives a Curie
law susceptibility corresponding to an electron density $2-\rho$ for $\rho>1$
and ρ for $\rho<1$, where ρ is the number of electrons per site[35]. Physically this means that the susceptibility is simply that of the sites
with unpaired electrons, assumed to be stationary and non-interacting.
So it's the same in this sense as the 3-electron, (or 1-electron)
2-site model, (for which U is irrelevant)[36].

Hence we have investigated a model that is a natural consequence of Fig. 1c. One can consider a zeroth-order ground state
corresponding to Fig. 1c as $(f^5 d_\ell)^{N(1-b)}(f^5)^{Nb}(d_k)^{Nb}$ where N is the
number of Sm ions; that is, there are N(1-b) Sm ions in a configuration with f^5 and a localized d,Nb Sm ions in f^5, and Nb conduction
electrons. An important point here is that we believe $f^5 d_\ell$ acts
essentially like $f^5 d_k$ or like Sm^{3+}, insofar as the volume, Mössbauer

isomer shift, and XPS are concerned (these are important measures of
"valence"). This point will be discussed in some detail in later sections. But then this zeroth-order state would give all Sm behaving
as Sm^{3+}, in disagreement with these experiments which suggest
~70% Sm^{3+} and ~30% Sm^{2+}. We proposed a mechanism[2,14,15] which mixes
f^5d_ℓ and f^6 and which can lead to the right order of mixing. It is
this mixing matrix element, proportional to a quantity W, that causes
the change from Fig. 1c to 1d; the Fermi level relative to the bottom of the band is thus reduced to ~.4 eV lowering b to about .13
electrons/Sm. The mean field ground state is of the form (symbolic-
ally) $(Af^6+Bf^5d_\ell)^{N(1-b)}(f^5)^{Nb}(d_k)^{Nb}$, with $B^2/A^2\cong 7/3$. Some very recent experimental results indicating rather directly (but, unfortun-
nately, not entirely conclusively) that the 5d-electrons emitted in
the transition $Sm^{2+}\rightarrow Sm^{3+}$ are localized[37,38,39] have, of course, been
encouraging to us.[40]

In section II we describe a highly simplified version of our
model indicating how it leads to a fairly satisfactory account of
volume, amount of mixing and isomer shift vs. pressure at T=0.
This is then extended to T≠0 leading to a phase diagram appropriate to this simplified model; this is characterized by both a first-
order and a second-order boundary. Noting that the shape of the first
order boundary depends sensitively on effects neglected in the simple
model, we proceed to consider this boundary in more general terms;
this analysis leads to what we believe are significant advances in
theoretical understanding. In section III we consider various experimental properties in the collapsed phase, showing in particular
how the metallic-like plasma edge frequency observed in optical
absorption, and the d.c. electrical conductivity are consistent with
our model despite the localized nature of most of the d-electrons
in this model. The low-energy absorption is also discussed, contact being made with the very recently observed[9] anomaly. Finally,
our speculation as to how nonmagnetic character can occur, and
general conclusions are given in section IV.

II. VOLUME AND DEGREE OF MIXING VS. PRESSURE; PHASE DIAGRAM

We first present our model in a mean-field approximation, neglecting the small number of conduction electrons. We begin with
the ground state and work for simplicity within a two-level model[14]
involving an f^2 spin-singlet and fd spin-singlet at each Sm-site
(in place of f^6 and f^5d, respectively). In a mean-field approximation (not Hartree-Fock) we write the crystal wave function

$$\Phi = \pi_i[A\, a^+_{fi\uparrow}a^+_{fi\downarrow}+ \frac{B}{\sqrt{2}}\, (a^+_{fi\uparrow}a^+_{di\downarrow} - a^+_{fi\uparrow}a^+_{di\downarrow})]|0\rangle\equiv\pi_i\phi_i\ ; \qquad (1)$$

$a^+_{\mu i\sigma}$ creates an electron in Wannier function μ at site i with spin
σ, and $|A|^2+|B|^2 = 1$. We calculate $(\Phi,H\Phi)$ where H is the usual non-

relativistic Hamiltonian minus the ionic kinetic energy, assuming
that (A) the ionic positions always form the NaCl lattice, (B) we
can retain in the Sm-Sm interactions only terms of leading order
in (Sm-Sm separation)$^{-1}$, and (C) we may expand all matrix elements
to second order in $V-V_2$ where V is the volume per Sm and V_2 is the
value of V at p=0. We thus find the energy per Sm atom,

$$E(n,V)=(1-n)\tfrac{1}{2}k_2(V-V_2)^2+n[\tfrac{1}{2}k_3(V-V_3)^2+c]-W(V)\ n(1-n)\ . \qquad (2)$$

Here $n=|B|^2$, the average number of d electrons per Sm in (1), V_3 is
the location of the energy minimum when n=1, k_2 and k_3 are constants
and $c = E(1,V_3)-E(0,V_2)$. W is the quantity $4\ z|\xi|$, estimated to be
≈ tenths of an eV in Ref. 14: W is proportional to the 2-site
Coulomb matrix element that corresponds to the transition $(f_if_j)\rightarrow$
(d_id_j), for neighboring sites i and j, and gives rise to the mixing
in (1). Clearly this gives $E(0,V)=(1/2)k_2(V-V_2)^2$ and $E(1,V)=$
$c+(1/2)k_3(V-V_3)^2$, i.e. the harmonic approximation has been assumed
for each pure valence state.

It is useful to consider this type of formula graphically as
shown in Fig. 2. In Fig. 2a, the curves E(0,V) and E(1,V) are
labelled respectively f^6 and f^5d. For volume V, let the value n*
of the variational parameter n minimize the energy in Eq. (2).
The corresponding energy E(n*,V) is the minimum energy curve in
Fig. 1a excluding the dotted segment LK. n*=0 on the solid part
lying to the right of M, and 0<n*<1 on the dashed part of this
minimum energy curve [which is the region where $|\Delta\varepsilon(V)|<W(V)$].
Similar discussion holds for Figs. 2b and 2c, which correspond to
different parameter values. Of course, if W=0, E(n*,V)=Min[E(0,V),
E(1,V)], in which case n* is either 0 or 1. In Fig. 1a, E(n*,V)
is not always convex; the dotted line KL is the double tangent, so
that the physical ground state energy $E_0(V)$ is the lowest curve in
the figure, and $p=-\partial E_0(V)/\partial V$. There is thus a first order transition
from point L(n*=0) to K(0<n*<1) as p increases through a critical
value p_c (given by the negative slope of LK) as is observed in SmS.
In Fig. 1(b), $E(n*,V)=E_0(V)$ is convex and the transition is smooth
(a second order transition occurs at N) as observed in SmSe and
SmTe. Similar smooth behavior can also occur when E(1,V)>E(0,V)
for all V. Figure 1(c) shows a possible picture of SmB_6, which is
in the mixed-valence phase at atmospheric pressure. The f^5d para-
bola (E(1,V)) is lower, but there is appreciable mixing near its
minimum so that even at p=0 (at R) SmB_6 is in the mixed phase. Presumably on applying negative pressure one would eventually reach
the unmixed phase; the transition can either be first-order (as
indicated in the figure) or smooth.

Using Eq. (2) we calculate V and n* vs. p, where n* minimizes
E(n,V) for fixed V. Details of the calculation and the results are
given in ref. 15. It turns out[15] that all the parameters except W

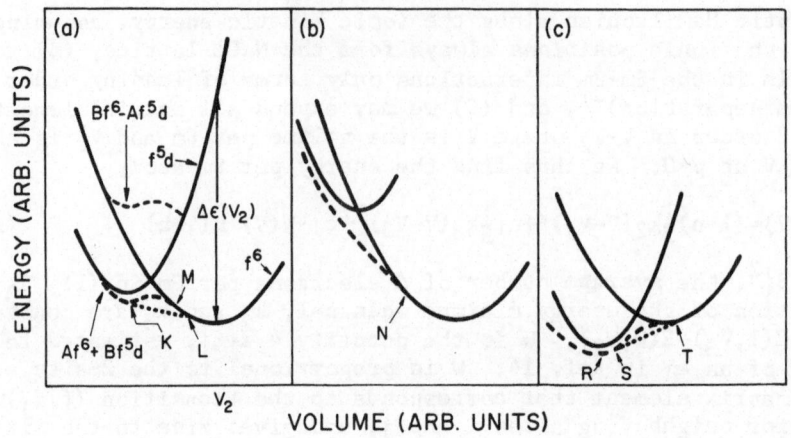

Figure 2.   Energy E(n,V) vs. Volume V (see eq. 2 of text) for dif-
            ferent values of the parameters.

are obtainable from presently available experiments. Taking essentially mean values of these parameters (without attempting refinement), and using values of W, assumed V-independent for simplicity, of the expected order of magnitude, we found a 1st order
transition at p_o=12 kbar for SmS, while for SmSe and SmTe we found
relatively smooth transitions occurring at much higher pressures,
in reasonable agreement with experiment. In the mixed phase our
model predicts the existence of atomic electric dipoles[14] ordered
in an unknown and presumably complicated way. In addition to the
above transition, which is n=0→n=n*, 0<n*<1 (i.e. f^6→Af^6+Bf^5d_ℓ),
there is another transition from the mixed phase Af^6+Bf^5d_ℓ to
f^5d_ℓ, which is a second order transition. (It marks the disappearance of the ordered electric dipoles.) It occurs when
$\Delta(V) \equiv E(1,V) - E(0,V) = -W$, and the parameters from above give for SmS,
a critical pressure p_o' of ~70 kbar. Recently a second phase transition was claimed to have been seen experimentally[41] at about
20 kbar. We have made a slight refinement in our parameters in
order to improve our value of p_o, which turns out to be very sensitive. Reducing $\Delta(V_2)$ from .6 eV to .55 eV reduced p_o from 12 to
5 kbar. Recognizing the fact that the critical pressure of about
6 kbar observed[3,4,10] refers to the nonequilibrium transition with
increasing pressure, an equilibrium value of p_o (which is what is
calculated) of about 5 kbar is roughly right. With this adjustment, the value of p_o' we obtain is about 40 kbar, close enough to
the value found[41] to strongly suggest that our second transition
is related to that observed[41,42,43] .

We note here that our calculated V vs. p is linear for p<p_o
with bulk modulus B_o characteristic of p=0 (which we chose as
500 kbar). For p=p_o+0 the bulk modulus jumps to a value about
20% less than B_q, and then increases gradually with increasing pres-

sure until $p=p_0'-0$ (at which it reaches the value ~800 kbar for
$p_0'=40$ kbar), suddenly jumping at $p_0'+0$ to 1050 kbar (about the bulk
modulus for Sm^{3+}). A discontinuity in B at p_0' was observed[41].
Unfortunately the experimental situation for $p_0<p<p_0'$ is unclear.
Jayaraman et al.[3] find behavior similar to our calculation in that
the value of B for p slightly above p_0 is about 20% reduced from
B_0, and B increases with increasing p; whereas Guntherodt et al.[41]
found, for $p_0<p<p_0'$, that B is constant at a value about 10% larger
than B_0. Clearly, it is essential to resolve these experimental discrepancies (see also ref. 42).

   To extend our results to T>0 within the simple two-state model
it is convenient to work with the spin-hamiltonian representation[14]
of that model, and apply mean-field theory. (The zero-T limit of
the Helmholtz free energy then agrees with (2), as it should.)
Although we will omit the details of the calculation, we note the
somewhat surprising fact that for $|\Delta E(V)|<W$ the excited-state single-
site energy at zero temperature, shown in Fig. 2a lies above the
corresponding ground state energy by W, independent of V. For SmS,

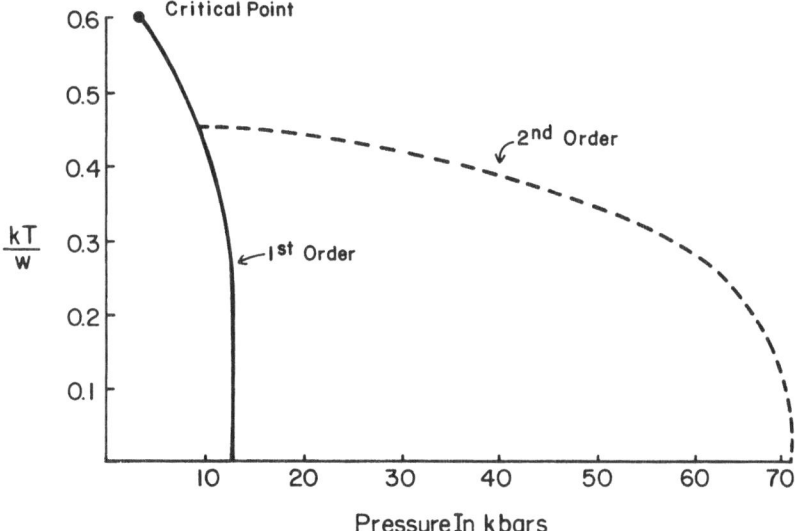

Figure 3.   Energy vs. $\vec{k}$ for a simplified version of the present
            model. Band 2 is a free-electron-like band whereas
            band 1 is a narrow band associated with the motion of
            a d_ℓ hole which is strongly correlated with the f-elect-
            rons.

we finally obtain the phase diagram shown in Fig. 3; we kept the original unrefined parameters for this calculation, so the 2nd order boundary approaches T=0 at ~70 kbar. It is seen that the 1st order boundary terminates in a critical point at $T \tilde{=} 1500°K$ (W= .3 eV), which is in the ballpark of the experimental estimates based on extrapolation[3,10]. While the order of magnitude of T_c in the essentially localized model is probably fairly estimated by our simple 2-state model, the shape of the 1st order boundary (within the essentially localized model) is clearly very sensitive to factors beyond those considered in the 2-state model. We now proceed with a more general discussion of these factors.

We will limit ourselves to $T \tilde{<} 500°K$ where much of the interesting behavior is observed. In the semiconductor phase the situation is simple. Aside from phonons, there is a Sm^{2+} J=1 level lying about 400°K above the J=0 ground level[44], the next level being over 10^3 °K above J=0. The metallic phase is more complicated.

Consider first the essentially localized model. The ground state in the collapsed phase, as discussed qualitatively above, consists of a small number, b of holes in Φ (Eq.(1)) plus the same number of electrons occupying the free-electron-like Bloch states near the bottom of the conduction-band. Formally this would be obtained by destroying Nb (N=number of Sm's) localized d-electrons in Φ and creating the same number of nearly free electrons. The destruction of a d_i in Φ_i (Eq.(1)) leaves f_i (renormalized to 1). This could have been done at any site; and symmetry demands that we multiply Φ by $a_{dk\sigma}=(1/\sqrt{N})\Sigma_j \exp(-ik \cdot R_j)a_{dj\sigma}$ rather than $a_{dj\sigma}$. This gives rise to a hole band, with energies ε_{1k}. Note that these states have correlation built in. We expect this band to be very narrow (of the order of several 100°K) - sufficiently so to give rise to the large observed[7] specific heat at low T, and arguments leading to such a narrow width will be given below. Thus we have a few electrons in a broad band (band 2) and a few holes in a narrow band (band 1) - this is depicted in Fig. 4 where we have taken cutoff effective-mass expressions for each band energy $\varepsilon_{k\nu}$. In addition to the positional degeneracy of the holes in Φ, which as we just explained, is removed by hole-hopping, there is the degeneracy of the f^5 configuration (which constitutes a hole). We have speculated[15] that this degeneracy is removed by Kondo-like interaction of the broad-band conduction electrons with the f^5 electrons (although at a concentration of ~10%, interactions between sites might be important) — we expect the characteristic temperature to be $\tilde{<}10^2$ °K. Thus at $kT \sim D_1$, the width of the narrow band, the electronic entropy in our model for the collapsed phase is about S_α where

$$S_\alpha/k = \alpha \ln 6 - \alpha \ln \alpha - (1-\alpha) \ln(1-\alpha),$$

and α is the value of b at this temperature. (This neglects the broadband contribution which is small at these T.) ln6 is essen-

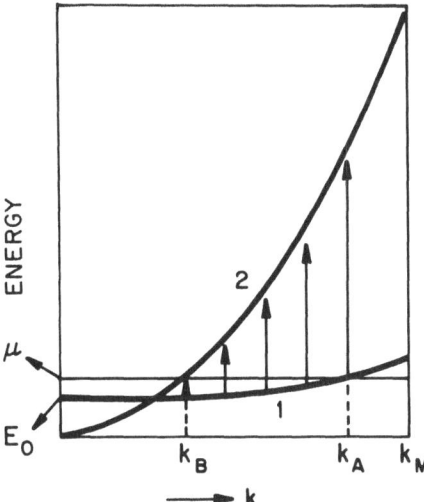

Figure 4. Phase diagram for SmS in the T-p plane in the simple
two-state model.

tially the entropy of a hole at $T \gtrsim 100^{\circ}K$: the low-lying doublet and
quartet of f^5 are separated by $\sim150^{\circ}K$[7], the next level being
$\sim10^3$ $^{\circ}K$ higher. We expect that α will not change drastically as long
as $kT \ll D_2$, the broad-band width (which is $\gtrsim 10^{4}{}^{\circ}K$). Hence, in our
essentially localized (and quite highly correlated) model, we expect
the entropy to start linearly at low T and saturate at $kT \approx D_1$ to a
value of about S_b, which ranges from .50k to .86k as b changes from
.1 to .2. We wish to simulate this behavior by a simply calculable
model, which consists of non-interacting electrons occupying two
overlapping bands (qualitatively as in Fig. 4). The broad-band
width will be ~4 eV; [50] the width of the narrow band D_1 is chosen
to give the correct low-T behavior, and its location is chosen to
give a value b' of the number of holes in the narrow band ($\approx.1$)
such that this noninteracting model gives an entropy of about .7k
at $kT \approx D_1$. We find b'=.15 at $T \cong 0$ and D_1=.047 ev. Within our essentially localized model, where $\approx90\%$ of the sites are in a singlet
ground state $\phi_0 = A_0 f^6 + B_0 f^5 d_\ell$, there will be additional localized
excitations (to $A' f^6 + B' f^5 d_\ell$). However, since we expect ϕ_0 to be
split from a degenerate level $f^5 d_\ell$ by the rather strong interaction
$W(\sim.3$ eV$)$, we guess that the first excitation above ϕ_0 will be
$\approx.1$ eV and therefore will not contribute appreciably at the temperature of interest here.

In addition to the electronic entropy, there is the lattice
contribution. This cannot be calculated a priori since we don't
know the phonon dispersion curves near the transition. We therefore treat it phenomenologically, using the simple Debye model.

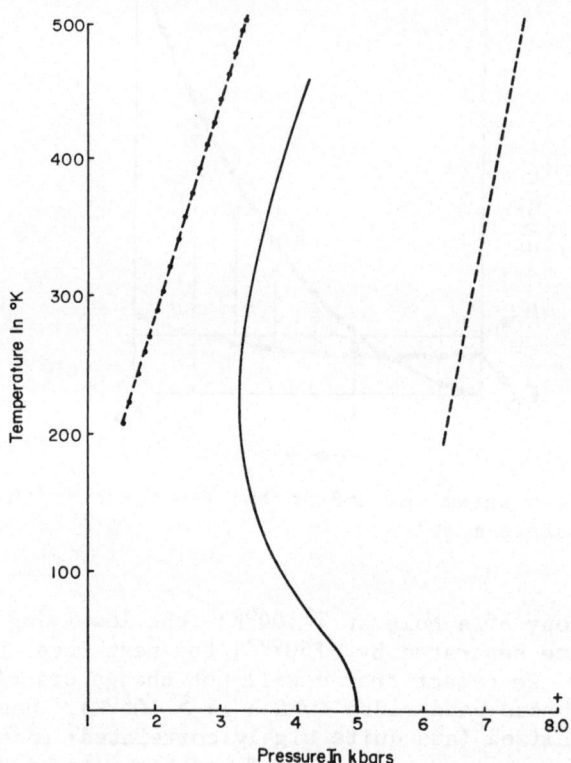

First Order Phase Boundary For T < 500 °K (SmS)

Figure 5.   First order phase boundary for SmS for low T in the T-
            p plane. The solid line is our theoretical curve. The
            dashed line is the experimental transition observed with
            increasing pressure, while the dash-dotted line is the
            corresponding curve found under decreasing pressure[3,10].
            (The dots are not experimental points). + represents a
            possible value[7] of the first order transition pressure at
            4°K (with p increasing).

    The first-order boundary satisfies the well-known Clausius-
Clapeyron equation $dp/dT \doteq \Delta S/\Delta V$. To calculate p vs. T we need to
know ΔV as a function of T. Since we expect ΔV to be slowly varying with T for the lower T of interest, we take it to be constant
at its room-temperature value. We chose p_0, the zero-T intersection of the boundary to be 5 kbar. The values of the Debye
temperature used are 266°K for the semiconductor and 233°K for the
metal. The result is shown in Fig. 5. The low-T behavior is a

general consequence of the fact that the entropy of the metal is
$\sim \gamma T$ (=specific heat) which dominates that of the semiconductor
(which is $\sim T^3$); it follows that as $T \to 0$, the phase boundary is given
by

$$T = \sqrt{\frac{2|\Delta V|}{\gamma}} \; (p_o - p) \tag{3}$$

Hence $dT/dp \to -\infty$ as $T \to 0$. The essential reasons for the turnaround
(i.e. the change from a negative to a positive dT/dp) are (i) the
rapid rise in the semiconductor contribution from the J=0 to J=1
excitations at $\sim 200°K$ plus (ii) the rapid approximate saturation
at $T \sim 150°K$ of the narrow-band entropy at a rather small value (due
to the small number of holes). Because of (ii), the semiconductor
entropy increases beyond that of the metal, causing a change in
sign of dT/dp.[44a]

Actually the phonon entropy is quantitatively important — there
is a small softening (i.e. entropy increase) in going from semiconductor to metal; if it were neglected, the turnaround would have
occurred at a somewhat lower temperature ($\sim 150°K$), and the slope at
high T would be reduced by a factor ~ 2.5. The only evidence we have
been able to find concerning the phonon contribution is in $Sm_x Y_{1-x} S$,
for which ultrasonic[45] and X-ray diffraction[45a] measurements have
been made. Both of these indicate softening[45,45a]. Further, the
entropy due to the acoustic modes that we estimated roughly from the
sound velocites[45] is of the same order-of-magnitude as, but appreciably smaller than, what we found phenomenologically. This plus
the fact that the optical mode softening[45a] will also give an increase in entropy is clearly supportive of our phenomenology -
particularly in light of the suggestion[45,45a] that the softening is
due to the mixed valence, suggesting in turn that a similar softening will occur in SmS under pressure (and in $Sm_x Gd_{1-x} S$).

The band model would appear to have a very difficult task of explaining the negative ΔS observed[3] in the high-T region for SmS under
p (and for $Sm_x Gd_{1-x} S$). In the band model, for $T \gtrsim 100°K$, the electronic entropy in the metal will be[46] $\sim .7k\ln 6 + .3k\ln 4$ (from 70% f^5
and 30% f^6) which is $> k\ln 4$, the approximate electronic entropy in
the semiconducting phase. This alone gives $\Delta S > 0$. The phonons
will likely increase ΔS further, as discussed above, and it is
difficult to see a significant negative contribution.

It thus seems obvious that there is an urgent need for determining the phonon dispersion curves in the neighborhood of the
first-order boundary, particularly for SmS under p and for $Sm_x Gd_{1-x} S$.

It is important to discuss the assumption made[15] that the vol-
ume V_3 associated with $f^5 d_\ell$ (or fd_ℓ in our simplified model) is the
experimental Sm^{3+} value. (A similar assumption was made[16] in connection with SmB_6.) Arguing empirically, it is found that the Sm^{3+}

radius as determined[46a] from the insulator[47] Sm_2O_3 agrees very closely (to within about 1%) with the extrapolation from[11] the series
RS, R=La, Ce,... . The latter is interpreted as the size appropriate to metallic Sm^{3+} S^{2-} with ~1 conduction electron per Sm, since
the neighboring members of the series (NdS and GdS) have this property. Hence the presence of Sm 5d-conduction electrons does not
seem to alter the size of Sm^{3+}. But it's known that even for nearly-
free electrons the conduction electron charge density is not uniform, showing appreciable peaking near the atomic sites. This
suggests that localized 5d electrons will not appreciably alter
the size of Sm^{3+}. From a more basic point of view, we note that
the outermost filled Sm shells are 5s and 5p, and they are thought
to provide the repulsive force in the cases of $Sm^{2+}(4f^6)$. It is the
shrinkage in size of these shells on going from $4f^6$ to $4f^5$ (the 4f
radius is less than the 5s or 5p radii) that causes Sm^{3+} to be smaller
than $Sm^{2+}(4f^6)$. We expect that the shrinkage of the 5s 5p shells
will not be appreciably changed on adding a 5d electron to $4f^5$ since
the 5d radius is larger than the 5p radius. The basic question remaining is the effect of this 5d on the repulsive forces. As far
as the interaction between neighboring Sm's, the extra 5d electron
would appear to give an <u>attractive</u> force, tending to reduce the
effective size of the Sm. On the other hand, this 5d electron would
presumably increase the Sm-S repulsion. However, we feel that this
will not alter the situation wherein the ionic size is essentially
determined by the Sm 5p electrons because (a) the 5d shell is only
1/10 filled (the 5p being completely filled), so the 5d electron
can avoid the sulphur p-electron, and (b) the σ-overlap between
$Sm5d(t_{2g})$ and S3p is zero by symmetry, leaving only the relatively
small π-overlap to contribute to the repulsion resulting from the
exclusion principle.

We also note here that the addition of a localized 5d to $4f^5$
should not appreciably change the isomer shift (as has already been
recognized[16]) because this effect is due mainly to the 5s electrons.

III. OPTICAL PROPERTIES, XPS, SPECIFIC HEAT AND ELECTRICAL

CONDUCTIVITY IN THE COLLAPSED PHASE

An obvious difficulty facing the present approach is in understanding the apparent metallic nature of the d.c. conductivity[7]
σ and the plasma frequency ωp, in the collapsed phase, in view of
the small number of conduction electrons. We show here that this
difficulty is only apparent.

Consider σ first. Values of the electron mobility at room temperature found[19] in semiconducting SmS range from about 5 to 15
cm^2/volt-sec. A value of 10 cm^2/volt-sec with b=.15 electrons/Sm
gives $\rho \cong 200\mu\Omega$-cm, the observed value[7]. It seems reasonable to

expect that the mobility will not change appreciably in going from the semiconducting to the collapsed phase since the scattering is presumably dominated by phonons at 300°K. Hence the above estimate would appear to be justified, leading to the conclusion that b~.1 is probably the correct order of magnitude.

The optical reflectivity shows a plasma-edge at an energy of roughly several volts[8]. In fact, the steep rise in reflectivity with decreasing energy follows very closely Drude-type behavior[48]. Making corrections for the high frequency dielectric constant ε_∞, Battlogg et al.[8] found a bare value of plasma frequency ω_p that corresponds, through $\omega_p^2 = 4\pi be^2/m^*$, to b≈1 (taking $m^* \sim m$). And in fact Battlogg et al.[8] conclude that there must be about 1 nearly free electron per Sm in collapsed SmS. This conclusion, if correct, would clearly invalidate our approach. We now show that this conclusion does not follow the data, and in fact our model (with b≈.1), is probably consistent with the data.

As we have already pointed out while discussing the low-T thermal properties, the ground state within the essentially localized model consists of a few electrons in a broad band (effective mass m_2) and a few holes in a strongly correlated narrow band (effective mass m_1). Optical absorption will be the sum of three contributions: the two ordinary intraband terms, giving the familiar Drude expressions for the conductivity, $\sigma_\nu(\omega) = (\omega_{p\nu}^2/4\pi) \times (i\omega + 1/\tau)^{-1}$ ($\nu = 1, 2$) and an interband term resulting from transitions from the occupied narrow-band states between k_B and k_A (Fig. 4) to the empty broad band states above. Since $m_1 \gg m_2$, the intraband terms would indeed give a plasma edge at $\omega_p = (4\pi be^2/m_2)^{1/2}$ and as pointed out above, would disagree seriously with experiment for b~.1 and $m_2 \sim$ a free-electron mass. We now argue however that the interband contribution also gives rise to a Drude behavior, and can have a large coefficient.

Intuitively one might suspect an important contribution of this sort from the interband conductivity $\sigma_{12}(\omega)$, because of a possibly small excitation energy at $k \cong k_B$. To get a rough picture, we consider a two-band model of non-interacting electrons corresponding to Fig. 4. The neglected correlation effects, characteristic of our model, will be discussed below; we feel they will not alter the essential idea. At the least, this simple calculation will prove rigorously that the conclusion of Battlogg et al.[8], namely b~1, cannot be made unambiguously on the basis of the observed plasma frequency.

A straightforward calculation using the Kubo formula yields, in the T→0 limit,

$$\sigma_{12} = \frac{e^2}{2\pi^3 m^2} \int d^3k \underset{k_B < k < k_A}{} \frac{|(\psi_{k1}, p\,\psi_{k2})|^2}{\varepsilon_{k2} - \varepsilon_{k1}} \left[i\left(\omega - \frac{\varepsilon_{k2} - \varepsilon_{k1}}{\hbar}\right) + \frac{1}{\tau}\right]^{-1} \quad (4)$$

where e and m are the electron charge and mass, $\Psi_{k\nu}$ is a Bloch function for band ν, $p = -i\hbar \, \partial/\partial z$, and τ is presumed to $\to \infty$. We choose $\varepsilon_{k2} = \hbar^2 k^2/2m_2$, $\varepsilon_{k1} = \frac{m_2}{m_1} \varepsilon_{k2} + E_o$ ($m_2 \approx m$, the free-electron mass, m_1 being the effective mass for the narrow band). To get an order-of-magnitude estimate of (4) we have assumed that in the matrix element we can put the broadband wave function $\Psi_{k2} \cong \Omega^{-1/2} \exp(i\vec{k} \cdot \vec{r})$ (Ω=crystal volume). Then $(\Psi_{k1} \, ' p \, \Psi_{k2}) = \hbar k_z d_k$ where $d_k \equiv \sqrt{N/\Omega} \int \exp(i\vec{k} \cdot \vec{r}) d(\vec{r}) \, d^3 r$ and $d(\vec{r})$ is the Wannier function for the narrow band (Ψ_{k1}). We also take $d(r) = Ce^{-\alpha r}$, $\alpha^{-1} \cong 1\text{Å}$, (the maximum in r d(r) for the $\text{Sm}^{2+}(f^5 d)$ ion[49]) which gives

$$d_k^2 = (N/\Omega) \; 64\pi\alpha^5/(\alpha^2 + k^2)^4 \tag{5}$$

Then (4) can be written, replacing $1 - m_2/m_1$ by 1, and for $m_2 = m$,

$$\sigma_{12} = \frac{2(ed_o)^2}{3\pi^2 m} \left(\frac{2m}{\hbar^2}\right)^{3/2} \int_{z_B}^{z_A} \frac{dz}{z}(z+E_o)^{3/2}[1 + \frac{z+E_o}{\varepsilon_c}]^{-4}[i(\omega - \frac{z}{\hbar}) + \frac{1}{\tau}]^{-1} \tag{6}$$

where $z = \varepsilon_{k2} - \varepsilon_{k1}$ and $\varepsilon_c = (h^2\alpha^2/2m_2)$. To estimate the various quantities we choose $b = .15$, $D_1 \equiv \varepsilon_{1k_m} - E_o \cong .047$ eV (the width of band 1) and $D_2 = \varepsilon_{2k_m}$ (the width of band 2) = 100 D_1.[50] This gives $\mu \cong .79$ eV and $k_m \cong 1 \text{ Å}^{-1}$ and the density of states at μ for the narrow band correctly reproduces the observed[7] low-T specific heat ($m_1/m \cong 94$). It follows that $E_o = .75$ eV, $z_A = 3.4$ eV and $z_B = .037$ eV. Thus for $\hbar\omega \sim$ several eV and $\varepsilon_c \cong 4$ eV, z_B is much smaller than the other energies appearing in the integrand. Hence, the asymptotic value of the integral as $z_B \to 0$ will give an adequate approximation. Putting $\omega_o^2 = 4\pi be^2/m_2$, the total conductivity is then

$$\sigma = (\omega_o^2/4\pi) \; (i\omega + 1/\tau)^{-1} \; (1+\nu) \tag{7}$$

where
$$\nu = 2d_o^2 \left(\frac{E_o}{\mu}\right)^{3/2} (\ln \frac{z_A}{z_B}) \; (1+E_o/\varepsilon_c)^{-4} \tag{8}$$

The above numbers give $\nu \cong 19$. We must conclude that observation of a plasma edge at frequency ω_p (after correction for ε_∞) should be interpreted according to the formula $\omega_p^2 = \omega_o^2(1+\nu)$ and therefore the number of band electrons per unit volume is $b = m_2\omega_p^2/[4\pi e^2(1+\nu)]$ rather than $m_2\omega_p^2/4\pi e^2$, the latter formula possibly being seriously in error. The value of ν obtained in our model is somewhat too large (by ~ a factor of 2) to give agreement with experiment, but is clearly of the right order of magnitude. In fact it is easy to see that our correlated state with one hole namely $\sqrt{2} \, a_{bk\sigma}^+ a_{dk\sigma}\Phi$, gives a momentum matrix element which is $1/\sqrt{2}$ of the uncorrelated value. [Here $a_{dk\sigma}$

was defined as a Bloch sum of $a_{di\sigma}$ which destroys an electron in
the excitonic d-state at site i, i.e. in the presence of f_i^5 (rather
than f_i^6); $a_{bk\sigma}$ corresponds to a free-electron-like Bloch state (b
for band-like) near the bottom of the conduction band.]

In order to obtain the low frequency behavior of $\sigma(\omega)$, we have
to keep the frequency dependent factor $[i(\omega-z/\hbar)+1/\tau]^{-1}$ inside the
integrand of Eq. 6. However, to a good approximation, the weakly
z dependent factors $(z+E_o)^{3/2}$ and $[1 + \frac{z+E_o}{\epsilon_c}]_{-4}$ can still be taken
out of the integrand. The resulting integration can be carried out
analytically. We have calculated $\epsilon_1(\omega)$, the real part of the dielectric function using a value of τ obtained from d.c. conductivity
measurements. We find that $\epsilon_1(\omega)$ follows the usual Drude-like behavior for $\hbar\omega>1$ eV and deviates from it for $\hbar\omega<1$ eV, the deviation
being drastic as $\hbar\omega\to0$. For example, $\epsilon_1(\omega)$ rises rapidly to a large
positive value as $\hbar\omega\to\sim.05$ eV. from above, a feature which is in remarkable qualitative agreement with experiment[9]. In addition to
this anomalous behavior, our model also predicts low-energy optical
absorption in the range of tenths of an eV. It results from the
intraatomic transition between ground and excited mixed states,
represented symbolically by $Af^6 + Bf^5d \to Bf^6 - Af^5d$; the corresponding momentum matrix element in our simplified 2-electron model
is $\langle Af^2 + Bfd|p|Bf^2 - Afd\rangle = \langle fd|p|f^2\rangle$, the same as that in the semi-
conducting phase. Here we have assumed A and B are real, appropriate
to the mean field state[14], and the single-particle matrix element
of momentum between localized f and d is imaginary. Thus the intensity is expected to be appreciable.[50a]

We now discuss the XPS experiments[6], where the spectral intensity
corresponds to $\sim70\%$ Sm^{3+} character, the remainder looking like Sm^{2+}.
Sm^{3+} character corresponds to energies for $f^5 \to f^4$ transitions while
Sm^{2+} involves $f^6 \to f^5$ transitions. The ground state in our model for
a single site, is $Af^6 + Bf^5d$ which we simplify to $Af^2 + Bfd \equiv \phi$. If
the final state is f + plane wave ($\vec{k}$) (idealization of $f^5 + \vec{k}$), then
the matrix element for the transition is, to within a sign, $M(\phi\to f,\vec{k})=$
$A\langle\vec{k}|p|f\rangle + (B/\sqrt{2})\langle\vec{k}|p|d\rangle$, while for a final state, d+ plane wave
(idealization of $f^4d + \vec{k}$), we have $M(\phi\to d,\vec{k}) = (B/\sqrt{2})\langle\vec{k}|p|d\rangle|$. For
the optical experiment $|\langle\vec{k}|p|d\rangle|>>\langle\vec{k}|p|f\rangle$ (since $\alpha_f>>\alpha_d$ and
$k^2<<\alpha_d^2$ – see Eq. (5)). If this held in the XPS case, then most of
the observed weight would be in the Sm^{2+} spectrum, in disagreement with experiment. However, for the XPS experiment $k^2>>\alpha_f^2$;
this gives the opposite result, $|\langle\vec{k}|p|d\rangle|<<|\langle\vec{k}|p|f\rangle|$; which leads
to $M(\phi\to fk)\cong A M(f^2\to f\vec{k})$ and $M(\phi\to dk)=B M(df\to d\vec{k})$, and this is observed:
$|M(\phi\to d\vec{k})/M(\phi\to f\vec{k})|^2 = B^2/A^2$ x ratio of intensities for the unmixed
cases.

Finally we argue the plausibility that the width of the narrow
band ϵ_{1k} be $\sim.05$ eV. As a start we consider the calculated[20] bandwidth of ~5 eV associated with the t_{2g} states. The band calculation[20] determines 5d-states in the presence of $4f^6$ whereas our

narrow band involves d-states in the presence of $4f^5$, giving an obvious tendency toward band-narrowing. A crude guess as to the amount can be obtained as follows: one-electron hopping integrals t are usually of the order of 1 Ryd. x an overlap integral Δ, and bandwidths are roughly t x the number of nearest-neighbors giving width $\approx 10^2\Delta$ eV, thus for a band of~10 eV width, $\Delta \sim .1$; putting $\Delta \approx e-\alpha R$ where R is the interatomic distance and $\alpha \propto$ ionic charge seen by an electron in an atomic state, α would increase by a factor of 3/2 giving $\Delta' \approx (.1)^{3/2} \approx \Delta/3$ a reduction factor of about 3. An important effect has to do with the form of our state ϕ ; putting $\phi_{i\sigma}^{3+} \equiv a_{fi\sigma}^{\dagger}|0>$, the hopping integral of interest is (in terms of ϕ_i of Eq. (1)') $T_{ij} = <\phi_i \phi_{j\sigma}^{3+}|H|\phi_{i\sigma}^{3+} \phi_j>$, which gives $T_{ij} = -(B^2/2)t'_{ij}$, where we put the effective one-electron part of $H = \sum_{ij} t'_{ij} a_{di\sigma}^{+} a_{dj\sigma}$, t'_{ij} being the hopping integral between d-states calculated in the presence of f^5 ions on i and j. Since $B^2 \approx .7$ our band width at this stage is ~1/2 eV. But there is another wellknown effect that we have neglected, namely the small-polaron effect; due to the term A $a_{fi\downarrow}^{+}$, there will be appreciable ionic size difference between a site with ϕ_i and one with $\phi_{i\sigma}^{3+}$. In view of the large reduction factors that are known to occur from this effect[51], it does not seem unreasonable to expect a further reduction of the order of 1/10 giving a final band width $\approx .05$ eV, which is of the order needed.

IV. DISCUSSION

Understanding the magnetic behavior of SmS (and SmB_6) is very difficult on the basis of the conventional model, as discussed in Sec. I. There is considerable hope that our model will provide such understanding, as discussed earlier[15]. A most promising possibility is the following: Working within our mean-field (single-site) approximation, the basic mixing interaction W, which is an intersite interaction, provides an effective mean (electric) field at the site of interest, causing mixing of the lowest $f^5 d\ell$ multiplet with the higher-lying f^6 J=0 state. If this mixing dominates that between other $f^5 d_\ell$ -f^6 pairs, it can be seen that the single-ion ground state must be nondegenerate (see Appendix 2); and this would then have to be nonmagnetic because of time reversal invariance of the mean-field Hamiltonian. However, this theorem does not apply if the $f^5 d_\ell$ multiplet mixes equally strongly with other multiplets. Whether the dominant mixing in SmS is indeed with the f^6 (J=0) singlet can be known only after carrying out detailed calculations. Such calculations are presently in progress.

To summarize, the determination that in the collapsed phase there is a small number of conduction electrons, (by extrapolation from the semiconducting phase) plus what we see as the unlikelihood of explaining the magnetic properties in the "conventional" models has compelled us to investigate the "unconventional" model where most

of the 5d-electrons in the collapsed phase are localized. We have
investigated the consequences of our model, in a simplified version,
for a number of experiments, and so far have not found any inconsistency. The necessary calculations for a more realistic version
of our model are in progress.

ACKNOWLEDGMENTS

We wish to thank Drs. A. Jayaraman, D.B. McWhan, J.W. Allen,
E. Carlson and M. Thorpe for helpful discussions, and Prof. S.
Doniach for useful correspondence.

APPENDIX I   STABILITY REQUIREMENT

We derive a simple stability requirement for the semiconductor
phase at zero pressure and show that it is difficult for Falicov-
Kimball-type (FK) models to satisfy this given the experimental
parameters for SmS. We contrast our model with the FK-type models.

Letting n = (average) number of conduction electrons and V =
volume per Sm, we can write the energy per Sm of a state characterized by n and V rather generally as

$$E(n,V) = E(0,V) + \Delta(V)n + g(n,V) \qquad (A.1)$$

where

$$\frac{g(n,V)}{n} \to 0 \quad \text{as} \quad n \to 0 \qquad (A.2)$$

In FK models, $\Delta(V)$ is the minimum excitation energy for one electron
from the semiconducting state (i.e. the band gap), and $g(1,V)$ is
negative for $V=V_c^+$, the volume at which a first order transition
occurs with decreasing V. Thus $\Delta(V_2)$, the value at zero pressure,
is between limits of about .1 and .2 eV.[8,11] In the explicit
Falicov-Kimball model

$$g(1,V) = - (G - \alpha W_B) \qquad (A.3)$$

where G>0 is the f-d intrasite Coulomb repulsion, W_B is the bandwidth, and α is a numerical constant, $\sim 1/2$, (which depends on
whether the density of states is rectangular, semielliptical or a
similar featureless function). G is considered V-independent and
$dW_B/dV<0$.

Clearly stability of the semiconductor phase (n=0) at zero
pressure requires

$$E(1,V_3) > E(0,V_2) \qquad (A.4)$$

where V_3 minimizes $E(1,V)$ (the equilibrium volume in $Sm^{3+}S$). This can be written

$$E_R \equiv E(1,V_2) - E(1,V_3) < E(1,V_2) - E(0,V_2) \qquad (A.5)$$

Here E_R is the energy change if the system initially finds itself in the $Sm^{3+}S$ phase but at volume V_2 and then relaxes to volume V_3 (it is closely related to the familiar Franck-Condon shift, but probably somewhat larger since the latter allows relaxation only locally around one Sm^{3+} in the presence of N-1 Sm^{2+}'s). The right side of (A.5) is the difference in energy between $Sm^{3+}S$ and $Sm^{2+}S$ at volume V_2.

Using (A.1) and (A.2), (A.5) requires

$$E_R < \Delta(V_2) + g(1,V_2) \qquad (A.6)$$

The important fact to consider now is that $g(1,V_2)$ is either negative or very small: $g(1,V_c^+)$ must be negative and $(V_2-V_c^+)/V_2$ is very small ($\sim.01$). For the explicit FK model, using the usual and reasonable assumption that $dW_B/dV < 0$, it follows that

$$g(1,V_2) < 0 \qquad \text{(FK model)} \qquad (A.7)$$

so that

$$E_R < \Delta(V_2) \cdot \qquad (A.8)$$

An independent estimate of the relaxation energy can be made knowing the bulk modulus B_3 in $Sm^{3+}S$ plus V_2 and V_3: $B_3=901$ kbar (ref.41). $V_i=R_i^3/4$ with $R_2=5.97Å$, $R_3=5.62Å$. We find $E_R=.49$ eV using the harmonic approximation, $E_R=\frac{1}{2}\frac{B_3}{V_3}(V_2-V_3)^2$. We tried two

other approximations, both of the form $E(V)=E_{rep}(V) - A/V$; in both we took A as the appropriate Madelung constant for the NaCl structure with charge 2e on the Sm, -2e on the S. For $E_{rep}(V)=Ce^{-R/\rho}$, where $R=(4V)^{1/3}$ is the lattice constant, we found $E_R=.37$ eV ; while for $E_{rep}(V)=C'/R^{\nu}$ we obtained $E_R=.34$ eV. Actually our use of the Madelung constant appropriate to $Sm^{2+}S$ underestimates the binding of the conduction electrons and so underestimates E_R.

Thus with these estimates plus the fact that the maximum value of $\Delta(V_2)$ quoted is .25 eV (Batlogg et al.[8] claim $\Delta(V_2)$ is between .06 eV and .25 eV) we conclude that the stability requirement (A.8) is violated, suggesting that FK type models of the first order transition are not applicable to SmS. Unfortunately this conclusion is weakened by the uncertainty in the experimental parameter $\Delta(V_2)$ and of the quantity E_R. A possibility for pinning E_R down is to use thermal expansion data to obtain additional information;

also as stated above, measurement of the Franck-Condon shift should give a lower bound on E_R. It would seem that careful measurement of the resistivity could pin down $\Delta(V_2)$.

We also wish to note that most of the explicit FK type of theories don't take into account the fact that B_3 is appreciably larger than B_2; neglect of the fact that $B_3 > B_2$ (incorrectly) underestimates E_R, and therefore might lead to internal consistency with (A.6), allowing one to miss a real physical inconsistency.

A crucial difference between FK-type models and our localized model is that $g(1,V)$ is positive and appreciable in the latter (also, in the latter, n is not the number of conduction electrons, but rather the number per Sm of $f^5 d_\ell$-sites). Thus (A.6) is not critical in this localized model.

APPENDIX 2

Consider an N-dimensional matrix M, whose elements are given by

$M_{ii} = 0$, for $1 \leq i \leq N-1$

$M_{ii} = \varepsilon$, for $i = N$

$M_{i,N} = M^*_{N,i} = W_i$, for $1 \leq i \leq N-1$

$M_{ij} = 0$, for all other i and j.

The eigenvalues of the matrix M are given by the solutions of the equation $\det(M - \lambda I) = P(\lambda) = 0$, where I is the N-dimensional identity matrix. It can be easily seen that the polynomial $P(\lambda)$ is given by

$$P(\lambda) = \lambda^{N-2} \left[\lambda(\lambda - \varepsilon) - \sum_{i=1}^{N-1} W_i^2 \right]$$

The N eigenvalues of M are then

$\lambda_i = 0$ for $1 \leq i \leq N-2$

$\lambda_{N-1} = \frac{1}{2} \left(\varepsilon + \sqrt{\varepsilon^2 + 4\alpha^2} \right)$

$\lambda_N = \frac{1}{2} \left(\varepsilon - \sqrt{\varepsilon^2 + 4\alpha^2} \right)$

where $\alpha^2 = \sum_{i=1}^{N-1} W_i^2$.

Thus for nonzero α, the lowest eigenvalue is λ_N and is nondegenerate.

REFERENCES

1. Valence Instabilities and Related Narrow-Band Phenomena, Edited by R.D. Parks, Plenum Press 1977.
2. T.A. Kaplan, S.D. Mahanti and Mustansir Barma, ref. 1, p. 153.
3. A. Jayaraman, V. Narayanamurti, E. Bucher, and R.G. Maines, Phys. Rev. $\underline{B11}$, 2783 (1975); and ref. 1 p. 61.
4. J.M.D. Coey, S.K. Ghatak, and F. Holtzberg, AIP Conf. Proc. $\underline{24}$, 38 (1974); and ref. 1, p. 211.
5. M.B. Maple and D. Wohlleben, Phys. Rev. Letters $\underline{27}$, 511 (1971).
6. M. Campagna, E. Bucher, G.K. Wertheim, and L.D. Longinotti, Phys. Rev. Letters $\underline{33}$, 165 (1974). The X-ray photoemission studies were not made on SmS under pressure, but rather on $Sm_{1-x}R_xS$ where R = Y or Gd. The concentration x acts similarly to pressure (see e.g. ref. 3).
7. S.D. Bader, N.E. Phillips, and D.B. McWhan, Phys. Rev. $\underline{B7}$, 4786 (1973).
8. E. Kaldis and P. Wachter, Sol. St. Comm. $\underline{11}$, 907 (1972); B. Batlogg, A. Schlegel, and P. Wachter, Phys. Rev. $\underline{B14}$, 5503 (1976).
9. B. Batlogg and P. Wachter, ref. 1, p. 537; J. W. Allen, ref. 1, p. 533 found a similar result in the mixed valence compound SmB_6.
10. E. Yu. Tonkov and I.L. Aptekar, Sov. Phys. Solid State $\underline{16}$, 972 (1974).
11. C.M. Varma, Rev. Mod. Phys. $\underline{48}$, 219 (1976).
12. Some examples in addition to those discussed in ref. 11 are: L.L. Hirst, ref. 1, p. 3; T. Penney, ref. 1, p. 86; S.K. Ghatak and M. Avignon, ref. 1, p. 229; B. Alascio, ref. 1, p. 247; B. Coqblin, A.K. Bhattacharjee, J.R. Iglesias-Sicardi, and R. Jullien, ref. 1, p. 365; J.H. Jefferson and K.W.H. Stevens, ref. 1, p. 419. For others consult ref. 1.
13. According to the titles of the papers by S. Doniach, ref. 1, p. 169 and in Proc. of the Oxford Conference on Itinerant Electron Magnetism, Sept. 1976, to appear in Physica, his work would appear to fall under this band-model category. However, the specific model Hamiltonian studied is rigorously a localized model; whereas the connection to the band model is not rigorous.
14. T. A. Kaplan and S.D. Mahanti, Phys. Lett. $\underline{51A}$, 265 (1975).
15. S.D. Mahanti, T.A. Kaplan and M. Barma, Phys. Lett. $\underline{58A}$, 43 (1976).
15a. J. Schweitzer, Phys. Rev. $\underline{B13}$, 3506 (1976), studied stability properties of the localized model.
16. A localized picture for most of the d-electrons in connection with SmB_6, was discussed earlier by R.L. Cohen, M. Eibschütz and K.W. West, Phys. Rev. Lett. $\underline{24}$, 382 (1970)and in more detail by J.C. Nickerson, R.M. White, K.N. Lee, R. Bachmann, T.H. Geballe, and G.W. Hull, Jr., Phys. Rev. $\underline{B3}$, 2030 (1971). A spatially homogeneous mixing was not considered, in contrast to ref.'s 2, 14, 15, and 15a.

17. F. Holtzberg and J. Torrance, AIP Conf. Proc. no. 5, 860 (1971).

18. The interpretation of ref. 17 followed the earlier work on the analogous situation in Eu compounds, by M.J. Freiser, S. Methfessel, and F. Holtzberg, J. Appl. Phys. 39, 900 (1968).

19. J.W. McClure, J. Phys. Chem. Sd. 24, 871 (1963); A. V. Golubkov, E.V. Goncharova, V.P. Zhuze and I.G. Manilova, Sov. Phys. Sol. St. 7, /1963/ (1966).

20. H.L. Davis, Proc. 9th Rare Earth Res. Conf., Va. Poly. Inst. and State Univ., Blacksburg, VA (1971).

21. K. Lendi, Phys. Cond. Matt. 17, 215 (1974).

22. D.L. Greenaway and G. Harbeke, Optical Properties and Band Structure of Semiconductors, Pergamon Press (1968), p. 114.

23. A. Jayaraman, P.D. Dernier, and L.D. Longinotti, High Temperatures, High Pressures 7, 1 (1975).

24. The lowering of 0.6 ev. follows from T. Penney and F. Holtzberg, Phys. Rev. Lett. 33, 165 (1975), with $\Delta V/V = 12\%$ across the transition.

25. Estimates were also made with a circular density of states, $D(e) \propto [1 - (\frac{2e}{e_c} - 1)^2]^{1/2}$, resulting in only minor changes.

26. C.M. Varma and V. Heine, Phys. Rev. B11, 4763 (1975).

27. One might worry about a possible discontinuous electron rearrangement as a function of V, such as that occurring in the Falicov-Kimball theory (Phys. Rev. Lett. 22, 997 (1969)). It is argued in Appendix I that this band theory is probably inconsistent with the stability of Sm^{2+} at atmospheric pressure.

28. L.L. Hirst, J. Phys. Chem. Solids 35, 1285 (1974).

29. C.M. Varma and Y. Yafet, Phys. Rev. B13, 2950 (1976).

30. J. Hubbard, Proc. Roy. Soc. (London) A276, 238 (1963).

31. D. Penn, Physics Lett. 26A, 509 (1968).

32. A marked difference between Ni and SmS is that the narrow band (3d) in Ni is much broader than that (4f) in SmS. This suggests that interactions neglected in the Varma-Yafet[29] Hamiltonian (e.g. interband Coulomb interactions) which, if properly treated, give a tendency for exciton formation might be crucial in understanding the difference. Haldane (ref. 1, p. 191) seems to have taken a small step in the direction of including such interactions (still within a single-impurity model).

33. Like those considered by P.W. Anderson and A. Hasegawa, Phys. Rev. 100, 675 (1955).

34. C.M. Varma, ref. 1, p. 201.

35. J.B. Sokoloff, Phys. Rev. B2, 779 (1970); G. Beni, T. Holstein, and P. Pincus, Phys. Rev. B8, 312 (1973).

36. Lest one be misled, one should realize that the same ($U \to \infty$) Hubbard model, for certain 3-dimensional lattices, is probably ferromagnetic (Y. Nagaoka, Phys. Rev. 147, 392 (1966)), at least for $\rho \cong 1$.

37. J.M. D. Coey, S.K. Ghatak, M. Avignon, and F. Holtzberg, Phys. Rev. B14, 3744 (1976); ref. 1, p. 211.

38. I. Nowik, ref. 1, p. 261.

39. I. Nowik, ref. 1, p. 167.

40. Nowik's result is that the electron released in $Sm^{2+} \rightarrow Sm^{3+}$ is not felt at the Eu^{153} probe. His conclusion[38] that this implies that the electron is in a d-band seems strained in light of the fact that the electron released in $R^{2+} \rightarrow R^{3+}$ where R = Y, Gd in $Sm_{1-x}R_xS$ is felt at the Eu^{153} and presumably goes into the same (d) band as the electron in $Sm^{2+} \rightarrow Sm^{3+} + e$. The remarkable result is the difference (observed at the Eu^{153} nucleus) between the electron in $Sm^{2+} \rightarrow Sm^{3+} + e$ and that in $R^{2+} \rightarrow R^{3+} + e$. As noted by Nowik[38,39], consistent with his result is localization of the extra Sm electron at that Sm site (postulated by us, and recently claimed to be found by Coey et al., ref. 37). Understanding the difference between Sm^{3+} and Gd^{3+} or Y^{3+} is a fundamental problem. A possible source is the charge asphericity of the Sm^{3+} core $(4f^5)$, not present in the other ions.

41. G. Güntherodt et al., ref. 1, p. 321.

42. Unfortunately, the low-p transition is found at 10 kbar in ref 41, in disagreement with the consensus of earlier results,[3,4,10] introducing some quantitative uncertainty in the experimental situation. It is important to have this resolved.

43. We note here that anharmonicity, particularly in $E(0,V)$, that might begin to become appreciable at these higher pressures goes in the direction of reducing p'_0.

44. S.M. Shapiro, R.J. Birgeneau and E. Bucher, Phys. Rev. Lett. 34, 470 (1976).

44a. Obtaining the qualitatively correct picture (fig. 5) within the simple theoretical model we've presented suggests the non-existence of a new semiconducting phase (which had been suggested, and labelled B'[3]). We add the comment that the calculations of this phase boundary were also carried out for rectangular and elliptical densities of states — this resulted in no important changes.

45. R.L. Melcher, G. Güntherodt, T. Penney, F. Holtzberg, 1975 Ultrasonics Symposium Proc. IEEE Cat. #75 CHO 994-45A, pg. 16.

45a. P.D. Dernier, W. Weber, L.D. Longinotti, Phys. Rev. B14, 3635 (1976).

46. If one neglects the term $-k(.3 \ln.3 + .7 \ln.7) \simeq .6k$ which arises from the different spatial configurations of $30\% f^6$ and $70\% f^5$. The latter term is appreciable in the difference ΔS, and increases the difficulty for the band theories.

46a. D.H. Templeton and C.H. Dauben, J. Am. Chem. Soc. 76, 5237 (1954).

47. N.P. Bogoroditskii, V.V. Pasynkov, Rifat Rizk Basili, and Yu. M. Volokobinskii, Sov. Phys. Doklady 10, 85 (1965).

48. J.W. Allen, private communication.

49. This was found by E. Carlson (private communication) in a Hartree-Fock Calculation for Sm^{2+} in the excited configuration $4f^5d$.

50. The numerical value of the broad-band width used in Sections II
     and III is rather larger than that (~2.5 eV) used in Sec. 1.
     The essential reason for the difference is as follows: The
     smaller value is based on the picture from the band calculation
     (Davis, ref. 20) where the conduction-band minimum is at the
     X-point, with a local curvature corresponding to a nearly-free-
     electron mass; due to the symmetry-related minima, the density-
     of-states is 3 times that of a Γ-minimum with the same mass.
     The difference is completely unimportant for the thermodynamic
     properties considered in Sec. II. Although it wasn't necessary
     to choose the (less realistic) larger value for these considera-
     tions, we did so to obtain symmetry with the simplified model
     used in Sec. III to crudely estimate optical properties. It
     will also be noted that the width of the narrow band, .047 eV,
     is rather larger than that quoted in ref. 2 (.033 eV); this
     discrepancy is due essentially to an error made in the earlier
     work. The value .047 eV is correct and depends only on the
     low-T specific heat (γT) and the value of b; for parabolic
     density of states it is easy to see that
     $D_1 = (1-b/2)^{1/3} \pi^2 k^2 N_a/\gamma$ where k=Boltzmann's constant, N_a =
     Avogadro's no. and γ is per mole. None of our qualitative con-
     clusions are affected by these changes.

50a. Comparison with experiment[9] of $\varepsilon_1(\omega)$, as calculated above (i.e.
     from Eq.(6)),indeed suggests that there are large additional
     contributions in the range .1 to .8 eV (roughly), which might
     be due to these intraatomic excitations.

51. D. Sherrington and S. von Molnar, Sol. St. Comm. <u>16</u>, 1347 (1975).

QUESTIONS AND COMMENTS

M.B. Maple: I just wanted to say that our interpretation of the
                 nonmagnetism in the nonintegral state is simple.
                 Let us allow for fluctuations between the two con-
                 figurations which are quantum mechanical. Then, if
                 your temperature is much lower than the character-
                 istic temperature, the magnetic moment can't undergo
                 an appreciable fraction of a whole period of a Zeeman
                 precession in a magnetic field before an electron is
                 emitted or absorbed by the 4f electron shell. There-
                 fore, it can't be aligned in the magnetic field. That
                 is the essence of it-- it's not based upon theoreti-
                 cal considerations.

T.A. Kaplan: Yes, well, I can understand that, for example, for
                 the Kondo Hamiltonian, where the ground state is a
                 singlet. Thus when the temperature is much lower
                 than the characteristic temperature, the behavior
                 will be non-magnetic. The problem is if I put two

impurities in they will interact. We know that, at least in some cases, when you increase the concentration you start to get local moments and spin glass behavior at very small concentrations. So it's very troublesome going from one impurity to many, where your fluctuation argument might fail as illustrated by this example.

THE α-γ PHASE TRANSITIONS IN Ce-Th ALLOYS UNDER HIGH PRESSURE[*]

C.Y. Huang, J.L. Smith
Los Alamos Scientific Laboratory, University of California
Los Alamos, NM 87545

C.W. Chu[†][‡][§]
Physics Department, Cleveland State University
Cleveland Ohio 44115

and

P.H. Schmidt
Bell Laboratories, Murray Hill, NJ 07974

I. INTRODUCTION

In the past few years a great deal of research has been done
on rare earth metals and compounds of them which exhibit non-integral valences.[1-3] The rare earth ion in these materials undergoes a change of valence in response to a temperature, pressure,
or alloying variation. In the Sm mixed valence systems the atomic-
like $4f$ state is usually situated energetically below the $5d$-$6s$
conduction band. As the temperature, pressure, or composition is
varied, the $5d$-$6s$ bands broaden and move down in energy relative
to the $4f$ level and finally, at some critical value, the $4f$ level
is pinned at the Fermi level. In this latter state, the system is
defined as a mixed valence material.[2] Historically, Ce was the
first material found to exhibit a valence change. X-ray studies
revealed that when Ce was compressed to modest pressures (> 8 **kbar**
at 300 K)[4] or cooled to low temperature (< 100 K at 1 atm)[5] a large

---

[*]Work performed under the auspices of the USERDA.
[†]Visiting Staff Member at Los Alamos Scientific Laboratory, Los
 Alamos, NM 87545
[‡]Supported in part by NSF.
[§]Present address: University of Houston, Houston, Texas, 77004

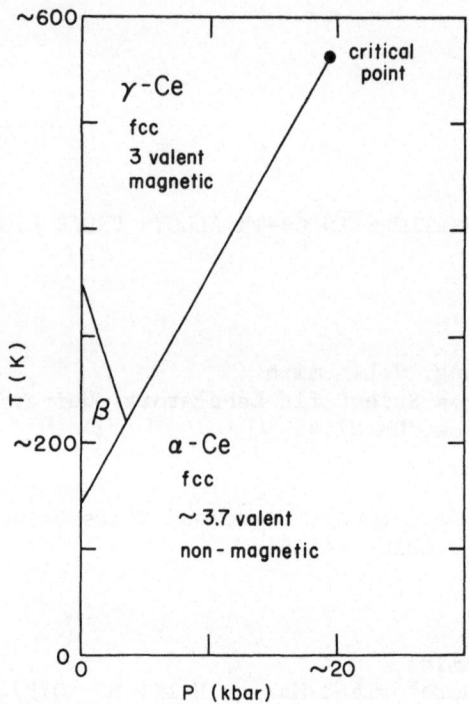

Figure 1.   Cerium Phase diagram.

(> 12%) volume contraction (the γ–α transformation) took place
while the crystal structure (fcc) remained unchanged. Zachariasen[6]
and Pauling[7] independently suggested that this volume contraction
corresponded to a valence change from 3 for γ-Ce(the $4\underline{f}^1 5\underline{d}^1 6\underline{s}^2$ configuration) to 4 for α-Ce (the $4\underline{f}^0 5\underline{d}^2 6\underline{s}^2$ configuration), that is,
the $4\underline{f}$ electron of γ-Ce was promoted to the valence band, leaving
the $4\underline{f}$ level of α-Ce empty. Subsequently, based on very strong
empirical correlations between valence and metallic radii in the
periodic table, Gschneidner and Smoluchowski[8] pointed out that
only about 0.67 of an electron per atom is transferred from the
$4\underline{f}$ level to the valence band. As a consequence of this valence
change, there exists the first-order phase-transition line between
the α- and γ- phases in the pressure-temperature (P-T) phase diagram, and it ends[9-11] in a critical point (Fig. 1). This critical
point is in most respects analogous to a liquid-gas critical point
and, by properly choosing a path in the P-T diagram it is possible
to go continuously from a large-spacing γ-phase to a more dense
α-phase. To date, Ce is the only element known to have a critical
point in a solid phase. A brief survey of the theoretical work on
this subject is given in Section II.

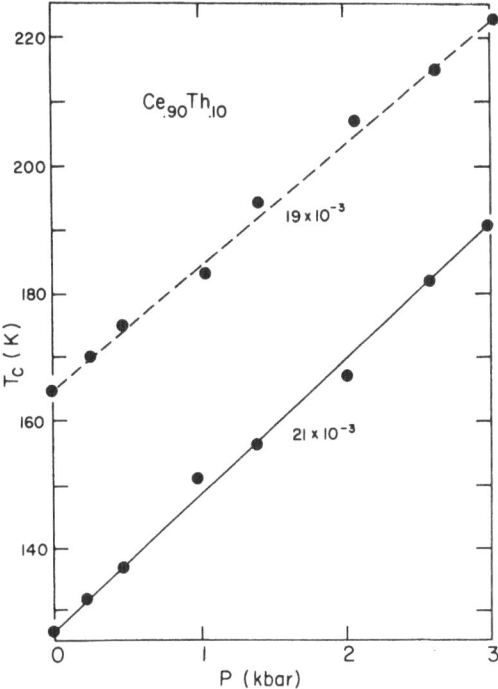

Figure 2(a). P–T phase diagram of $Ce_{.90}Th_{.10}$. The dashed and
solid lines represent the data taken when warming and
cooling, respectively. The numbers by these lines
are their slopes.

As early as 1962, Gschneidner, Elliott and McDonald[12] showed
that the first-order γ–α transitions can also be driven second-
order at a critical point by intra-rare-earth alloying. In particular, by analogy to the P–T plane, Lawrence et al. in their work
in the x–T plane of the $Ce_{1-x}Th_x$ alloy system found from their
electrical resistance data that at P = 0 the line of first order
transitions terminated at a critical point $\{x_0 = .265, \ T_0 = 148 \ K\}$
beyond which only continuous transitions were observed. In this
work, these authors pursued the analogy to liquid–vapor isobars by
considering the isoconcentration curves in the critical region to
have the form

$$A(\Delta R)^\delta + B \ \Delta x \Delta R \quad = T - T_0, \eqno(1)$$

where for volume they substituted the resistance R and for pressure
the concentration x, and where $\Delta R = R - R_0$, $\Delta x = x - x_0$, R_0 and x_0
being the critical values. A and B are constants and δ is the critical exponent. Their data yielded $\delta \simeq 3.0$, which is the value for

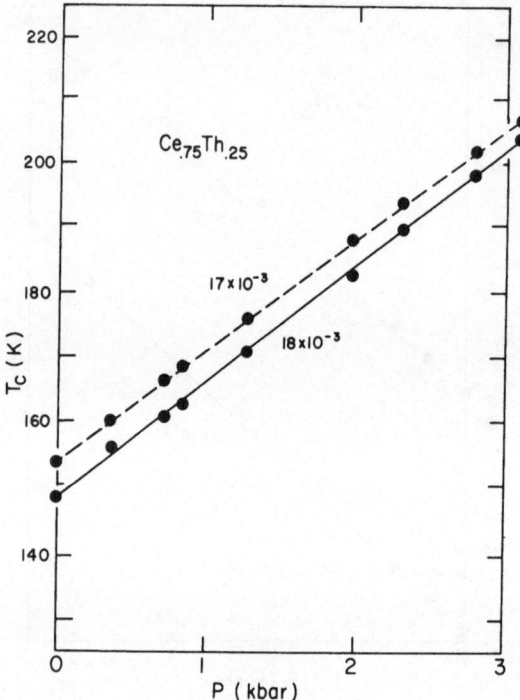

Figure 2(b). P-T phase diagram of Ce .80Th .20. The dashed and
solid lines represent the data taken when warming and
cooling, respectively. The numbers by these lines
are their slopes.

mean-field behavior. It was a very unusual result, since $\delta \sim 5.2$
for Ising-model calculations and ~ 4.0 for liquid-gas transitions
and magnetic transitions.[14] This particular mean-field behavior
obtained by Lawrence et al.[13] stimulated our interest in looking
into the critical behavior of $Ce_{1-x}Th_x$ under pressure. In this
manner the substitution of P by x as done in Fig. 1 is avoided.
In Section III we will present our experimental results.

II. BRIEF THEORETICAL SURVEY

To date there have been several models for the mechanism of the
$\gamma \leftrightarrow \alpha$ transition and the nature of α-Ce, but none of these adequately
explain all of the data, that is, magnetic as well as thermal.[15]
As pointed out in Section I, Zachariasen[6] and Pauling[7] first proposed that the electronic configuration of the Ce atom in the solid
changed from $4\underline{f}^1(\underline{sd})^3$ to $4\underline{f}^0 (\underline{sd})^4$ during the transition. This
promotion of an electron from the localized 4f state to the ($\underline{sd}$)

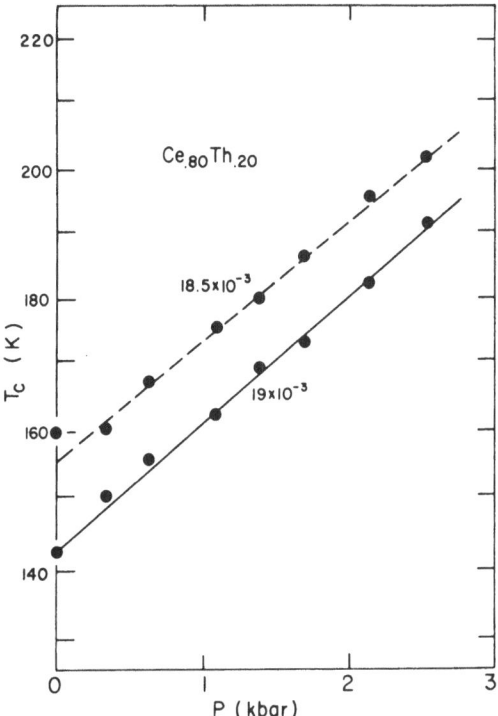

Figure 2(c). P-T phase diagram of Ce$_{.75}$Th$_{.25}$. The dashed and solid
lines represent the data taken when warming and cool-
ing, respectively. The numbers by these lines are
their slopes.

conduction band transforms the trivalent solid to a tetravalent
one and simultaneously the magnetic moment associated with the
4f state is lost. From the correlations between valence and metallic radii[8] and from the high observed electronic-specific-heat constant[15] for α-Ce, it is believed that only a fraction (∿0.7 at 1 atm)
of the 4f electron is transferred during the γ-α transition.

   Another viewpoint of this promotion idea has been given by
Ramirez and Falicov[16] who included an electron-electron matrix between the localized and the conduction-band states. They assumed
that the f level in Ce lies in a very narrow, highly correlated
band slightly above the Fermi level, and that all interactions are
short range. Taking into account three classes of quasiparticles:
(a) electron excitations in the conduction band; (b) hole excitation
in the conduction band; and (c) localized electrons in the narrow
f band, and also including the interaction between f electrons and
conduction electrons, they obtained an expression for the free
energy of the system. By minimizing this expression with respect

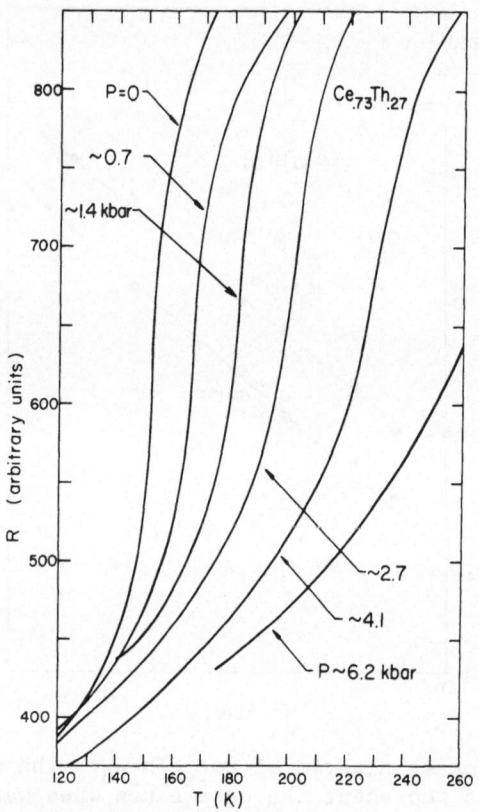

Figure 3.   The resistance versus temperature curves for $Ce_{.73}Th_{.27}$
            for various pressures.

to the occupation numbers of the conduction electrons, conduction
holes, and localized electrons, they found that the number of f
electrons varies discontinuously as a function of temperature
and pressure. They also predicted a critical point. The short-
coming resulting from some simplifying assumptions is that this
model predicts α-Ce to be tetravalent at low temperatures in disagreement with experiment. In order to improve this model, Alascio,
Lopez, and Olmedo[17] considered the hybridization of the localized
4f states with the sd conduction band and were able to predict non-
integral valences for α-Ce. However, in order to agree with magnetic
susceptibility data they had to assume a shift in the 4f level with
temperature.

A somewhat different picture of the γ-α transition has been put
forward by Coqblin and Blandin.[18] This model, the so-called virtual
bound state model, essentially takes into account the compensation
of the spin of the 4f electron by the conduction electrons. In this
manner, the 4f density of states has a Lorentzian shape with a width

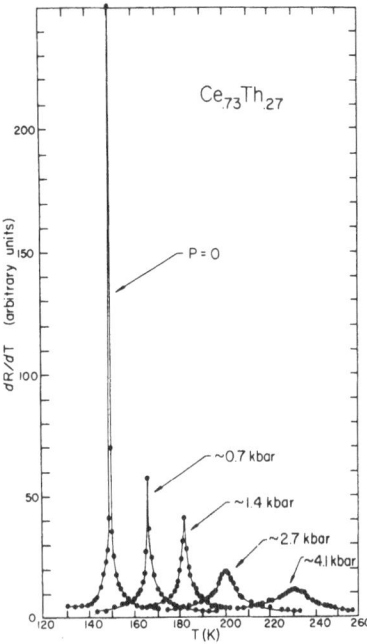

Figure 4. dR/dT versus T curves for Ce$_{.73}$Th$_{.27}$ at various pressures.

of order 0.01 eV. By assuming that the energy of the 4f̱ state is shifted upwards compared to the Fermi energy when pressure is applied and downward with increasing temperature, they obtained various degrees of spin compensation. Thus, in the γ-phase there is essentially no compensation of spin and in the α-phase the spin is partially compensated. However, these authors as in Ref. 17 did not give the reason for the shift in the 4f̱ level.

Both theories presented above involve the assumption of the localized nature of the 4f̱ state in both α- and γ-Ce. However, recent APW band calculations presented by Kmetko[19] revealed that the 4f̱ electrons are somewhat itinerant. In partial agreement with this observation, Johansson[20] proposed the use of the Mott transition[21] to explain the γ-α transition. In this model, the γ-phase is on the low-density side of the Mott transition. In the α-phase the 4f̱ band was assumed to overlap the Fermi level and a fractional number of electrons (near 0.7) would occupy the band. However, this model cannot account for the high observed magnetic susceptibility.[15]

Another promotional idea is the interconfigurational-fluctuation (ICF) model of Hirst.[22] According to him, integral values of the 4f̱ occupation have a stability due to Hund's rule correlations. However, the stability of the non-integral valency could arise from

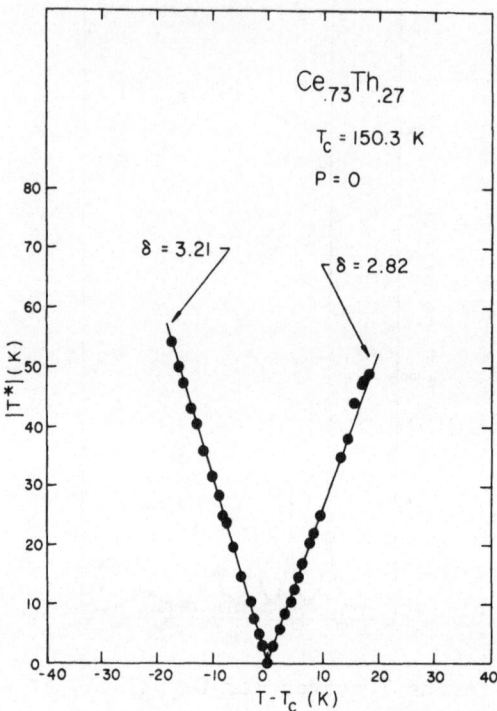

Figure 5(a) Reduced temperature $|T^*|$ versus $(T-T_c)$ plots for
             $Ce_{.73}Th_{.27}$ at P = 0. $T^* \equiv (R-R_c)/(dR/dT)$ where T_c
             and R_c are deduced from the peak in dR/dT.

an interplay of conduction electrons with the unstable 4$\underline{f}$ shell;
that is, over a certain pressure-temperature range the two levels
(4$\underline{f}^1$ and 4$\underline{f}^0$) are tied to the Fermi level in such a way that mixing
between the 4$\underline{f}$ electrons and conduction electrons is sufficiently
strong to overcome the Hund's rule correlations. This is an ICF
state in which the valence on individual ions fluctuates between
integral values with lifetimes on the order of 10^{-13} sec. Again,
this model cannot explain the temperature dependence of the observed magnetic susceptibility[15] of α-Ce.

Based on a possible similarity of α-Ce to spin-compensated systems exhibiting a Kondo effect, Edelstein[23] predicted a $T^{-1/2}$ dependence of the magnetic susceptibility of α-Ce. However, no Kondo
effect has been observed in α-Ce.

In addition to these, recently Kincaid and Stell[24] tackled the
γ-α allotropic phase transition of $Ce_{1-x}Th_x$ from the phenomenological
point of view of solid mixtures. In this model, the particles are
assumed to interact with each other in a manner that can be repre-

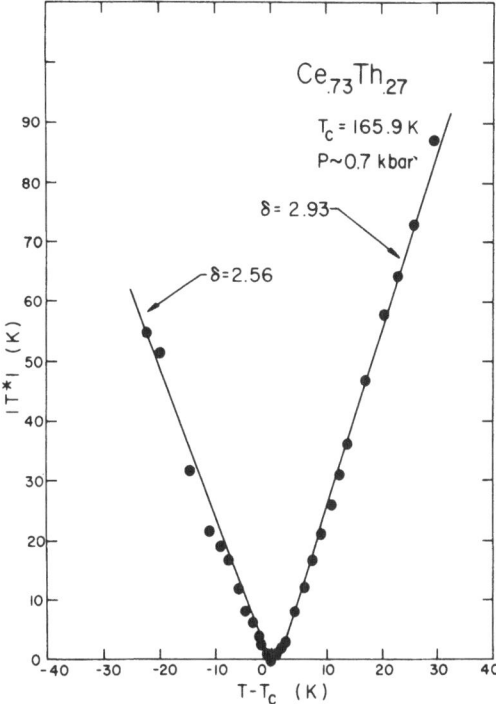

Figure 5(b)   Reduced temperature $|T^*|$ versus $(T-T_c)$ plots for
               Ce$_{.73}$Th$_{.27}$ at P$\sim$0.7 kbar. $T^* \equiv (R-R_c)/(dR/dT)$ where
               T_c and R_c are deduced from the peak in dR/dT.

sented by a sum of two-body potentials ϕ_{CC}, ϕ_{CT}, and ϕ_{TT}, where C
and T stand for Ce and Th, respectively. By assuming ϕ_{TT} to be a
simple hard-sphere pair potential, ϕ_{CC} and ϕ_{CT} to be the shouldered
hard-sphere pair potentials, they were able to predict the first-
order transition line terminated by a second-order critical point
in the P-T as well as x-T planes. Nonetheless, this model is in-
capable of dealing with any magnetic properties.

    In conclusion, no existing theory is capable of interpreting all
the physical properties of Ce. The fascinating properties of this
substance point to the need for more experimental as well as theoretical investigations.

III.   EXPERIMENTAL RESULTS

    Our samples were prepared from 99.9% pure Ce and 99.995% pure
Th by arc melting in argon followed by vacuum annealing at 600ºC for
two weeks. Some of the polycrystalline samples were then checked by

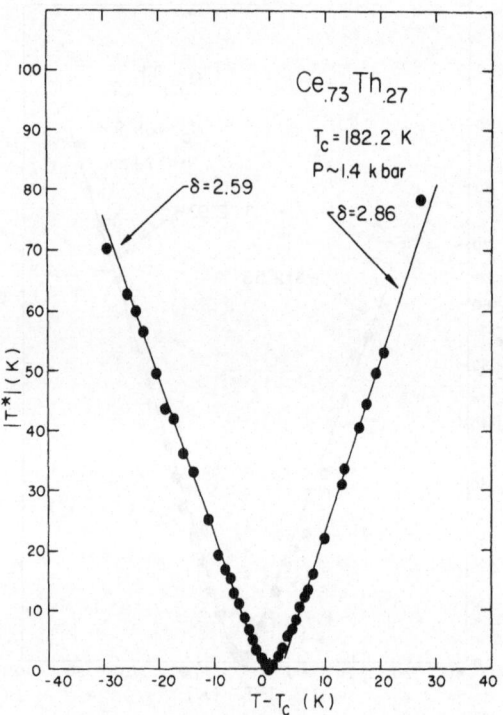

Figure 5(c)   Reduced Temperature $|T^*|$ versus $(T-T_c)$ plots for
Ce$_{.73}$Th$_{.27}$ at P~1.4 kbar. $T^* \equiv (R-R_c)/(dR/dT)$ where
T_c and R_c are deduced from the peak in dR/dT.

chemical analysis and x-rays in order to guarantee the correct compositions and crystal structure (fcc). For P $\lesssim$ 3 kbar, argon was used as a pressure transmitting medium in a conventional Be-Cu high pressure cell. For P $\gtrsim$ 3 kbar, the self-clamp technique[25] was employed. A four-point technique was utilized to measure the sample resistance, and in the first-order region a differential thermal analysis (DTA) technique was used to obtain the transition temperatures in order to avoid difficulty from the large volume contraction at the transition.

Figure 2a shows our data on Ce$_{.90}$Th$_{.10}$ obtained by DTA in argon pressure up to 3 kbar. The dashed and solid straight lines represent the fits to the γ-α transition temperature data taken when warming and cooling the sample, respectively. The transition temperature, T_c, is apparently linear in pressure. From the slopes of these straight lines, we obtain $dT_c/dP = 19 \times 10^{-3}$ K-bar^{-1} and 21×10^{-3} K-bar^{-1} for warming and cooling, respectively. As shown, the hysteresis indicating a first-order transition is very pronounced.

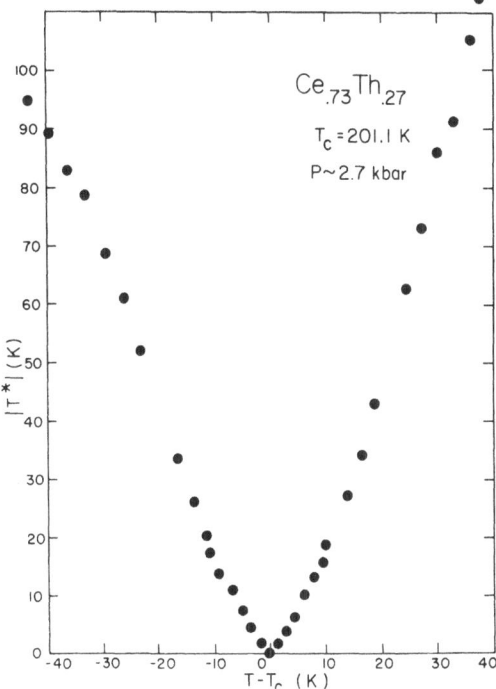

Figure 5(d)   Reduced Temperature $|T^*|$ versus $(T-T_c)$ plots for
Ce$_{.73}$Th$_{.27}$ at P∿2.7 kbar. $T^* \equiv (R-R_c)/(dR/dT)$ where
T_c and R_c are deduced from the peak in dR/dT.

Figures 2(b) and (c) display the data obtained by the same technique
for Ce$_{.80}$Th$_{.20}$ and Ce$_{.75}$Th$_{.25}$.   Again T_c is linear in P.   It is clear
to see that the hysteresis and dT_c/dP decrease with the thorium concentration.   All these figures clearly demonstrate that the warming
and cooling lines will meet at a critical point.   In particular, by
extrapolating these straight lines (solid and dashed) in Fib. 2(c)
for Ce$_{.75}$Th$_{.25}$, we obtain the critical point {P$_0$∿4.8 kbar and T$_0$∿
273 K}.   This was verified later (see Figs. 6 and 7).

We have also investigated Ce$_{.73}$Th$_{.27}$ in which the thorium concentration is very close to the critical one (x_0=0.265 at P=0).
Figure 3 shows the resistance variations taken at various pressures
for Ce$_{.73}$Th$_{.27}$.   Their derivatives, dR/dT, are shown in Fig. 4, and
we find that increasing pressures broaden the peak (make it less
divergent) and shift it to higher temperatures.   This is quite different from the results of Ref. 13 in which they obtained T_c∿150 K
at P=0 for the samples wih thorium concentrations within a range
about x_0 while our T_c varies drastically with pressure, where T_c

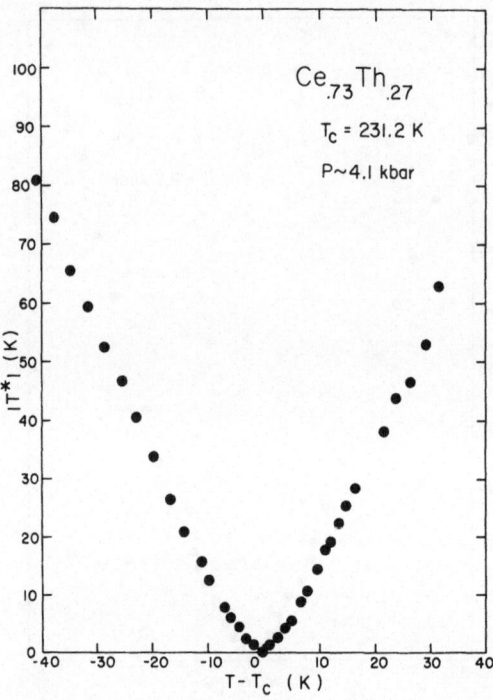

Figure 5(e)   Reduced Temperature $|T^*|$ versus $(T-T_c)$ plots for
              $Ce_{.73}Th_{.27}$ at $P \sim 4.1$ kbar. $T^* \equiv (R-R_c)/(dR/dT)$ where
              T_c and R_c are deduced from the peak in dR/dT.

is the temperature at which dR/dT peaks. Thus x and P can be seen
not to be equivalent.

Following Ref. 13 we define a reduced temperature

$$T^* \equiv (R - R_c)/(dR/dT).$$

It can be shown that

$$T^* = \delta(T - T_c) + B (1 - \delta) (R - R_c) \Delta x, \qquad (2)$$

where R_c is the value of R at T_c. Note that $\Delta x = (x - x_0)$ now de-
pends on pressure. Figure 5(a) through (e) shows our $|T^*|$ versus
$(T-T_c)$ plots for $Ce_{.73}Th_{.27}$ at various pressures. In this case
$\Delta x \simeq 0$ when $P = 0$, and Eq. (2) predicts that $|T^*|$ is linear in T
with a slope, δ. Thus from Fig. 5(a), we obtain $\delta = 3.21$ and 2.82
for $T < T_c$ and $> T_c$, respectively. These values are in good agreement
with those given in Ref. 13. For $P \neq 0$, Δx is no longer nearly zero
and hence the second term of Eq. (2) contributes resulting in the

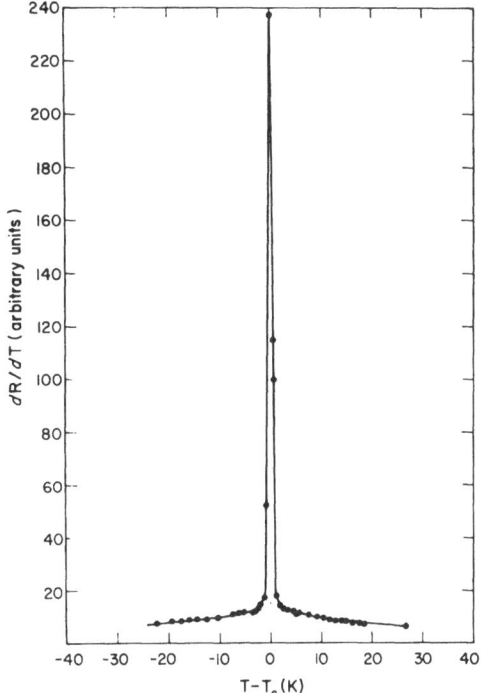

Figure 6. dR/dT vs T-T$_c$ for Ce$_{.75}$Th$_{.25}$ at P∿4.8 kbar.

rounding near T$_c$ shown in Fig. 5(b)-(e). It is interesting to note that the slopes of the linear parts in Fig. 5(b) are different from the mean-field values in Fig. 5(a). Hence, as low as P∿0.7 kbar, the mean field behavior does not seem to govern the transition.

Following the result of Fig. 2(c), we have investigated Ce$_{.75}$Th$_{.25}$ at P∿4.8 kbar. The sharp peak at T$_c$ = 272.7 K shown in Fig. 6 demonstrates that the critical point is close to P$_0$ = 4.8 kbar and T$_0$ = T$_c$ = 272.7 K. We have also obtained |T*| versus T-T$_c$ for the composition. As shown in Fig. 7 |T*| is again linear in T-T$_c$ within our experimental error. The slopes obtained are δ = 2.60 and 2.85 for T<T$_c$ and >T$_c$, respectively, which clearly deviate from the mean-field value.

We have also made measurements on Ce$_{.71}$Th$_{.29}$ which show that the dR/dT versus T curves are not as sharp as those given for Ce$_{.73}$Th$_{.27}$. This is understandable, since the thorium concentration of the former is higher than the critical concentration, and Δx is not zero even for P=0.

Figure 8 shows the temperatures of the peaks in dR/dT(i.e.,T$_c$)

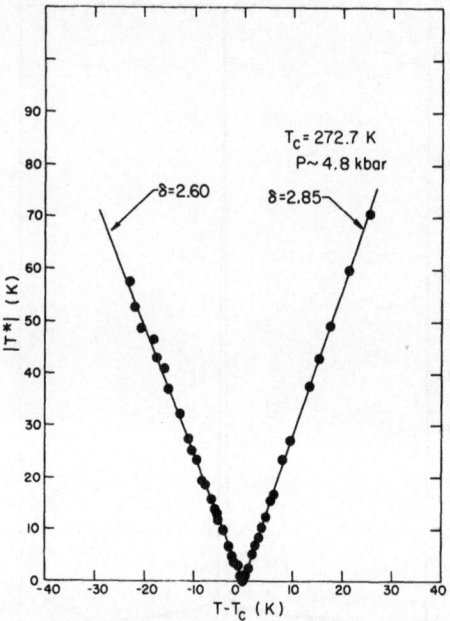

Figure 7.   Reduced Temperature, $|T^*|$, versus $(T-T_c)$ for $Ce_{.75}Th_{.25}$
at $P \sim 4.8$ kbar.

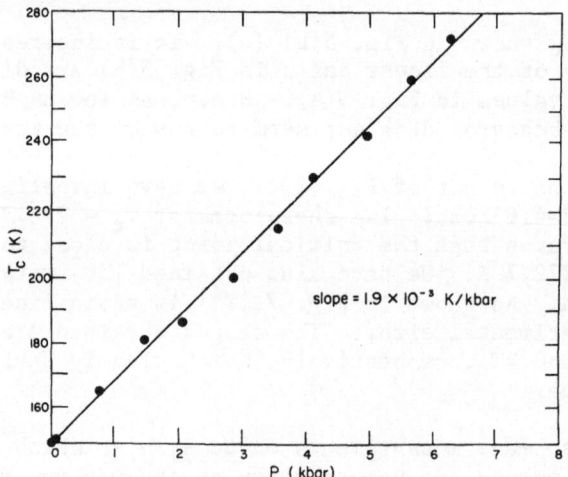

Figure 8.   The temperature variations of the peak of dR/dT versus
pressure for $Ce_{.73}Th_{.27}$ and $Ce_{.71}Th_{.29}$ which behave
identically.

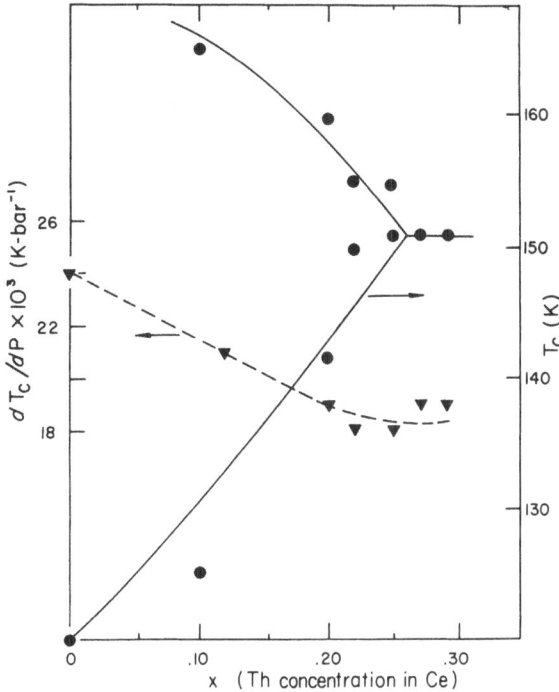

Figure 9. dT_c/dP (triangles) and T_c (circles) versus thorium concentration x.

for both $Ce_{.73}Th_{.27}$ and $Ce_{.71}Th_{.29}$. Within our experimental error T_c is linear in P for both samples, and the data points fall on the same straight line.

We summarize our results for the dependence of the thorium concentrations, x, in Fig. 9. As shown by the circles the hysteretic first-order transitions at P = 0 terminate at the thorium concentration of $x_0 \sim 0.26$, in agreement with that obtained in Ref. 13. The dT_c/dP data (triangles) show that dT_c/dP decreases with x and becomes constant for $x \gtrsim x_0$.

IV. SUMMARY

The important new results concerning the α-γ phase transitions in $Ce_{1-x}Th_x$ alloys under pressure can be summarized. For a given x, the temperature of the transition increases linearly with pressure. For a given x, larger than x_c, the divergence of dR/dT decreases with increasing pressure, causing broader peaks. The critical exponents, where they could be extracted, were always less than the mean field value of 3 under any applied pressure. Finally, the shift of the transition temperature with pressure (dT_c/dP) decreases with increasing x until it levels off near the critical concentration, x_c. These results point out that x and P cannot be taken to be thermodynamically equivalent and that the transition does not appear to be a mean field type for $P > 0$.

In this paper we have not attempted to support or disprove any of the theories presented in Section II. With the availability of new results, such as ours, we hope that the theories and their assumptions can be reexamined. The deviation of our results from the mean-field behavior does suggest that the true order parameter, the volume, might be measured rather than the resistance as in our work.

ACKNOWLEDGMENTS

We would like to thank J.M. Lawrence, E.A. Kmetko and J.M. Kincaid for stimulating discussions. We are grateful to V.O. Struebing for much of the sample preparation and to R.J. Laskowski for aid in the data acquisition and reduction.

REFERENCES

1. D.K. Wohlleben and B.R. Coles in Magnetism, Ed. H. Suhl (Academic Press, New York, 1973) Vol. V, p.3.
2. C.M. Varma, Rev. Mod. Physics 48 219 (1976).
3. See papers in "Valence Instabilities and Related Narrow-Band Phenomena," Ed. R.D. Parks (Plenum Press, 1977).
4. A.W. Lawson and T.-Y. Tang, Phys. Rev. 76 301 (1949).
5. A.F. Schuch and J.H. Sturdivant, J. Chem. Phys. 18, 145 (1950).
6. W.H. Zachariasen, unpublished information quoted in Ref. 4.
7. L. Pauling, unpublished information quoted in Ref. 5.
8. K.A. Gschneidner, Jr., and R. Smoluchowski, J. Less-Common Metals 5 374 (1963).
9. R.I. Beecroft and C.A. Swenson, J. Phys. Chem. Solids 15, 234 (1960).
10. E. Poyatovskii, Dokl. Akad. Nauk SSSR 120 1021 (1958). [Soviet Phys. Doklady 3 498 (1958).]
11. A. Jayaraman, Phys. Rev. 137 A179 (1965).
12. K.A. Gschneidner, Jr., R.O. Elliott, and R.R. McDonald, J. Phys. Chem. Solids 23 555, 1191, 1201 (1962).
13. J.M. Lawrence, M.C. Croft and R.D. Parks, Phys. Rev. Letters 35 289 (1975).
14. P. Heller, Rep. Progr. Phys. 30 31 (1967).
15. K. Gschneidner, Ref. 3, p. 89: D.C. Koskimaki and K.A. Gschneidner, Jr., Phys. Rev. B11 4463 (1975).

16. R. Ramirez and L.M. Falicov, Phys. Rev. B3, 1225 (1971).

17. B. Alascio, A. Lopez, and C.F.E. Olmedo, J. Phys. F3, 1324
     (1973); C.F.E. Olmedo, A. Lopez, and B. Alascio, Solid State
     Commun. 12 1239 (1973).

18. B. Coqblin and A. Blandin, Adv. Phys. 17 281 (1968): B. Coqblin,
     J. Phys. Suppl. 32 C1-599 (1971): B. Coqblin, A.K. Bhattacharjee,
     J.R. Iglesias-Sicardi, in Ref. 3, p. 365.

19. E.A. Kmetko, Bull. Am. Phys. Soc. 22 445 (1977) H.H. Hill and
     E.A. Kmetko, J. Phys. F5, 1119 (1975).

20. B. Johansson, Phil. Mag. 30 469 (1974): J. Phys. F7, 877 (1977).

21. For example, see N.F. Mott, Phil. Mag. 30 403 (1974).

22. L.L. Hirst, J. Phys. Chem. Solids 35, 1285 (1974); Phys. Kondens.
     Mater. 11 255 (1970); AIP Conf. Proc. No. 24 11 (1975); Phys.
     Rev. B15 1 (1977); Ref. 3, p.3.

23. A.S. Edelstein, Phys. Rev. Letters. 20 1348 (1968); Solid State
     Commun. 8 1849 (1970).

24. J.M. Kincaid and G. Stell, J. Chem. Phys. 67 420 (1977).

25. C.W. Chu, this conference.

QUESTIONS AND COMMENTS

D. Debray: Have you done any optical absorption measurements
                 on this sample?

C.Y. Huang: We didn't do that. I don't know who has done that.

C. Vettier: You mentioned that you could make the transition from
                 pressure to concentration.

C.Y. Huang: The Rochester group did that.

V.H. Schmidt: In your exponent, delta, what is your ordering field
                 and your order parameter?

C.Y. Huang: Well, that's something related to the previous
                 question. Ordinarily in the equation of state, the
                 order parameter is volume. For some reason the
                 Rochester group was able to show a linear relation-
                 ship between volume and resistivity. The origin
                 for this linear relationship is not clear to us.
                 Our data seem to show that the relationship is
                 torn apart. In fact we have just started the
                 work to measure the volume change directly,
                 rather than resistivity change, with temperature
                 under pressures.

I.L. Spain: I just wanted to make a comment about the labeling
                 of the transitions. Back to the slides; they show

those discontinuities in resistance, and so on. It
is true to say that the first two slides that you
showed were first order transitions. By analogy with,
say, a gas/liquid transition there comes a point
which you get in a critical transition, and you
called it a second order transition. And then,
further on when you went to greater fraction of
thorium you also called it a second order transition. This is not correct usage of the terminology.
Perhaps continuous transition is more appropriate.

C.Y. Huang: It is certainly a continuous transition at a pressure above the critical value. However, I don't
know whether it is fair to say that the continuous
transition is not second-order.

THE PRESSURE DEPENDENCE OF THE MAGNETIC SUSCEPTIBILITY OF RARE-

EARTH SUBSTITUTED SAMARIUM SULFIDE

C. Cordero-Montalvo[*], K. Vedam and L.N. Mulay[†]

Materials Research Laboratory, The Pennsylvanvia State

University, University Park, PA 16802

ABSTRACT

Magnetic susceptibility (χ) has been measured for SmS alloyed with GdS, LaS and YS as a function of pressure (P) and of composition (x) of the dopant at 295 K. A sharp decrease in χ similar to that of SmS when going into the Intermediate Valent (I.V.) state is obtained. χ of $Sm_{0.85}Gd_{0.15}S$ which retains the high pressure phase metastably has been measured at 1 atm for several temperatures below ambient. χ deviates from the simple additivity law for SmS and dopant when (x) is varied. This has been attributed to several mechanisms, among which conduction electron enhancement of Sm-Sm exchange interaction and an increase in valence due to alloying are the most important. A gradual increase in the valence of Sm with pressure is also observed. It is concluded that the amount of the Gd ion has no role in the transition to the I.V. state resulting from either pressure or alloying and that these two parameters are not equivalent for these alloys.

INTRODUCTION

The magnetic properties of many solids may be described by a localized model in which a magnetic moment is assigned to certain ions in the solid.[1] For rare earth compounds, this moment arises from the unpaired electrons occupying the partially filled 4f shell

---

[*]Now at Colegio Universitario de Cayey, Universidad de Puerto Rico, Cayey, Puerto Rico 00633

[†]Address inquiries to this author.

of the rare earth ion, which is well localized in the interior of
the latter and is shielded by the outer shells.[1,2] A rare earth
ion may exhibit different valences depending on the integral occupation of its 4f shell. Each valence state will thus be expected to
have different magnetic properties.[2]

Sm is divalent in SmS, which crystallizes in the NaCℓ structure
with a lattice constant[3] of 5.99Å. A semiconductor to metal transition without change of crystal structure was discovered in this compound by the application of external pressure.[4] This is accompanied
by a dramatic change in color from black to golden yellow,[5] and a
large decrease ($\sim$10%) in volume. This was attributed to the promotion of a 4f electron to the conduction band.[3] The discontinuous
nature of this transition as opposed to the continuous character of
that for the similar compounds SmSe and SmTe, as well as its low
transition-pressure of 6.5 kbar and its electronic nature have attracted the attention of workers in the field of semiconductor to
metal transition. The Sm ion is believed to change at the transition from divalent state to trivalent state, which is a magnetic
state.[6] However, Maple and Wohlleben[7] measured the magnetic susceptibility under pressure and computed a valence intermediate be-
bween Sm^{2+} and Sm^{3+} at room temperature with an absence of ordering
at low temperatures. The existence of a nonmagnetic state in a concentrated rare-earth system and its easy accessibility at low pressures called the attention of workers in the field of local moment
formation in dilute alloys.[8] Based on the theories for the local
moment problem, Maple and Wohlleben[9] proposed the existence of an
intermediate valent state due to a dynamical fluctuation between the
Sm^{2+} and Sm^{3+} configurations.* It was found later that the semiconductor-metal transition can be induced by alloying with other
rare earth ions, and the intermediate valence of the metallic state
so obtained was shown by magnetic susceptibility and lattice parameter measurements with temperature.[3,11] Temperature was also found
to induce similar changes between valence states.[3,11]

Although earlier workers[11] studied the effects of doping SmS
with nonmagnetic ions such as La^{3+} and Y^{3+}, magnetic studies involving a paramagnetic ion such as Gd^{3} remained to be done both as
a function of concentration and of pressure. These studies were of
particular interest in order to find out whether the magnetic moment
of the Gd ion plays a role in the induced transition. Magnetic susceptibility is expected to be a very useful probe in this regard
since it yields clearly different values of the ionic moment corres-

---

*Another interesting area of mixed valent systems, but without dy-
namical fluctuations is found in the oxides of titanium (Ti_nO_{2n-1})
which contain Ti^{3+} and Ti^{4+} ions; these have been extensively studied by Mulay and co-workers.[10]

ponding to different integral valences and allow comparisons to be made of the results obtained for magnetic and nonmagnetic dopants.[2] These studies were also expected to aid in elucidating the relationship between pressure and alloying, which were initially thought to be equivalent in their effects.[3]   A further interest of high pressure magnetic studies lies in the expectation that a transition from the nonmagnetic intermediate valent state to a magnetic integral valent state will be achieved at high enough pressures.[8,12]

In view of the above studies, the present work was undertaken to further elucidate the semiconductor-to-metal transitions in SmS induced by either pressure or by alloying with selected dopants.   In this paper we discuss results on the pressure dependence of the magnetic susceptibility of SmS alloyed with different concentrations (x) of not only non-magnetic dopants such as La^{3+} and Y^{3+} but also with a magnetic dopant, Gd^{3+}.

EXPERIMENTAL ASPECTS AND RESULTS

The $Sm_{1-x}La_xS$ (x = 0, 0.20, 0.30, 0.35, 0.40, 0.50 and 1), $Sm_{1-x}Y_xS$ (x = 0, 0.15, 0.17, 0.20, 0.21, 0.23, 0.25, 0.31, 0.35 and 0.75), and the $Sm_{1-x}Gd_xS$ (x = 0.13, 0.15, 0.16, 0.18, 0.20, 0.23, 0.30, 0.40, 0.50 and 1.0) used in this study were kindly given to us by Dr. A. Jayaraman of the Bell Telephone Laboratories, Murray Hill, NJ.   A Faraday magnetometer and a pressure cell were constructed for this study essentially according to the design of Wohlleben and Maple.[13]   The magnetometer was adapted for easy use at room temperature and could also be used for measurements at a single pressure at lower temperatures obtained with refrigerant baths.   The samples, originally in the form of single crystals, were broken into small granules in order to fit them in a small teflon cup which was filled with a mixture of n-pentane and isoamyl alcohol and placed inside the pressure cell.

The results for the susceptibility per mole (χ_M) of SmS at 295 K for the nonmagnetic $Sm_{1-x}La_xS$ and $Sm_{1-x}Y_xS$ systems are shown in Figs. 1 and 2.   In order to obtain this curve the Pauli paramagnetic susceptibility of the fraction x of LaS was subtracted, that is, χ for LaS was considered to be independent of concentration. Selected values of Tao and Holtzberg[11] are also included in these figures.   For the La doped system the susceptibility is enhanced above the value expected for undoped SmS (x = 0) below 25 at % La and is reduced below this value (x = 0) thereafter.   The reduction is observed to occur gradually and after 40 at % La the rate of decrease becomes small and approximately constant; thus the susceptibility reaches a value which is intermediate between that of undoped divalent and collapsed SmS, and the system is in an intermediate valent state, as shown by Tao and Holtzberg.[11]

The results for doping with Y also show an enhancement of χ

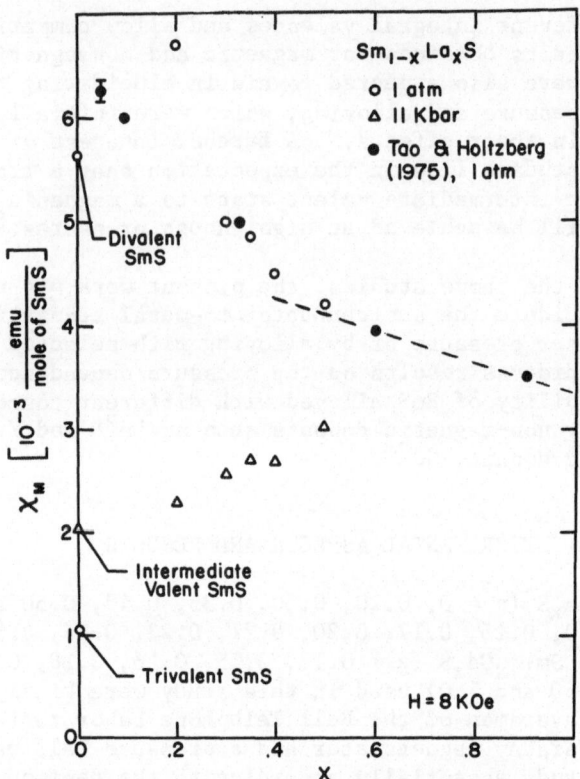

FIG. 1 Magnetic susceptibility χ_M per mole of SmS vs composition x
for $Sm_{1-x}La_xS$ at 295 K both at 1 atm and at 11 kbar. The broken line
is a guide to Tao and Holtzberg's results above x ~0.40 in the collapsed state.

below 20 at % Y and a reduced χ thereafter for the intermediate valent state. In contrast to La doping, the transition from one region
to the other is sharp at a critical concentration, x_{cr}. However, our
value of x_{cr} ~20% is higher than that observed by Tao and Holtzberg,
which we attribute to a difference in the stated composition of the
alloy.

The pressure dependence of χ in units of the χ_M of SmS for several concentrations of La and Y are shown in Figs. 3 to 8. It is observed that the initial value of χ decreases for increasing x as expected from the effect of the dilution of the Sm ions, since χ for
LaS and YS is small. The slope $d\chi/dP$ is initially negative, in contrast to the positive slope obtained for SmS by Maple and Wohlleben[7]
and corroborated by us. For samples below 30 at % La or 20 at % Y a
sharp decrease in χ is obtained with increasing pressure, as for SmS,
at a transition pressure P_{tr}. Upon reduction of pressure these sam-

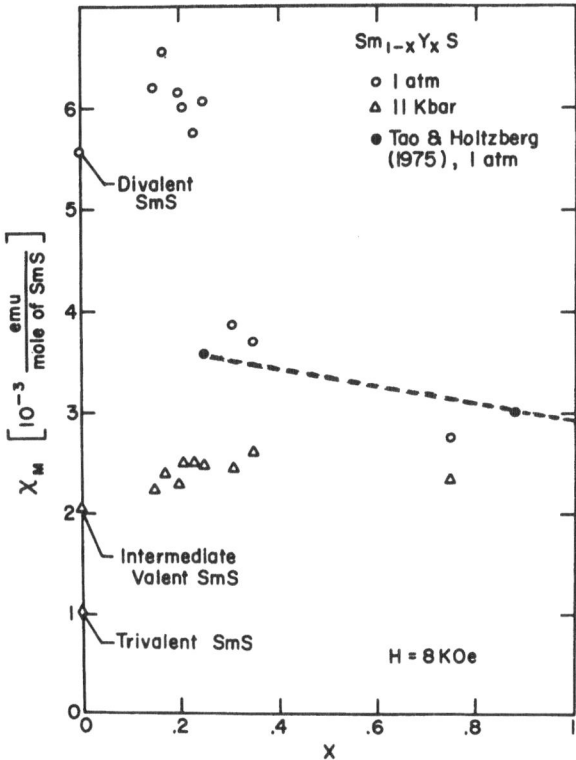

FIG. 2 Magnetic susceptibility χ_M per mole of SmS vs composition x
for $Sm_{1-x}Y_xS$ at 295 K both at 1 atm and at 11 kbar. The broken line
is a guide to Tao and Holtzberg's results above x~0.25 in the collapsed state.

ples exhibit a hysteresis and a sharp reverse transition in the region 1 to 2 kbar. The hysteresis becomes smaller and P_{tr} decreases
with increasing x, in agreement with Pohl's lattice contraction[14]
and reflectivity results. Above 30 at % La or 20 at % Y only a gradual decrease in χ is observed, which becomes less for x → 1. Figures 1 and 2 also summarize the results for χ vs composition x at
the fixed pressure of 11 kbar for these systems. The susceptibility
χ is observed to increase starting from the collapsed undoped SmS
value towards the 1 atm collapsed value due to alloying. This suggests that alloying is effectively acting in opposition to pressure
under these conditions.

The results for the molar susceptibility (χ_M) for the formula
unit vs x for the SmS system, doped with the paramagnetic Gd^{3+} are
shown in Fig. 9. This graph directly shows the effect of alloying
as x increases from the SmS value to that of GdS which is higher.

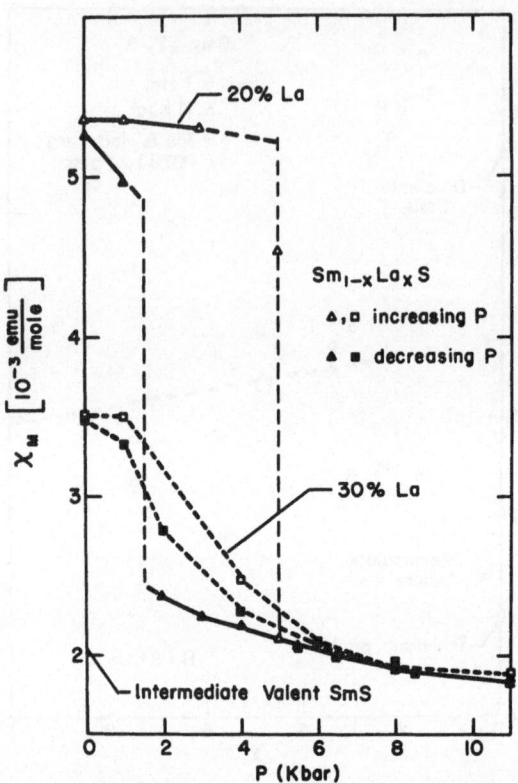

FIG. 3 Pressure dependence of the molar magnetic susceptibility,
χ_M of two samples in the system $Sm_{1-x}La_xS$ at 295°K.

The broken straight lines show the expected value of χ if a simple
additivity law were followed. However, in contrast to the La and Y
cases, the magnetism of the Gd ion now has to be taken into account.[3] The figure clearly shows the sharp transition induced by
alloying in this system at 16 at % Gd.[3,15] Figures 10 and 11 show
the results for the dependence of χ on pressure. For samples below
15 at % Gd, χ has an initially negative slope as a function of P, as
obtained for the La and Y alloys and in contrast to undoped SmS.
The susceptibility for these samples also shows a sharp decrease with
increasing pressure and a small dependence on P thereafter. A hysteresis is obtained with decreasing P which, in contrast to the La and Y
systems, does not become smaller and is retained metastably at 1 atm.
The initially dark granules are observed to have a golden yellow color
when taken out of pressure cell. For $x > x_{cr}$ a decrease in χ is obtained with pressure, which becomes smaller as $x \to 1$, as in the La
and Y systems.

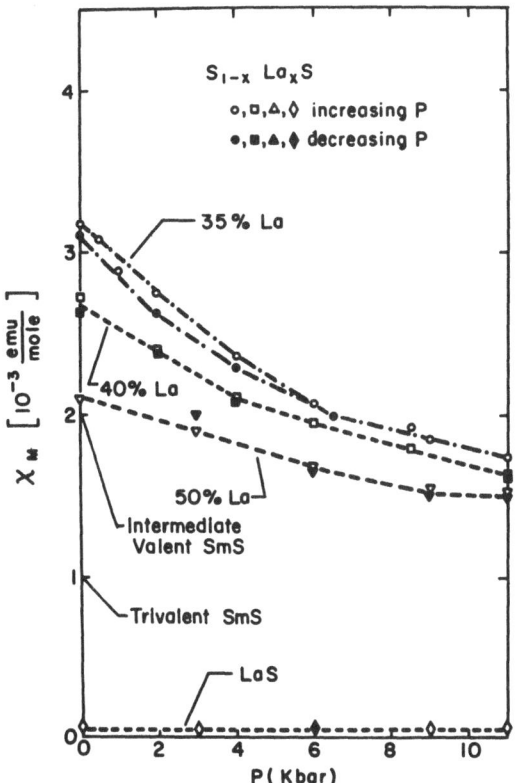

FIG. 4 Pressure dependence of the molar magnetic susceptibility χ_M at 295°K of three samples in the system $Sm_{1-x}La_xS$ compared with that of LaS.

Figure 12 summarizes the results of χ vs x at the fixed pressure of 11 Kbar. χ increases with x again due to the increased dominance of the Gd moment but the values are depressed below the 1 atm values due to the pressure induced transition. However, the values increasingly deviate from the straight dotted line for the additivity of susceptibility from collapsed undoped SmS towards the 1 atm values of the alloys. These trends are similar to the ones observed for La and Y doping.

DISCUSSION

The Sm ion in SmS is divalent and has a J = 0 ground state[16] as given by Hund's rules and its susceptibility at a temperature T can be obtained from Van Vleck's expression, which is dependent on the spin-orbit splitting Δ between the J = 0 and J = 1 states. Birgeneau et al.[17] have found that χ of SmS is enhanced at T = 0 K due to the

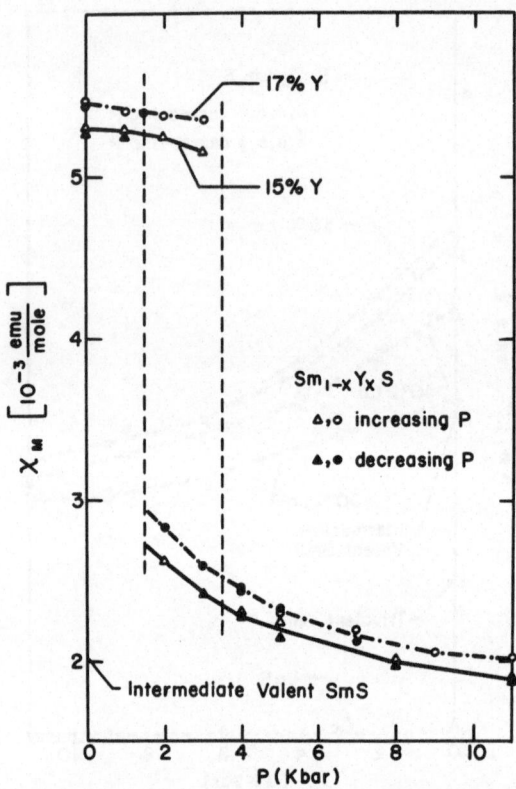

FIG. 5 Pressure dependence of the molar magnetic susceptibility χ_M
at 295 K for two samples in the system $Sm_{1-x}Y_xS$.

reduction of Δ by the presence of Sm-Sm interactions, and the pressure dependence of χ obtained by Maple and Wohlleben[7] can be explained
in terms of the dependence of this exchange on the cation-cation distance. The effect of La doping has been considered by several workers as giving rise to a number of mechanisms. First of all, since
La^{3+} is nonmagnetic its substitution for Sm^{2+} in SmS effectively reduces the total number of Sm ions contributing to the susceptibility
of the sample. In addition, dilution reduces the number of Sm ions
available for the Sm-Sm exchange interactions.[18] Further effects
must be taken into account due to the smaller size of the La^{3+} ion
(the lattice constant $a_o = 5.85$Å for LaS) relative to that of Sm^{2+}
($a_o = 5.97$Å for Sm^{2+}S), although higher than that expected for Sm^{3+}
($a_o = 5.62$Å for Sm^{3+}S). This smaller size induces a positive lattice pressure on the Sm ions which both enhances the Sm-Sm exchange
interaction, which increases χ, and induces the conversion of Sm
ions to the intermediate valent state, which reduces χ towards its
value for collapsed undoped SmS. A further and very important mech-

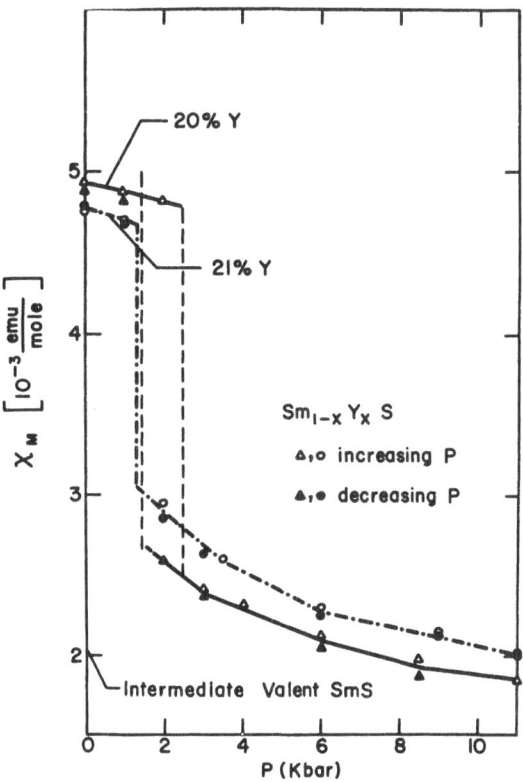

FIG. 6 Pressure dependence of the molar magnetic susceptibility χ_M at 295 K of two samples in the system $Sm_{1-x}Y_xS$.

anism is the enhancement of the Sm-Sm exchange interactions by the mediation of the conduction electrons added by La, also acting to enhance χ.

Assuming these interactions and their effect on Δ to be valid with the same strength at room temperature we calculated χ as explained above and obtained a value, for the sample with 20 at % La, of 5.5×10^{-3} emu/mole, which is lower than the observed value of 6.8×10^{-3} emu/mole. Similar low results were obtained for χ in the enhanced region. The discrepancy may be attributed to the presence of magnetic impurities unaccounted for and to the assumption that exactly the same mechanisms are operative at 295 K as at T = 0 K. It should be noted that high values of χ for undoped SmS have been observed by Tao and Holtzberg[11] and attributed to deviations from stoichiometry and different Sm-Sm exchange.

In the intermediate valent state χ should be given approximately by the relation

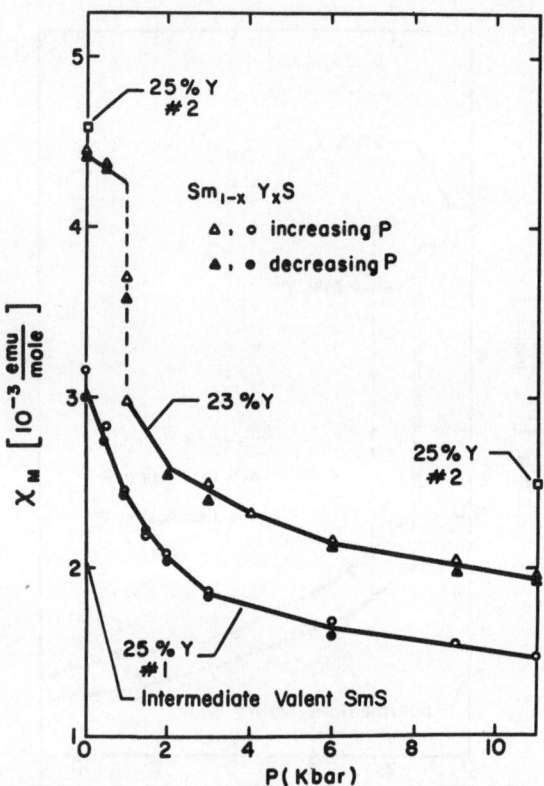

FIG. 7 Pressure dependence of the molar magnetic susceptibility χ_M at 295 K of three samples in the system $Sm_{1-x}Y_xS$. Two of the samples have the same composition (25 at % Y) but different susceptibilities at 1 atm and at 11 kbar presumably due to deviations from stoichiometry.

$$\chi = (1-\varepsilon)\chi_{Sm^{2+}S} + \varepsilon\chi_{Sm^{3+}S},$$

where $(2+\varepsilon)$ is the valence, according to the model proposed by Maple and Wohlleben.[9] Using the measured susceptibility for the alloys and the value of $\chi_{Sm^{2+}}$ obtained by means of the calculation explained above (but now for the range $x > 0.40$) and the value expected for $\chi_{Sm^{3+}S}$ (~1 x 10^3 emu/mole) a value of ε ~2.4 is obtained. This value is lower than that obtained (2.5 → 2.7 as χ → 1) from lattice of the intermediate valence susceptibility relation, since room temperature lies in a region of transition between two limiting regions in which moment-fluctuations of the Sm ion in the intermediate valent state are dominated by different mechanisms. These are respectively the interactions with the conduction electrons at low

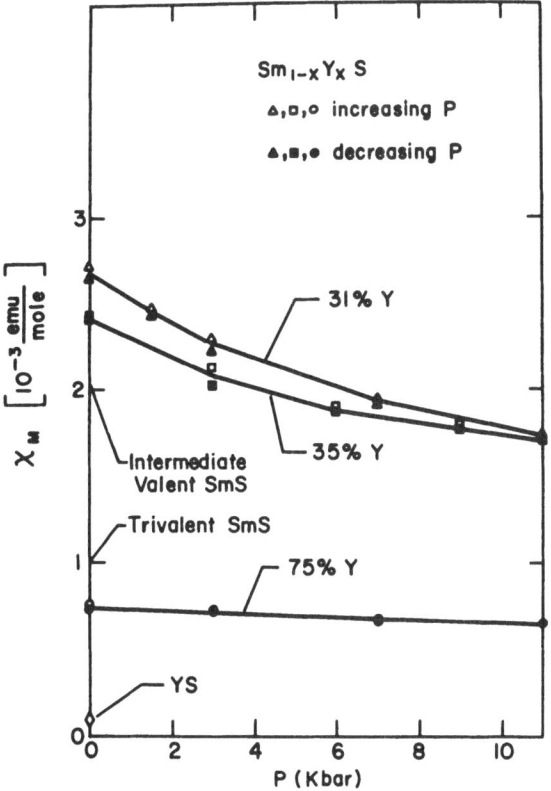

FIG. 8 Pressure dependence of the molar magnetic susceptibility χ_M at 295 K of three samples in the system $Sm_{1-x}Y_xS$ compared to the susceptibility of YS at 1 atm (quoted by Steiger and Carter[26]).

temperatures and thermal fluctuations at high temperatures; in the latter range the above relation should be accurate.[9,20] Similar results are also expected for the nonmagnetic Y alloys, except that the smaller size of Y^{3+} (a_0 = 5.49Å for YS),[15] induces a sharp transition to the intermediate valent state as compared to the gradual transition induced by La.[12]

The Gd doped samples will also be expected to behave likewise (a_0 = 5.55Å for GdS)[3] if the magnetism of the Gd moment can be taken into account and does not take part in the transition. The total susceptibility of the Gd doped samples was calculated by adding the susceptibilities of the fraction x of GdS in the solution and for the (1-x) fraction of SmS. Since GdS is an antiferromagnet at low temperatures[11] its susceptibility is given by a Curie-Weiss expression $C/(T-\theta)$. For GdS this expression was appropriately modified by considering the effects of the concentration (x) to account for the expected Curie-Weiss behavior in the dilute limit as $x \rightarrow 0$. Again we obtained calculated values which lie below the measured values, for

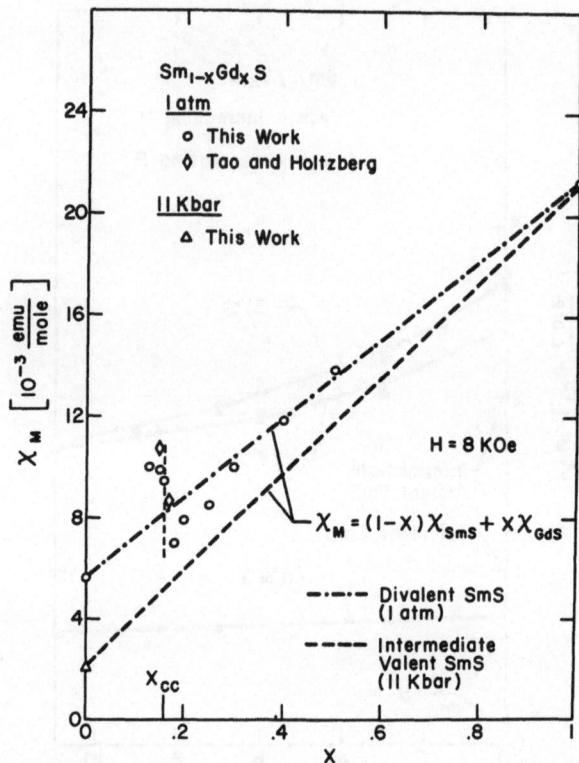

FIG. 9 Molar magnetic susceptibility χ_M vs composition (x) for
$Sm_{1-x}Gd_xS$ at 295°K at 1 atm. The vertical dotted line at x_{cc} shows
the critical concentration (~16 at % Gd) for the collapsed state.

x > 16 at % Gd as for La doping, and we attribute this to the same
reasons as before. In the intermediate valent state the results obtained are similar to the La case.

The initial negative slope $d\chi/dp$ obtained below P_{tr} for samples
below x_{cr} in the three systems is attributed to the increasing valence of the Sm ions due to pressure. This effect is also present
in SmS but is overshadowed by the dependence of the Sm-Sm exchange
interaction on distance. However, the similarities in the samples
studied with regard to chemical effects makes the valence effect
dominant. The transition at P_{tr} is of first order as shown by the
hysteresis and as discussed by earlier workers.[3,12] It is attributed,
as in the case of undoped SmS, to the promotion of a localized 4f
electron from the Sm ion to the 5d conduction band.[6] The latter is
split by crystal field effects; pressure increases such splitting,
which in turn lowers the bottom of the conduction band relative to

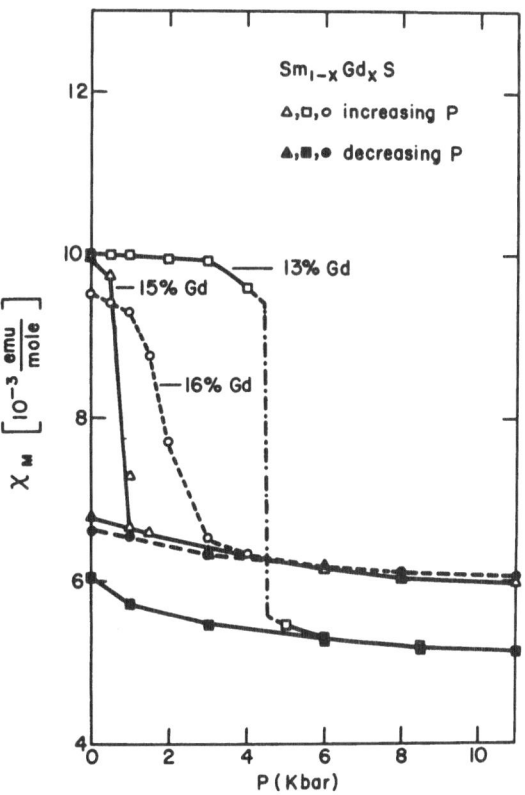

FIG. 10 Pressure dependence of the molar magnetic susceptibility χ_M at 295 K of three samples in the system $Sm_{1-x}Gd_xS$. The samples with 16 at % Gd shows a spread out transition which may be due to strains in the sample granules.

the 4f level, thus reducing the 4f-5d gap and giving rise to the delocalization.[21] The increasingly small dependence of χ on P above P_{tr} observed in every case is due to the increasing rigidity of the lattice. A decelerating mechanism for this transition is suggested by Hirst[22] and by Penney and Melcher.[23]

The softened transition for the 30 at % La and the 35 at % La samples is consistent with Pohl's suggestion[14] of a critical concentration of x = 0.27 % above which pressure induced transitions become second order. As shown in Fig. 10 for a black sample with 15 at % Gd we measured an initial value of χ of 9.9 x 10^{-3} emu/mole at 1 atm. This value decreased to 6.95 x 10^{-3} emu/mole on the application of 11 kbar, and after releasing the pressure the collapsed phase with a golden yellow color could be retained metastably with a value for χ of 6.8 x 10^{-3} emu/mole. Upon cooling this sample below

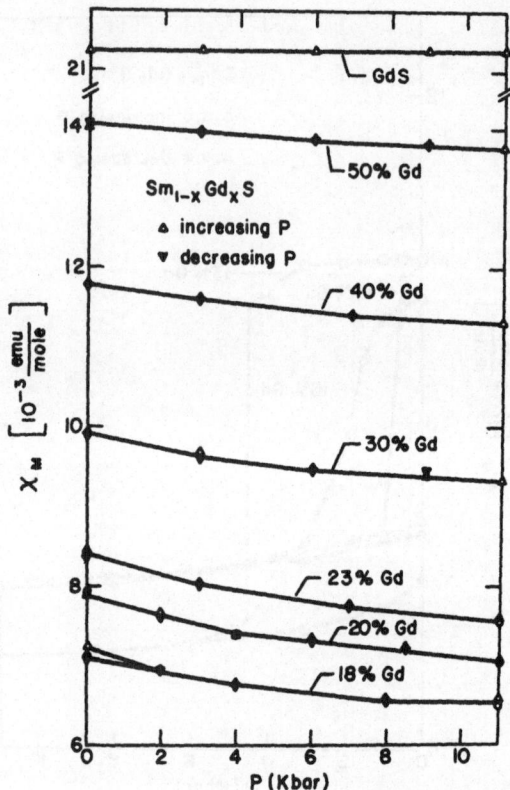

FIG. 11 Pressure dependence of the molar magnetic susceptibility χ_M at 295 K of six samples in the $Sm_{1-x}Gd_xS$ system compared to that obtained for GdS.

ambient, χ increased to 6.82×10^{-3} emu/mole at 273°K, 9.6×10^{-3} emu/mole at 195°K and 34.7×10^{-3} emu/mole at 77°K. After letting it warm up to the ambient temperature, the sample was black and had a value of χ as 10.02×10^{-3} emu/mole, so that the sample had reverted back to the stable semiconducting state. The mechanisms involved in the pressure hysteresis, and especially in the metastability observed in the Gd alloys, have not yet been elucidated but are expected to be found in the balance between the lattice and electronic energies.[12]

The otherwise overall similarity of the behavior of the susceptibility of the Sm ion in the La and Y doped systems compared with that of the Gd doped system, both as a function of concentration and of pressure is evidence of the absence of a discernible role for the paramagnetic Gd ion in the pressure or alloying induced transition.

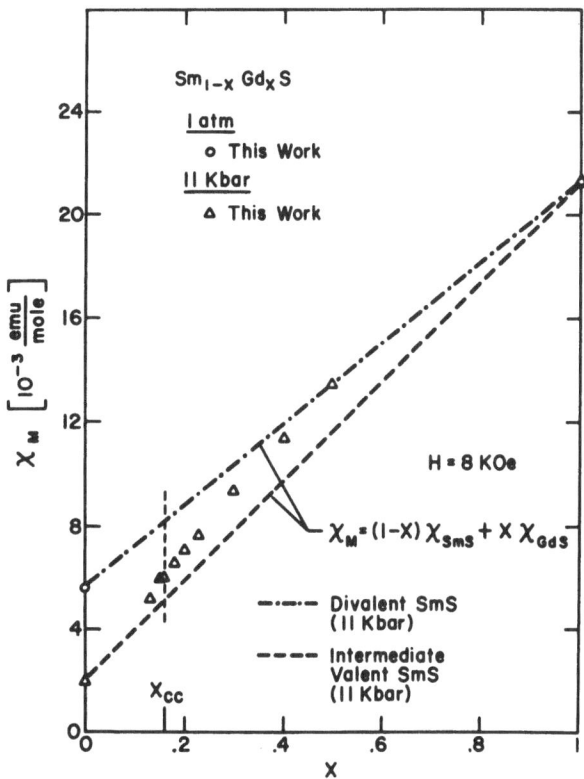

FIG. 12 Molar magnetic susceptibility χ_M vs composition (x) for $Sm_{1-x}Gd_xS$ at 295 K at 11 kbar. The vertical dotted line at x_{cc} shows the critical concentration (~16 at % Gd) for the collapsed state.

The same conclusion was arrived at by Hedman et al.[24] who measured the susceptibility as a function of temperature for several Gd samples near x_{cr}. It is of interest to mention, however, that such a role was deduced indirectly by Gronau and Methfessel,[25] who aligned the 4f spins of Gd in a magnetic field, and thus reduced the energy of its 5d states. This improved their hybridization with the 4f-5d states and shifted downwards the transition temperature between semiconducting and metallic states for a sample $Sm_{0.79}Gd_{0.21}S$.

The χ vs x values at 11 kbar for the three systems show a definite trend to increase as x increases. These systems may be looked at as starting with undoped collapsed SmS and gradually substituting the intermediate valent Sm ions by the trivalent rare earth ions. It has been shown previously that the I.V. state due to alloying has a lower valence than that exhibited by SmS.[21] The effect of alloying

in increasing the value of the intermediate valence susceptibility
is thus correlated with a decrease in valence. This in turn illustrates the nonequivalence of alloying and pressure since alloying
is shown to reduce the effect achieved by pressure. It is interesting to note that the electronic structure of the dopant plays a
significant role in the transition.[3] The relationship between pressure and alloying involves several mechanisms. The effect of pressure is to contract the lattice as the ions move closer to each
other, and in doing so the crystal field splitting of the 5d band of
SmS increases and the 5d bands overlap.[21] The bottom of the band is
thus lowered to the 4f level and the 4f-5d gap is reduced, facilitat-
ing the delocalization of a 4f electron. Alloying as discussed
earlier, also has a lattice contraction effect but, in addition, it
adds a partially occupied conduction band to the system. The addition of conduction electrons as x increases gradually fills the conduction band which moves the Fermi level away from the 4f level,[12]
increasing the energy needed by the electron to delocalize. The
position of the bottom of the conduction band and its density of
states may also need to be considered,[23] as well as the possible
formation of clusters[22] and deviations from stoichiometry of the
resulting ternary system. We expect that the net effect of alloying will be given by the balance between the lattice contraction and
the addition of the d band of the dopant. Since both collapsed SmS
and the rare earth sulfides have high bulk moduli, the effect of the
lattice contraction is small and that due to the d-band dominates at
11 kbar as x increases.

CONCLUSION

The systems $Sm_{1-x}La_xS$, $Sm_{1-x}Y_xS$ and $Sm_{1-x}Gd_xS$ show changes in
magnetic behavior at 295 K under pressure indicating a transition
to the intermediate valent state which is essentially similar to
that of undoped SmS. However, the proximity of some of the samples
to the chemically induced collapse manifests itself in a negative
initial slope $d\chi/dP$ with the increase in valence arising from pressure becoming more important than the Sm-Sm exchange enhancement
stemming from lattice contraction obtained for SmS.[7] Although the
Gd doped samples with concentrations x = 0.13, 0.15 and 0.16 show
metastability upon the release of pressure, a comparison of the
χ vs x as well as χ vs P results for these and other concentrations
of Gd doped alloys with those for the nonmagnetic La and Y alloys
showed no difference that could be attributed to the moment of Gd
playing a discernible role in the transition. While the magnetism
of the dopant does not appear to play a significant role in semiconductor to metal transitions in these alloys, their electronic configurations do play an important role; thus pressure and alloying
are not equivalent and can even act in opposite directions.

ACKNOWLEDGMENT

One of us (C.C.) would like to acknowledge a scholarship granted by the Economic Development Administration of Puerto Rico during part of this work. We are grateful to Professor Maple of the University of California at San Diego for helpful advice regarding the construction of the high pressure cell. We thank Dr. Jayaraman of the Bell Telephone Laboratories, Murray Hill, NJ, for the samples.

REFERENCES

1. D.H. Martin, Magnetism in Solids. Iliffe Books, Ltd., London (1967).
2. J.H. Van Vleck, The Theory of Electric and Magnetic Susceptibilities. Oxford University Press, London (1932).
3. A. Jayaraman, P. Dernier and L.D. Longinotti, Phys. Rev. (B) 11, 2763 (1975); High Temp. High Pressure 7,(1975).
4. A. Jayaraman, V. Narayanamurti, E. Bucher and R. G. Maines, Phys. Rev. Lett. 25, 1430 (1970).
5. J.L. Kirk, K. Vedam, V. Nayanaramurti, A. Jayaraman and E. Bucher, Phys. Rev. B 6, 3023 (1972).
6. E. Bucher, V. Narayanamurti and A. Jayaraman, J. Appl. Phys. 42, 1741 (1971).
7. M.B. Maple and D. Wohlleben, Phys. Rev. Lett. 27, 511 (1971).
8. D. Wohlleben and B.R. Coles, in Magnetism, edited by H. Suhl, Vol. 5, p. 3. Academic Press, New York (1973).
9. M. B. Maple and D. Wohlleben, Am. Inst. Phys. Conf. Proc., edited by Wolfe, 18, 447 (1973).
10. L.N. Mulay and W.J. Danley, J. Appl. Phys. 41, 877 (1970); Mat. Res. Bull. 7, 739 (1972); L. N. Mulay and J.F. Houlihan, phys. stat. solidi (b) 61, 647 (1974); Mat. Res. Bull. 11, 307 (1976), etc.
11. L.J. Tao and F. Holtzberg, Phys. Rev. B 11, 3842 (1975).
12. J.A. Wilson, in Structure and Bonding. Springer-Verlag, Berlin (1976).
13. D. Wohlleben and M.B. Maple, Rev. Sc. Instr. 42, 1573 (1971).
14. D.W. Pohl, Phys. Rev. B 15, 3855 (1977).
15. C.M. Varma, Rev. Mod. Phys. 48, 219 (1976).
16. J.B. Torrance, F. Holtzberg and T.R. McGuire, Am. Inst. Phys. Conf. Proc., edited by Wolfe 10, 1279 (1973).
17. R.J. Birgeneau, E. Bucher, L.W. Rupp, Jr. and W.M. Walsh, Jr., Phys. Rev. B 5, 3412 (1972).
18. F. Mehran, J.B. Torrance and F. Holtzberg, Phys. Rev. B 8, 1268 (1973).
19. F. Holtzberg, Am. Inst. Phys. Conf. Proc., edited by Wolfe 18, 478 (1974).
20. B.C. Sales, J.L. Temp. Phys. 28, 107 (1977).

21. B. Battlog, E. Kaldis, A. Schlegel and Y. Wachter, Phys. Rev.
     B 14, 5503 (1976).
22. L.L. Hirst, J. Phys. Chem. Solids 35, 1285 (1976).
23. T. Penney and R.L. Melcher, J. Physique Colloq. 37, C4-273
     (1977).
24. L. Hedman, B. Johansson and K.V. Rao, Physica 86-88B,221
     (1977).
25. M. Gronau and S. Methfessel, Physica 86-88B, 218 (1977).
26. R.A. Steiger and E. Cater, High Temp. Sci. 7, 204 (1975).

QUESTIONS AND COMMENTS

D. Bloch: When you dissolve the system, you have simultan-
                 eously local effects and overall effects. Is
                 that true for Ce-Th? Can you distinguish directly
                 between localized effects and homogeneous effects?

C. Cordero: As far as I have seen, the local effects are evi-
                 dent in the gadolinium-doped samarium sulfide al-
                 loys for very low concentrations of gadolinium.
                 As suggested by ESR g-shift measurements both at
                 atmospheric and higher pressures (Birgeneau,
                 et al., 1972; Walsh, et al., 1974) this is a dif-
                 ferent situation for which, up to about one per-
                 cent of gadolinium, the additional electron con-
                 tributed by the gadolinium ion remains localized
                 around this ion. The susceptibility results appear
                 to be due to the whole sample, that is, as a bulk
                 property. Earlier workers have also reported that
                 gadolinium-doped samarium sulfide already shows
                 antiferromagnetic order for a seven percent
                 gadolinium concentration.

PRESSURE DEPENDENCE OF CRYSTAL FIELD PARAMETERS IN RARE EARTH

INTERMETALLIC COMPOUNDS

H.R. Ott

Laboratorium für Festkörperphysik

ETH-Hönggerberg, 8093 Zurich, Switzerland

ABSTRACT

Pressure derivatives of crystal field parameters at zero pressure can be derived from measurements of the low temperature specific heat and thermal expansion of compounds exhibiting CEF-effects in their thermal properties. We demonstrate this for the singlet ground state systems PrSb and TmSb and for the two-level systems SmSb and CeTe. The results are in disagreement with expectations from a simple effective point charge model. They demonstrate the importance of other contributions to the crystal electric field than that expected from the charges on the ligand sites.

Crystal electric field (CEF) effects in metals and alloys have recently become of increasing interest[1,2] . Almost exclusively substances containing rare earth ions have been investigated. The local character of the 4f electrons of the rare earth ions offers the possibility that an ionic approach may successfully describe the behaviour found in these materials. Although this is not a priori to be expected in metallic substances many experimental results could be explained in this way. In compounds of rare earths with the Vth and VIth column of the periodic system (pnictides and chalcogenides) the rare earth ions usually adopt the 3+ configuration. The Hunds rule ground state of the 4f electrons is then split by a crystal electric field with the symmetry of the crystal lattice. This splitting mainly determines the physical properties of these materials below room temperature and measurements of magnetic, thermal, and transport properties give evidence for the existence of crystal electric fields in metals and alloys. For the rare earth monopnictide series it has been shown[3] that the CEF parameters follow the predictions of an effective-point-charge model rather closely but it is still not quite clear what contributions determine the actual crystal field in such materials. Additional information on the nature of crystal electric fields can be obtained by investigating its change under external influences as e.g. pressure induced volume change. The most straightforward way of doing this is by means of inelastic neutron scattering experiments where the variation of CEF excitation energies with increasing pressure can be measured directly. Another possibility is given by means of measurements of the thermal expansion from which the pressure or volume dependence of the energy level splitting at zero pressure may be derived[4,5]. In this work we describe such measurements on relatively simple systems, namely a) PrSb and TmSb (singlet ground state systems) and b) SmSb and CeTe (two level systems).

If we consider a system of non interacting rare earth ions, the free energy density F is given by

$$F = - kTN \Sigma \exp. - (E_{\Gamma_n^i}/kT) \qquad (1)$$

where N is the number of rare earth ions per cm^3 and Γ_n^i is the representation index for CEF levels. In our case it is sufficient to consider the CEF split levels of the Hunds rule ground state multiplet since we are only interested in low temperature properties. The specific heat C and the volume thermal expansion β are given as derivatives of F

$$C = - T \frac{\partial^2 F}{\partial T^2} = \frac{N}{kT^2} [<E^2> - <E>^2] \qquad (2)$$

$$\beta = -\varkappa \frac{\partial^2 F}{\partial V \partial T} = \frac{\varkappa N}{kT^2} [<E^2\gamma> - <E\gamma><E>] \qquad (3)$$

$\varkappa$ is the compressibility and the statistical averages over the energy levels are defined as

$$< X > = \frac{\Sigma_i X_i e^{-E_i/kT}}{\Sigma_i e^{-E_i/kT}}$$

The CEF Grüneisenparameter γ introduced in eq. (3) is defined as

$$\gamma_{CEF}^i = - \partial \ln E_i/\partial \ln V \qquad (4)$$

and describes the volume dependence of E_i with respect to the ground state level. In the simple case of a two level system the

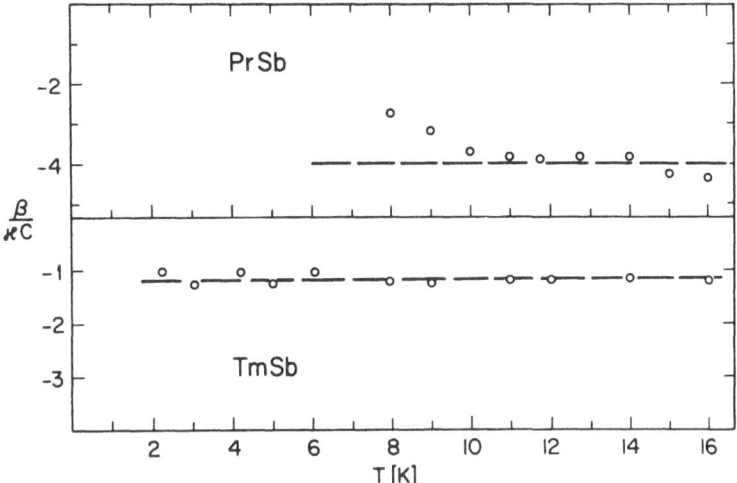

Figure 1.   a) β/(𝝒C) for PrSb
            b) β/(𝝒C) for TmSb
            only CEF contributions to the thermal expansion and
            the specific heat are considered.

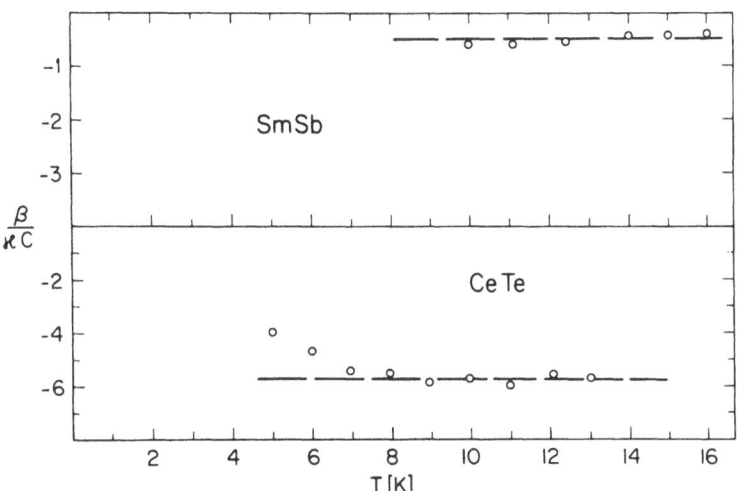

Figure 2.   a) β/(𝝒C) for SmSb
            b) β/(𝝒C) for CeTe
            for CeTe, no background corrections have been taken
            into account.

CEF Grüneisenparameter reduces to

$$\gamma_{CEF} = \frac{\beta}{C} \frac{1}{\varkappa} \tag{5}$$

Note that C is the specific heat per unit volume. The same relation obviously holds when in a multilevel system the volume dependence of all the levels is the same.

The γ_{CEF} are related to the magnetoelastic coupling constants entering the Hamiltonian describing such systems. If we consider only the part involving volume strain coordinates this Hamiltonian can be written[6]

$$\varkappa(C_B) = - G_{10} N\varepsilon_v - G_{11} \sum_i \varepsilon_v O_4 - G_{12} \sum_i \varepsilon_v O_{6_i} \tag{6}$$

The G_{ii} are the magnetoelastic coupling constants and O_4 and O_6 are the cubic CEF equivalent operators. These magnetoelastic coupling constants can also be written as

$G_{11} = - \frac{\partial B_4}{\partial \varepsilon_v}$ and $G_{12} = - \frac{\partial B_6}{\partial \varepsilon_v}$, where $\varepsilon_v = \varepsilon_{xx} + \varepsilon_{yy} + \varepsilon_{zz}$ and B_4 and

B_6 are the 4th and 6th order CEF parameters respectively. γ_{CEF}^i is now given by

$$\gamma_{CEF}^i = \frac{G_{11}(\mu_i - \mu_o) + G_{12}(\nu_i - \nu_o)}{E_i - E_o} \tag{7}$$

where $\mu_i = \langle \Gamma_i | O_4 | \Gamma_i \rangle$ and $\nu_i = \langle \Gamma_i | O_6 | \Gamma_i \rangle$.

We first consider the singlet ground state systems PrSb and TmSb. Both compounds have a Γ_1 ground state and are nearly ideal Van Vleck paramagnets. Therefore there are no complications due to magnetic or structural phase transitions at low temperatures. In Figs. 1a and 1b we show the temperature dependence of the ratio $\beta/(\varkappa C)$ as derived from experiment. For both β and C only the magnetic parts (CEF contribution) are considered. In PrSb, the splitting between the ground state Γ_1 and the first excited state Γ_4 is 73 K[7]. Therefore only $\gamma_{CEF}(\Gamma_4)$ may be determined from our experiment (see table I). In TmSb the three lowest levels $\Gamma_1 - \Gamma_4$ (25 K) $- \Gamma_5^2$ (56 K)[8] have to be considered in our temperature range. From our measurements we find $\gamma_{CEF}(\Gamma_4) = \gamma_{CEF}(\Gamma_5^2)$.

In Fig. 2a and 2b we show $\beta/(\varkappa C)$ for SmSb and CeTe. In these compounds the rare earth ions adopt a J = 5/2 ground state and the cubic crystal electric field splits this ground state into a Γ_7 doublet and a Γ_8 quartet. Hence only one Grüneisenparameter has to be considered. The CEF effect is rather weak in SmSb but much

TABLE I. CEF-Grüneisenparameters as derived in this work

	γ_{CEF}
PrSb	-4^1
TmSb	-1.2^1
SmSb	-0.5
CeTe	-5.8

1) γ_{CEF} for the two lowest levels

stronger in CeTe. For both compounds the variation of γ at the lowest temperatures have to be attributed to the phase transitions occurring at approximately 2.1 K.

As a first result we note that all the CEF anomalies in the thermal expansion coefficients of these materials obviously appear to be negative (negative γ's). This leads to the conclusion that the energy splitting of the CEF levels decreases with decreasing volume, just opposite to what one would expect from a simple point charge model. The explicit volume dependence of the fourth order CEF parameter B_4 in the point charge model approximation leads to $\gamma_{CEF} = + 5/3$. The experimental values are between -0.5 (SmSb) and -5.8 (CeTe). They show that a microscopic description of the origin of crystal electric fields in metals has to take into account also contributions other than those stemming from electron charges on the ligand site ions.

In the case of PrSb and TmSb our results can be used for the interpretation of the pressure dependence of the magnetic susceptibility χ. At low temperatures these singlet ground state systems show the expected Van Vleck-type susceptibility given by

$$\chi = 2g_J^2 \, \mu_B^2 \, \frac{<\Gamma_4|J_z|\Gamma_1>}{E_4 - E_1} \tag{8}$$

where E_1 and E_4 are the ground state (Γ_1) and the first excited state (Γ_4) respectively $(E_1 = 0)$. $<\Gamma_4|J_z|\Gamma_1>$ is the dipole matrix element between these states. Since $\gamma_{CEF} = - \partial \ln E_4 / \partial \ln V$, the volume dependence of χ in this case is given by $\partial \ln \chi / \partial \ln V = \gamma_{CEF}$. As may be seen from Table 1 the values for $\partial \ln \chi / \partial p$ as calculated from the appropriate Grüneisenparameters account very well for the observed values in real pressure experiments[9]. Thus the volume dependence of χ and probably also that of the Knight shift are mainly determined by CEF effects and the volume dependence of the exchange contribution seems to be of minor importance in these materials.

Very recently, inelastic neutron scattering techniques have been used to measure the excitation energies directly as a function of pressure in $PrSb^{10}$. From these experiments a value of γ_{CEF} (Γ_4) = $-11/3$ has been deduced, in very good agreement with our value from thermal expansion measurements.

ACKNOWLEDGEMENTS

It is a pleasure to acknowledge the generous supply of single crystal specimens from E. Bucher, D.B. McWhan, L.D. Longinotti and F. Hulliger. Moreover the author is grateful to B. Lüthi and K. Andres for numerous discussions concerning several aspects of this work. Financial support of the Schweizerische National-fonds zur Förderung der Wissenschaft is also gratefully acknowl-edged.

REFERENCES

1. Proc. 1st Int. Conf. on CEF effects in metals and alloys, Montreal 1974.
2. Proc. 2nd Int. Conf. on CEF effects in metals and alloys, Zürich (Plenum Press, New York 1977).
3. R.J. Birgeneau, E. Bucher, J.P. Maita, L. Passell and K.C. Turberfield, Phys. Rev. B8, 5345 (1973).
4. H.R. Ott and B. Lüthi, Phys. Rev. Letters 36, 600 (1976).
5. H.R. Ott and B. Lüthi, to appear in Z. Physik B.
6. E.R. Callen and H.B. Callen, Phys. Rev. 129, 578 (1963).
7. K.C. Turberfield, L. Passell, R.J. Birgeneau and E. Bucher, Phys. Rev. Letters 25, 757 (1970).
8. R.J. Birgeneau, E. Bucher, L. Passell and K.C. Turberfield, Phys. Rev. B4, 718 (1971).
9. R.P. Guertin, J.E. Crow, L.D. Longinotti, E. Bucher, L. Kupferberg and S. Foner, Phys. Rev. B12, 1005 (1975).
10. D.B. McWhan, this conference.

QUESTIONS AND COMMENTS

T.F. Smith: A couple of points I wanted to ask. First of all, you said at the beginning that the point charge model failed to explain your thermal expansion, and in fact you get completely the opposite sign. Can you explain this? We heard this morning that the point charge model does work.

H.R. Ott: What I think now is that one has d-electron con-
                   tributions to the crystal field. I mean, these
                   measurements should also help to find out what the
                   origin of the crystal field is, in these compounds.
                   Up to now, one has assumed that it's just the ligand
                   ions which produce a crystal field and act on the
                   rare earth ions. But, it could also be that, as Hirst
                   pointed out several years ago, that in virtual bound
                   states like d-electrons which are around the 4f-elec-
                   tron of the 4f-ion, could also give a quite appreci-
                   able contribution to the crystal field. The pressure
                   dependence of this contribution can have the opposite
                   sign, and they add up to a pressure dependence which
                   is probably not so easily predicted by simple argu-
                   ments leading to the lattice constant dependence of
                   the crystal field parameters, of a^{-5} and a^{-7}.

T.F. Smith: I just would like to make the comment that you can
                   get away with a very simple crystal field model for
                   Fe^{2+} ions in zinc sulfide. There, you go through
                   exactly the same sort of analysis of the experimental
                   data, but you get contributions to the thermal ex-
                   pansion which are exactly what you expect your crys-
                   tal field contribution to be.

H.R. Ott: Well it's a different case here, because here the
                   crystal field splitting delta goes in proportion.
                   In the case you're talking about, it goes inverse.
                   That makes the sign incorrect.

T.F. Smith: It makes the sign difference, but it's still cor-
                   rect. There again we're talking about insulators.

H.R. Ott: Right.

L.N. Mulay: Is it correct to assume that there is magnetic or-
                   dering at very low temperatures in all of the com-
                   pounds that you discussed?

H.R. Ott: No, I would say it's only in the cerium telluride
                   and the samarium antimonide. Those which have a
                   ground state doublet order at 2K, and the others
                   do not. What they might do is to show nuclear
                   ordering induced by the high Van Vleck suscepti-
                   bility and the high internal field due to the
                   electrons.

L.N. Mulay: Towards the end, you made a comment that the effect
                   of pressure on the exchange interaction J is very
                   small. What is the magnitude of the exchange in-

teraction? How difficult is it to measure the small
pressure dependence?

H.R. Ott: No, I mean the pressure dependence might still be
                rather high, but if the exchange is small, you have
                this prefactor, and that's what can make it small.
                I cannot say the pressure dependence is small, but
                I can say it's unimportant to describe the pressure
                dependence of the susceptibility.

J. Wittig: How does the thermal expansion look for lanthanum
                antimony?

H.R. Ott: It's positive, and more or less normal. It's
                Grüneisen constant, from the specific heat, is of
                the order of one to two. That's more or less nor-
                mal.

PRESSURE INDUCED SPIN-GLASS→SINGLE-IMPURITY TRANSITION IN LaCe,

SUPERCONDUCTIVITY IN LaCe AND YCe UNDER PRESSURE

F. Zimmer and J.S. Schilling*

Institut für Experimentalphysik IV, Universität Bochum

463 Bochum, Germany

The results of high pressure electrical resistivity measurements on classical spin-glass systems such as AuFe, CuMn, and AuMn are briefly reviewed. It is shown how such studies give information about the nature of the interactions leading to the well known spin-glass freezing phenomena at a temperature T_O. LaCe, at dilute Ce concentrations a Kondo system, shows typical spin-glass behavior in the resistivity above 1K for 10-40% Ce. Upon application of moderate pressures (10-20 kbar) this system undergoes an apparent spin-glass→single-impurity transition as predicted by Larsen and Doniach. Whereas the superconducting transition temperature T_c is less than in pure La or Ce at all pressures, for $Y_{.6}Ce_{.4}$ T_c=1.6K for 50 kbar and T_c=2.7K for 110 kbar, values which are higher at those pressures than in either Y or Ce alone.

SPIN-GLASSES UNDER HIGH PRESSURE

In recent years the application of high pressure has become an increasingly useful tool in probing the magnetic properties of solids.[1] Using this technique one now attempts to explore the appropriateness of the band or local moment picture for a given system, as well as to extract the volume dependence of the basic interactions leading to magnetism. Such studies are of obvious importance to the general problem of the development of a microscopic theory of magnetism in solids. Perhaps one of the more spectacular effects of the application of high pressure is the apparent

*Research supported in part by the Deutsche Forschungsgemeinschaft

destruction of magnetism observed, for example, both in the weak
itinerant ferromagnet $ZrZn_2$[2] and in the local moment systems LaCe
and pure Ce.[3] Our attention in this paper will be concentrated
on LaCe in a concentration range where spin-glass like behavior
is observed. Under pressure, LaCe is rapidly pushed toward single-
impurity like behavior, a transition of the type predicted by
Larsen[4] and Doniach.[5]

Before embarking on a discussion of the properties of LaCe
and YCe under pressure, it would be well to first briefly review
the principle results of previous pressure studies of classical
spin-glass systems such as AuFe, AuMn, and CuMn. Spin-glasses,
which have been defined[6] as random, metallic, magnetic systems
characterized by a random freezing of the moments at a temperature T_O without long range order, are presently the objects of
extensive experimental[6] and theoretical[7] activity. Perhaps the
most characteristic physical property of such systems is the
presence of a sharp cusp in the a.c. susceptibility at the temperature T_O, as shown schematically in Fig. 1. The exact nature
of the interaction mechanisms responsible for this cusp is not yet
clear, but is expected to be basically either a cooperative freezing
of spins interacting via the indirect RKKY-interaction or a blocking
of magnetic clusters by an anisotropy field which could originate,
for example, from spin-spin dipolar interactions.[6] The RKKY-interaction between two spins at a separation R_{12} is given by:

$$\mathcal{J}_{RKKY} = 6\pi z J^2 n \left(\frac{\cos 2k_f R_{12}}{(2k_f R_{12})^3} - \frac{\sin 2k_f R_{12}}{(2k_f R_{12})^4} \right) , \qquad (1)$$

where z is the number of electrons per atom, k_f the Fermi wavevector, n the density of states at the Fermi energy, and J the
effective itinerant-electron to local-spin exchange parameter.
The dipolar anisotropy interaction is given by:

$$\mathcal{J}_{DA} = \frac{\overline{m}_1 \cdot \overline{m}_2 - 3(\hat{r}_{12} \cdot \overline{m}_1)(\hat{r}_{12} \cdot \overline{m}_2)}{(R_{12})^3} , \qquad (2)$$

where m is the magnetic moment and $\hat{r}_{12} = \overline{R}_{12}/R_{12}$. The spin-glass
freezing temperature T_O is determined by a sum over the interac-
tions between all pairs of spins. From Eqs. 1 and 2 it can be seen
that both RKKY- and dipolar interactions are proportional to $1/V$
for dilute systems which leads in a certain approximation to the
so-called scaling law $T_O \propto c$.[8] Important deviations from this law
have been pointed out by Larsen.[9] On the basis of the concentration dependence of T_O alone, therefore, it is not possible to
determine which of the above spin freezing mechanisms is appropriate.
Pressure experiments are, however, useful for this purpose due to

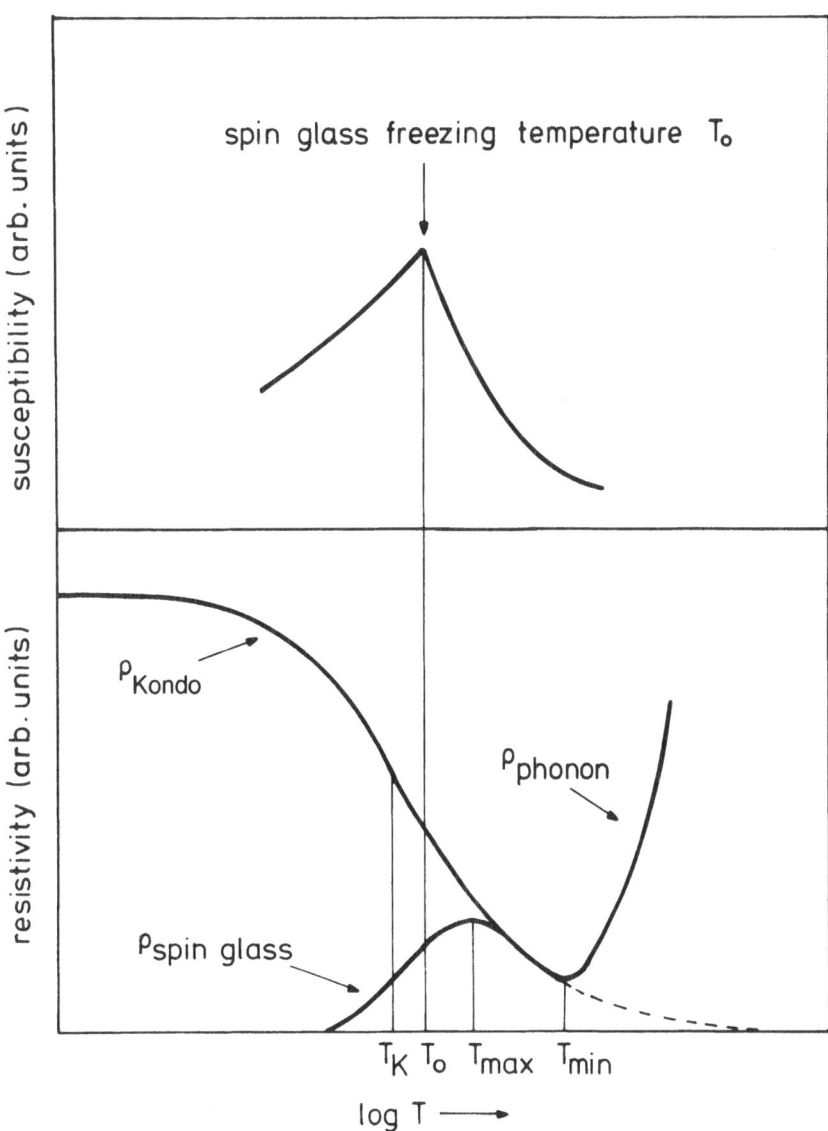

Figure 1. a.c. magnetic susceptibility of a spin-glass and Kondo and spin-glass resistivities versus log T.

the differing pressure dependences of $\mathcal{J}_{RKKY}$ and $\mathcal{J}_{DA}$. In Eq. 1, since $k_f r_{12}$ is not volume dependent within the free electron model ($k_f \propto V^{-1/3}$ and $R_{12} \propto V^{+1/3}$), the terms in the brackets are pressure independent and thus $d\ln T_o/d\ln V = d\ln(nJ^2)/d\ln V$. For noble metal host systems like $\underline{Au}Fe$ and $\underline{Cu}Mn$, n has only a relatively small volume dependence,[10],[11] so that $d\ln T_o/d\ln V \simeq 2d\ln|J|/d\ln V$. The volume dependence of J is thus needed. On the other hand, for the dipolar interaction $\mathcal{J}_{DA}$, the only pressure dependent term in Eq. 2 for cubic systems is $R_{12}^{-3} \propto V^{-1}$ so that if the spin blocking model is correct, then $d\ln T_o/d\ln V = -1$. The problem that remains, therefore, is to determine the intrinsic volume dependence of the effective exchange parameter J. Fortunately, most known spin-glass systems are Kondo systems at low dilution so that the volume dependence of J is "exponentially amplified" into the volume dependence of the Kondo temperature T_K where $T_K \propto e^{-1/n|J|}$. The pressure dependence of T_K has been determined for a variety of very dilute systems and one finds universally $dT_K/dP > 0$.[10-12] For noble metal host systems with Mn impurities one finds $d\ln|J|/d\ln V \simeq -2.5$. If the RKKY-cooperative spin freezing model is appropriate in spin-glasses then one expects $d\ln T_o/d\ln V \approx -5$. This volume dependence of T_o is significantly different from that of the spin blocking model where one expects $d\ln T_o/d\ln V = -1$.

High pressure susceptibility measurements of the volume dependence of T_o would be useful to help resolve the above questions. Such investigations are underway at the present time in this laboratory. Another way to gain information about volume dependences of interactions in spin-glasses is by high pressure resistivity studies. As is indicated in Fig. 1, the interactions leading to the spin-glass freezing phenomena at T_o cause the rising Kondo resistivity to fall off at lower temperatures, creating a resistivity maximum at a temperature T_{max}. T_{max} is therefore a function of both T_o and the Kondo temperature T_K. The pressure dependence of T_{max} has been investigated for a number of systems[10],[12],[13] and both the sign and magnitude of dT_{max}/dP have been found to vary markedly from one system to another. It was, however, established that systems with relatively high Kondo temperatures T_K show $dT_{max}/dP < 0$. These and other aspects of the pressure and concentration dependence of T_{max} have been successfully accounted for by a recent theory of Larsen.[4],[9],[14] In this theory an analytical expression was derived expressing T_{max} as a function of T_K and the r.m.s. interaction strength Δ_c. In Fig. 2, T_{max}, T_K, and Δ_c are plotted as a function of $|J|$. Recalling that universally $|J|$ increases with pressure in Kondo systems (direction of arrows), the complex pressure dependence of T_{max} becomes understandable. For low T_K systems such as $\underline{Ag}Mn$[15] and $\underline{Au}Mn$, $|J|$ is small and T_{max} should increase with pressure, as observed. As $|J|$ increases in magnitude, either by increasing the pressure or changing the system, dT_{max}/dP decreases, finally becoming negative as observed for $\underline{Au}Fe$, $\underline{Mo}Fe$, and $\underline{Cu}Cr$.[13] Since $\Delta_c \propto J^2$ and $T_K \propto e^{-1/n|J|}$, for small

$|J|$ one has $\Delta_c > T_K$ and one is in the interaction dominated spin-glass regime. With increasing $|J|$, T_K increases rapidly, eventually catching up with and surpassing Δ_c. When $T_K > \Delta_c$ one is in the single-impurity Kondo regime. The curve in Fig. 2 was drawn for a single impurity concentration; with increasing concentration the T_{max} "mound" is stretched to the right, pushing the phase boundary to larger values of $|J|$. The length of the arrows in Fig. 2 correspond to the change in $|J|$ for approximately 100 kbar applied pressure. We see that only the very compressible system LaCe can be pushed by pressure from well below to well above the phase boundary at $\Delta_c = T_K$. Because of the relatively high T_K value of LaCe (~.1K), one would also expect T_{max} to decrease rapidly with pressure. As will be seen, the measurements confirm these expectations.

The pressure dependence of the average interaction strength $\Delta_c(P)$ can now be derived from $T_{max}(P)$ and $T_K(P)$ using Larsen's formula.[4] For Mn in noble metal hosts one finds that $d\ln\Delta_c/d\ln V \simeq -5$ to -8, depending on the system and the initial value of T_K assumed.[10,12] These values agree reasonably well with the value -5 expected if Δ_c is determined by the RKKY-interaction but are far larger in magnitude than the value -1 expected for the dipolar interaction. It should, however, be pointed out that in the dipolar model discussed here it was tacitly assumed that the number of spins in a cluster was pressure independent. More complicated cluster-blocking models in fact assume that the cluster size increases with the RKKY-interaction,[16] precluding the possibility of distinguishing this model from the RKKY-cooperative spin freezing model by the present pressure experiments. The precise relationship between Δ_c and T_o is at present unclear. Studies of the concentration dependence of these two quantities, however, indicate that they are at least proportional over a considerable range.[9,14]

We now would like to present a graph which we feel represents a reasonable estimate of the resistivity of dilute and spin glass alloys over a wide range of temperature, T_K, and Δ_c. In Fig. 3 the resistivity per impurity is plotted versus the relative temperature $\ln T/T_K$ for various values of the parameter Δ_c/T_K. The uppermost curve is for $\Delta_c = 0$ and was taken from a calculation by Larsen.[17] The lower curves represent increasingly larger values of Δ_c. At first the inter-impurity interactions only serve to disturb the full Kondo resistivity and lower the unitarity limit value; such effects are, for example, seen by Daybell and Yeo on CuCr[18] and by Silverman and Briscoe on quenched CuFe films.[19] For larger values of Δ_c (or smaller values of T_K) a resistivity maximum appears which with increasing Δ_c/T_K becomes more symmetric about T_{max}. Such effects are seen in pressure studies on MoFe and CuCr[13] where Δ_c/T_K decreases as T_K increases rapidly with pressure. For $\Delta_c > T_K$ one is in the spin-glass regime and T_{max} is related to T_K and Δ_c according the Larsen theory.[4] In Fig. 3, the

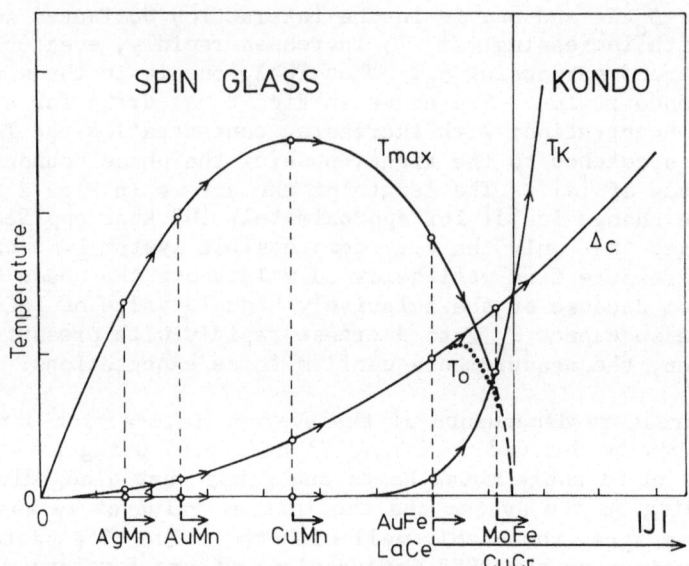

Figure 2. Expected functional dependence of T_{max}, Δ_c, T_K, and
           T_o on $|J|$ at fixed impurity concentration in both the
           spin-glass $\Delta_c > T_K$ and Kondo $\Delta_c < T_K$ regimes.
           $\Delta_c \propto J^2$, $T_K \propto e^{-1/n|J|}$, T_{max} is from Larsen,[4] and the T_o
           dependence is conjectured. Pressure can push LaCe
           over the phase boundary at $\Delta_c = T_K$. The arrows correspond
           to an increase of pressure of about 100 kbar.

shape of a curve is only a function of Δ_c/T_K. This means that an
increase of Δ_c is equivalent to a decrease of T_K and a shift on
the temperature axis, as is found in Larsen's theory.[4] Let us consider one of the curves for $\Delta_c > T_K$ with a well-defined maximum and
ask how it behaves with pressure. If T_K is not too small T_{max}
will decrease with increasing pressure as seen in Fig. 2. In addition, since Δ_c/T_K decreases with increasing pressure, from Fig.
3 we see that the low temperature (to the left of T_{max}) part of
the resistivity curve rises up leading to an asymmetrical curve
about T_{max}. At still higher pressures, the $\Delta_c = T_K$ phase boundary
is crossed and the low temperature resistivity rises up further,
the maximum disappears, and the resistivity looks very much Kondo-
like. In the next section we will compare these expectations with
the high pressure measurements on LaCe and YCe.

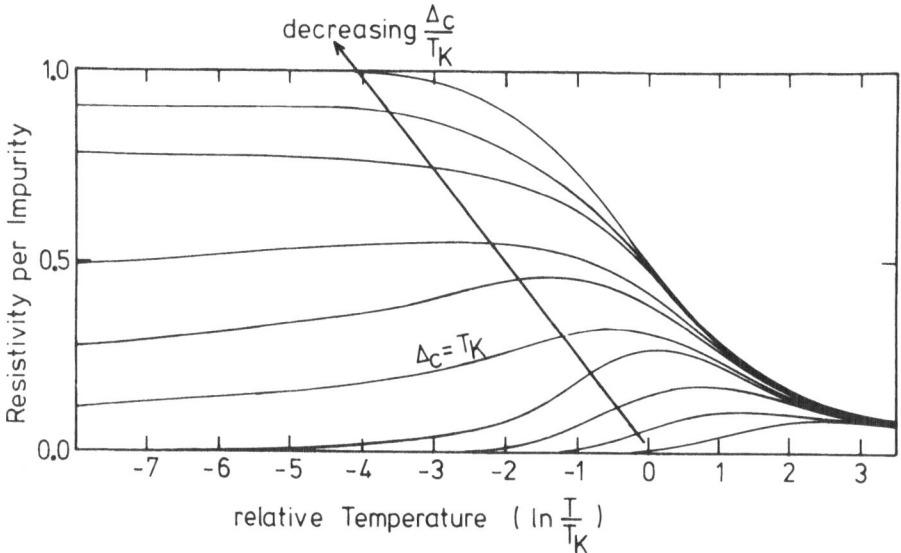

Figure 3. Proposed functional dependence of the resistivity per impurity on the relative temperature T/T_K and on the relative interaction strength Δ_c/T_K. The uppermost curve is for $\Delta_c=0$, the Kondo limit.[17] Pressure always decreases Δ_c/T_K so that the resistivity is pushed from spin-glass to Kondo-like behavior.

EXPERIMENTAL

The samples were prepared from 99.9% pure La, Y, and Ce by induction melting the components on a water-cooled copper boat. A sample was held in the molten state for a total period of 15 minutes, being rotated several times and agitated continually. The mass loss was well under 0.1%. Thin discs were cut out of the buttons and cold rolled to 50-80 micron thick foils. That LaCe samples prepared in this way are predominantly in the dhcp-phase is evidenced by the superconducting transition temperature $T_c \approx 5.2K$ of a pure La sample (for dhcp La T_c =4.9K and for fcc La T_c =6.1K). The value of T_{max} for a La-10%Ce sample was found to be the same for unannealed, fcc-annealed, or dhcp-annealed samples. The present cold-worked samples are the most appropriate kind to use in a quasihydrostatic pressure cell where a limited amount of sample deformation during the initial preloading procedure[11] is unavoidable.

High pressure measurements to 40 kbar were carried out in a preloaded[11] belt-type pressure cell.[20] For measurements to 140 kbar a Wittig-type pressure cell was used.[21] Two LaCe or YCe samples were measured simultaneously in the same pressure cell in addition to a lead manometer for the pressure determination.[11] The high pressure clamp is constructed from hardened CuBe alloy and is a design similar to that of Eichler and Wittig.[21] Up to 40 tons of force can be generated at any temperature with the clamp resting in the cryostat by successively tightening six bolts. The electrical resistivity can be measured in the temperature range 1-300K. Due to the massive CuBe clamp and well shielded leads, a resolution of 1 nanovolt is easily attained using an Amplispot galvanometric DC-amplifier connected to a Solatron LM 1490 digitalvoltmeter with 2 sec. integration time. The linearity of the amplifier is $1:10^3$ and the long time stability of the entire measuring system is about $1:10^5$ over 10 hours. Temperatures are determined with a calibrated Germanium resistor.

RESULTS OF EXPERIMENT

Spin-Scattering Resistivity

The results of the high pressure resistivity measurements on the alloys La-3%, 10%, 20%, and 40% Ce are shown in Fig. 4. For P=0 the 3%Ce Data show the typical Kondo logT behavior whereas at higher Ce concentrations a resistivity maximum appears which shifts with increasing concentration to higher temperatures. The maximum for the 40%Ce data is buried in the phonon scattering which rises rapidly above about 8K. If the resistivity of pure La is subtracted off from the data a resistivity maximum appears at about 8K. The present values of T_{max} for P=0 lie about 10% higher than those found by Edelstein on evaporated LaCe films;[22] presumably the large amount of defect scattering in such quenched alloys damps the RKKY-interactions and lowers T_{max}. The increase of the slope of the resistivity for the La-3%Ce alloy has been extensively studied and is due to an increase of the Kondo temperature T_K with pressure.[23] One should imagine that the uppermost ($\Delta_c=0$) curve in Fig. 3 is shifting through our temperature window, from 1 to 8K, as pressure increases. When T_K is within our temperature window, the slope is a maximum. The resistivity maximum at 26 kbar is not due to spin-spin interactions but is the onset of superconductivity. Such an effect is expected in Kondo systems by the theory of Müller-Hartmann et al.[24] where the pair-breaking parameter falls off for temperatures below T_K, allowing superconductivity to reappear.[3] The same effect is also seen in the other higher concentration systems. The resistivity maximum for the La-10%Ce alloy is seen to shift with increasing pressure to lower temperatures,

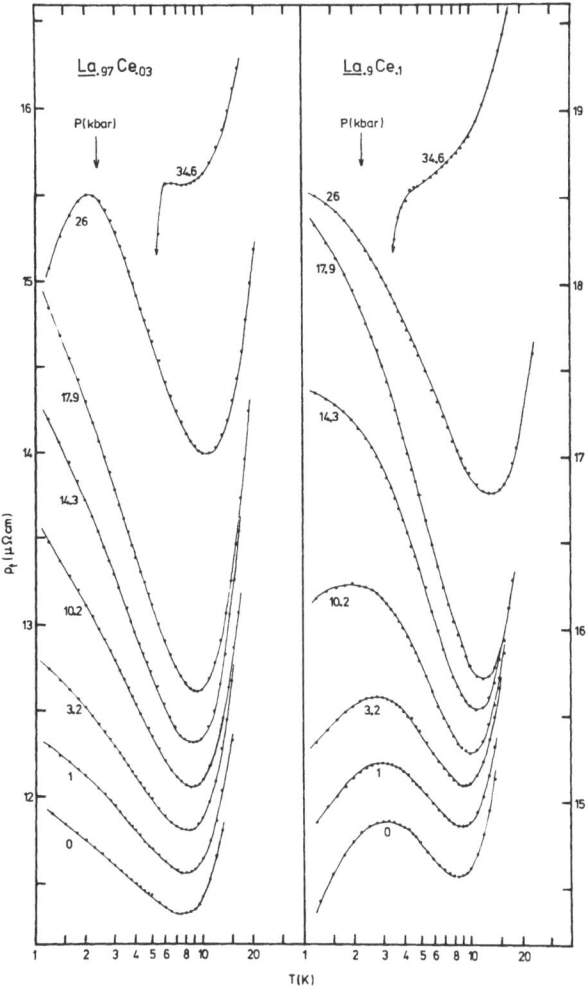

Figure 4a. Total measured electrical resistivity of various L̲aCe
            alloys versus logT at different pressures. The high
            pressure data is normalized to the absolutely measured
            P=0 data by matching slopes at low temperature. The
            resistivity scales apply to the P=0 data. The high
            pressure data is shifted arbitrarily vertically for
            clarity with the relative resistivity the same as for
            P=0. The value of the resistivity at 8K for the high
            pressure data is from low to high pressure: for La-
            3%Ce, 11.26, 11.88, 13.44, 15.42, 17.60, 23.62, and
            30.18; for La-10%Ce, 14.42, 15.63, 18.56, 22.27, 26.33,
            34.27, and 34.84 with all values in $\mu\Omega$cm. The La-3%
            Ce results are in reasonable agreement with the hydro-
            static measurements of Kim and Maple.[23]

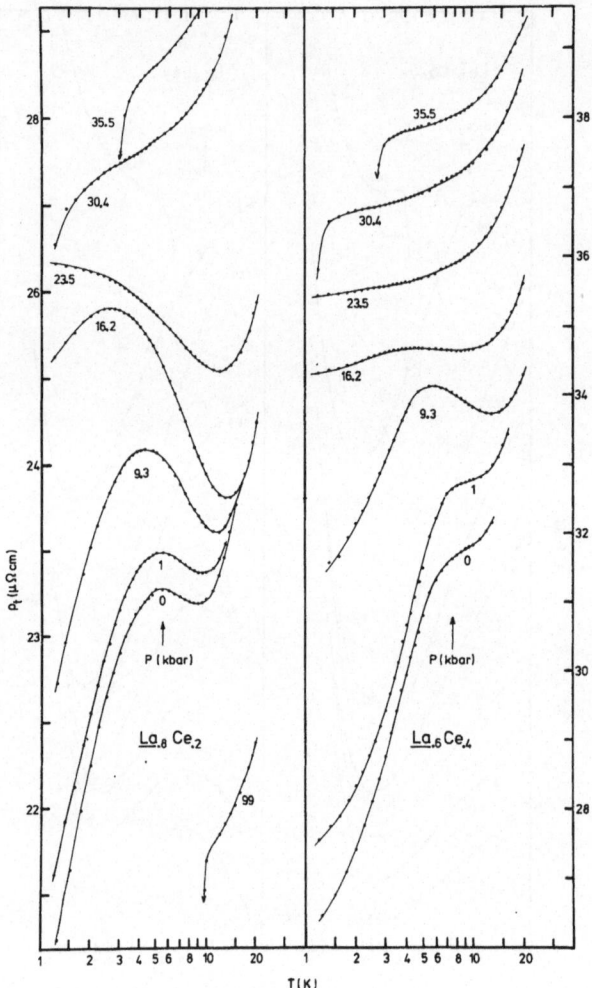

Figure 4b. Total measured electrical resistivity of various L̲a̲Ce
alloys versus logT at different pressures. The high
pressure data is normalized to the absolutely measured
P=0 data by matching slopes at low temperature. The
resistivity scales apply to the P=0 data. The high
pressure data is shifted arbitrarily vertically for
clarity with the relative resistivity the same as for
P=0. The value of the resistivity at 8K for the high
pressure data is from low to high pressures: for
La-20%Ce, 21.00, 25.45, 33.45, 39.62, 41.77, and 39.88;
for La-40%Ce, 35.72, 48.10, 55.38, 53.19, 49.64, and
50.10 with all values in μΩcm.

becoming increasingly asymmetric. At 17.9 kbar the resistivity looks
remarkably Kondo-like and at 26 kbar the slope has decreased strongly as the resistivity bends over and heads for the unitarity limit
value. This behavior is exactly what was anticipated in the last
section with Fig. 3 for the case where application of pressure induces a spin-glass to single-impurity phase transition. In addition, the fact that T_{max} behaves under pressure like a high T_K
spin-glass (AuFe, for example) is good evidence that LaCe is indeed
a spin-glass itself and will show a susceptibility cusp at a lower
temperature. Such a cusp has, in fact, been observed in (LaCe)
Al_2.[25] The behavior under pressure of the resistivity of the two
higher concentration LaCe alloys is analogous to the La-10%Ce results. In the 40%Ce data the phase transition boundary $\Delta_c = T_K$ has
been pushed to a high enough temperature that the resistivity maximum can be seen to "die out" within our temperature window. Measurements were also carried out upon unloading the pressure from 35 kbar
and some pressure hysteresis was observed. This is not surprising
since above 25 kbar a dhcp-fcc phase transformation takes place. The
superconducting transition temperatures of the two phases differ by
approximately 1.2K which corresponds to a pressure difference of
about 8 kbar. The data shown in Fig. 4, which were taken with increasing pressure, are certainly representative of the dhcp-phase
for pressures less than about 25 kbar.

The first step in an analysis of the data is the determination
of the volume dependence of T_K or $|J|$. Estimating this using the
second Born approximation result $d\rho/d\log T \propto n^2 |J|^3$ is invalid because
of the relatively large value of the Kondo temperature. A simple
and natural way to estimate dT_K/dP is to assume that the La-3%Ce
high pressure data are pieces of a universal resistivity curve and
that the spin degeneracy and potential scattering phase shift δ_V
are independent of pressure in our pressure range. LaCe is believed to have a doublet ground state which is well ($\sim$100K) separated from excited states.[3,27] Previous pressure measurements on
very dilute CuFe and AuV alloys also confirm the validity of the
above assumptions for these systems as well as the existence of a
"universal" resistivity curve.[11] To determine $T_K(P)$, therefore,
we shift the La-3%Ce data on the temperature axis until the curves
overlap. Assuming T_K at 17.9 kbar is approximately at 10K we find
that at P=0, $T_K \simeq$ 15mK. In fact, other estimates place T_K for LaCe
at temperatures well below 1K.[3,23,28] For the initial volume depen-
dence of T_K we find $d\ln T_K/d\ln V \simeq -80$. For P=0, $n|J| \simeq .065$ and initial-
ly $d\ln(n|J|)/d\ln V \simeq -6$. Between 0 and 18 kbar $n|J|$ thus approximate-
ly doubles in magnitude. Due to the presence of s- and d- bands at
the Fermi surface of La there is some uncertainty as to the appropriate magnitude and pressure dependence of the density of states
n. From measurements of the pressure dependence of the thermodynamic critical field of La, Takata and Oshida[29] estimate that

Table I. Pressure dependence of various parameters for L̲a̲Ce alloys.
Parameters are defined in text. T_{max} values are corrected
for influence of phonon resistivity. V_o is unit cell volume in pure La at P=0. Δ_c only calculated for $T_K < T_{max}$.

System	P(kbar)	T_K(K)	V/V_o	$n\lvert J\rvert$	T_{max}(K)	Δ_c(K)
La-3% Ce	0	.015	1.0000	.0646	---	---
	1	.018	.9960	.0653	---	---
	3.2	.045	.9874	.0695	---	---
	10.2	.302	.9618	.0801	---	---
	14.3	1.78	.9479	.0933	---	---
	17.9	10.0	.9364	.111	---	---
La-10% Ce	0	.024	.9941	.0666	3.15	1.052
	1	.031	.9902	.0677	2.95	1.044
	3.2	.061	.9816	.0710	2.75	1.128
	10.2	.664	.9562	.0855	2.00	1.813
La-20% Ce	0	.044	.9858	.0694	5.10	1.741
	1	.061	.9819	.0710	4.90	1.790
	9.3	1.15	.9513	.0897	4.1	3.473
	16.2	30.0	.9284	.1268	2.6	-----
La-40% Ce	0	.141	.9715	.0755	8.0	3.129
	1	.202	.9676	.0776	7.8	3.324
	9.3	6.89	.9375	.1068	5.5	-----
	16.2	407	.9150	.1894	4.15	-----

Table II. Pressure dependence of the superconducting transition
temperature T_c for L̲a̲Ce and Y̲Ce alloys.

System	P(kbar)	T_c(K)
La-3% Ce	35	4.3
La-10% Ce	35	2.1
La-20% Ce	35	1.9
La-20% Ce	99	8.5
La-40% Ce	35	2.1
Y-40% Ce	50	1.6
Y-40% Ce	86	2.7
Y-40% Ce	110	2.7

$dlnN(E_f)/dlnV \simeq dln|J|/dlnV \simeq -4.5$, so that $dlnN(E_f)|J|/dlnV \simeq -9$, which is larger than our estimate. However, whereas both s- and d- electrons are important in the superconductivity of La, the s-electrons presumably play a larger role in the Kondo resistivity. With this assumption and since the pressure dependence of the s-band density of states n is small,[11] we find $dln|J|/dlnV \simeq -6$. If we use the full pressure dependence of $N(E_f)$, then $dln|J|/dlnV \simeq -1.5$.

From the pressure dependence of $T_K(P)$ and $T_{max}(P)$ for the more concentrated LaCe systems, it is possible to estimate $\Delta_c(P)$ from the theory of Larsen.[4] The values of the parameters are given in Table I. It is important to note that the parameters had to be corrected for the fact that LaCe contracts as the Ce concentration increases. V_0 is thus the volume of the unit cell of pure La. Δ_c can be seen in all cases to increase with pressure. In the above the compressibility is taken from the data of Syassen and Holzapfel.[30]

Resistivity measurements were also carried out on the alloys Y-0.5, 10, 20, and 40% Ce under high pressures. This system has a relatively high Kondo temperature ($T_K \simeq 50K$) and thus no resistivity maximum is expected according to Fig. 3. A clear suppression of the low temperature resistivity per impurity is observed as the Ce concentration is increased, as expected. Application of pressure shifts T_K rapidly to higher temperatures[26] and reduces drastically the negative slope and curvature in the resistivity in our temperature window, as expected from Fig. 3. Since for this system $|J|$ at P=0 is already to the right of the phase boundary in Fig. 2, no phase change is possible.

Superconductivity

The effect of pressure on the superconducting transition temperature T_c of LaCe alloys up to 16% Ce has been extensively studied.[3,23,31] The present work extends these studies to higher Ce concentrations and also includes the system YCe. The results are given in Table II. For P=35 kbar, and probably at higher pressures, T_c is essentially the same for the three most concentrated alloys of LaCe. It is interesting to note that whereas adding Ce to La always decreases T_c at any pressure, for $Y_{.6}Ce_{.4}$ above 50 kbars, T_c is higher than in either Y or Ce alone.

ACKNOWLEDGEMENTS

The authors would like to thank S. Methfessel for his continued support of this research. Stimulating discussions with P.J. Ford, J.A. Mydosh, and especially U. Larsen are gratefully acknowledged. The authors are also grateful to A. Zimmer and E. Havenstein for superb technical assistance.

REFERENCES

1. For a review of the work prior to 1969 see: D. Bloch and
    A.S. Pavlovic, Ch. 2, Vol. 3, in "Advances in High Pressure
    Research", R.S. Bradley ed., Academic Press (1969).
2. T.F. Smith, J.A. Mydosh and E.P. Wohlfarth, Phys. Rev. Letters
    27, 1732 (1971).
3. See, for example, M.B. Maple and D. Wohlleben, AIP Conf. Proc.
    No. 18, Magnetism and Magnetic Materials, p. 447 (1974);
    B. Coqblin, M.B. Maple and G. Toulouse, Intern. J. Magnetism
    1, 333 (1971).
4. U. Larsen, Phys. Rev. B 14, 4356 (1976).
5. S. Doniach, Conference on Itinerant Electron Magnetism, Oxford,
    Sept. 1976.
6. For a review see: J.A. Mydosh in Amorphous Magnetism II edited
    by R.A. Levy and R. Hasegawa (Plenum Press, NY, 1977) and
    J.A. Mydosh, Proceedings of the International Conference on
    Magnetic Alloys and Oxides, 15 - 18 Aug. 1977, Haifa, Israel,
    Published in: Journal of Magnetism and Magnetic Materials.
7. See references in Ref. 6.
8. J. Souletie and R. Tournier, J. Low Temp. Physics 1, 95 (1969).
9. U. Larsen, Solid State Commun. 22, 311 (1977).
10. J.S. Schilling, P.J. Ford, U. Larsen and J.A. Mydosh, Phys.
    Rev. B 14, 4368 (1976).
11. J.S. Schilling and W.B. Holzapfel, Phys. Rev. B 8, 1216 (1973).
    J. Crone and J.S. Schilling, Solid State Commun. 17, 791 (1975).
    J. Crone, Doctoral Thesis, Technische Universität München(1975).
12. J. S. Schilling, P.J. Ford, U. Larsen, and J.A. Mydosh, to ap-
    pear in the "Proceedings of the Second International Symposium
    on Amorphous Magnetism" (Plenum, NY, 1977).
13. J.S. Schilling, J. Crone, P.J. Ford, S. Methfessel and J.A.
    Mydosh, J. Phys. F 4, L116 (1974); P.J. Ford and J.S. Schilling,
    J. Phys. F 6, L285 (1976); P.J. Ford, J.S. Schilling, U. Larsen
    and J.A. Mydosh, Physica 86 - 88 B, 848 (1977); J. Willer,
    H. Olijnyk, J. Crone and E. Lüscher, Proceedings of the Inter-
    national Conference on Magnetic Alloys and Oxides, 15 - 18 Aug.
    1977, Haifa, Israel, published in: Journal of Magnetism and
    Magnetic Materials.
14. U. Larsen, J.S. Schilling, P.J. Ford and J.A. Mydosh, Physics
    86 - 88 (B + C), 846 (1977); Amorphous Magnetism, Vol. 2,
    R.A. Levy and R. Hasegawa ed. (Plenum Press, N.Y., 1977), to
    appear; U. Larsen, Proceedings of the International Conference
    on Magnetic Alloys and Oxides, 15 - 18 Aug. 1977, Haifa, Israel,
    published in: Journal of Magnetism and Magnetic Materials.
15. H. Olijnyk, J. Crone and E. Lüscher (to be published).
16. J.L. Tholence, Conference on Glasses and Spin Glasses, Aussois,
    1977 (to be published).
17. U. Larsen, Thesis, University of Copenhagen, 1974 (unpublished);
    Proc. of the 14th Intern. Conf. on Low Temp. Physics, Otaniemi,
    Finland, 14 - 20 Aug. 1975.

18. M.D. Daybell and Y.K. Yeo, Bull. Am. Phys. Soc. 16, 840 (1971).
19. P.J. Silverman and C.V. Briscoe, Phys. Rev. B 15, 4336 (1977).
20. J.S. Schilling, U.F. Klein and W.B. Holzapfel, Rev. Sci. Instr.
     45, 1353 (1974).
21. A Eichler and J. Wittig, Z. Angew, Phys. 25, 319 (1968).
22. A.S. Edelstein, Phys. Letters 27 A, 614 (1968).
23. K.S. Kim and M.B. Maple, Phys. Rev. B 2, 4696 (1970); W. Gey and
     E. Umlauf, Z,Physik 242, 241 (1971).
24. E Müller – Hartmann, B. Schuh and J. Zittartz, Solid State
     Commun. 19, 439 (1976).
25. W. Felsch, K. Winzer and G.V. Minnigerode, Z. Physik B 21,
     151 (1975).
26. M.B. Maple and J. Wittig, Solid State Commun. 9, 1611 (1971).
27. A.S. Edelstein, Phys. Rev. Letters 20, 1348 (1968).
28. J. Flouqet, Phys. Rev. Letters 27, 515 (1971).
29. M. Takata and S. Oshida, J. Phys. Soc. Japan 30, 1640 (1971).
30. K. Syassen and W.B. Holzapfel, Solid State Commun. 16, 533
     (1975).
31. M.B. Maple, J. Wittig and K.S. Kim, Phys. Rev. Letters 23,
     1375 (1969); T.F. Smith, Phys. Rev. Letters 17, 386 (1966);
     B. Coqblin and C.F. Ratto, Phys. Rev. Letters 21, 1065 (1968).

QUESTIONS AND COMMENTS

T.A. Kaplan: What are the good theories for handling and fitting
                 your data in the spin glass region to determine the
                 value of delta?

J.S. Schilling: Larsen's theory is valid in the spin glass regime
                 where Δ_c is greater than T_K. There is, as far as I
                 know, no good theory around the transition region
                 and lower. That's the most difficult region to han-
                 dle. In the region $\Delta_c \gg T_K$ we feel Larsen's theory
                 of T_{max} yields reliable values of Δ_c. The high pres-
                 sure data was taken first, and Larsen was able to ac-
                 count quantitatively for the observed $T_{max}(P)$ de-
                 pendence which is a rather complicated function.
                 He was able to predict for a different concentration
                 on a given system how the pressure dependence should
                 change, and this was subsequently confirmed by ex-
                 periment.

M.B. Maple: I think Doniach very recently proposed a competition
                 between the Kondo effect and magnetic ordering in con-
                 centrated systems.

J.S. Schilling: I forgot to mention that Larsen, as well as Doniach,
                 had predicted such a magnetic phase transition. I

guess Doniach did it for a one-dimensional chain
Kondo system including interactions. I think that
perhaps these results support those people who look
upon valency fluctuations from this point of view.
We went to concentrations as high as 40 percent Ce
and it still looks very much like a Kondo system plus
interactions.

STRESS EFFECTS ON CRITICAL BEHAVIOUR

D. Bloch

Laboratoire Louis Néel, C.N.R.S., 166X,

38042 Grenoble Cedex, France

ABSTRACT

We give a review of high pressure or uniaxial stress effect on magnetic phase transitions, with particular emphasis on manganese oxide.

INTRODUCTION

Although the renormalization group approach[1] has led recently to major progress in the analysis and description of phase transitions, various aspects of phase transitions are still not fully understood. Among these aspects are the effects of internal stresses, such as spin-lattice coupling, or external stresses, such as those associated with uniaxial stress or hydrostatic pressure, on the character of phase transitions.

PHASE DIAGRAMS

Phase diagrams of ferromagnets and antiferromagnets (figure 1) are characterized by a line of transitions which ends at a critical point, the Curie point for a ferromagnet, and the Néel point for an antiferromagnet. At low temperatures, the magnetic moment of a ferromagnet flips discontinuously when the transition-line is crossed, that is when the uniform applied magnetic field changes its sign. A similar feature is observed for an antiferromagnet, where the reversal of the staggered field on each atom reverses discontinuously the direction of the atomic moments. Above the critical temperature the flipping is continuous. Thus turning around

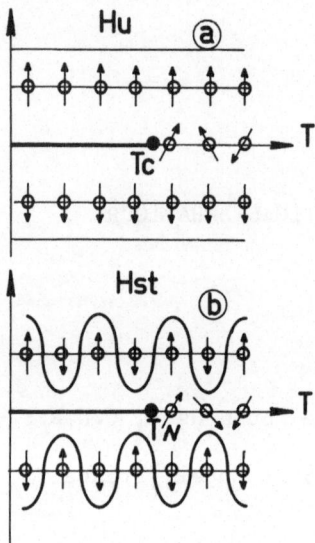

Figure 1. Phase diagram of a ferromagnet (a) and of an anti-
           ferromagnet (b).

T_C or T_N permits any points of the H–T diagram to be connected
through a continuous transformation.

    The usual role assigned to high pressure or stress is to change
internal parameters, such as exchange interactions between atomic
moments. Thereby, pressure or stress modifies the Hamiltonian and
leads to a change in the values of the Curie or Néel temperature
which are related to the magnitude of the interactions.

    In some cases, however, which are not covered by the present
discussion, the atomic moment itself can disappear or can appear
at high pressure or high stress, and the $T \rightarrow 0$ K phase diagram
should be considered as well as the thermal behaviour.

ORDER PARAMETER

    The exchange energy favours order, whereas the magnetic entropy
favours disorder. The Curie (or Néel) temperature separates the
ordered magnetic state from the disordered (paramagnetic) one.
The degree of order is characterized by an order parameter σ which
is zero in the disordered temperature range. For a ferromagnet
σ will be the $M(T)/M(T = 0)$ ratio, where M is the magnetic moment
at temperature T ; for an antiferromagnet it will be $(M_A - M_B)_T/$
$(M_A - M_B)_{T=0}$, where $M_A - M_B$ is the difference between the magnetic

moments of up and down magnetic sublattices. The dimensionality of
the order parameter for a ferromagnet is n = 1 for an Ising system,
that is for moment aligned in a well defined direction. It is
n = 2 for a X - Y system with magnetic moments in a plane and
n = 3 for the classical isotropic Heisenberg ferromagnet.

As demonstrated by Alben[2], the dimensionality of the
order parameter can be larger than 3 for antiferromagnets or heli-
magnets. Let us consider manganese oxide (MnO) which possesses
a fcc structure. The magnetic structure consists of ferromagnetic
(111) plane, that is planes where the magnetic moments are all
parallel together. The successive (111) planes are stacked anti-
ferromagnetically. As four equivalent body diagonals exist, there
are four different propagation vectors for this magnetic structure.
For a chosen propagation direction, the magnetic moment is de -
fined from its three components and thus 12 components are to be
specified in order to describe the actual situation. When the
magnetic moments lie in a (111) plane, which is the case for MnO,
due to dipolar forces, the dimensionality is reduced to 8. It is
4 when the spins point along a unique direction[3].

LANDAU THEORY - CRITICAL AND TRICRITICAL BEHAVIOUR

In the calculation of thermodynamical quantities, a first approximation is to neglect all fluctuations in the magnetization.
The difference between the actual free energy for zero magnetization can be expanded near the critical point as a function of
the magnetic moment :

$$F_M - F_{M = 0} = AM^2 + BM^4 + CM^6 - K(\vec{\nabla M})^2 \tag{1}$$

The actual magnetization is that which minimizes the free energy
density. The coefficient K is positive to insure that a uniform
magnetization exists and C is positive to insure a finite magnetic
moment. B is a smooth function of T, whereas A changes its sign
at temperature T_c following the simple law:

$$A = A' (T - T_c) \tag{2}$$

Two cases are to be considered, depending on the sign of B.
When B > 0, for A > 0, no magnetic moment exists, since F is mini-
mized for M = 0. A magnetic moment exists for A < 0 and thus T_c
when A = 0 is the critical temperature. In the vicinity of the
ordering temperature, the magnetization depends on the square root
of the temperature. The critical classical exponent for the description of the magnetization close to the ordering temperature is
1/2. When B < 0, magnetic ordering occurs discontinuously. Suppose
B is pressure or stress dependent. An interesting situation occurs,
if pressure or stress changes the sign of B, since simultaneously,
the ordering transition changes from continuous to discontinuous at

a tricritical pressure and a tricritical temperature. When B = 0
in the vicinity of the tricritical temperature, M varies as the
1/4 power of the temperature; this is the classical tricritical
exponent for the description of the magnetization close to the
tricritical temperature.

FLUCTUATIONS

In the vicinity of the critical temperature, the ordered and
disordered phases have similar free energy. For temperatures
slightly above the transition temperature fluctuations occur from
the disordered phase to ordered domains, of dimension ξ and life-time
τ ; ξ is called the correlation length and τ the correlation time.
If $\xi \to \infty$, when $T \to T_c$, the transition is continuous, whereas, if
the transition occurs too early, at finite ξ, it is discontinuous.
The renormalization group approach has been able to treat the fluctuations and to describe the approach of the system to the critical
point with exact exponents for various quantities such as magnetization, susceptibility, specific heats, correlation length. These
exponents depend on the dimensionality of the lattice space as
well as on the dimensionality of the order parameter. Note that,
in 3 dimensional space, the critical exponents for systems with
n = 3 at tricritical transition are the classical Landau exponents.

EXPERIMENTS

We will describe neutron scattering experiments on a single
crystal of MnO. The intensity of the magnetic peaks, at low temperature, is proportional to the square of the magnetic moments,
and thus from diffraction experiments performed at various temperatures, one can obtain the temperature dependence of the magnetic moments, and therefore the temperature dependence of the order
parameter.

At zero (or small) stress, one notes a discontinuous jump of
the sublattice magnetization at $T_1 \sim 120$ K associated with the
discontinuous character of the antiferro-paramagnetic transition
in MnO[4]. The crystallographic structure of MnO is cubic above
120 K ; it undergoes a rhombohedral distortion α below 120 K, the
(111) antiferromagnetic planes move closer together and the ferromagnetically aligned atoms within (111) plane move further apart.

If we apply a stress along the (111) diagonal, this rhombohedral
distortion is increased by an amount readily calculated from the
values of the elastic coefficients. If we look at the temperature
dependence of the order parameter at various applied stresses[5],
one notices that the discontinuity of the order parameter decreases
when the stress increases. The antiferro-paramagnetic transition

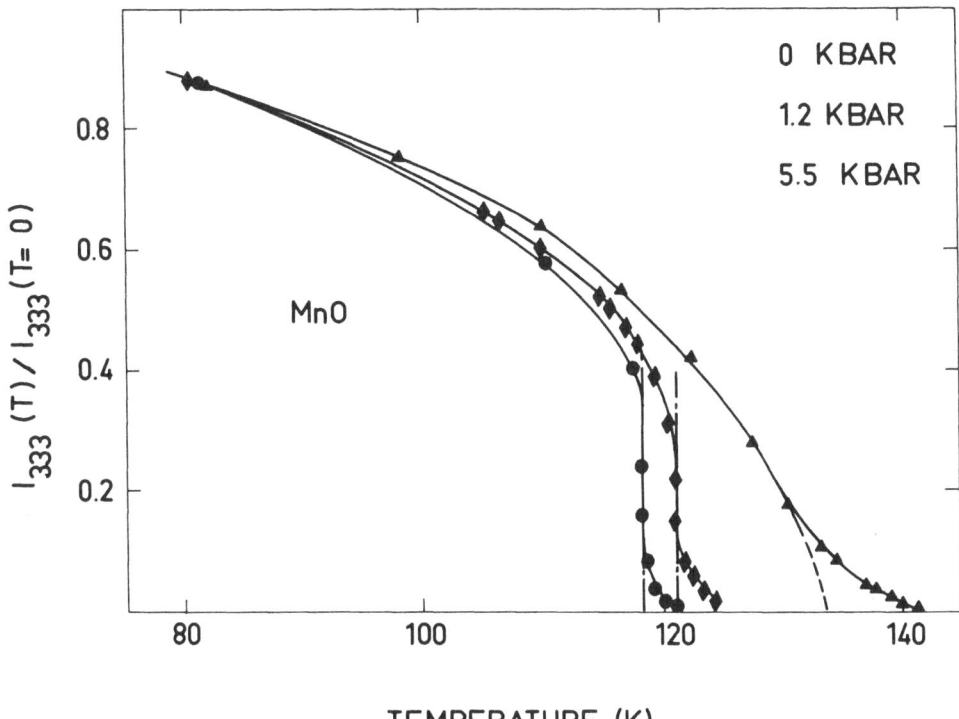

Figure 2.   Variation of the integrated intensity of the (333) lines
            with temperature for several applied stresses.

temperature itself increases with stress at a rate of approximately 3 K/kbar ; this value is about 10 times larger than the
variation with hydrostatic pressure. At 5.5 kbar, the antiferro-
paramagnetic transition is clearly continuous (figure 2).

MAGNETOELASTIC LANDAU THEORY

     We can include the effects of elastic distortion and associated
modifications of exchange interactions in the Landau free energy
expansion[5]. The magnetic moment and strain dependent parts of the
free energy can be written as :

$$\Delta F = 6NS^2\sigma^2 \ (J_2^0 - jJ_i^0 \ \delta\alpha) - NkT \ \left[\log(2S+1) - B_S^{-1}(\sigma') \ d\sigma'\right]$$
$$+ \ \frac{3}{2} V_m \ C_{44} \ (\delta\alpha)^2 \tag{3}$$

where N is the Avogadro's number, $S = \frac{5}{2}$ the spin of the magnetic
atoms, J_1^0 and J_2^0 the nearest and next nearest neighbour exchange

constants, V_m the molar volume ; C_{44} the cubic elastic constant, B_S^{-1} the inverse Brillouin function for spin S and

$$j = \frac{2\delta \log \left| J_1^0 \right|}{\delta \alpha}$$

The minimization of ΔF with respect to $\delta\alpha$ gives :

$$\delta\alpha = 12.5 \ N\sigma^2 \ jJ_1^0/C_{44} \ V_m + \tau/C_{44} \qquad (4)$$

This is the spontaneous magnetoelastic stress (τ) induced contribution to the distortion. The order parameter σ is found by minimizing ΔF with respect to σ .

The transition is discontinuous when it occurs at $T_1 < T_{3c}$, with

$$kT_{3c} = 966(jJ_1^0)^2/C_{44} \ V_m \qquad (5)$$

It is continuous when occurring at $T_2 > T_{3c}$, but with renormalized exponents.

The jJ_1^0 value can be obtained from the measured value of the tricritical temperature, from the spontaneous distortion, or from the stress dependence of the transition temperature. The agreement between these values, within 50% is rather unsatisfactory.

Nonetheless, the magnetoelastic Landau theory allows for a qualitative understanding of magnetoelastic coupling and tricritical phenomena. The absence of a quantitative agreement can eventually be attributed to the molecular field approximation, which neglects fluctuation effects.

COMPRESSIBLE MODELS – DIMENSIONALITY EFFECTS

As the Landau model does not take fluctuations in account, the magnetoelastic coupling should be treated using more appropriate models. This has been the subject of numerous but conflicting studies within the last few years[6]. The effects of this coupling can give rise to bare or renormalized critical exponents, or to first order transitions. First order transitions occur depending on the rigidity of the lattice, and therefore high pressure can change the character of the transition. However, the main effects of fluctuations on magnetoelastic Hamiltonians have not yet been definitively assessed, and no quantitative effects can be unambiguously predicted.

The effect of the dimensionality of the spin system on the character of the transition has been noticed in the last two years independently by Brazovskii and Dzyaloshinskii[7] and Bak, Krinsky, and Mukamel[8], which have demonstrated that in the space of parameters, no stable fixed point exists for n > 4, which is indicative of a discontinuous transition, as observed at zero stress in manganese oxide. This argument has been used recently by Bak et al.[9], to give an interpretation of the tricritical properties of MnO, where the applied (111) compressive stress reduces from 8 to 2 the dimensionality of the order parameter[2] and therefore changes the character of the antiferro-paramagnetic transition from discontinuous (n = 8) to continuous (n = 2). The treatment however does not give an account of the tricritical behaviour observed in MnO. The effects of fluctuations for propagation vectors out of the stress direction are in fact not entirely suppressed for small stresses and a more accurate description of the critical region should be used in order to explain the observed phase diagram[10]. One can notice that intermediary values can be obtained through a judicious orientation of the compressive or tensile stress.

CONCLUSION

The use of high pressure or high uniaxial stress as a tool to study ordering mechanisms in solids has been up to now quite infrequent. However, as we have seen, such experiments are of prime interest as they provide an experimental data base on which to test the predictions of the various theories of the effects of dimensionality and lattice distortion on critical phenomena. In this way we hope to help resolve conflicting views in the present theories and provide new directions for future investigations.

REFERENCES

1. For a review on renormalization group theory, see for instance G. Toulouse and P. Pfeuty "Introduction au groupe de renormalization et á ses applications", Presses Universitaires de Grenoble, 1975.
2. R. ALBEN
   C.R. Acad. Sc. Paris, 279, 111 (1974).
3. P. BAK, S. KRINSKY and D. MUKAMEL
   Phys. Rev. Lett., 36, 52 (1976).
4. D. BLOCH, R. MAURY, C. VETTIER and W.B. YELON
   Phys. Lett., 49A, 354 (1974).
5. D. BLOCH, D. HERRMANN-RONZAUD, C. VETTIER, W.B. YELON and R. ALBEN
   Phys. Rev. Lett., 35, 963 (1975).
6. M.E. FISHER
   Phys. Rev., 176, 257 (1968).

G.A. BAKER and J.W. ESSAM
Phys. Rev. Lett., 24, 447 (1970).
A. AHARONI
Phys. Rev. Lett., 138, 4314 (1973).
Y. IMRI, O. ENTIN, W. OHZMAN and D.J. BERGMAN
J. de Physique, C 6, 2846 (1973).
Y. IMRI
Phys. Rev. Lett., 33, 1304 (1974).
7. S.A. BRAZORSKII and I.E. DZYALOSHINSKII
J.E.T.P. Lett., 21, 164 (1975).
8. P. BAK, S. KRINSKY and D. MUKAMEL
Phys. Rev. Lett., 36, 52 (1976).
9. P. BAK, S. KRINSKY and D. MUKAMEL
Phys. Rev. Lett., 36, 829 (1976).
10. E. DOMANY, D. MUKAMEL and E. FISHER
Phys. Rev., B 15, 5432 (1977).

TRICRITICAL POINT IN KDP*

V. Hugo Schmidt

Department of Physics, Montana State University

Bozeman, Montana 59715

ABSTRACT

Static dielectric measurements on the ferroelectric crystal
KH_2PO_4 (KDP) at pressures to 3 kbar indicate a tricritical point
at 2.4±0.2 kbar. The exponent δ at this pressure agrees well with
the value of 5 obtained from mean-field theory for the tricritical
exponent.

INTRODUCTION

Pressure effects are important in hydrogen-bonded crystals because of the large compressibility of the hydrogen bonds. In hydrogen-bonded ferroelectrics, pressure usually decreases the ferroelectric Curie temperature T_c.[1] In the potassium dihydrogen phosphate family of ferroelectrics and antiferroelectrics, pressure reduces the transition temperature T_c, in some cases to 0 K. For example, Samara[2] showed that in KH_2PO_4 (KDP), a drop in T_c from 123 K
at ambient pressure to 0 K at 17 kbar occurs.

The structure of KDP consists of a square array of $K-PO_4$ chains
running parallel to the ferroelectric (c) axis. Each phosphate group
is connected by four hydrogen bonds to the four neighboring chains.

---

*Work supported by National Science Foundation Grant DMR76-11828
A01.

The hydrogens tend to obey the "ice rules" that each bond contains one proton, and that the protons are in off-center positions such that two protons are close to a given PO_4. Above T_c the protons occupy random positions subject to the above rules, but below T_c they order in one of two ways, in which two protons are near the top (or bottom) of each PO_4.

The proton ordering is accompanied by phosphorus and potassium ion displacement responsible for the polarization in the ferro-electric phase, and by a shear which changes the unit cell shape from tetragonal to orthorhombic. Near T_c a soft transverse optic mode and a soft shear mode are associated with these motions.

While pressure tends to shorten the hydrogen bond, and also brings the proton closer to the center of the bond, deuteration has the opposite effect because the greater deuteron mass de-creases the zero-point motion in the asymmetric off-center poten-tial wells. As a result, the differences between the energies of the various hydrogen configurations become greater upon deuteration, thereby increasing the transition temperature T_c as indicated in Fig. 1.

For KD_2PO_4, Sidnenko and Gladkii[3] found that the ferroelectric transition, which is first order at zero electric field, ends at a critical field E_{cr} of 7100 V/cm. For KH_2PO_4, E_{cr} is much lower, about 200 V/cm at zero pressure from results of several workers.[4]

Because halving the hydrogen mass almost halves T_c and causes a much more drastic drop in E_{cr}, and because applying pressure de-creases T_c also, it was natural to infer that pressure would also reduce E_{cr}, perhaps to zero.[5] In a three-dimensional p-T-E (pressure, temperature, electric field) space the two lines of $E \neq 0$ critical points would then meet, and continue to higher pressure in the E=0 plane as a line corresponding to a second-order ferro--electric phase transition. Such a location where three critical lines meet is termed a tricritical point.[6]

EXPERIMENTAL PROCEDURE

We searched for a tricritical point in KDP and found it to be located near 2.4 kbar as shown in Fig. 2.[4] We accomplished this by measuring the static polarization P as a function of pressure, temperature, and electric field. To create the necessary pressure we used the apparatus shown in Fig. 3, consisting of a hand pump, remote head, and intensifier, capable of attaining pressures to 100,000 psi. The pressure vessel is made of beryllium copper. The pressure medium was helium, which remains fluid in the pressure-temperature range we used. The pressure was measured with an

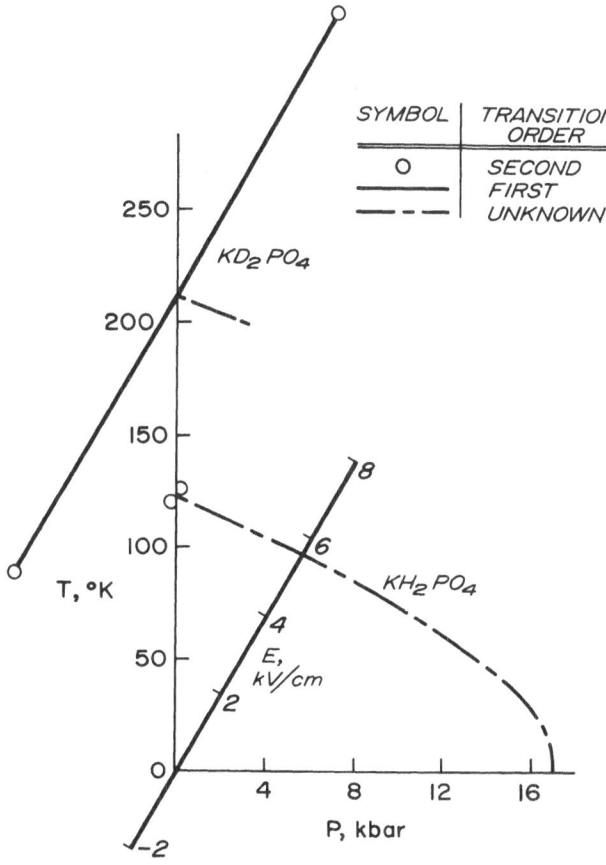

Figure 1.   Phase diagrams of KH_2PO_4 and KD_2PO_4 in pressure – tem-
            perature-electric field space, showing features known at
            the outset of this work, namely the wing first-order
            transition lines which end in critical points in the
            zero-pressure plane, and the pressure dependence of the
            ferroelectric Curie temperatures in the zero-field plane.

accuracy of ±0.4 bar by means of a thermostated manganin cell.

    To obtain the accurate temperature control and stability needed,
we used two concentric evacuated cans as shown in Fig. 4.   The
outer can was immersed in liquid nitrogen which was maintained within a given height range by an automatic filler.   The inner can
served as a radiation shield and for this purpose was kept about
1 K colder than the pressure vessel by means of a heater and servo
system.   The pressure vessel temperature was monitored and maintained constant within ±2 mK by a Lakeshore CSC 400 Capacitance

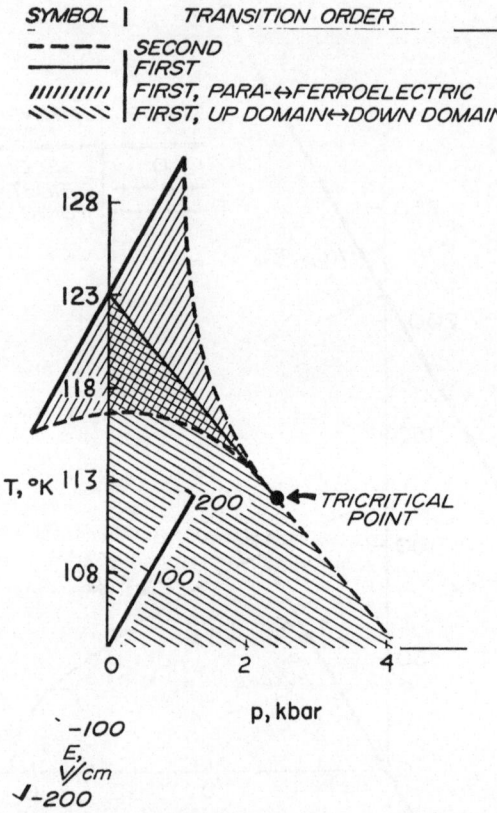

Figure 2.   Phase diagram of KH_2PO_4 in pressure-temperature-electric
field space, showing the tricritical point located at
the confluence of the two wing critical lines, the
zero-field critical line, and the line of triple points.
The first-order transition surfaces bounded by these
lines are shown also.

Temperature Controller and associated heater.   Thermocouples on the
can and pressure vessel were also used to monitor temperature.

The crystals used were supplied by Interactive Radiation (No. 1)
and Cleveland Crystal (Nos. 2 and 3), and were 1x1x0.2 cm in size,
the thin dimension along the c axis.   The crystal surfaces were
coated with evaporated gold.   To minimize anisotropic strains the
crystal was suspended by wires which were attached to the electrodes
by daubs of silver paint.   These electrical leads to the crystals
were supplied by a Lakeshore Type A Ultraminiature Coaxial Cable
which ran through the pressure tubing.

The polarization measurements were made by applying a voltage

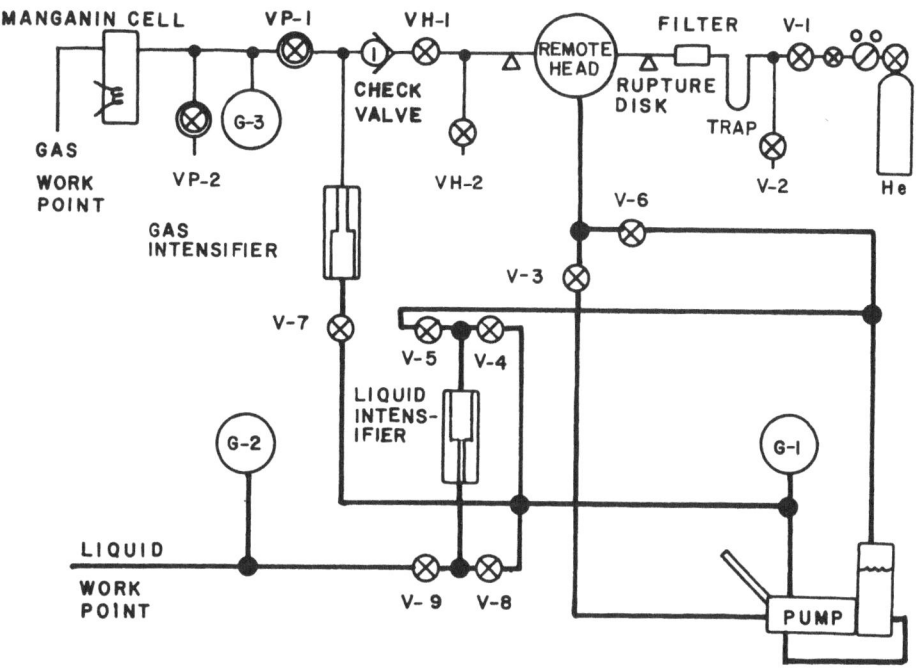

Figure 3.   Schematic of pressure system, showing both the liquid
            system for testing pressure vessels and the helium
            system used in the measurements.

from a battery-driven voltage divider across a series circuit consisting of the crystal and an 8 μf polystyrene capacitor. Digital
voltmeters monitored the voltage across the crystal and the capacitor voltage which was proportional to the polarization.

Initially we made runs in which temperature and electric fields
were varied simultaneously while holding polarization and pressure
constant, with the thought that hysteresis effects would be min-
imized by keeping polarization fixed. We found that the slow rate
of approaching thermal equilibrium using this technique made it
advantageous to instead hold temperature constant and vary polarization, obtaining data in regions in which KDP exists only in one
phase to avoid hysteresis effects.

By this time we had realized the advantage of analyzing data
for lines of constant polarization, so we obtained all data as
selected polarizations. When changing polarization at a given Temperature, we would wait several minutes for thermal and dielectric
transients to die away before recording data.

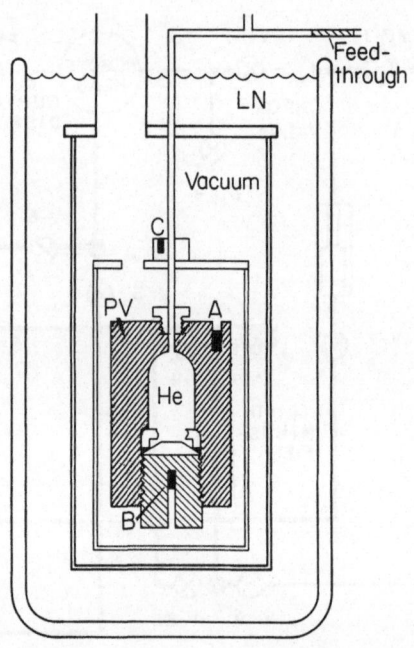

Figure 4.   Pressure vessel and its temperature control provisions.
The fine control heater is wound around the top of the
pressure vessel near A, and is controlled by a cap-
acitive temperature sensor located at A. The coarse
control heater is wrapped around a small cylinder at
the junction of the pressure tubing and the radiation
shield can, and is controlled by a thermocouple located
at C. The sample temperature is monitored by a cap-
acitive sensor and thermocouple located near B.

THEORY

We now discuss the reason for analyzing data in terms of isopols,
or lines of constant polarization. The static critical behavior of
a uniaxial ferroelectric such as KH_2PO_4 is expected to obey mean-
field theory except in an extremely narrow temperature region, perhaps 10^{-4} K wide, near the transition. Any mean-field theory can
be cast in terms of a Landau expansion.

$$F = \Sigma_n A_n P^n, \tag{1}$$

of the free energy F in powers of the order parameter P (polarization for KDP). The A_n are coefficients which in general depend on
temperature, pressure, and other parameters. The symmetry of KDP
excludes odd powers of P. Although Reese[7] found reason to include

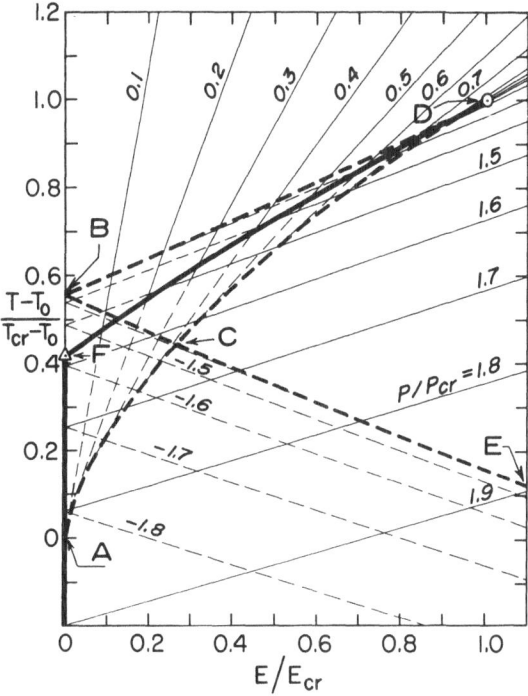

Figure 5.   Dimensionless isopol plot for equation of state given
            in Eq. 3, shown for a negative Landau coefficient B, for
            which the transition is of first order. The ferroelectric
            transition temperature or Curie point T_c is at F, the
            Curie-Weiss temperature T_0 at A, and the wing critical
            point at D.   Free energy minima corresponding to para-
            electric and "up domain" ferroelectric phases both exist
            in the region BCD. Minima corresponding to "up domain"
            and "down Domain" phases both exist in region ACE.   All
            three minima occur in region ABC.   The first-order para-
            electric-"up domain" transition line is indicated by DF,
            while the first-order "up domain"-"down domain" transition
            line extends downward from F.

a P^8 term to fit his electrocaloric results for KDP, our results
are fit equally well if the series is truncated at the P^6 term,
as follows:

$$F = \frac{1}{2} A_0 (T-T_0)P^2 + \frac{1}{4} BP^4 + \frac{1}{6} CP^6. \tag{2}$$

From the thermodynamic relation $E = \partial F/\partial P$ we obtain the equation
of state

$$E = A_0(T-T_0)P+BP^3+CP^5 \tag{3}$$

which for fixed P is the isopol equation giving a linear relation
between electric field E and temperature T.

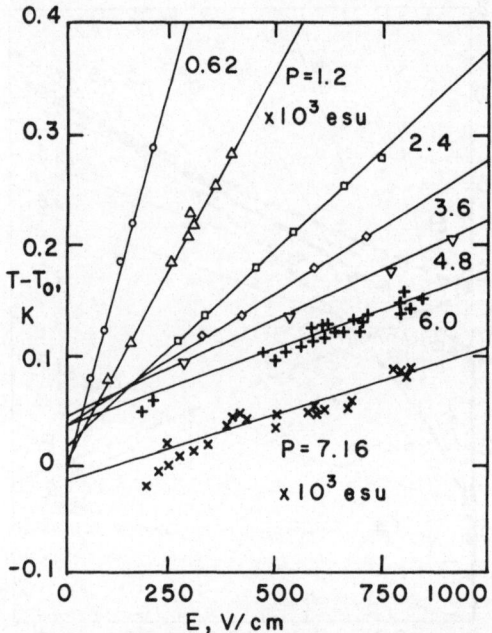

Figure 6.   Isopol plot for Crystal No. 2 at ambient pressure.

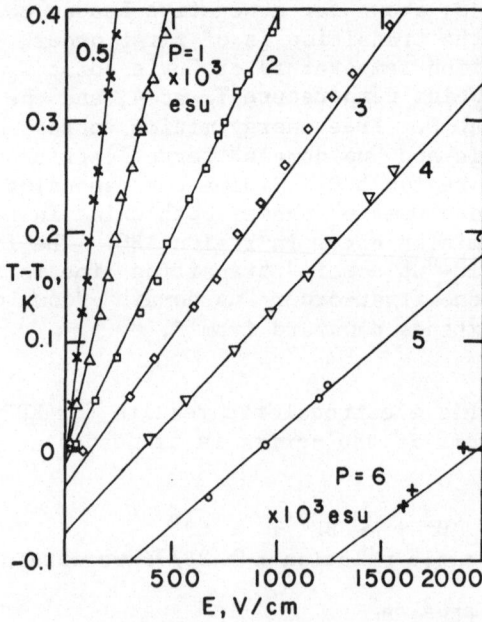

Figure 7.   Isopol plot for Crystal No. 2 at 3 kbar pressure.

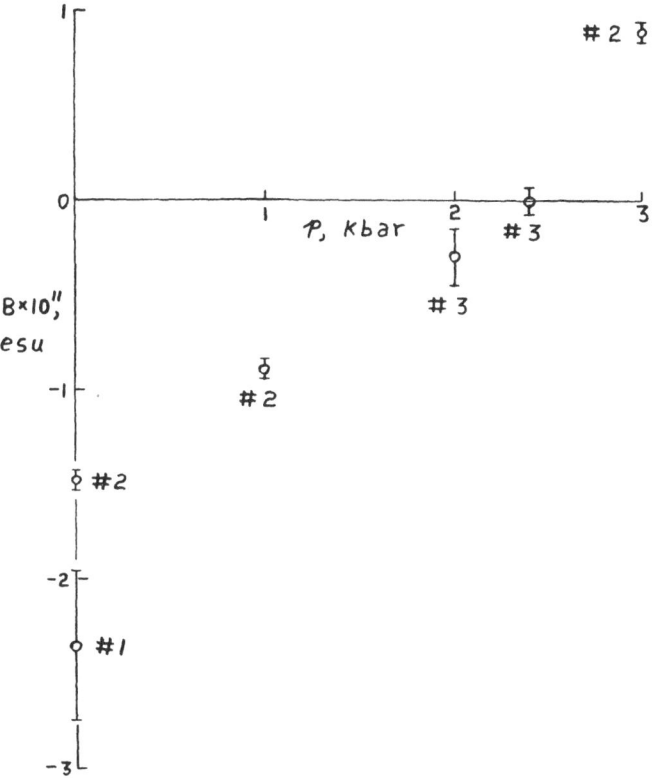

Figure 8. Plot of Landau coefficient B as a function of pressure,
from results for Crystals No. 1, 2, and 3 as indicated.

The interpretation of experimental results using Eq. (3) is now
explained with the aid of Fig. 5. Here isopols for negative as
well as positive P are shown. Isopols are seen to intersect, so
that it is necessary to determine which polarization actually exists
for a given (E,T) combination. To analyze the isopol overlap region,
one examines the "free energy at constant field", F-EP, for minima
as P is varied for fixed E and T. Where isopols do not intersect,
there is one minimum at the P value of the isopol going through
(E,T). If one follows the isopol into a region of isopol intersection, other local minima appear. As the isopol is traversed
farther into the intersection region, its value of P at some point
ceases to give an absolute minimum of F-EP and gives merely a local
minimum. Where the isopol becomes tangent to a caustic curve formed
by a family of neighboring isopols, F-EP for that isopol's value
of P changes from a minimum to a maximum.

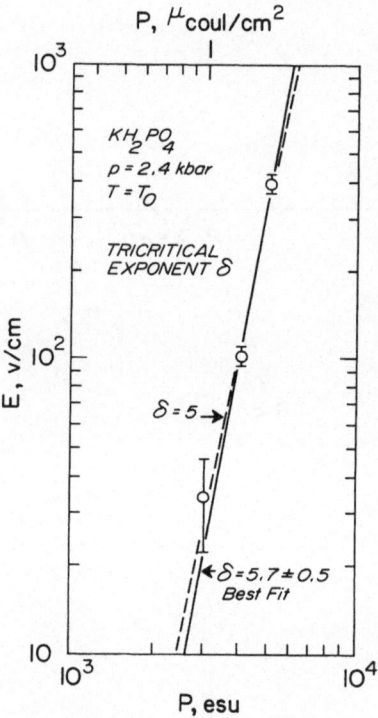

Figure 9. Plot of electric field E vs. polarization P measured
at the Curie-Weiss temperature T_o, for one of the runs
at 2.4 kbar pressure. Both the best-fit line and the
line corresponding to the mean-field-theory value of
5 for the tricritical exponent δ are shown.

EXPERIMENTAL RESULTS

In analyzing the data obtained at several values of P, namely
isopols corresponding to 500, 1000, 2000, 3000, 4000, 5000, and
6000 esu polarization, we first obtained for the data points for
each isopol the mean values of E and T, the slope, and the standard
deviations of T and of the slope. From the weighted average of
$P^{-1}\partial E/\partial T$ and deviations from this average we determined A_o and its
standard deviation. Then, given the mean E for the data points for
each isopol, we found the (T_o, B, C) combination giving the best fit
to the corresponding mean T's.

The isopols for 0 and 3 kbar are presented in Figs. 6 and 7 respectively. Isopol plots for pressures of 1, 2, and 2.4 kbar have
been[4] or will be[8] presented elsewhere. At zero pressure the isopols
overlap, indicating negative Landau coefficient B and a first-order
transition. At 3 kbar no overlap is present, B is positive, and

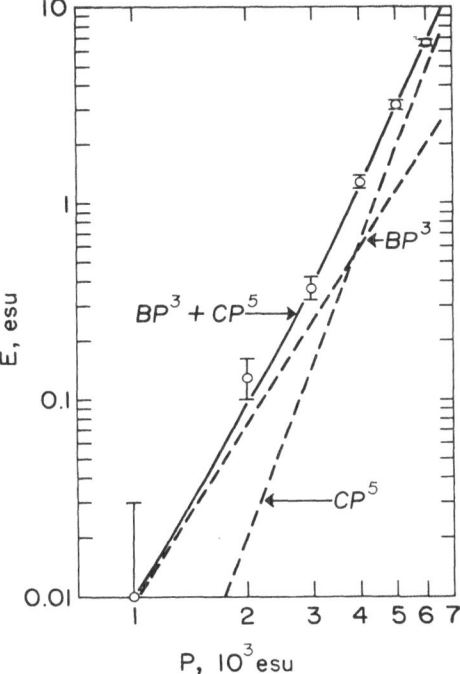

Figure 10. Plot of electric field E vs. polarization P measured
            at the Curie-Weiss temperature T_o (which is also the
            Curie temperature T_c because the transition is of sec-
            ond order), at 3 kbar pressure. The BP^3 and CP^5 con-
            tributions to $E(P,T_o)$ are shown both separately and
            summed. The crossover of δ between its critical and
            tricritical value is apparent.

a second-order transition exists. The plot of B vs. p shown in
Fig. 8 indicates that the tricritical point where B=0 occurs at
2.4±0.2 kbar.

     Further evidence for a tricritical point near 2.4 kbar is given
by the behavior of the exponent δ relating P and E according to
$E \propto p^\delta$, evaluated at $T=T_o$. From Eq. (3) it is seen that near the
critical line in the $E = 0$ plane, for which B>0, the exponent δ
has the mean-field critical value of 3. However, at the tricritical point, B = 0 and the exponent δ becomes 5, the mean-field tricritical value. The plot of E vs. P for p=2.4 kbar shown in Fig. 9
indicates that δ is near the tricritical value of 5. This is the
first observation of the tricritical exponent δ in any system. A
similar plot in Fig. 10 for p=3.0 kbar exhibits crossover of δ from

3 to 5 as P increases. Additional evidence for a tricritical point
in KDP is presented by Vettier et al.[9] elsewhere in these proceedings.

CONCLUSIONS

Pressure provides another dimension to parameter spaces in which
higher order critical points can be examined. In particular,
ferroelectrics are of interest because the neighborhood of the tricritical point can be studied in three dimensions, temperature,
pressure and electric field, all of which are experimentally ac-
cessible. Evidence for a tricritical point in KH_2PO_4 at 2.40 ± 0.2
kbar has been presented in this paper.

ACKNOWLEDGEMENTS

This work was performed in collaboration with, at various
times, Charles R. Bacon, Alan G. Baker, Richard J. Pollina, and
Arthur B. Western.

REFERENCES

1. V.H. Schmidt, in The Hydrogen Bond, edited by P. Schuster, G.
    Zundel, and C. Sandorfy (North-Holland, Amsterdam, 1976),
    Vol. III, Ch. 23.
2. G.A. Samara, Phys. Rev. Lett. 27, 103 (1971).
3. E.V. Sidnenko and V.V. Gladkii, Kristallografiya 17, 978 (1972)
    [Sov. Phys. Crystallogr. 17, 861 (1973)].
4. V.H. Schmidt, A.B. Western, and A.G. Baker, Phys. Rev. Lett. 37,
    839 (1976).
5. V.H. Schmidt, Bull. Am. Phys. Soc. 19, 649 (1974).
6. R.B. Griffiths, Phys. Rev. Lett. 24, 715 (1970).
7. W. Reese, Phys. Rev. 181, 905 (1969).
8. A.B. Western, A.G. Baker, C.R. Bacon, and V.H. Schmidt, to
    appear in Phys. Rev. B.
9. C. Vettier, P. Bastie, and M. Vallade, Proceedings of this
    Conference.

QUESTIONS AND COMMENTS

P.S. Peercy: Do you have any reason to think that these
                            coefficients (A_o, B, C) should be linear
                            with pressure on both sides of the tri-
                            critical pressure?

V.H. Schmidt: Well, of course, over a long range they
                             wouldn't be linear with pressure, but I
                             don't see why there should be any change
                             in slope right at the tricritical point.
                             You were thinking that maybe there actually
                             is a change which would explain why the
                             3 kbar value of B lies above the line through
                             the other B points?

P.S. Peercy: No, I thought perhaps the B coefficient
                             might not be strictly linear with pressure.

V.H. Schmidt: It is certainly possible that there would
                             be curvature, there's no reason that there
                             shouldn't be over that large a pressure
                             range.

INVESTIGATION OF THE TRICRITICAL POINT OF KDP UNDER PRESSURE

C. Vettier

Laboratoire Louis Néel, C.N.R.S. 166X
38042 Grenoble Cedex, France

P. Bastie, M. Vallade

Laboratoire de Spectrométrie Physique, Université
Scientifique et Médicale, B.P. 53, 38041 Grenoble Cedex,
France

C.M.E. Zeyen

Institut Laue-Langevin, 156X
38042 Grenoble Cedex, France

KDP was investigated recently by Schmidt et al.[1], using static dielectric measurements under pressures up to 3 kbar. The ferroelectric phase transformation which is first order at ambient pressure was found to be second-order at a pressure of 3 kbar.

We present preliminary results on direct measurements of the spontaneous shear u_{xy} using γ-ray diffractometry and of the variation u_{zz} of the c lattice parameter using neutron diffraction in the vicinity of the ferroelectric transition of KDP at pressures up to 3.5 kbar at E = 0.

The pressure assembly consisted of an autofrettaged high strength aluminium alloy cell[2] and a helium gas pressure generator. The pressure was measured by a manganine gauge. We performed slow temperature sweeps through the transition temperature T_c. The temperature was measured by a platinum resistor which was attached to the cell body.

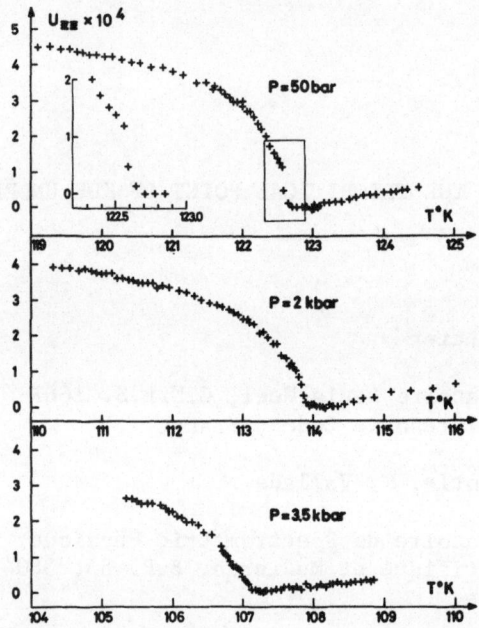

Figure 1.   Temperature dependence of the spontaneous strain u_{zz}
            for various pressures.

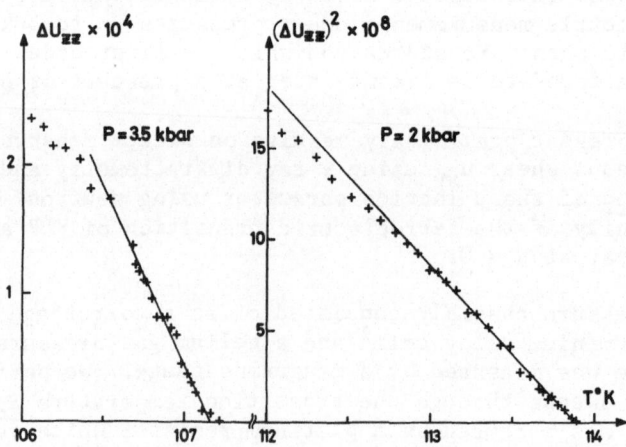

Figure 2.   Temperature dependence of powers of the polarization in-
            duced strain Δu_{zz}.

γ- ray scattering proves to be a useful tool for investigation of systems under pressure because of the very low absorption of this radiation. Using the I.L.L. γ-ray diffractometer[3] at a wavelength of $\lambda = .03$ Å and an angular collimation in the scattering plane of 30 seconds of arc, we measured directly[4] the spontaneous shear angle at different pressures as a function of the sample temperature. The shear u_{xy} is connected to the polarization by the relationship:

$$u_{xy} = b_{36}P_z$$

where b_{36} is a piezoelectric constant.

Our measurements show that the step at T_c characteristic of a first order transition decreases with increasing pressure and vanishes at 2.15 kbar. These results are compatible with the existence of a TCP at a pressure close to 2 kbar.

The neutron scattering experiments were carried out on the D10 spectrometer at I.L.L. A neutron high resolution parallel crystal arrangement was used[5]. The monochromator was a deuterated KD_2PO_4 single crystal plate. At a wavelength of 3.2 Å the angular position of the (004) reflexion was monitored as a function of temperature to yield the corresponding variation of the c-lattice constant.

The temperature dependence of u_{zz} near T_c can be written down as follows:

$$u_{zz}(T) = \alpha_3 T + Q_{33}P_z^2$$

where $\alpha_3 T$ is the usual thermal expansion and Q_{33} is an electrostriction coefficient.

Analysis of our results (figure 1) leads to the following conclusions:
(i). the shift of T_c as a function of pressure is given by:
    $\partial T_c/\partial P = -4.46 \pm 0.08$ K.kbar^{-1} in good agreement with the most recent result.[6]
(ii). the step in u_{zz} at T_c disappears completely at the pressure of 3.5 kbar. An extrapolation of α_3 for temperatures below T_c yields the pure electrostrictive part, $\Delta u_{zz}(T) \sim P_z^2(T)$.

The Landau theory leads to classical $(T_c - T)^\beta$ asymptotic temperature dependence for the order parameter with $\beta = 1/2$ at a critical point and $\beta = 1/4$ near a tricritical point.

In figure 2 we have plotted our results, $\Delta u_{zz} = Q_{33}P_z^2$ for a pressure of 3.5 kbar and $(\Delta u_{zz})^2 = Q_{33}^2 P_z^4$ for a pressure of 2 kbar. The linearity of these plots near T_c is in good agreement with the existence of a TCP at a pressure very close to 2 kbar.

In conclusion the present measurements[7] of the spontaneous shear u_{xy} and strain u_{zz} confirm the existence of the pressure induced tricritical point in KDP.

REFERENCES

1. W.H. Schmidt, A.B. Western, A.G. Baker, Phys. Rev. Lett., 37, 839 (1976); A.G. Baker, C.R. Bacon, V.H. Schmidt, Bull. Am. Phys. Soc., 22, 324 (1976).
2. J. Paureau, C. Vettier, Rev. Sci. Instr., 46, 1484 (1975).
3. J.R. Schneider, J. Appl. Cryst., 7, 541 (1974) and 7, 547 (1974).
4. P. Bastie, J. Bornarel, J. Lajzerowicz, M. Vallade, J.R. Schneider, Phys. Rev., B 12, 5112 (1975).
5. C.M.E. Zeyen, Thesis, Munich (1975) unpublished; C.M.E. Zeyen, H. Meister, W. Kley, Sol. Stat. Commun., 18, 621 (1976).
6. H. Umebayashi, B.C. Frazer, G. Shirane, W.B. Daniels, Sol. Stat. Commun., 5, 591 (1967); G.A. Samara, Phys. Rev. Lett., 27, 163 (1971).
7. B. Bastie, M. Vallade, C. Vettier, C.M.E. Zeyen, to be published.

HIGH PRESSURE STUDIES OF SOFT MODE TRANSITIONS IN SOLIDS*

G. A. Samara

Sandia Laboratories,

Albuquerque, New Mexico 87115

ABSTRACT

An overview of the soft mode concept and of high pressure studies of soft mode transitions in solids is given. Typical results on displacive ferroelectric (SbSI) antiferrodistortive [$Gd_2(MoO_4)_3$ and $BaMnF_4$] transitions as well as on coupled proton-phonon soft mode transitions (KH_2PO_4) are used to demonstrate that hydrostatic pressure can strongly modify the balance of forces responsible for soft mode behavior. Emphasis is placed on the physical implications of the results and on the general trends of the available data on the different classes of soft mode transitions, for it is such trends that have led to a greater understanding of the nature of the forces involved.

I. INTRODUCTION

Ever since it was introduced in 1960 to describe certain displacive ferroelectric transitions,[1,2] the concept of soft phonon modes has greatly enhanced our understanding of the static and dynamic properties of many structural phase transitions in solids. By a soft mode we simply mean a normal mode of vibration whose frequency de-

---

* Work supported by the U.S. Energy Research and Development Administration under Contract AT(29-1)789.

creases (and thus the lattice literally softens with respect to this
mode) and approaches zero as the transition point is approached. From
a lattice dynamical point of view, the limit of stability of a crystal lattice is approached as the frequency of any one mode decreases
and approaches zero, for then, once the atoms are displaced in the
course of the particular vibration, there is no restoring force to
bring them back, and they assume new equilibrium positions determined
by the symmetry (eigenvector) of this soft mode.

The soft mode concept was first developed to describe continuous
or second-order phase transitions.[1] For such transitions, symmetry
considerations require that only one normal mode of a certain polarization and wavevector be involved.[1] The frequency of this mode vanishes precisely at the transition and there is a one-to-one correspondence between the symmetry properties of the eigenvector of the
soft mode and the static ionic displacements accompanying the transition. The eigenvector describes the pattern of atomic displacement
associated with the phonon relative to the equilibrium lattice sites.
The crystal structure of the new phase is determined by the superposition of this pattern of atomic displacements on the structure of
the old phase. For first-order phase transitions, on the other hand,
there are generally no symmetry requirements, and the description
of such transitions in terms of soft modes, if at all valid, is quite
involved. However, there are many known cases where, while the transition is first order, it appears nearly second-order from a lattice
dynamical viewpoint (i.e. there is substantial pre-transition mode
softening), and for such cases the considerations obtained from
treatments of second-order transitions apply.

In addition to its frequency and eigenvector, the soft mode (or
any phonon for that matter) must also be characterized by its wavelength (or wavevector q, where $q = 2\pi/\lambda$). The wavelength determines
how the atomic displacements in one unit cell relate to those in adjacent unit cells. In ferroelectric crystals, identical ions in adjacent unit cells undergo identical displacements but with negative
and positive ions undergoing opposite displacements and thereby producing a macroscopic polarization. The soft mode leading to such a
circumstance is thus necessarily a long wavelength (or Brillouin zone
center, $q = 0$) transverse optic (TO), infrared-active phonon (Fig.
1a). In antiferroelectric crystals, on the other hand, identical
ions in adjacent unit cells undergo opposite displacements and there
is no net polarization. The characteristic wavelength of a phonon
leading to such displacements is thus typically twice the lattice
parameter $\underline{a}$, and the wavevector of the soft mode is then $q = 2\pi/2a$,
i.e., a Brillouin zone boundary phonon (Fig. 1b). More generally,
the soft mode can be either an optic or acoustic phonon and of either
long or short wavelength. Furthermore, in more complicated cases
the soft mode can be a coupled mode involving the coupling of two
or more fundamental excitations. Examples of these various possi-
bilities are known,[3,4] and we shall discuss a few examples in this
paper.

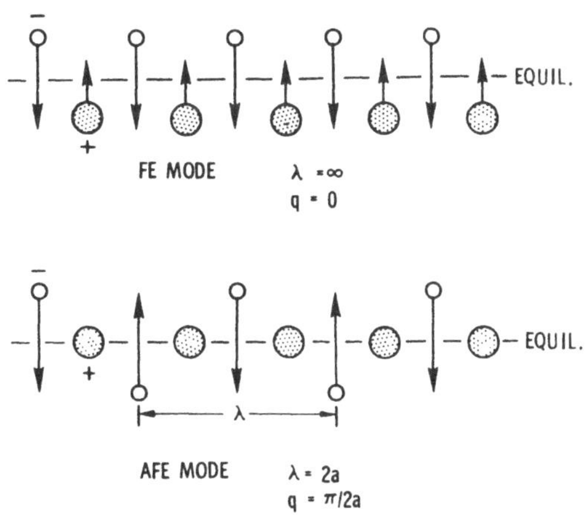

Figure 1. Schematic representation of the ionic displacements as-
           sociated with a ferroelectric (FE) mode and an antiferro-
           electric (AFE) mode in a diatomic linear chain.

It is generally accepted that the decrease in the frequency of
the soft mode (ω_s) as the transition temperature (T_c) is approached
is caused by the cancellation between competing lattice forces.
One might suspect that pressure can significantly affect the balance between such forces and thereby strongly influence, or even
induce, soft mode behavior. This has indeed turned out to be the
case[5] and there now exists a substantial body of literature illustrating that high pressure studies have been important --sometimes
essential, to the understanding of the phenomena involved.

In the space available for this review it is not possible to
give anything like a detailed account of the pressure results on
materials which exhibit soft mode behavior. Rather we shall enumer-
ate the various classes of soft mode transitions that have been investigated under hydrostatic pressure, illustrate some of the important features by a few specific results, and comment on the physical
implications of these results. The reader is referred to the cited
references for details. The following paper in this Conference Proceedings (by Peercy) presents specific pressure results on soft modes
in some ferroelectrics.

For the present purpose we have, somewhat arbitrarily, divided
the soft mode transitions investigated at high pressure into the
following classes:
   1 - Displacive ferroelectric soft mode transitions
   2 - Displacive antiferrodistortive soft mode transitions
   3 - Coupled proton-phonon soft mode transitions
   4 - Other soft mode transitions.

In the fourth class we include several interesting, but generally unrelated, soft mode transitions where high pressure studies have led
to important results. Following some brief theoretical considerations we shall deal with each of the above classes.

II. THEORETICAL CONSIDERATIONS

Inherent in the soft mode picture of phase transitions is the
premise that the crystal is unstable in the harmonic approximation
with respect to the soft mode.[1] Specifically, the square of the
harmonic frequency, ω_o^2, is presumed to be sufficiently negative
(i.e., ω_o is imaginary) that this mode cannot be stabilized by
zero-point anharmonicities alone. Thermal fluctuations then re-
normalize ω_o and make it real at finite temperature, thereby stab-
ilizing the lattice. The transition temperature is the temperature
where the renormalization is complete. Formally, the anharmonic
crystal potential can be written as a series expansion in the displacements (or normal mode coordinates), and perturbation or self-
consistent treatments are then used to solve for the normal modes.
Considering only quartic anharmonicities for the purposes of the
discussion to follow, self-consistent treatments[6] yield for the
renormalized frequency of mode j with wavevector q

$$\omega_T^2 \ (jq) = \omega_o^2(jq) + \sum_{\mu k} g_{j\mu}^{(4)} \ (qk) \ \frac{1}{2\omega(\mu k)} \ \coth \frac{\omega(\mu k)}{2k_B T} \ , \tag{1}$$

where $g_{j\mu}^{(4)}$ are effective fourth-order coupling constants and the
summation is over all modes μ and wavevectors k. At suitably high
temperature the second term on the right-hand side of Eq. 1 can be
expected to vary linearly with T so that

$$\omega_T^2 \ (jq) = \omega_o^2 \ (jq) + \alpha \ T \tag{2}$$

where α is a positive constant. Experimentally, the soft-mode frequency ω_s is found to vary with temperature as

$$\omega_s^2 \equiv \omega_T^2(j'q') = K(T - T_o) \ , \tag{3}$$

where K is a positive constant and T_o is the (second-order) transition temperature. Comparing Eqs. 2 and 3 we then note that $T_o =
- \omega_o^2/K$.

In ferroelectric crystals the temperature dependence of ω_s
(Eq.3) is responsible for the observed Curie-Weiss temperature
dependence of the static dielectric constant, ε, and the divergence
of ε at the transition, namely

$$\varepsilon = C/(T- T_o) \ , \tag{4}$$

where C is a constant. This is because ε and ω_s are related by the

Lyddane–Sachs–Teller relationship which, for example, for a diatomic crystal, has the form

$$\frac{\varepsilon}{\varepsilon_\infty} = \frac{\omega_{LO}^2}{\omega_{TO}^2} \equiv \frac{\omega_{LO}^2}{\omega_s^2} \quad , \tag{5}$$

where ε_∞ is the high (optical) frequency dielectric constant and ω_{LO} and ω_{TO} are the long wavelength longitudinal and transverse optic frequencies of the polar phonons. ε_∞ and ω_{LO} exhibit very weak temperature dependence so that to a very good approximation

$$\varepsilon(T)\omega_s^2(T) = \text{constant} \tag{6}$$

When dealing with a soft mode, $\omega_o(j'q')$ is purely imaginary and stabilization to $\omega_T(jq)$ is provided by the αT term (neglecting any small stabilization provided by zero-point fluctuations). Even so, it should be strongly emphasized that the interesting and unusual features of the soft mode are not so much a consequence of the large anharmonicities as they are a result of the extremely small (or imaginary) harmonic frequencies. This point has been emphasized earlier.[1,7] Furthermore, there is evidence suggesting that the changes in the harmonic interactions dominate the pressure effects. Thus, although the anharmonic interactions will change with pressure, it will suffice for the purposes of qualitative arguments to consider only the harmonic interactions.

What causes $\omega_o(jq)$ to be small or imaginary and what determines its pressure dependence? Some answers are provided in the following Sections.

III. DISPLACIVE FERROELECTRIC SOFT MODE TRANSITIONS

Conceptually displacive ferroelectric transitions are the simplest and best understood soft mode transitions. As indicated earlier, in this case the soft mode is a zone center TO phonon, and its harmonic frequency [dropping the (jq) notation] can be expressed as[1]

$$\mu \, \omega_o^2 = R - \frac{(Ze*)^2}{v}[C(0)] \quad , \tag{7a}$$

$$\propto (\text{S.R. interaction}) - (\text{L.R. interaction}) \quad , \tag{7b}$$

where μ is the appropriate reduced mass, $(Ze*)$ is the effective ionic charge, v is the unit cell volume, $C(0)$ is the electrostatic force constant for the $q = 0$ mode and R is the sum of the appropriate short-range force constants. Thus it is seen that ω_o^2 is given by the difference between a short-range (S.R.) interaction and a long-range (L.R.) interaction, and the negative value of ω_o^2 (for ferroelectrics with finite transition temperatures) results from the overcancellation of the positive S.R. interactions by the L.R. interactions.

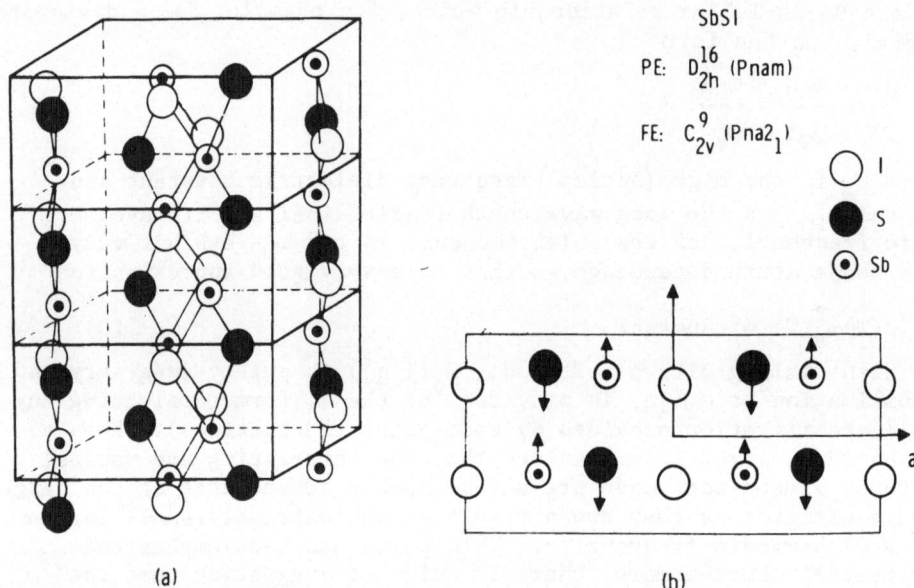

Figure 2.   (a) structure of SbSI.
            (b) eigenvector of the ferroelectric soft mode showing
                the Sb and S displacements relative to I.

With pressure the interionic distance r is decreased and this
increases the S.R. forces ($\sim r^{-n}$, where $n \simeq 10$) much more rapidly
than the L.R. forces ($\sim r^{-3}$) leading to an increase in ω_0^2 which in
turn leads to a decrease in T_o. This effect can be most readily
seen from Eq. 3. A larger ω_s at constant T necessitates a lower
T_c. Alternatively, as ω^2 becomes less negative, we do not need to
go to as high a temperature ($\equiv T_o$) to stabilize ω_s and make it real.
Ultimately, at sufficiently high pressure, ω_o can be expected to
become real. In such a circumstance the crystal will be stable with
respect to this mode at all temperatures, i.e., the transition vanishes.

These theoretical predictions are confirmed by the available experimental evidence. The decrease in T_c with pressure has been observed experimentally for many displacive ferroelectrics[5,8] and the
ultimate vanishing of ferroelectricity has been observed in perovskite ferroelectrics[5] and SbSI.[9] The SbSI results will serve to
illustrate the main features of the pressure effects.

Antimony sulpho-iodide, SbSI, is a widely studied ferroelectric.
The crystal is orthorhombic both above (space group D_{2h}^{16} - Pnam) and
below (space group C_{2v}^9 - Pn22_1) the ferroelectric transition temperature $T_c \simeq 290K$. The structure, shown in Fig. 2a, consists of
doubly-linked chains extending along the ferroelectric c-axis. The

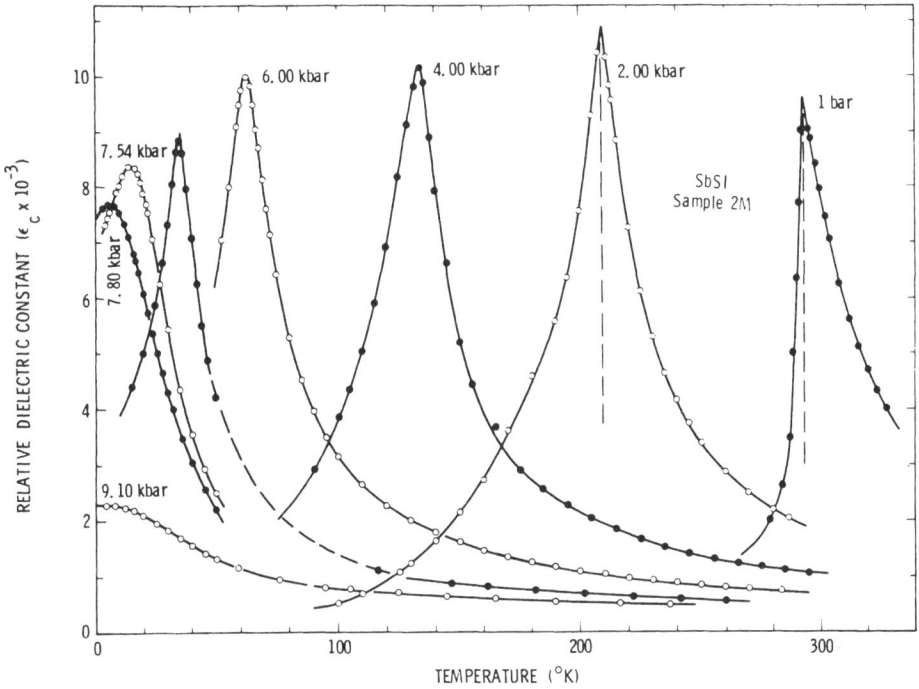

Figure 3. Temperature dependence of the static dielectric constant
of SbSI at different pressures showing the vanishing of
the ferroelectric state.

transition is driven by the softening of a long wavelength TO phonon[10]
whose eigenvector is depicted in Fig. 2b and consists of opposite
displacements of the Sb and S ions relative to the I ions.

Figure 3 shows the temperature dependence of the c-axis dielectric constant ε_c at different pressures.[9] The transition temperature, T_c,[11] defined by the peak in $\varepsilon_c(T)$, decreases rapidly with increasing pressure and ultimately the low temperature ferroelectric
phase vanishes. At 7.80 kbar there is a clearly defined transition
at ∼ 6 K, but the transition completely vanishes at slightly higher
pressure as shown by the 9.1-kbar isobar.

In the high temperature paraelectric phase the $\varepsilon_c(T)$ data in
Fig. 3 obey at each pressure the Curie-Weiss law, Eq. 4, as is illustrated by the linear $\varepsilon_c^{-1}(T)$ responses in Fig. 4. The data in
Fig. 4 also represent the $\omega_s^2(T)$ response at each pressure, as can
be readily deduced from Eq. 6. Representing this linear ω_s^2 response
by Eq. 2, we see that the extrapolation of the straight lines back

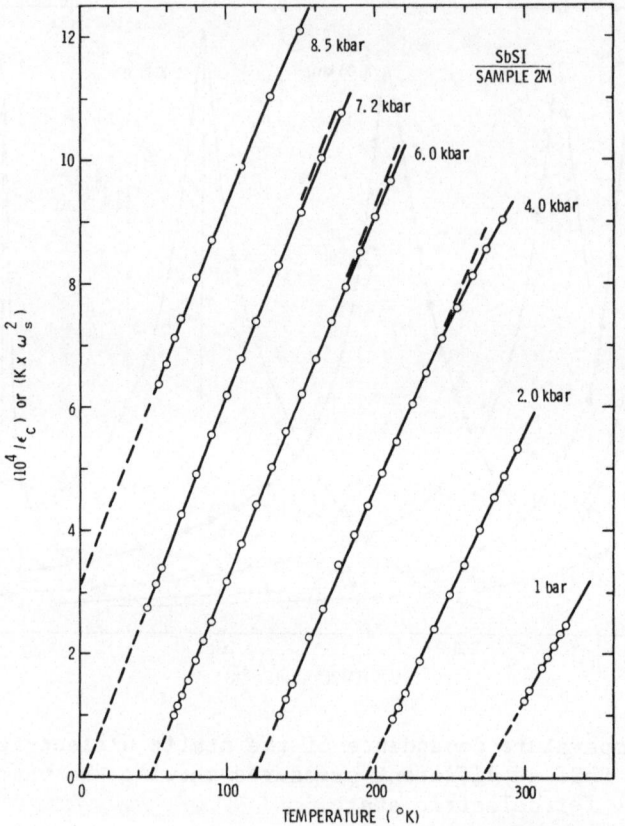

Figure 4. Temperature dependence of the inverse static dielectric
           constant (or the square of the soft mode frequency) in
           the high temperature paraelectric phase of SbSI at dif-
           ferent pressures.

to $T = 0$ K yields the harmonic frequency, ω_o, squared. Examination
of the results in Fig. 4 shows that for pressures $\lesssim 7.2$ kbar ω_o is
imaginary. It increases with pressure and ultimately becomes real
and finite (see the 8.5-kbar isobar). This implies that the lattice
then becomes stable with respect to this mode at all temperatures,
i.e. the ferroelectric phase vanishes, as we have mentioned.

Figure 5 shows the pressure dependence of T_c. The initial slope
$dT_c/dP = -39.6 \pm 0.6$ K/kbar, appears to be the largest pressure effect
for any ferroelectric.[12] The initial shift of T_c is linear but becomes somewhat nonlinear at high pressure. This nonlinearity is
most likely a result of the nonlinear pressure dependences of the
lattice parameters (or unit cell volume) of SbSI. Note that the
slope dT_c/dP is finite as $T_c \rightarrow 0$ K. This behavior is similar to

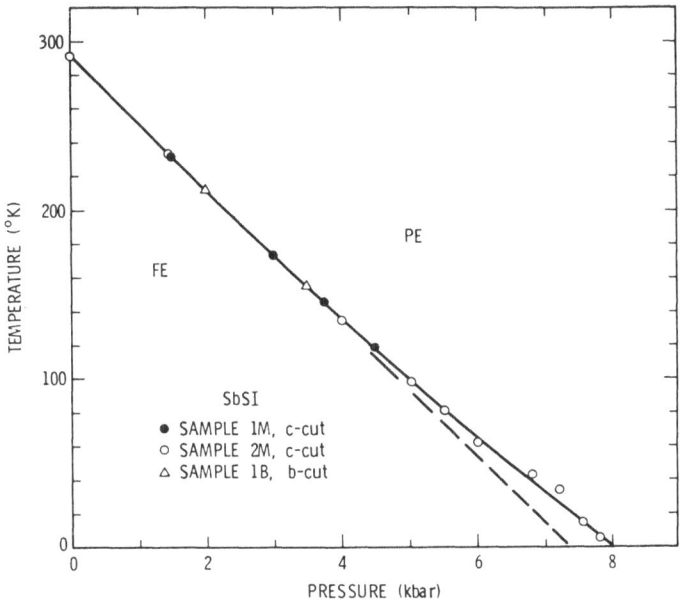

Figure 5. Pressure dependence of the ferroelectric transition
          temperature of SbSI.

that observed for the perovskites[5] and can be readily understood in
terms of the soft mode picture.

An important feature of the results in Fig. 5 is that the value
of dT_c/dP is about the same when $T_c \simeq 290$ K (the 1 bar value) as it
is when $T_c \lesssim 10$ K. Since the magnitude of the anharmonic contributions must have decreased drastically between 290 K and ~ 10 K,
the experimental evidence indicates that dT_c/dP is dominated by
changes in the harmonic interactions, i.e. by changes in ω_o, as
suggested earlier.

Another feature of the SbSI data that is worth noting is the
pressure-induced change in the character of the transition (compare
the 1-bar and 2.0-kbar isobars in Fig. 3). The transition is first
order at 1 bar but appears second order at ≥ 2 kbar. This change,
which is observed in most dielectric constant measurements,[9,13]
actually occurs between 1 and 2 kbar and is now definitely known
to be associated with the occurrence of a Curie critical point
(a tricritical point) at ~1.4 kbar and ~ 235 K.[14] This is the point
where the transition changes from first (at low pressure) to second
order. There has been much recent interest in tricritical points
and other multicritical points, and ferroelectrics at high pressure
are uniquely suited for exploring such critical points. In addition
to SbSI pressure-induced tricritical points occur or are expected

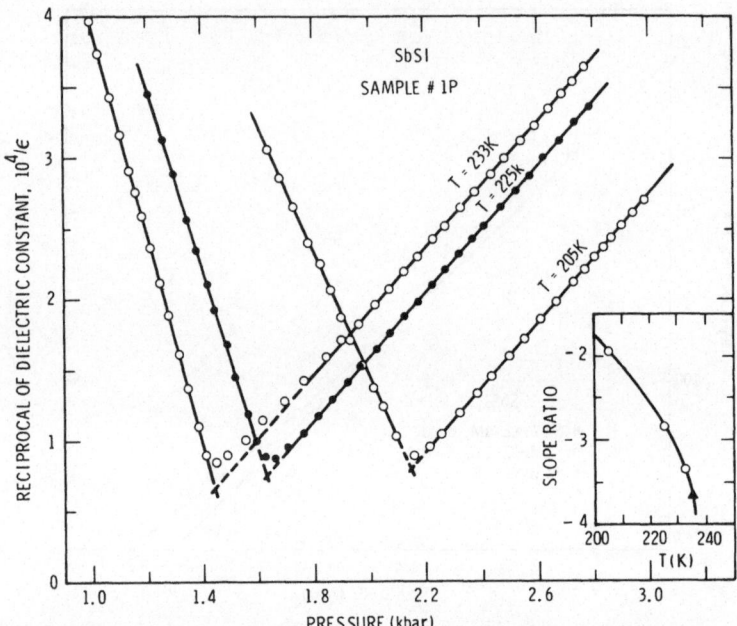

Figure 6. Pressure dependence of the inverse static dielectric
constant of SbSI at different temperatures showing the
variation of the slope ratio of the inverse dielectric
constant in the ferroelectric phase to that in the
paraelectric phase on approaching the tricritical point
at ~ 1.4 kbar and ~ 235 K.

to occur in ferroelectrics such as KH_2PO_4[15] and the perovskites.[16,17]
The tricritical points in KH_2PO_4 and SbSI are discussed in the papers
by V.H. Schmidt[18] and P.S. Peercy,[19] respectively, in the Proceedings
of this Conference.

One of the manifestations of a tricritical point in a ferroelectric is that slope ratio of the inverse susceptibility (or
dielectric constant) in the ferroelectric phase to that in the
paraelectric phase should change from 2 for a second-order transition to 4 at the tricritical point.[9,14] Recent, yet unpublished,
data by the author on an SbSI sample from the same batch of material as used by Peercy in his Raman scattering work[14] confirm this
expected change. The results are shown in Fig. 6. A more detailed
account and interpretation of these results will be published elsewhere.

Finally we note that there are many crystals which have a q =
0 TO phonon which softens with decreasing temperature (i.e. a
ferroelectric soft mode) but for which the softening is not complete

down to the lowest temperatures, and thus the crystals remain stable
with respect to this mode. Such crystals have become known as incipient ferroelectrics. Several such crystals have been investigated
at high pressure, and the results exhibit the expected relatively
large increase in soft mode frequency and decrease in the extrapolated transition temperature T_0 (see Eq. 3) with increasing pressure.
Examples include $SrTiO_3$,[12,20] $KTaO_3$,[20,21] TiO_2,[22] PbF_2,[23]
$KMnF_3$,[24] and $Sn_{1-x}GeTe$ alloys.[25] The reader is referred to the
cited papers for details.

IV. DISPLACIVE ANTIFERRODISTORTIVE SOFT MODE TRANSITIONS

Whereas for ferroelectric transitions the physical origin of the
soft mode behavior has been understood in terms of a cancellation
of the positive short range forces by the negative Coulomb forces,
the origin of the soft mode behavior in so-called antiferrodistortive transitions has not been known. However, some recent
results [8,26,27] based on high pressure data and lattice dynamical calculations have shed considerable light on this question.
By antiferrodistortive transitions we refer to those transitions
driven by soft zone boundary (or short wavelength) optic phonons.
Displacive antiferroelectric transitions are a sub-set of these.

There are now many known examples of displacive antiferrodistortive transitions and most of these have been studied at high
pressure. In all cases the transition temperature _increases_ with
pressure.[26] This is in marked contrast with the behavior of the
soft mode ferroelectrics where T_c always _decreases_ with pressure.
We illustrate the behavior of the antiferrodistortive transitions
by two examples: $Gd_2(MoO_4)_3$ and $BaMnF_4$.

Gadolinium molybdate, $Gd_2(MoO_4)_3$, and its rare-earth isomorphs
crystallize in a tetragonal structure (space group $D_{2d}^3 - P\bar{4}2_1m$) at
high temperature and transform on cooling to an orthorhombic phase
(space group $C_{2v}^8 - Pba2$). The transition is driven by a soft zone-
boundary optic phonon at the $M(\frac{1}{2},\frac{1}{2}, 0)$ point of the Brillouin zone
whose frequency obeys Eq. 3.[28] The ionic displacements associated
with this mode consist primarily of counter rotations of adjacent
MoO_4 tetrahedra. The orthorhombic phase is actually ferroelectric,[29]
but this ferroelectricity is accidental being a manifestation of
anharmonic coupling between the soft mode displacements and zone-
center optic and acoustic mode displacements.[28] The ferroelectricity, but not the transition, disappears when the crystal is clamped.[29]

Figure 7 shows the temperature dependence of the static dielectric constant of $Gd_2(MoO_4)_3$ in the vicinity of the transition temperature measured along the orthorhombic c-axis. The transition is
first order and its temperature T_c exhibits a large increase with
pressure, the initial slope being $dT_c/dP = 33.5$ K/kbar. This is

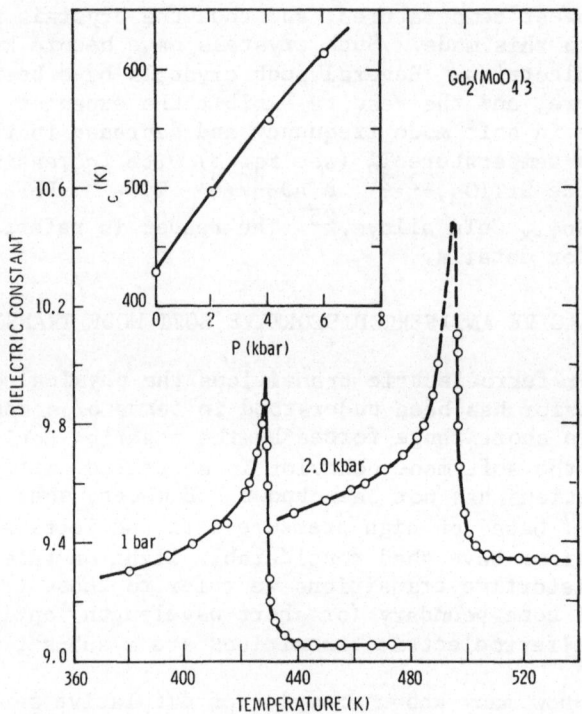

Figure 7.   Isobars of the temperature dependence of the dielectric
constant of $Gd_2(MoO_4)_3$ along the tetragonal c-axis showing the effect of pressure on the antiferrodistortive
transition temperature.

in close agreement with that reported by Shirokov et al.[30]   The
inset shows that $T_c(P)$ is somewhat nonlinear over the 6-kbar range
of the data.

Barium maganese fluoride, $BaMnF_4$, is orthorhombic (space group
C_{2v}^{12} - $A2_1am$) at room temperature, the structure consisting of layered sheets of linked MnF_6 octahedra with the Ba ions located between the layers.[31]   On cooling it undergoes an apparently second-
order structural transition at 247 K.[31]   The structure of the low
temperature phase is still unresolved.   Raman scattering results
indicated that the transition is driven by a soft zone boundary
optical phonon,[32] but more recent inelastic neutron scattering
data[33] indicated that the wavelength of the mode is actually incommensurate with the lattice.   Despite this complication, the soft
mode is a short wavelength optic phonon which most likely involves
rotations of adjacent MnF_6 octahedra.

Figure 8 shows several isobars of the static dielectric constant in the vicinity of T_c measured along the orthorhombic a-axis

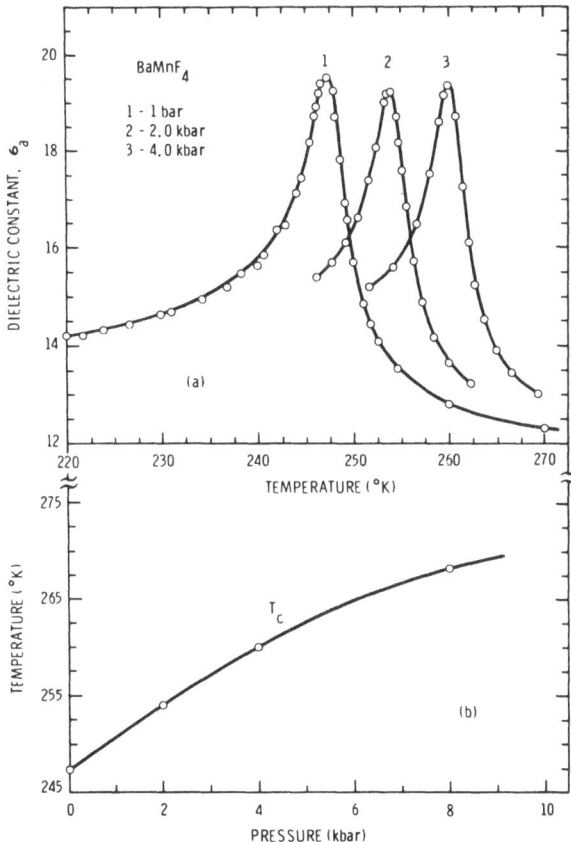

Figure 8.   (a)   Isobars of the dielectric constant of BaMnF$_4$ vs.
                  temperature.
            (b)   Pressure dependence of the antiferrodistortive tran-
                  sition temperature.

and the shift of T$_c$ with pressure.[31] The qualitative similarity of
these results to those on Gd$_2$(MoO$_4$)$_3$ in Fig. 7 is evident.

    Table 1 summarizes results on a variety of antiferrodistortive
transitions including the best understood and studied cases SrTiO$_3$
and KMnF$_3$. For contrast, results on a variety of soft mode ferroelectric transitions are given. The Table emphasizes what was mentioned earlier: in all cases where the transition is driven by a
soft zone center (z.c.) TO mode (i.e. a ferroelectric mode) T$_c$
decreases with pressure, whereas in all cases where the transition
is driven by a soft zone boundary (z.b.) or short wavelength mode
T$_c$ increases with pressure. This contrast is most striking in crystals like SrTiO$_3$, PbZrO$_3$ and PbHfO$_3$ which have both kinds of soft
modes in the same crystal and where the associated transition tem-

Table I. Pressure dependence of the transition temperature T_C for a variety of crystals which exhibit displacive structural phase transitions. z.b. and z.c. stand for Brillouin zone boundary and zone center, respectively. (Table taken from Ref. 26)

Crystal	Symmetry of Hi Temp Phase	Soft Mode (Sym.)	T_C^a (°K)	dT_c/dP (°K/kbar)
$BaMnF_4$	Ortho.$-C_{2v}^{12}$	z.b.	47.3	3.3
$Gd_2(MoO_4)_3$	Tetrag.$-D_{2d}^3$	z.b.	430	33.5
$(NH_4)_2Cd_2(SO_4)_3$	Cub. $-T^4$	z.b.	92	3.3
$SrTiO_3$	Cub.$-O_h^1$	z.b. (R_{25})	110	1.7
"		z.c. (Γ_{15})	36	-14.0
$KMnF_3$	Cub.$-O_h^1$	z.b. (R_{25})	186	3.0
$CsPbCl_3$	Cub.$-O_h$	z.b. (M_3)	320	7.6
"		z.b. $(R_{25})^b$	315	5.2
"		z.b. $(R_{25})^b$	311	5.4
$PbZrO_3$	Cub.$-O_h^1$	z.b.	507	4.5
"		z.c. (Γ_{15})	475	-16.0
$PbHfO_3$	Cub.$-O_h^1$	z.b.	434	5.9
"		z.c. (Γ_{15})	378	-10.0
$BaTiO_3$	Cub.$-O_h^1$	z.c. (Γ_{15})	393	-5.2
$PbTiO_3$	Cub.$-O_h^1$	z.c. (Γ_{15})	765	-8.4
$SbSI$	Ortho.$-D_{2h}^{16}$	z.c.	292	-37.0
$Pb_5Ge_3O_{11}$	Hexag.$-C_{3h}^1$	z.c.	450	-6.7

a – T_C is either the actual transition temperature or the Curie-Weiss temperature deduced from the static susceptibility.
b – These modes derive from the R_{25} mode of the cubic phase.

peratures exhibit opposite effects. This appears to be a general result for ferroelectric and antiferrodistortive soft mode transitions
to which we know of no exception at present. What are the lattice
dynamical implications of these results?

As can be deduced from Eq. 3, the increase of T_c with pressure
for the soft z.b. phonon transitions results from the softening of
the soft mode frequency with pressure. The clue for the explanation of this result was suggested by the results of some lattice
dynamical calculations on $SrTiO_3$.[34,35] These results show that for
the z.b. modes at the $(\frac{1}{2} \frac{1}{2} \frac{1}{2})R$ and $(\frac{1}{2} \frac{1}{2} 0)M$ points of the Brillouin
zone the short-range interaction is negative (i.e. attractive) where-
as the electrostatic interaction is positive. These results and
their implications do not appear to have been appreciated until recently.[8,26,27] They indicate that, for the modes in question, the
balance of forces leading to an imaginary harmonic frequency, and
therefore soft mode behavior, is due to the overcancellation of the
positive electrostatic interactions by the negative short-range interactions. This situation is opposite to that in ferroelectrics. In
other words, for the z.b. soft modes the roles of the short-range
and electrostatic interactions are reversed from what they are for
the ferroelectric case. It has been suggested that this reversal
in the roles of forces is a general phenomenon applicable to all
the crystals in Table I.[26] This generalization then provides a
ready explanation for the increase of T_c with pressure for the soft
z.b. phonon transition. By analogy to Eq. 7b we write

$$\mu\omega_o^2 \propto [(\text{Electrostatic interaction}) - (\text{S.R. interaction})]. \quad (8)$$

Since pressure increases the magnitude of the short-range forces
more rapidly than the magnitude of the electrostatic forces, ω^2 becomes more negative and a higher temperature is needed to provide
the necessary stabilization of the high temperature phase by anharmonic interactions, i.e. a higher T_c.

An interesting observation based on the above discussion is
that with increasingly higher pressure, and as the ions get closer
and closer, the short-range interaction could not be expected to
continue to be attractive. Ultimately they should become repulsive.
If this were the case, then a reversal in the sign of dT_c/dP might
be expected at sufficiently high pressure. Such a reversal has
been observed [31] in $BaMnF_4$ at ~ 10 kbar. This observation might
reflect the expected reversal in the sign of the short-range interactions.

V.   COUPLED PHONON-PROTON SOFT MODE TRANSITIONS

Potassium dihydrogen phosphate (KH_2PO_4 or KDP) and its isomorph
are an important class of soft mode hydrogen-bonded ferroelectrics

in which the protons play an essential role in determining the transition and related physical properties. This role is most evidenced by the unusually large effects on both the static and dynamic properties observed on deuteration.[36] For example the Curie temperature T_c of KDP increases from 122 K to ~ 230 K on replacing the two protons by deuterons. Pressure studies on these materials have turned out to be crucial to the understanding of the nature of soft mode and the transition. Among the important pressure results are: (i) the vanishing of the ferroelectricity at high pressure,[37] (ii) the demonstration that the soft mode, which is overdamped at all temperatures at 1 bar, can be made underdamped at high pressure.[38,39] This resolved a long-standing question as to whether the mode is a collective, propagating excitation or a single particle-like diffusive excitation, and (iii) the demonstration that the proton motion remains coupled to an optic phonon in the low temperature phase.[40] Items (ii) and (iii) are considered in Peercy's paper.[19] Here we comment briefly on the pressure dependence of T_c and the vanishing of the ordered state. Specifically, we consider KDP.

In the high temperature phase KDP is tetragonal, the structure consisting of tetrahedral PO_4 groups connected by a network of O-H...O hydrogen bonds which lie very nearly in the basal plane with proton motion predominantly perpendicular to the c-axis. The polarization in the ferroelectric state is determined by the displacements of the $K-PO_4$ groups along the c-axis. The presently accepted lattice dynamical picture for the transition is as follows.[41] In the high temperature phase the protons tunnel between two minima of a double-well potential along the O-H...O bonds with the proton motion coupled to the $K-PO_4$ TO phonon with polarization along the c-axis. At the transition the protons order preferentially in the potential minima and consequently $K-PO_4$ displacements become frozen in. The picture is then one of a coupled proton-phonon motion. A detailed approximate treatment of this coupled motion was given by Kobayashi,[41] who on solving the coupled system Hamiltonian for the coupled proton tunneling-optic mode frequencies and requiring that the lower frequency branch ω_- vanish at T_o, obtained the following expression for T_o

$$4\Omega/\tilde{J} = \tanh(\Omega/kT_o) , \qquad\qquad (9)$$

where Ω is the proton tunneling frequency and $\tilde{J}$ accounts for the dipolar interaction among protons and for the proton-lattice coupling. This equation has solutions and therefore a finite T_o, only when $(4\Omega/\tilde{J}) < 1$. The physical picture represented by Eq. 9 is as follows: T_o is determined by the competition between two fields, (i) a dipolar field represented by $\tilde{J}$ which tends to order the protons and induce ferroelectricity, and (ii) a transverse tunneling field represented by Ω which tends to disorder the protons and stabilize the high temperature phase. At 1 bar the dipolar field dominates, so that

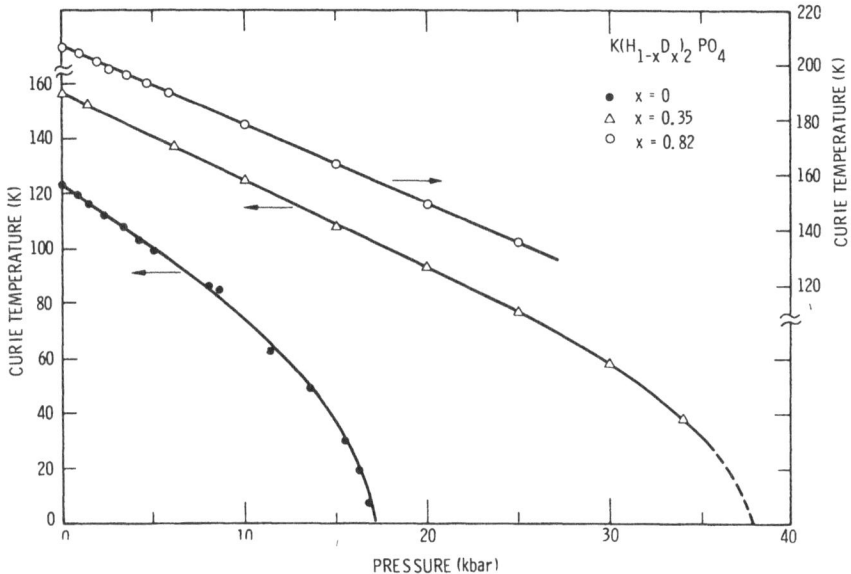

Figure 9. Pressure dependence of the ferroelectric transition
temperature of KH_2PO_4 and two partially deuterated crystals.

$(4\Omega/\tilde{J}) < 1$, and there is a finite T_o.

It is expected,[37,42] and has been confirmed experimentally,[40,42]
that Ω increases and $\tilde{J}$ decreases with pressure. This enhances the
disordering tendency of the tunneling field with respect to the
ordering tendency of the dipolar field and leads to a lowering of
T_o. At sufficiently high pressure the ratio $(4\Omega/\tilde{J})$ should become
≥ 1 so that no ordering is possible and the transition vanishes.

These predictions are confirmed by experiments as can be seen
for KDP in Fig. 9. In fact the $T_c(P)$ data in this figure are well-
fitted by Eq. 9 (solid curve).[43] The fit is quite good and Eq. 9
correctly predicts the observed result that $(dT_c/dP) \rightarrow -\infty$ as $T_c \rightarrow$
0 K. Figure 9 also shows results for two partially deuterated crys-
tals. The data for the 35% deuterated crystal are qualitatively
similar to those for KDP, although we were not able to completely
suppress T_c for this crystal because of the limited pressure range
of the apparatus used. The data on the 82% deuterated sample extend to only 25 kbar, and $T_c(P)$ is linear over this range. It would
be of considerable interest to extend the pressure measurements on
this and on completely deuterated crystals to much higher pressure
and observe how T_c vanishes.

Finally we mention briefly the case of the antiferroelectric
members of the KDP class of hydrogen-bonded crystals exemplified
by ammonium dihydrogen phosphate ($NH_4H_2PO_4$ or ADP) and its isomorphs.
The high temperature phase of these crystals is isomorphous with
KDP, and the crystals exhibit deuteration effects on T_c similar to
those for KDP; however, the ordering of the protons in the hydrogen bonds at T_c is such as to result in antiferroelectric low temperature phases.[37]   The effect of pressure on T_c of ADP has been
reported[37] and is qualitatively similar to that for KDP: T_c
<u>decreases</u> with pressure and ultimately vanishes with $dT_c/dP \rightarrow -\infty$.
Thus, for these hydrogen-bonded crystals T_c <u>decreases</u> with pressure
for both the ferroelectric and antiferroelectric transitions. This
is in marked contrast to the behavior of the displacive soft mode
transitions described in Secs. III and IV where T_c decreases with
pressure for ferroelectric transitions and increases for antiferroelectric transitions. This difference in behavior emphasizes the
fact that it is the proton motion in the hydrogen bond that plays
the key role in KDP-type crystals.[37]

VI.   OTHER SOFT MODE TRANSITIONS

In this section we merely want to draw attention to a few interesting though unrelated, soft mode transitions for which pressure
studies have proven to be very useful. Specifically we mention the
transitions in TeO_2, the rare-earth pentaphosphates and the β-tung-
sten superconductors.

Paratellurite, TeO_2, undergoes a pressure-induced second-order
transition at ~ 9 kbar and 295 K from the tetragonal structure
(D_4^4 - $P4_12_12$) to an orthorhombic structure (D_2^4 - $P2_12_12_1$).[44,45]
This was the first known example of a pure strain-induced transition. The transition is driven by a soft shear acoustic mode propagating along a <110> crystal direction and polarized along a <1$\bar{1}$0>.
The velocity associated with this mode softens slightly with decreasing temperature, but the transition cannot be induced by temperature
alone; pressure is thus essential for studying this transition. A
phenomenological theory for the transition has been developed,[46] but
the microscopic origin of the instability leading to the transition
remains unknown.

The rare earth pentaphosphates REP_5O_{14} (where RE is one of the
rare earth ions Nd, La or Tb) crystallize in an orthorhombic structure (point group D_{2h}) at high temperature and transform on cooling
to a monoclinic structure (point group C_{2h}).[19]   The monoclinic
phase is ferroelastic, i.e. its domains can be oriented by the application of a mechanical stress. The transition is accompanied
by a soft long wavelength Raman-active (nonpolar) optical phonon

which interacts strongly near T_c with an acoustic phonon governed by the elastic constant C_{55}. It is C_{55} which vanishes at T_c thereby precipitating the transition. The effects of pressure on the Raman spectra and T_c of some members of this group have been investigated by Peercy.[19] It is found that T_c increases with pressure, a behavior that may be typical of ferroelastic transitions, but it is not at present understood from a lattice dynamical viewpoint. More work is needed to better understand these materials.

The cubic β-tungsten (orA-15) are best known for their high temperature superconductivity. Also important is the fact that several members of this family exhibit, on cooling, a phase transition from cubic to tetragonal ($D_{4h}^9 - P4_2/mmc$) at a temperature T_L somewhat higher than the superconducting transition temperature T_c. The structural transition is associated with a soft shear acoustic mode propagating along <110> and polarized along <1$\bar{1}$0>[47]. There has been some evidence linking this lattice instability and the high superconducting T_c.[47]

Pressure studies have yielded some important results on V_3Si and Nb_3Sn, the two most investigated members of the A-15 family.[48,49] It has been found that for V_3Si, T_L decreases with pressure whereas T_c increases.[48] The opposite is true for Nb_3Sn.[49] These opposite pressure effects on T_L and T_c in the two isomorphous compounds have been interpreted in terms of a pressure-induced electron transfer from the s to the d bands.[49] A more detailed discussion of these results and other pressure effects on these compounds is given in the paper by Chu.[50]

VII. CONCLUDING REMARKS

In this paper we have given a brief overview of the soft mode concept and of high pressure studies of soft mode transitions in solids. It was mentioned that the decrease of the frequency of the soft mode as the transition is approached is caused by a balance between competing forces, and that this balance can be strongly influenced by hydrostatic pressure. The effects of pressure on soft mode behavior in different classes of soft mode transitions were discussed briefly with emphasis on the physical implications of the results. Comparisons of pressure effects among the different classes reveal general trends which have led to a better understanding of the nature of the competing forces involved.

The results presented and the literature cited emphasize the important — often essential, role of high pressure research in the study of soft mode transitions. Such research has not only yielded exciting new phenomena that occur at high pressure, but it has also improved our understanding of the response of materials at atmospheric pressure. There can be no doubt that such research will always be important in the study of phase transitions.

ACKNOWLEDGMENT

It is a pleasure to acknowledge the expert experimental assistance of B.E. Hammons.

REFERENCES

1. W. Cochran, Adv. Phys. $\underline{9}$, 387 (1960).
2. P.W. Anderson, in $\underline{\text{Fizika Dielektrekova}}$ (G.I. Shanski, Ed.), Acad. Sci. USSR, Moscow (1960).
3. See, e.g., J.F. Scott, Rev. Mod. Phys. $\underline{46}$, 83 (1974) and references therein.
4. P.S. Peercy and I.J. Fritz, Phys. Rev. Letters $\underline{32}$, 466 (1974).
5. G.A. Samara, Proc. 4th International Conference on High Pressure, The Physico-Chemical Soc. Japan, Kyoto (1974) p. 247.
6. E. Pytte, Comments Sol. St. Phys. $\underline{5}$, 41 (1973).
7. R.A. Cowley, Phys. Rev. $\underline{134}$, A981 (1964).
8. G.A. Samara, T. Sakudo and K. Yoshimitsu, Phys. Rev. Letters $\underline{35}$, 1767 (1975); G.A. Samara, Comments Sol. St. Phys. $\underline{8}$, 13 (1977).
9. G.A. Samara, Ferroelectrics $\underline{9}$, 209 (1975).
10. See e.g., D.K. Agrawal and C.H. Perry, Phys. Rev. B $\underline{4}$, 1893 (1971).
11. In this paper we use two transition temperatures T_o and T_c. T_o is the temperature where the soft mode frequency goes to zero or extrapolates to zero. T_c is the temperature where the structural transition actually takes place. For a second-order transition T_o and T_c are one and the same temperature.
12. G.A. Samara in $\underline{\text{Advances in High Pressure Research}}$, R.S. Bradley, Ed. (Academic Press, N.Y., 1969) Vol. 3, Chapt. 3.
13. T.R. Volk, E.I. Gerzanich and V.M. Fridkin, Bull. Acad. Sci. USSR, Phys. Ser. (USA) $\underline{33}$, 319 (1969).
14. P.S. Peercy, Phys. Rev. Letters $\underline{35}$, 1581 (1975).
15. V.H. Schmidt, A.B. Western and A.G. Baker, Phys. Rev. Letters $\underline{37}$, 839 (1976).
16. G.A. Samara, Ferroelectrics $\underline{2}$, 277 (1971).
17. R. Clarke and L. Benguigui, J. Phys. C: Solid State Phys. (to be published).
18. V.H. Schmidt, Proceedings of this Conference.
19. P.S. Peercy,Proceedings of this Conference and references therein.
20. R.P. Lowndes and A. Rastogi, J. Phys. C: Solid State Phys. $\underline{6}$, 932 (1973).
21. G.A. Samara and B. Morosin, Phys. Rev. B $\underline{8}$, 1256 (1973).
22. G.A. Samara and P.S. Peercy, Phys. Rev. B $\underline{7}$, 1131 (1973).
23. G.A. Samara, Phys. Rev. B $\underline{13}$, 4529 (1976).

24. We have just recently discovered the evidence for a soft zone center TO phonon in $KMnF_3$.

25. S.S. Kabalkina, N.R. Serebryanaya and L.F. Vereshchagin, Sov. Phys.-Solid State 9, 2527 (1968).

26. See Ref. 8 and G.A. Samara, Comments Sol. St. Phys. 8, 13 (1977).

27. K. Yoshimitsu, Prog. Theor. Phys. 54, 583 (1975).

28. J.D. Axe, B. Dorner and G. Shirane, Phys. Rev. Letters 26, 519 (1971).

29. L.E. Cross, A. Fouskova and S.E. Cummins, Phys. Rev. Letters 21, 812 (1968).

30. A.M. Shirokov, V.M. Mylov, A.I. Baranov and T.M. Prokhortseva, Sov. Phys.-Solid State 13, 2610 (1972).

31. See G.A. Samara and P.M. Richards, Phys. Rev. B 14, 5073 (1976) for references and details.

32. J.F. Ryan and J.F. Scott, Solid State Commun. 14, 5 (1974).

33. S.M. Shapiro, R.A. Cowley, D.E. Cox, M. Eibschutz and H.J. Guggenheim, Proc. Conf. on Neutron Scattering, Gatlingburg, Tenn. (1976).

34. W.G. Stirling, J. Phys. C: Solid State Phys. 5, 2711 (1972).

35. A.D. Bruce and R.A. Cowley, J. Phys. C: Solid State Phys. 6, 2422 (1973).

36. G.A. Samara, Ferroelectrics 5, 25 (1973).

37. G.A. Samara, Phys. Rev. Letters 27, 103 (1971).

38. P.S. Peercy, Phys. Rev. Letters 31, 380 (1973).

39. R.P. Lowndes, N.E. Tornberg and R.C. Leung, Phys. Rev. B 10, 911 (1974).

40. P.S. Peercy, Solid State Commun. 16, 439 (1975).

41. K. Kobayashi, J. Phys. Soc. Japan 24, 497 (1968).

42. P.S. Peercy and G.A. Samara, Phys. Rev. B 8, 2033 (1973).

43. The actual transition temperature T_c in KDP differs from T_0 in Eq. 9 by ~ 5 K because of optic-acoustic coupling. This coupling effect can be neglected for the purpose of this discussion.

44. P.S. Peercy, I.J. Fritz and G.A. Samara, J. Phys. Chem. Solids 36, 1105 (1975).

45. T.G. Worlton and R.A. Beyerlein, Phys. Rev. B 12, 1899 (1975).

46. I.J. Fritz and P.S. Peercy, Solid State Commun. 16, 1197 (1975).

47. L.R. Testardi, Comments Sol. St. Phys. 6, 131 (1975).

48. C.W. Chu and L.R. Testardi, Phys. Rev. Letters 32, 766 (1974).

49. C.W. Chu, Phys. Rev. Letters 33, 1283 (1974).

50. C. W. Chu, Proceedings of this Conference.

QUESTIONS AND COMMENTS

V.H. Schmidt: According to your results, deuteration increases the pressure at which the transition goes to zero, and finally it looks like you aren't going to get there at all. The coupling is a cooperative phenomenon. Is this possibly some kind of percolation phenomenon, that if you get too many deuterons in there, the tunneling can't percolate through the hydrogen network?

G.A. Samara: I haven't thought about that problem in this sense. There is no evidence for any percolation limit in the KDP-DKDP system.

B.T. Matthias: You said that there was no exception to your rule, however, the ammonium cadmium sulfate has the opposite sign.

G.A. Samara: I'm glad you raised this point, because I have an answer for that. There are two ferroelectrics on this list of zone boundary phonons: ammonium cadmium sulfate is one and gadolinium molybdate is the other one. Actually, it turns out that the ferroelectricity in these two cases is rather accidental. The phase transitions in these two crystals are driven by soft zone boundary phonons and the ferroelectricity is accidental, being a manifestation of anharmonic coupling between the soft mode displacements and zone-center optic and acoustic mode displacements.

B.T. Matthias: No, because you have many isomorphs which do the same.

G.A. Samara: Yes, but in both of these cases, the zone boundary phonons trigger the transformation. Gadolinium molybdate is really the classic example; its ferroelectricity is completely accidental.

B.T. Matthias: But this is wrong.

G.A. Samara: No, it's not (see Ref. 28). For example, if I clamp the crystals, the ferroelectricity goes away. However, the transition remains.

B.T. Matthias: The negative sign is the crucial feature of your presentation.

G.A. Samara: That's right. But the crucial question I think is this. What causes the transition? What soft phonon causes the transition?

B.T. Matthias: Then, let me ask you another question which I didn't understand. You quoted as ferroelectrics, potassium manganese fluoride and rutile. What was your evidence?

G.A. Samara: The evidence for that is a soft TO phonon. I
                    call these incipient ferroelectrics.

B.T. Matthias: But maybe I'm just too old fashioned, but to me
                    the definition of a ferroelectric is a dielectric
                    hysteresis loop.

G.A. Samara: Yes. I tried to qualify that by calling these
                    crystals incipient ferroelectrics, materials that
                    would like to become ferroelectrics, but they never
                    quite make it. In the same sense, of course,
                    strontium titanate and potassium tantalate al-
                    most make it. In fact, if you dope these latter
                    crystals slightly you can induce ferroelectric
                    transitions.

B.T. Matthias: Never mind. Potassium manganese fluoride will
                    never do it. We tried it many times.

G.A. Samara: But it exhibits a Curie-Weiss susceptibility which
                    is unlike normal dielectrics.

B.T. Matthias: Oh, many do it. Potassium manganese fluoride,
                    and TiO_2 particularly. Paul Chu tried everything
                    to make it ferroelectric and this was impossible!
                    So, I don't understand why today ferroelectricity
                    is mentioned for these two.

G.A. Samara: Well, I perhaps should have emphasized very strong-
                    ly what I really meant by incipient ferroelectrics.
                    I'm glad you raised this point. These involve
                    soft TO phonons, which behave unlike normal dielec-
                    trics. For example, normal dielectrics such as
                    sodium chloride, as you raise the temperature, the
                    phonon frequency decreases, the lattice becomes
                    softer, the opposite situation happens in all the
                    present crystals.

B.T. Matthias: They don't have anything to do with ferroelectric-
                    ity.

G.A. Samara: Well, I guess that revolves on whether you be-
                    lieve in the soft TO phonon description of ferro-
                    electric transitions. The crystals in question
                    are not ferroelectrics in the usual sense. I hope
                    everybody in the audience now knows what I mean
                    by incipient ferroelectrics.

PRESSURE DEPENDENCE OF RAMAN SCATTERING IN FERROELECTRICS AND

FERROELASTICS*

P.S. Peercy

Sandia Laboratories

Albuquerque, New Mexico 87115

ABSTRACT

Inelastic light scattering at high pressure from soft mode systems is briefly reviewed. Examples of the different types of soft-mode systems in which the pressure dependence of Raman and Brillouin scattering have been investigated are summarized. Special emphasis is given to three important systems: the coupled-proton-optic mode in hydrogen bonded ferroelectrics of the KDP class, the tricritical behavior in the ferroelectric SbSI, and optic-acoustic interactions in the ferroelastic rare-earth pentaphosphates.

INTRODUCTION

Hydrostatic pressure is a valuable complementary parameter to temperature for investigating phase transitions associated with soft modes. One of the first areas related to soft mode transitions in which pressure effects were investigated was that of static dielectric properties of ferroelectrics. In the past few years, measurements of the pressure dependence of inelastic light scattering has also been demonstrated to be a powerful technique for investigating soft mode systems. Light scattering permits one to directly measure the dynamic response of $k \approx 0$ soft modes as well as the order parameter of the phase transition. In this paper, pressure measurements of Raman scattering in soft mode systems are briefly reviewed. Table I lists examples of soft mode systems in which the pressure

*This work was supported by the United States Energy Research and Development Administration (ERDA) under Contract At(29-1)789.

Table 1: Types of soft-mode systems investigated by inelastic light
          scattering at high pressure.

- FERROELECTRICS -

$PbTiO_3$ - CLASSIC SOFT-MODE DISPLACIVE FERROELECTRIC
$BaTiO_3$ - PIEZOELECTRIC SOFT-OPTIC-ACOUSTIC COUPLING
KDP   - COUPLED PROTON-OPTIC MODE SYSTEM
SbSI   - TRICRITICAL BEHAVIOR; OPTIC-OPTIC INTERACTIONS

- FERROELASTICS -

TeO_2 - PURE STRAIN TRANSITION; SOFT ACOUSTIC MODE
REP_5O_{14} - NON-PIEZOELECTRIC SOFT-OPTIC-ACOUSTIC COUPLING

- UNIAXIAL STRESS -

$KTaO_3$ - INCIPIENT FERROELECTRICS; STRESS-INDUCED
$SrTiO_3$ FERROELECTRICITY
$Gd_2(MoO_4)_3$ - IMPROPER FERROELECTRIC

dependence of inelastic light scattering has been measured and indicates the primary points of interest in these systems. The list
is not meant to be complete but rather to represent the different
types of transitions investigated.

The pressure dependence of Raman scattering in $PbTiO_3$ was
measured by Cerdeira, et al.[1] $PbTiO_3$ is a ferroelectric perovskite
and represents a classic displacive ferroelectric transition in which
the interest was in investigating the soft mode. $BaTiO_3$ is another
ferroelectric perovskite. In $BaTiO_3$, the soft optic mode interacts
piezoelectrically with an acoustic mode near the tetragonal to orthorhombic phase transition.[2] The pressure dependence of this optic-
acoustic interaction was investigated by Peercy and Samara[3] using
both Raman and Brillouin scattering. (KDP and SbSI will be discussed
in more detail later). Paratellurite (TeO_2) and the rare-earth
pentaphosphates represent two types of ferroelastics in which the
order parameter is the spontaneous strain rather than the spontaneous polarization. TeO_2 does not exhibit a phase transition at

atmospheric pressure, but does undergo what appears to be an ideal second-order transition with pressure. The transition was first investigated by Peercy and Fritz[4] using Raman and Brillouin scattering combined with ultrasonic velocity measurements. No soft optic mode has been detected for this system so that the transition appears to be a pure strain transition[5] accompanied by a soft acoustic mode. The ferroelastic transition in NdP_5O_{14}, on the other hand, is accompanied by a soft zone center optic mode which is Raman active and underdamped in both phases.[6] This system will be discussed in more detail later.

For completeness, ferroelectrics for which uniaxial stress measurements have been performed are also listed. $KTaO_3$ and $SrTiO_3$ are incipient ferroelectrics in that they possess a $k{\sim}0$ optic mode whose frequency softens as the temperature is decreased; however, the soft mode frequency remains finite for all temperatures. Ferroelectricity can be induced by the application of uniaxial stress along appropriate symmetry directions and Uwe and Sakudo[7] have investigated Raman scattering from the soft mode for stress-induced ferroelectricity in these systems. $Gd_2(MoO_4)_3$ is an improper ferroelectric which undergoes a cell-doubling transition and the spontaneous polarization is a secondary order parameter. The uniaxial stress dependence of inelastic light scattering in this system was investigated by Ganguly and coworkers.[8]

These systems present an overview of the types of soft-mode systems for which pressure measurements of inelastic light scattering near the phase transition have been made to date. We now concentrate in more detail on KDP, SbSI and $Nd_{0.5}La_{0.5}P_5O_{14}$ to illustrate the types of information obtained from the pressure measurements of Raman scattering.

POTASSIUM DIHYDROGEN PHOSPHATE

Potassium dihydrogen phosphate (KH_2PO_4 or KDP) and its isomorphs comprise an important class of ferroelectrics which have been extensively studied. The soft mode of the phase transition is thought to be a coupled excitation of the proton-tunneling system and a transverse optic mode of the lattice.[9] At atmospheric pressure, the soft-mode response is overdamped for all temperatures in the paraelectric phase so the question arose as to whether the transition was an order-disorder transition governed by the proton motion or a soft-mode system reflecting the coupled response. The pressure measurements resolved this question as illustrated by the data in Fig. 1. At low pressure the lower frequency branch ω_- of the coupled system peaks at $\omega = 0$. As the pressure is increased, however, the ratio of the frequency ω_- to the damping Γ_- increases so that the response becomes underdamped, illustrating that the system exhibits soft-mode behavior.[10] The primary change in the spectra with pressure is due to a decrease in the damping Γ_- with pressure.

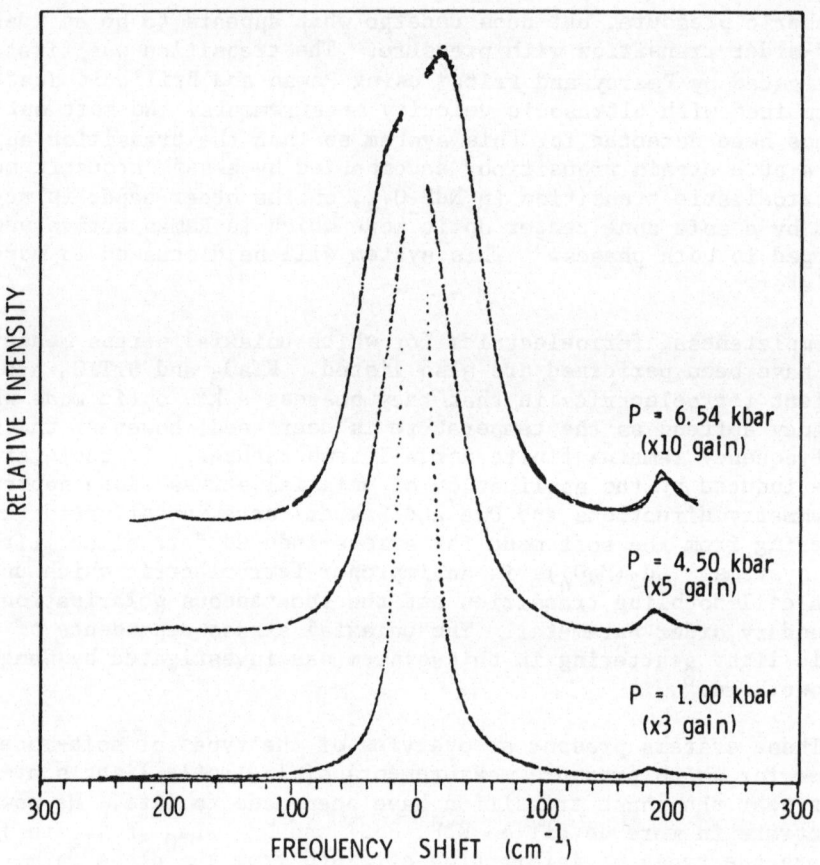

Figure 1.   Soft mode spectra of KDP for different pressures at
            140°K.

Another advantage of measurements at high pressure is that
because the response is underdamped, much more reliable fits to
the spectra can be obtained which permits a more accurate determination of the parameters for the coupled modes[11] which may be
compared with microscopic theories of the phase transition.   For
example, the measured temperature dependence of the proton-like
mode, obtained by decoupling the measured response assuming a
real interaction between the proton tunneling and optic mode, is
compared with the predicted temperature dependence in Fig. 2 for
p = 6.54 kbar.   The fit yields a proton-tunneling frequency of

89 cm^{-1} at this pressure and suggests that the proton system would undergo a transition at ~60°K in the absence of proton-lattice interaction.[12]

Similarly, the coupled response can be evaluated to obtain the more familiar soft mode behavior for these systems as shown in Fig. 3. These data, again for 6.54 kbar, illustrate that the coupled proton-lattice system exhibits mean-field behavior with $\omega_-^2 \propto (T_o - T)$ where the ~5° difference in T_o and T_c is produced by a piezo-electric optic-acoustic mode coupling[13c] analogous to that observed in $BaTiO_3$.

Pressure measurements have proven valuable in the ferroelectric phase of KDP where they were used to determine the soft mode of the system.[14] Even at high pressure, the response is overdamped near T_c so that it is not possible to follow the soft mode through T_c. However, at lower temperatures, the spectra become well-resolved as illustrated in Fig. 4. Since the transition temperature decreases with pressure, the system approaches T_c as p is increased for $T < T_c$. The soft mode, as well as any mode coupled to it, should thus "soften" as the pressure increases for $T < T_c$. This behavior is illustrated in Fig. 5 and demonstrates that the mode labeled ω_- is the soft mode for $T < T_c$ and the decrease in ω_+ with pressure indicates that the proton-motion remains coupled to the lattice in the ferroelectric phase.

The behavior of the phase transition in KDP is well described by the coupled proton-tunneling-optic-mode model for the transition, and the parameters obtained from Raman scattering at high pressure can be used to quantitatively evaluate the macroscopic dielectric properties within the framework of the model.[11]

Another question which a successful model must answer is the effect of deuteration on the phase transition in these materials. To investigate this effect, the pressure dependences of $\omega_\pm$ were measured versus deuteration for $T < T_c$.[15] The changes in these modes with deuteration are shown in Fig. 6 where it can be seen that both $\omega_\pm$ increase in frequency with increasing deuterium concentration while the proton-like branch ω_- of the coupled system decreases in intensity.

The changes in the pressure dependences of $\omega_\pm$ with deuteration are shown in Fig. 7. While both $\omega_\pm$ decreased in frequency with increasing pressure for KDP, for dKDP both modes display normal frequency increases with pressure. The solid curves in Fig. 7 are the pressure dependences calculated by Blinc et al.[16] who extended the coupled-mode model to include proton-proton, proton-deuteron and deuteron-deuteron interactions in partially deuterated

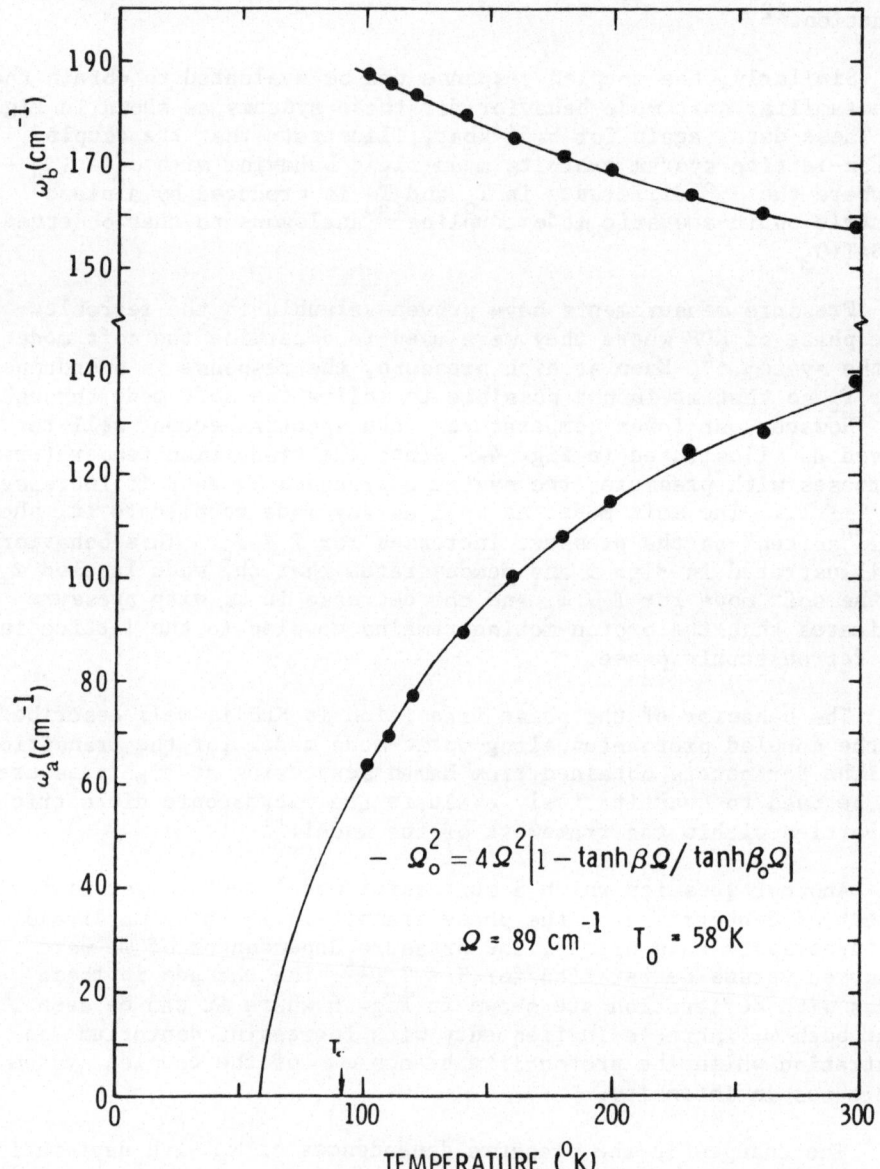

Figure 2.   Temperature dependences of the decoupled frequencies of
the coupled proton-optic mode system in KDP at 6.54
kbar.

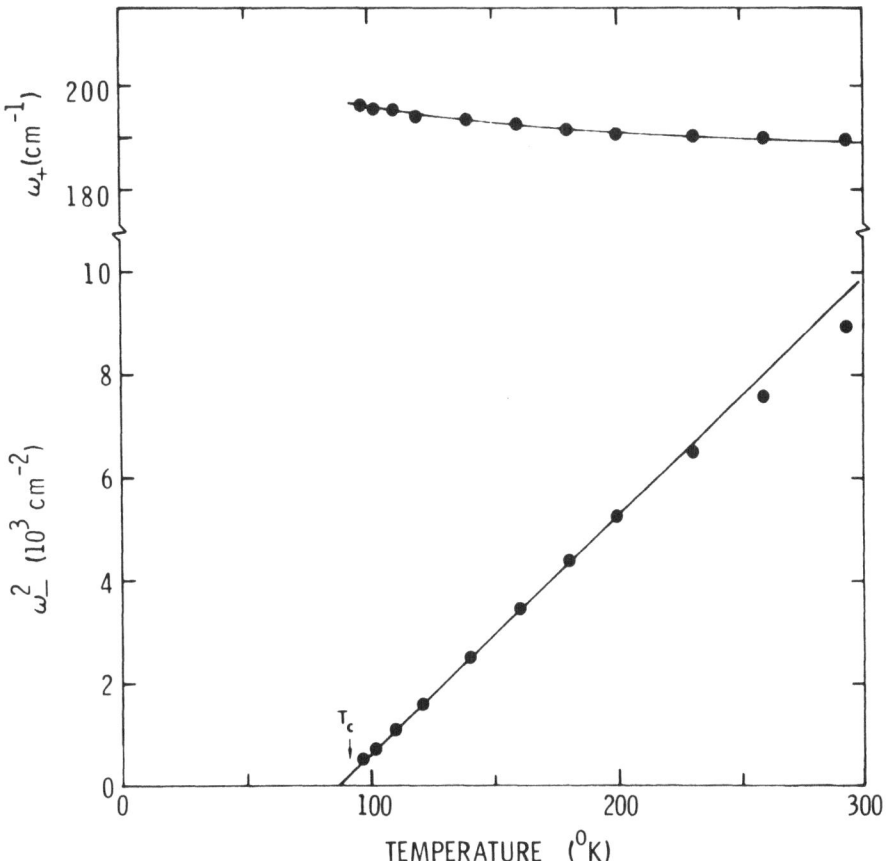

Figure 3.   Temperature dependence of the coupled response in the paraelectric phase of KDP at 6.54 kbar.   The soft mode is plotted as ω^2.

crystals. As can be seen from these results, the coupled-mode model can account not only for the behavior in KDP but also for the effects of deuteration on this system.

ANTIMONY SULPHOIODIDE

Antimony sulphoiodide (SbSI) is a semiconductor which undergoes a ferroelectric phase transition at 293°K accompanied by a soft transverse optic mode. Although by symmetry the transition is allowed to be second-order, at atmospheric pressure the transition

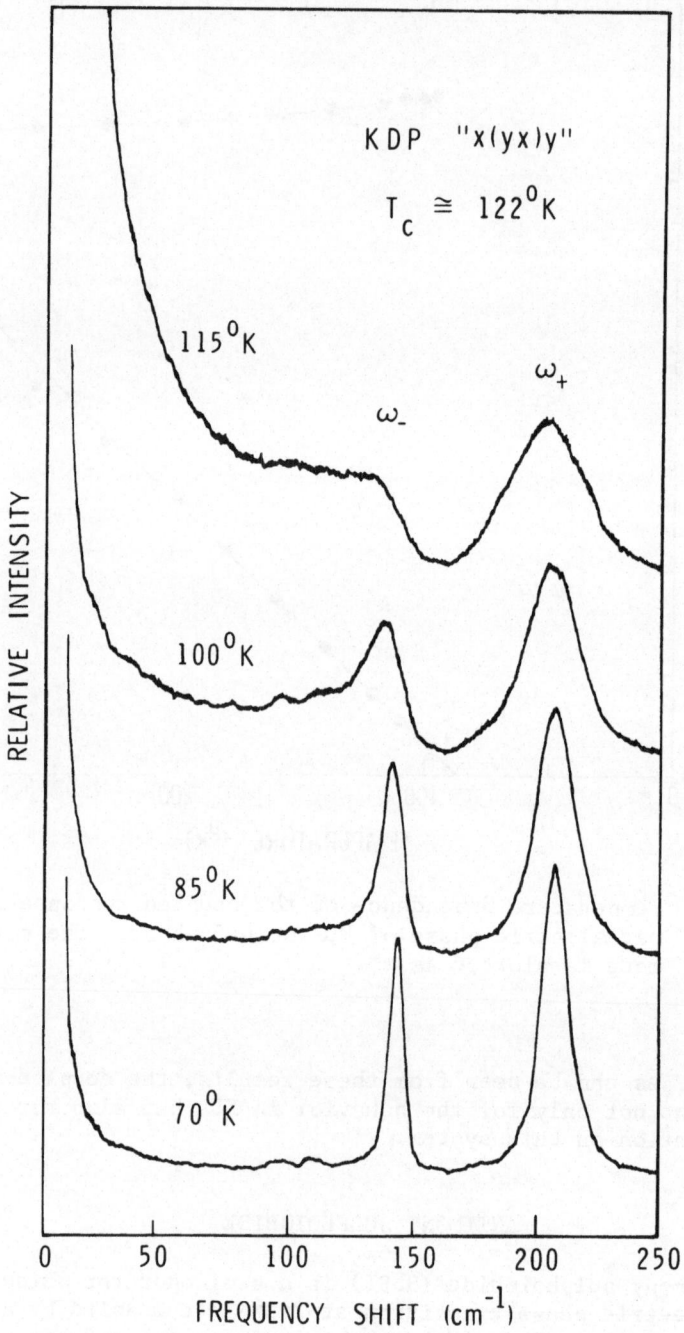

Figure 4. Temperature dependence of the coupled modes in KDP for
           T < T_c.

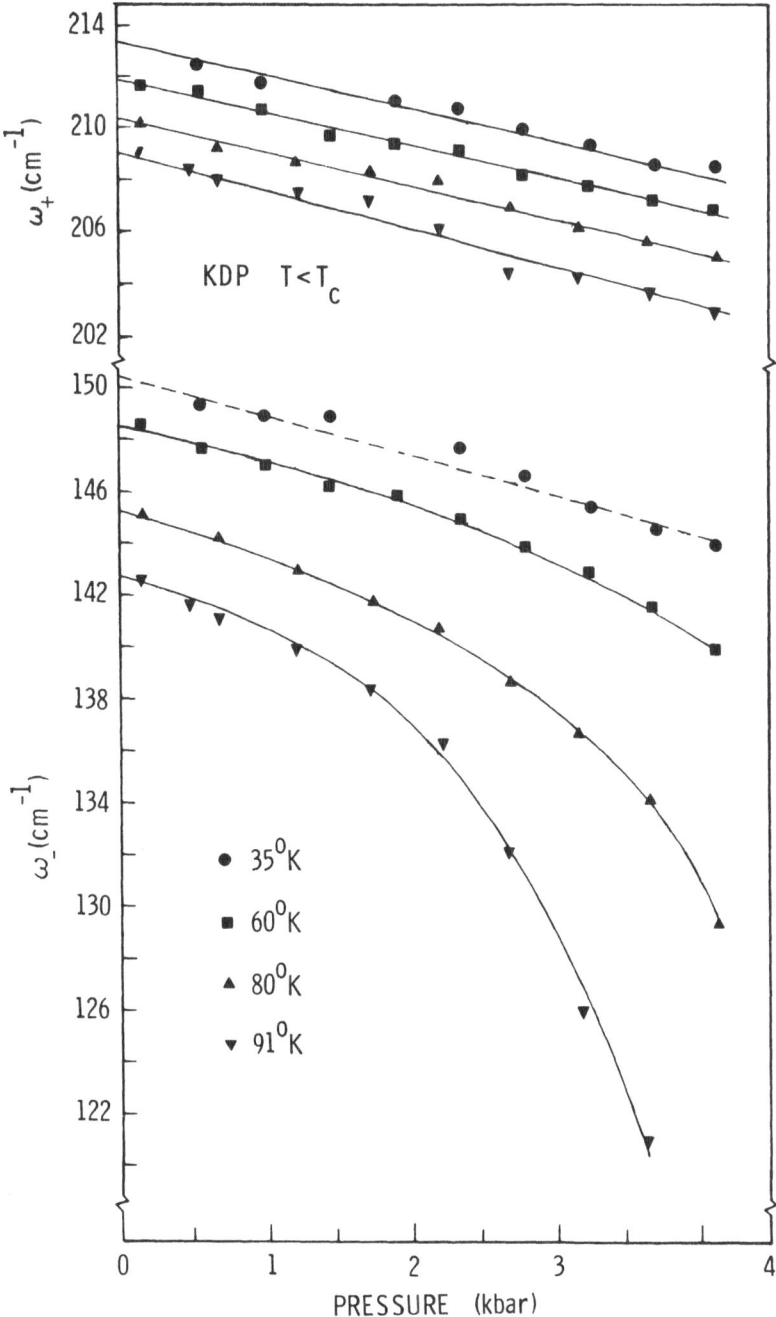

Figure 5. Pressure dependence of $\omega_{\pm}$ for various temperatures in
the ferroelectric phase.

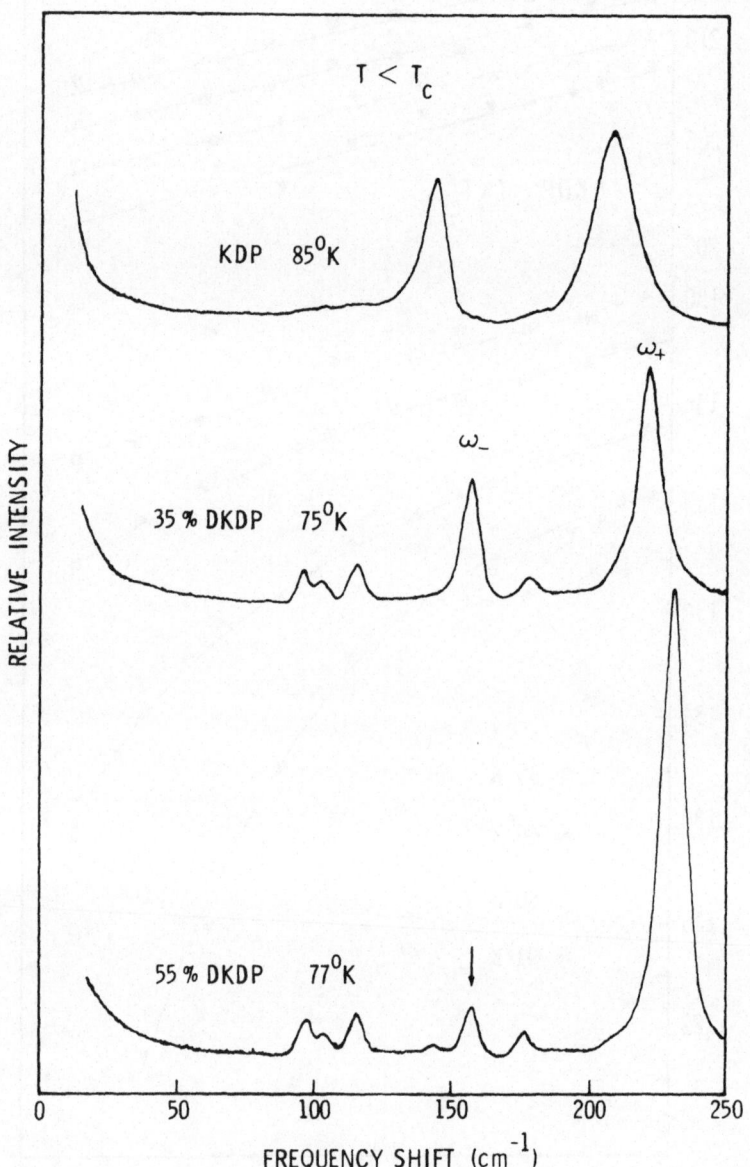

Figure 6.   Changes in $\omega_\pm$ with deuteration for $T \ll T_c$.

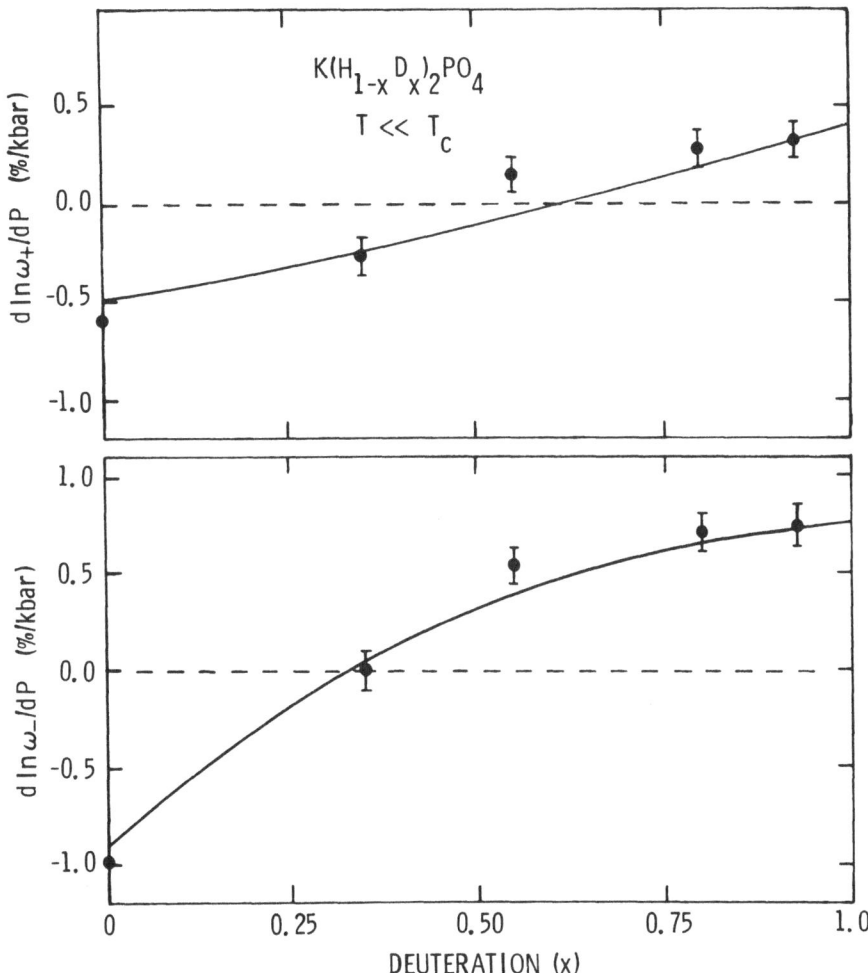

Figure 7. Changes in $(\partial \ell n \omega_{\pm}/\partial p)_T$ with deuteration for $T \ll T_c$.

is observed to be first order. From measurements of the pressure dependence of the Raman spectra, however, SbSI was found to exhibit a tricritical point[17] where the phase transition changed from first to second order. Such tricritical behavior has also been recently observed in KDP.[18]

The transition temperature in SbSI decreases very rapidly with pressure ($dT_c/dp = 40°K/kbar$) so that the ferroelectric phase may be completely suppressed by the modest pressure of <10 kbar.[19] Raman spectra taken at various pressures near the transition pressure p_c of 4.54 kbar at 119.4°K are shown in Fig. 8. The soft mode is not Raman-active in the paraelectric phase; by symmetry it is Raman active in the ferroelectric phase. As p is decreased below p_c, the soft mode appears as a shoulder on the laser line. On further decrease in p, the soft mode frequency increases and the mode becomes underdamped. Interactions between the soft mode and other optic modes[20] are observable near 32 cm^{-1} and 42 cm^{-1}.

The pressure dependence of the soft mode is shown in Fig. 9. From these data, the transition appears continuous at this temperature with the soft mode frequency extrapolating to zero at p_c. The system also appears to exhibit mean-field behavior with $\omega^2 \propto (p_c - p)$ in the pressure region below the optic mode interaction at ~32 cm^{-1}.

To further investigate the change in the order of the phase transition with pressure, we also monitored the intensity of the parasitic ($\omega = 0$) scattering and the intensity of optic modes which were Raman-active in both phases. Similar results were obtained for both methods and are illustrated in Fig. 10 where the pressure dependence of the intensity $I(0)$ of the $\omega = 0$ scattering is shown for two temperatures. Since the scattered light is collected from the bulk of the sample rather than the surface, $I(0)$ is dominated by the transmitted light.

At 272°K as p_c is approached $I(0)$ decreases as order parameter fluctuations increases and scatter light out of the laser beam. At p_c, $I(0)$ drops discontinuously as the crystal undergoes a first order transition and domains form which scatter the incident and collected light. Completely analogous results are obtained as a function of temperature at constant pressure and the transition exhibits considerable hysteresis in this temperature region.

As T_c is decreased with increasing pressure, the magnitude of the discontinuity in $I(0)$ and the hysteresis in p_c (or T_c) decreases until $T_c \simeq 235$°K. For transition temperatures below ~235°K, $I(0)$ changes continuously through the transition. This behavior is shown in the 234.2°K curve in Fig. 10. Here, $I(0)$ decreases smoothly as $p \to p_c$, attains a minimum at p_c, then increases smoothly as p moves away from p_c. No hysteresis is observed within experimental uncertainty (±0.002 kbar). SbSI thus appears to exhibit a tricritical point at $p_t \simeq 1.40$ kbar and $T_t \simeq 235$°K where the transition changes from first to second order. A more complete analysis of the Raman spectra[17] at various temperatures approaching T_t and of the dielectric constant[21] in this temperature and pressure region corroborate this conclusion. Unfortunately, detailed measurements of the spontaneous polarization have not yet

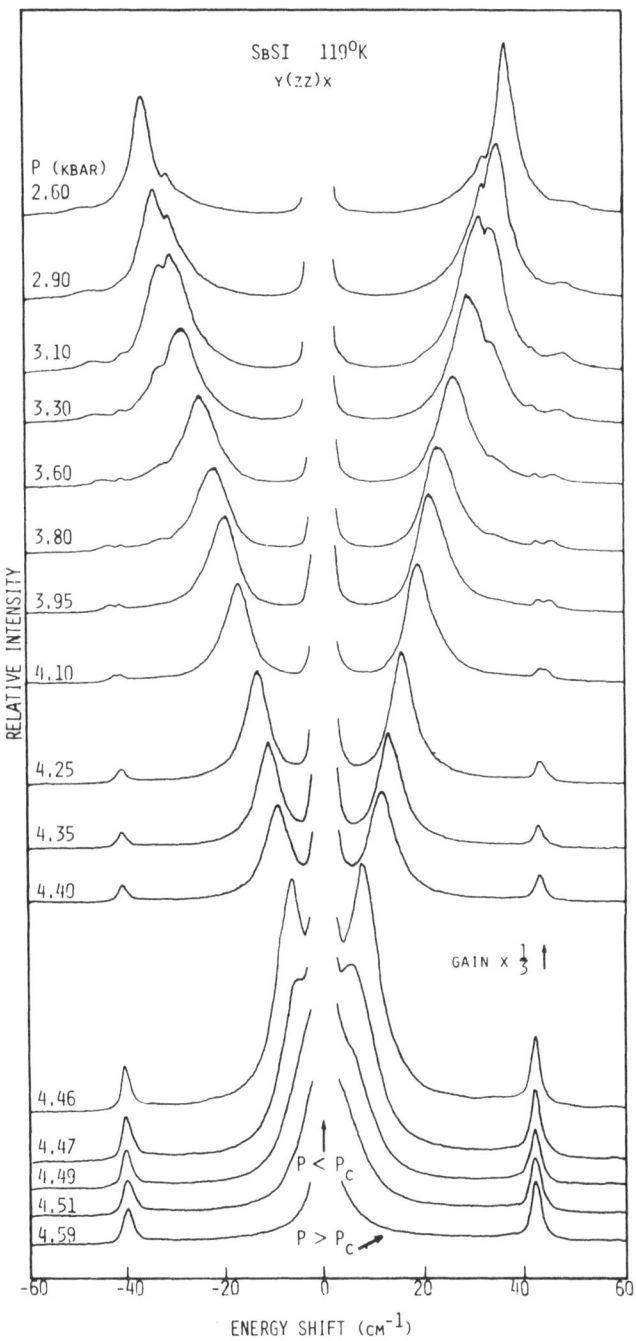

Figure 8. Raman spectra of the low frequency modes in SbSI at 119.4^{O}K for various pressures near p_c.

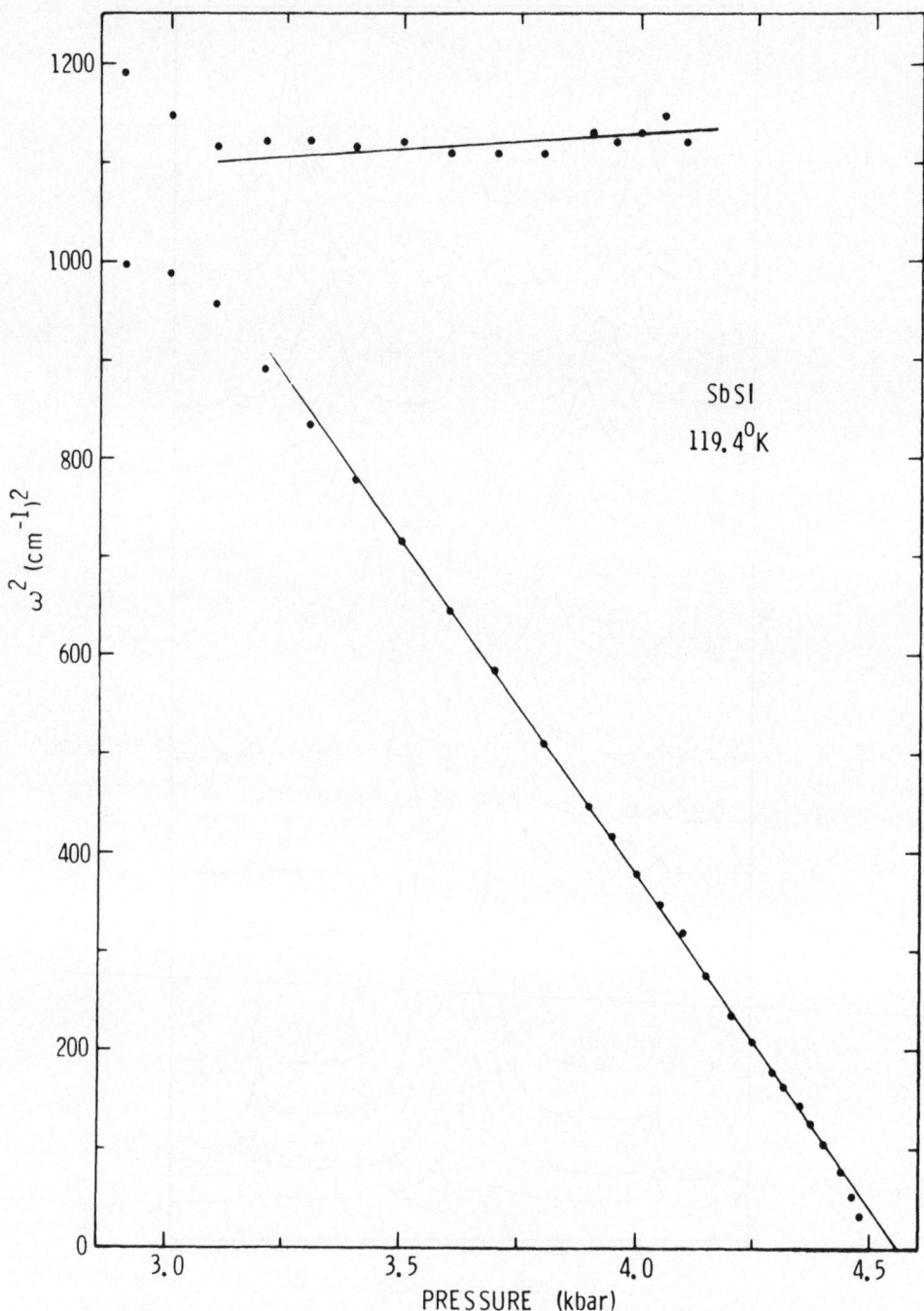

Figure 9.   Pressure dependence of the soft mode in SbSI at
            119.4°K.   The data are plotted as ω^2.

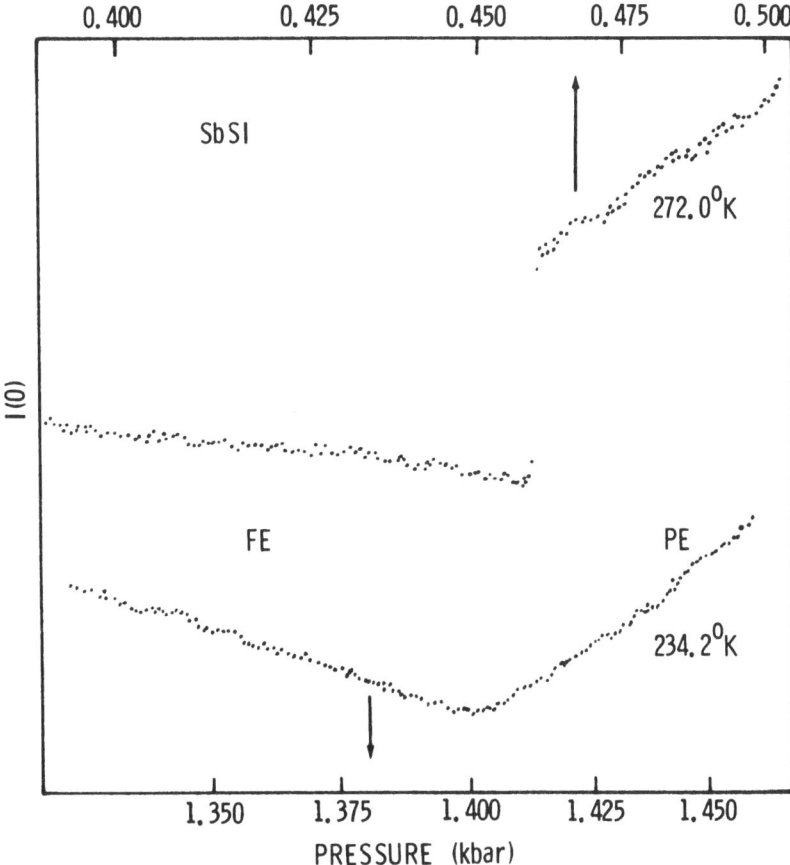

Figure 10. Change in the intensity of elastically scattered light
near the phase transition in SbSI for two temperatures.

been made because of the difficulties in obtaining suitable crystals. Such measurements would be of considerable interest for ex-
amining this system for deviations from mean-field behavior near
the tricritical point.

RARE EARTH PENTAPHOSPHATES, REP_4O_{14}

The rare earth pentaphosphates comprise a large class of crystals
with the formula REP_5O_{14} where RE designates rare earth. For

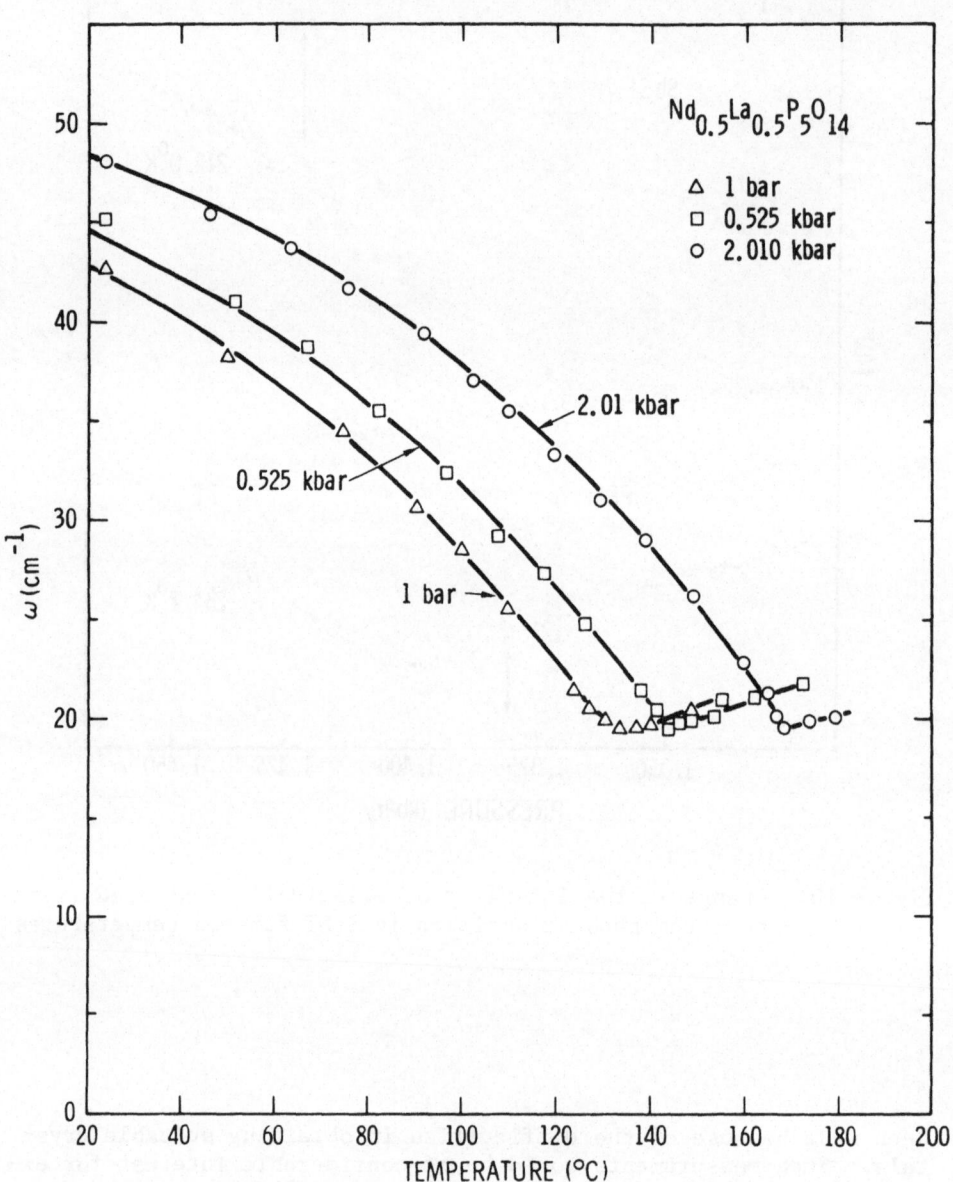

Figure 11.   Temperature dependence of the soft optic mode in
$Nd_{0.5}La_{0.5}P_5O_{14}$ through the phase transition at
different pressures.

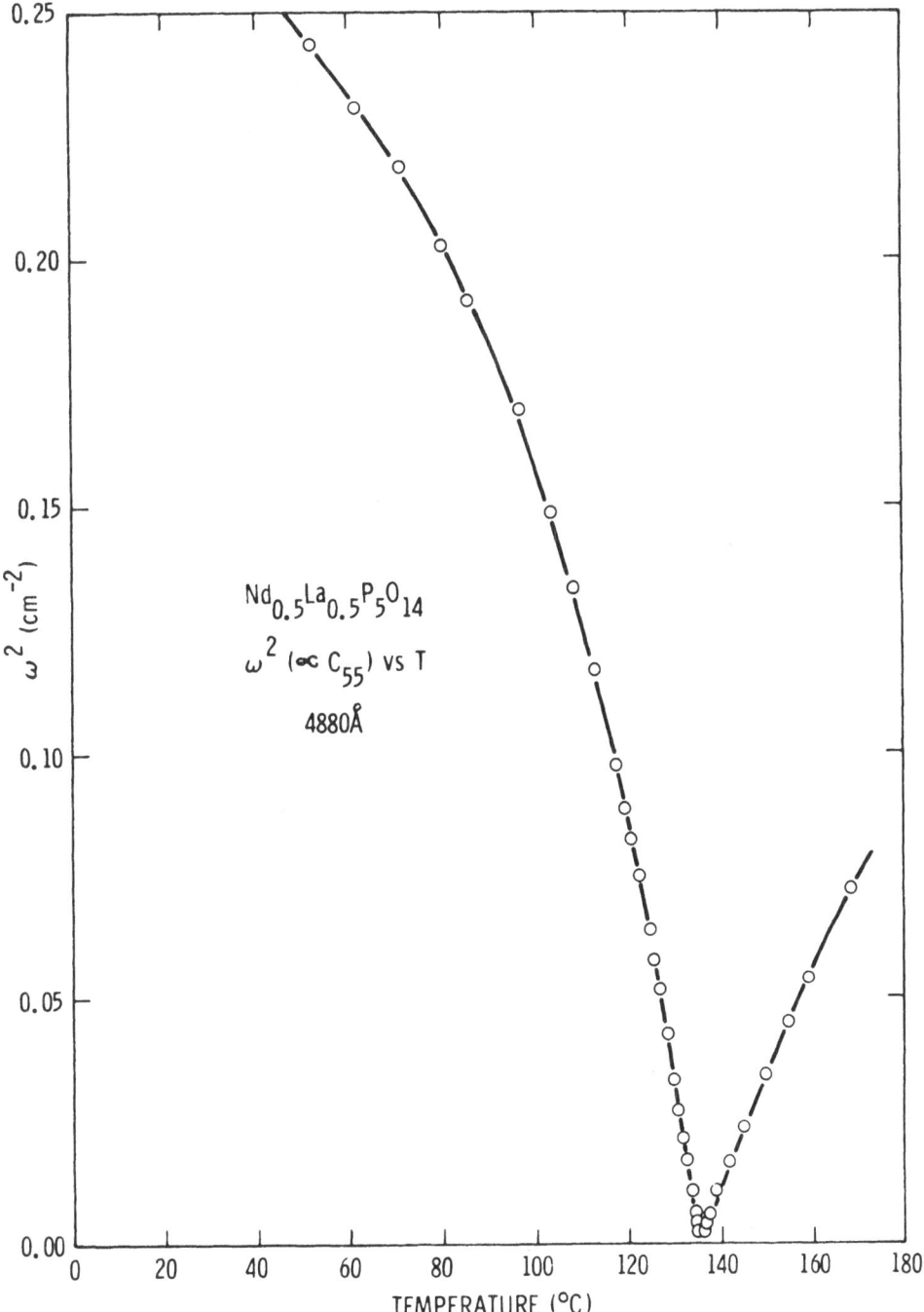

Figure 12. Temperature dependence of the Brillouin frequency associated with C_{55} near the phase transition in $Nd_{0.5}La_{0.5}P_5O_{14}$.

several rare earths, e.g., Tb, La and Nd, as well as mixed crystals of these constituents, the crystals undergo a phase transition from orthorhombic (D_{2h}) to monoclinic (C_{2h}) symmetry. The transition to the ferroelastic monoclinic phase is thought to be second order[22] and is accompanied by a soft zone-center optic mode which is Raman active in both phases.[6]

The transition exhibits a remarkable pressure dependence in that it is the only known zone-center soft optic mode system for which T_c increases with pressure.[23] The increase in T_c with pressure can be inferred from Fig. 11 where the temperature dependence of the soft-optic mode frequency is shown for different pressures for $Nd_{0.5}La_{0.5}P_5O_{14}$.

Because of the unusual pressure dependence of T_c and the fact that the soft mode frequency does not vanish at T_c, it is of interest to examine this ferroelastic system for optic-acoustic interactions. Although by symmetry there can be no piezoelectric coupling between optic and acoustic modes, there can be an internal strain[24] contribution which couples the soft-optic mode to the acoustic mode governed by C_{55}.

The effect of this interaction is illustrated by the data shown in Fig. 12 where the square of the Brillouin frequency, which is proportional to C_{55}, is plotted as a function of temperature. These data show that the elastic constant C_{55} essentially vanishes at T_c. The transition thus occurs not when the soft-optic mode frequency vanishes but rather when the frequency of the acoustic mode coupled to the soft-optic mode vanishes. The peculiar temperature dependence of the Brillouin frequency can be well understood within the framework of an internal strain contribution to the elastic constant.[25] However, the increase in T_c with pressure has not yet been explained. This behavior may be typical of ferroelastics and a complete understanding must await detailed calculations similar to those which have explained the pressure dependence in ferroelectrics and antiferroelectrics.[26]

CONCLUSIONS

We have briefly reviewed the results of pressure studies of Raman scattering in soft mode systems. Such studies have proven extremely valuable for investigating microscopic models for different types of phase transitions. Not only can the dynamic response of the soft mode be examined, but interactions between the soft mode and other degrees of freedom of the crystal can also be probed. This area of research is just beginning to be exploited, and, with the recent developments in the techniques using high pressure diamond cells, we can look forward to significant advances in the understanding of soft-mode phase transitions at high pressure.

REFERENCES

1. F. Cerdeira, W.G. Holzapfel and D. Bäverle, Phys. Rev. B11, 1188 (1975).
2. P.A. Fleury and P.D. Lazay, Phys. Rev. Lett. 26, 1331 (1971)
3. P.S. Peercy and G.A. Samara, Phys. Rev. B8, 2033 (1973).
4. P.S. Peercy and I.J. Fritz, Phys. Rev. Lett. 32, 466 (1974).
5. P.W. Anderson and E.I. Blount, Phys. Rev. Lett. 14, 217 (1965).
6. D.F. Fox, J.F. Scott and P.M. Bridenbaugh, Solid State Commun. 18, 111 (1976).
7. H. Uwe and T. Sakudo, Phys. Rev. B13, 271 (1976).
8. B.N. Ganguly, F.G. Ullman, R.D. Kirby and J.R. Hardy, Phys. Rev. B12, 3783 (1975).
9. K.K. Kobayashi, J. Phys. Soc. Japan 24, 497 (1968).
10, P.S. Peercy, Phys. Rev. Lett. 31, 380 (1971).
11. P.S. Peercy, Phys. Rev. B9, 4868 (1974).
12. P.S. Peercy, Phys. Rev. B12, 2725 (1975).
13. E.M. Brody and H.Z. Cummins, Phys. Rev. Lett. 21, 1263 (1968); 23, 1039 (1969).
14. P.S. Peercy, Solid State Commun. 16, 439 (1975).
15. P.S. Peercy, Phys. Rev. B13, 3945 (1976).
16. R. Blinc, R. Pirc and B. Zeks, Phys. Rev. B13, 2943 (1976).
17. P.S. Peercy, Phys. Rev. Lett. 35, 1581 (1975).
18. V.H. Schmidt, A.B. Western and A.G. Baker, Phys. Rev. Lett. 37, 839 (1976).
19. G.A. Samara, Ferroelectrics 9, 209 (1975).
20. M.K. Teng, M. Balkanski and M. Massot, Phys. Rev. B3, 1031 (1972).
21. G.A. Samara (Preceeding Paper).
22. See, e.g., H.P. Weber, R.C. Tofield and P.F. Liao, Phys.Rev. B11, 1152 (1975).
23. P.S. Peercy, J.F. Scott and P.M. Bridenbaugh, Bull. Am. Phys. Soc., Series II, Vol. 21, 337 (1976).
24. P.B. Miller and J.D. Axe, Phys. Rev. 163, 924 (1967).
25. P.S. Peercy, in "Proceedings of the Fifth Raman Conference," ed. by E.D. Schmid, J. Brandmuller, W. Kiefer, B. Schroder and H.W. Schrotter (Hans Ferdinand Schultz Verlag, Freiburg, 1976), p. 571.
26. See, e.g., G.A. Samara, T. Sakudo and K. Yoshimitsu, Phys. Rev. Lett. 35, 1767 (1975).

QUESTIONS AND COMMENTS

V.H. Schmidt: You said that it would be interesting to look for deviations from Landau theory in SbSI. I have heard people say that there aren't any expected deviations in ferroelectrics until you get to within 10^{-3} or 10^{-4} K of T_c.

P. Peercy: That is correct. Estimates made by Blinc, how-
                  ever, indicate that for SbSI you would expect
                  deviations over a larger range than, for example,
                  in KDP because of the linear-chain structure. It
                  would be interesting to see if the deviations are
                  experimentally accessible in SbSI. I don't know
                  that they are, and it would be a tough experiment.

B.T. Matthias: Whatever happened to the tungsten bronzes which
                  people thought might be ferroelectric? Have
                  you had a look for a soft mode?

P. Peercy: No, I have not. I think Prof. Scott looked for
                  soft modes in that system.

B.T. Matthias: Well, he saw it softening.

P. Peercy: He saw a softening, but not a vanishing, as I re-
                  call. I have not looked at the tungsten bronzes.

BRILLOUIN SCATTERING AT HIGH PRESSURES*

Hans D. Hochheimer

Fachbereich Physik, Universität Regensburg

8400 Regensburg, West-Germany

Walter F. Love[++] and Charles T. Walker

Department of Physics, Arizona State University

Tempe, Arizona 85281

ABSTRACT

Pressure and temperature dependences of the Brillouin spectra for KCN and NaCN are shown for pressures of 0 to 7 kbar and temperatures of 178 K to 295 K for KCN and of 285 K to 340 K for NaCN. Dispersion of c_{44} and $c' = (c_{11} + c_{12} + 2c_{44})/2$ and strong pressure dependence of T_o, the temperature where c_{44} would vanish, has been found for NaCN, whereas there is no dispersion for c_{44} and c' and no pressure dependence of T_o in KCN. On the other hand it is shown for both substances that multiphonon interactions, which increase approaching the phase transition, are the dominant anharmonic effect for the zone center acoustic phonons. A review of current theories for the order-disorder transition of the CN^--ion system is given.

*Supported by the National Science Foundation under Grant DMR 76-16793 and the Deutsche Forschungsgemeinschaft.

++Present address: Corning Glass Works, Corning, N.Y. 14830

INTRODUCTION

Though the alkali cyanides behave in many properties such as bulk compressibility, thermal expansion, cleavage, plasticity, and transformation into the CsCl structure under high pressure like an alkali halide of NaCl type, there are many important differences due to the molecular character of the CN^- ion.

In this paper we report on our results of Brillouin measurements of KCN and NaCN under hydrostatic pressures up to 7 kbar in the temperature range from 178 to 295 K for KCN and 285 to 340 K for NaCN. In these regions both materials undergo a phase transition from a high temperature fcc structure to a lower temperature orthorhombic structure.[1] In the fcc phase (space group O_h^5) the CN^- ions have directional disorder, while in the orthorhombic phase (space group D_{2h}^{25}) the ions are directionally aligned but still have head-to-tail randomness. The transition temperature, T_c, is strongly pressure dependent, increasing 2 K per kbar of hydrostatic pressure for KCN[2] and 4 K per kbar for NaCN. In both materials the phase transition is seen clearly in the elastic constants which have been studied by two different techniques; ultrasonic measurements at 15 MHz [3,4] and Brillouin measurements at gigahertz frequencies [4,5,6].

Ultrasonic measurements have shown that c_{44} softens logarithmically with temperature for KCN and linearly with temperature for NaCN as the phase transition is approached from above. But while the logarithmic softening continues until the phase transition occurs in the case of KCN it deviates from the linear behavior at about 300 K, 16 degrees above the phase transition, in the case of NaCN. At that temperature c_{44} becomes independent of temperature down to the phase transition temperature. Furthermore Brillouin measurements[4] have shown substantial dispersive effects for c_{44} and $c' = (c_{11} + c_{12} + 2 c_{44})/2$ at 3 GHz compared to the values determined ultrasonically at 15 MHz in NaCN, whereas there is no dispersion for c_{44} and c' in KCN.

To gain further insight into the phase transition we have undertaken a high pressure Brillouin study for both materials. As will be seen, this study reveals even more differences between NaCN and KCN.

In KCN c_{44} and T_o, the temperature where c_{44} would vanish are basically independent of pressure, whereas both show strong pressure dependence in NaCN. Besides these differences there are some similarities. In both substances the TA-mode associated with c_{44} will be seen to be strongly anharmonic, with large multiphonon contribution to the phonon self-energy. Furthermore, T_c depends on pressure linearly in both cases, supporting the idea of Julian and Lüty[7] that elastic dipole interactions are involved in this phase transition.

EXPERIMENTAL DETAILS

Pressure-dependent Brillouin scattering measurements below room temperature are somewhat difficult to effect, so we shall give a few experimental details here. A 20 kg maraging steel high pressure cell[8] with f/4 optics was used. The crystal was immersed in liquid pentene in order to match the indices of refraction and reduce surface scattering. The pentene thus served as the pressure transmitting medium and its freezing point ($\sim$ 178 K at 7 kbar) provided the lower limit on our temperature range. A standard high pressure apparatus was used, along with a fluid divider to keep the pentene out of the pressure intensifier. The pressure cell was mounted in a large styrofoam box and cooled with nitrogen gas. The large mass of the cell made it relatively easy to hold the temperature to within 1 K for the several minute integration times necessary for collecting good Brillouin spectra. Temperatures were measured with a copper-constantan thermocouple mounted in one of the plugs of the high pressure cell. Pressures were measured in the standard way with a manganin resistor mounted in the room temperature portion of the pressure line.

KCN single crystals of very high optical, mechanical and chemical quality were provided by Prof. Franz Rosenberger of the University of Utah. The experiments would not have been possible without the high optical quality which these crystals had. Even with these crystals, mounted in a medium with a matching index of refraction, the central elastic component of the scattered spectrum was very intense when the sample was in the pressure cell. Brillouin spectra were therefore analysed by using a piezoelectrically scanned, electronically-stabilized, triple-pass Fabry-Perot interferometer, preceded by a 5 Å bandpass interference filter. A multichannel analyser allowed the multiple scanning necessary for an acceptable signal-to noise ratio.

Incident and scattered light propagated along cubic axes and therefore the observed Brillouin spectra are from TA and LA phonons propagating in the (110)-direction, with velocities given by c_{44} and $\frac{1}{2}(c_{11} + c_{12} + 2c_{44})$ respectively. The sapphire windows in the pressure cell[12] caused the incident and scattered light to be depolarized and hence all spectra contain both LA and TA Brillouin peaks.

EXPERIMENTAL RESULTS

The quality of the Brillouin spectra for samples in the pressure cell is represented in Fig. 1, where the pressure and temperature dependences of the TA and LA modes of NaCN are shown. Large scale features of the pressure and temperature dependences of the TA and LA phonons can be seen. The strong central elastic peak (intensity of about 1 x 10^6 counts) overflows the multichannel an-

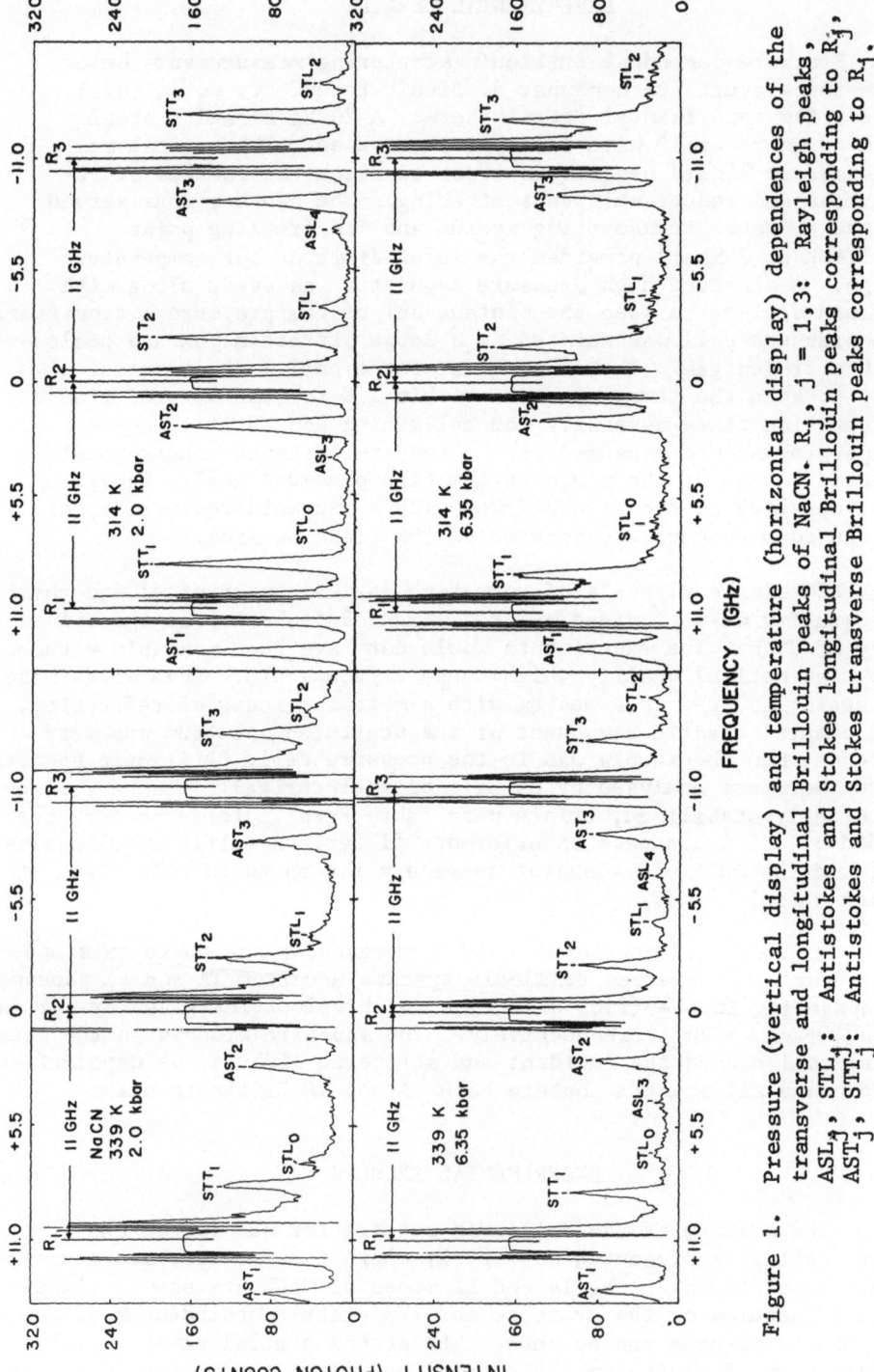

Figure 1. Pressure (vertical display) and temperature (horizontal display) dependences of the transverse and longitudinal Brillouin peaks of NaCN. R_j, $j = 1,3$: Rayleigh peaks, ASL_j, STL_j: Antistokes and Stokes longitudinal Brillouin peaks corresponding to R_j, AST_j, STT_j: Antistokes and Stokes transverse Brillouin peaks corresponding to R_j.

alyser and is therefore cut off several times. The TA peaks are clearly seen (intensity above background of about 10^2 counts) whereas the intensity of the LA peaks is much weaker. The stokes and antistokes components are indicated for each central elastic peak. The spectra were taken with a free spectral range of 11 GHz which allowed measurement of the TA- and LA-Brillouin shifts simultaneously.

It is seen that the TA-peak softens with decreasing temperature at constant pressure (horizontal display) as well as with increasing pressure at constant temperature (vertical display). At 2 kbar the TA mode shifts from 2.63 GHz at 339 K to 2.15 GHz at 314 K and at 6.35 kbar from 2.32 GHz at 339 K to 1.8 GHz at 314 K, i.e. the temperature shift at different pressures is the same within the error of the measurements. The pressure shift at constant temperature is 310 MHz at 339 K and 340 MHz at 314 K for the pressure difference of 4.35 kbar between 2 kbar and 6.35 kbar.

For the LA mode we have at 2 kbar a shift from 14.75 GHz at 339 K to 14.68 GHz at 314 K, whereas the pressure shift at constant temperature (339 K) is from 14.75 GHz at 2 kbar to 15.26 GHz at 6.35 kbar. As the intensity of the LA peak decreases rapidly with increasing pressure, most of the data above 5 kbar could not be analysed. But it is clearly seen that for the LA mode the pressure shift is much larger than the temperature shift.

The general quality of the Brillouin spectra for KCN is the same as seen in Fig. 1 for NaCN, although the pressure and temperature dependences of the Brillouin shifts are quite different. As we have shown previously[9] the LA peak shifts with pressure (at constant temperature (295 K)) from 12.9 GHz at 175 bar to 13.8 GHz at 6 kbar. The TA peak shifts downwards from 3.73 GHz to 3.62 GHz. At a constant pressure of 3.8 kbar the softening of c_{44} with decreasing temperature is clearly seen, starting with the overlapped TA peak at 240 K occurring at 3 GHz, down to the 1.69 GHz TA peak at 178 K.

The intensity of the TA peak increases with decreasing temperature, varying as $\frac{1}{c_{44}}$, if the elasto-optic coefficient P_{44} is constant.

From these and comparable spectra we have calculated the pressure and temperature dependences of the elastic constants c_{44} and c', using the relation between Brillouin shift, ν_B, and sound velocity, v_s,

$$\nu_B = \frac{2nv_s}{\lambda_o} \sin \frac{\theta}{2} \tag{1}$$

where n is the index of refraction, λ_o is the wavelength of the incident light and θ is the scattering angle.

Knowing the pressure and temperature dependence of the density, ρ, one can calculate the elastic constants from Eq. (1) assuming the constancy of the specific refraction, r, given by [10]

$$r = \frac{n^2 - 1}{\rho(n^2 + 2)} \tag{2}$$

This assumption is justified in the case of molecular crystals[11].

In the case of NaCN we have measured the temperature dependence of ρ[4], and for KCN we used the values given by Haussühl[3]. The pressure dependence of ρ was calculated from the bulk modulus.

For KCN we found the surprising result that the temperature dependence of c_{44} was the same at 1 bar, 56 bar, 3.8 kbar, and 7.0 kbar. Furthermore, all data could be fitted using the relation determined by Haussühl from his ultrasonic data

$$c_{44} = a \ln \left(\frac{T}{T_o} \right) \tag{3}$$

with a and T_o, the temperature where c_{44} would vanish, the same within the errors of our measurements for the ultrasonic data at 15 MHz and all Brillouin data at about 3 GHz, as seen in Table 1. In particular the agreement of our high pressure values with Haussühl's zero pressure values is astonishing, as at a pressure of 7 kbar the transition temperature, T_c, has been moved upward 14 K, while the same pressure has not moved T_o by any significant amount.

Fig. 2 shows the temperature dependence of c_{44} for NaCN at 1 bar, 3 kbar, and 6.3 kbar determined at about 3 GHz by Brillouin scattering together with the temperature dependence of c_{44} measured ultrasonically at 15 MHz. It can be seen that the temperature dependence of c_{44} determined by Brillouin scattering follows a linear relation, given by

$$c_{44} = a + b T \tag{4}$$

down to the phase transition temperature, whereas this holds true for the ultrasonically determined values of c_{44} only to about 300 K where c_{44} bends over. In Table 2 the pressure dependence of the coefficients a and b of Eq. (4) are listed together with the pressure dependence of T_o and T_c, the phase transition temperature. T_o changes about 3 K/kbar compared to 4K/kbar for T_c. The substantial differences to the case of KCN, such as dispersion and strong pressure dependence of T_o, are seen clearly. Furthermore c_{44} decreases with increasing pressure.

Table 1. Pressure dependence of T_o, the temperature where $c_{44} = 0$, and the constant a in Eq.(3) for KCN.

	Pressure (kbar)	T_o (K)	a (10^9 dyne/cm^2)
Ultrasonic[3]	0.001	153.7	21.9
Brillouin	0.056	155.9 ± 2.4%	23.1 ± 1.2%
Brillouin	3.8	156.3 ± 3.4%	22.4 ± 1.7%
Brillouin	7.0	157.1 ± 4.7%	22.7 ± 2.3%

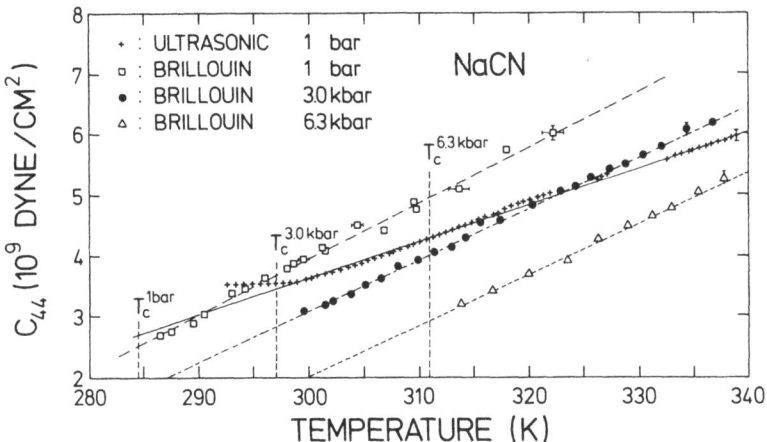

Figure 2. Isobaric temperature dependence of c_{44} determined at 3 GHz (Brillouin scattering) at 1 bar, 3 kbar, and 6.3 kbar and at 15 MHz (ultrasonically) at 1 bar.

Table 2. Pressure dependence of the coefficients a and b of Eq. (4), T_o, the temperature where $c_{44} = 0$. and T_c, the phase transition temperature for NaCN.

	Pressure	a	b	T_o	T_c
	[kbar]	[10^9 dyne/cm^2]	[10^{+7} dyne/cm^{2}K]	[K]	[K]
Ultrasonic	0.001	-14.4 ± 1.1 %	$6.01 \pm .85$ %	$240 \pm$ 2%	
Brillouin	0.001	-24.0 ± 2.4 %	9.3 ± 2 %	258 ± 4.4%	284.0/281.5
Brillouin	3.0	-22.4 ± 1 %	$8.5 \pm .9$ %	264 ± 1.9%	296.7
Brillouin	6.3	-23.2 ± 2.7 %	8.4 ± 2.3 %	276 ± 5%	310.6

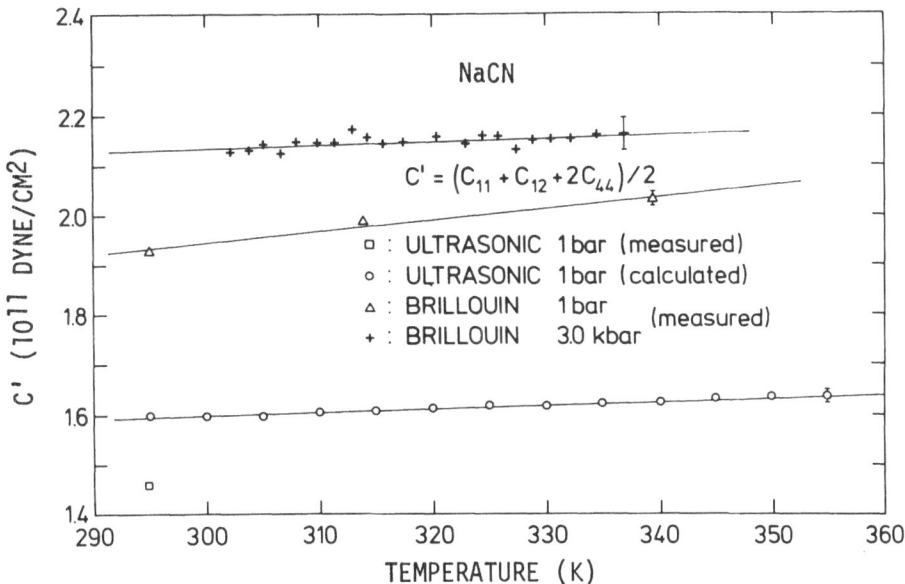

Figure 3. Isobaric temperature dependence of $c' = (c_{11} + c_{12} + 2c_{44})/2$ determined at 14 GHz (Brillouin scattering) at 1 bar and 3 kbar. The values for c' at 15 MHz are calculated from c_{11}, c_{12}, and c_{44} measured ultrasonically at 15 MHz.

The dispersion can also be seen in the LA mode which is connected with the elastic constant c', shown in Fig. 3. But unlike c_{44} the decrease of c' is small approaching the phase transition temperature from above. Furthermore c' increases with increasing pressure. Again no dispersion of c' is found in the case of KCN[5], though the pressure and temperature dependences of c' behave similarly for KCN and NaCN (Table 3).

We have seen that the TA and LA frequencies vary with temperature and pressure. These dependences can be discussed in terms of anharmonicity and related by

$$\left(\frac{\partial \ln\omega_i}{\partial T} \right)_P = - \frac{\beta}{\kappa_T} \left(\frac{\partial \ln\omega_i}{\partial P} \right)_T + \left(\frac{\partial \ln\omega_i}{\partial T} \right)_V \tag{5}$$

Table 3. Values obtained by fitting Eq. (5) to the temperature and pressure dependence of the Brillouin peaks of KCN and NaCN.

Crystal	Mode	T	ω_i	$\dfrac{1}{\kappa_T}\left(\dfrac{\partial \ln\omega_i}{\partial P}\right)_T$	$\left(\dfrac{\partial \ln\omega_i}{\partial T}\right)_P =$	$-\dfrac{\beta}{\kappa_T}\left(\dfrac{\partial \ln\omega_i}{\partial P}\right)_T +$	$\left(\dfrac{\partial \ln\omega_i}{\partial T}\right)_V$
		[K]	[GHz]		$[10^{-5}/K]$	$[10^{-5}/K]$	$[10^{-5}/K]$
KCN	TA	295	3.73	-0.82 ± 0.10	340.7	$+12.3$	328.4
	TA	188	2.12	-1.09 ± 0.16	3500.0	$+17.2$	3482.8
	LA	295	12.96	1.50 ± 0.03	13.9	-22.5	36.4
	LA	188	12.48	1.95 ± 0.07	33.5	-30.8	64.3
NaCN	TA	339.4	2.79	-4.05 ± 0.09	531.0	20.9	510.1
	TA	314.0	2.32	-5.54 ± 0.14	1022.2	28.1	994.1
	LA	339.4	14.48	1.30 ± 0.10	21.0	-6.5	27.5
	LA	314.0	14.35	1.40 ± 0.04	21.0	-7.1	28.1

where ω_i is the phonon frequency, β is the thermal expansion coefficient, and κ_T is the isothermal compressibility.

Maradudin and Fein[12] and Cowley[13] have shown that the terms on
the right hand side of Eq. (5) are related to the anharmonic frequency shifts of the real part of the phonon self-energy. The first
term can be related to the "thermal expansion" shift (i.e. a pure
volume effect) while the second term can be related to the frequency
shifts arising from three and four phonon interactions. The cubic
anharmonicities can only give a negative value for the second term
on the right hand side while quartic anharmonicity can give a negative or a positive value. Therefore a positive value for this term
is unambiguous evidence for four phonon interactions.

Table 3 lists the magnitudes of the two terms on the right
hand side of Eq. (5) for the TA and LA zone center phonons at
188 K and 295 K for KCN and 339.4 K and 314 K for NaCN. For the
LA modes in KCN and NaCN one sees that the thermal expansion and
multiphonon terms have about the same size. However, the multiphonon term is positive, indicating four phonon interactions. The
TA mode behaves dramatically differently. For KCN the multiphonon
term is 27 times larger than the thermal expansion term at room
temperature and 200 times larger at 188 K, 6 K above the phase
transition at 7 kbar. For NaCN the multiphonon term is 25 times
larger than the thermal expansion term at 339.4 K, and 50 times
larger at 314 K, 2 K above T_c at 7 kbar. Though the increase in
multiphonon interaction strength is not as large in NaCN as in KCN,
it is clearly the dominant anharmonic effect for the zone center
acoustic phonons in both materials. Furthermore the multiphonon
terms are also positive for the TA phonons in KCN and NaCN.

As we have argued previously for KCN[9], and as is now apparent
for NaCN, if one interprets our observed temperature and pressure
dependence in terms of anharmonic interactions in the usual way,
then one must conclude that with respect to the phonons quartic
anharmonicity is very important near the phase transition. This
conclusion is in good agreement with the conclusion drawn by Rowe
et al.[14] in their neutron scattering studies. The phonon anharmonic effects are probably more important in KCN than in NaCN.

On the other hand, the observed linear pressure dependence of
the transition temperature, T_c, is, as we have argued earlier[9],
fully compatible with Julian and Lüty's suggestion[7] that the phase
transition involves a "ferroelastic" ordering of the CN⁻ dipoles,
although that is not a sufficient explanation of their dielectric
measurements. This point has been considered more quantitatively
by Shuey and Beyeler[15].

DISCUSSION

At the moment there are three theories concerning the softening of c_{44} and c' with temperature for KCN. Of these theories one also offers an explanation for the pressure dependence in KCN.

Michel and Naudts[16] started with a Hamiltonian which included the lattice vibrational kinetic and potential energies, as well as an interaction between the ends of the CN^- ions and the surrounding K^+ ions. This latter interaction was represented by a parameterized Born-Mayer repulsive potential. When expanded in terms of small displacements from equilibrium this latter potential gave two terms, one of which was a purely rotational interaction with the other being an interaction between CN^- rotation and translational displacements of the ions. After making a transformation to the center-of-mass coordinates of the unit cell they found a term in the Hamiltonian which was bilinear in rotational and translational coordinates. The CN^- – CN^- interaction then occurs through this term, via a mechanism where rotation of one CN^- ion causes a lattice strain which couples to a neighbouring CN^-, making it rotate in turn. Their expression for the elastic constant was

$$c_{44} = c_{44}^o \left[1 - \frac{\delta(T)}{T} \right] \qquad (6)$$

where $\delta(T)$ is an eigenvalue of the susceptibility and is temperature dependent because of the pure rotation-rotation interaction. Their fit to the temperature dependence of c_{44}, which involves numerical evaluation of the susceptibilities, was satisfactory. One must note, however, that their theory is purely harmonic for the lattice vibrations, and therefore is not applicable to our analysis of our data in terms of strong multiphonon interactions. In fact, in its present form the theory of Michel and Naudts does not appear to offer an explanation of our pressure results even in terms of their translation-rotation coupling.

Rehwald et al.[17] have used a very different approach. They defined an order parameter which was related to the symmetrized occupation probabilities for the six equivalent (110) directions. (As the CN^- elastic dipole does not have a "head" a negative (110) direction is equivalent to a positive one.) The order parameter with T_{2g} symmetry was designated by η and the order parameter with E_g symmetry by ξ. In order to obtain a low temperature ordered state they assumed a ferrodistortive interaction between CN^- ions of the form

$$U = - \frac{J}{N} \sum_{i=1}^{6} N_i^2 \quad ,$$

where N_i is the occupation probability for a (110) direction, N is the total number of CN^- ions, and J is a positive interaction parameter. They also assumed a linear coupling between strain and the two order parameters η and ξ. The result of their calculation was

$$c_{44} = c_{44}^o \left(\frac{T - T_0}{T - T_0'} \right) \qquad (7)$$

where T_0 is the temperature at which c_{44} should vanish under the influence of both interactions and T_0' was the temperature at which c_{44} would vanish under the influence of the ferrodistortive term alone. By fitting Eq. (7) to the data they found $T_0 = 152$ K and $T_0' = -231$ K. That is, the ferrodistortive interaction alone could not produce ordering.

This theory, because it concentrates on the dynamics of the CN^- ions, is also not applicable to our analysis of our data in terms of strong lattice anharmonicity. It is also basically a zero pressure theory. Dultz et al.[18] have shown that at pressures above about 0.4 kbar the preferred CN^- orientation in KCN is (111). This means that if the theory of Rehwald et al. is to be applied to our high pressure data it must be recast in terms of a new order parameter which takes account of the (111) alignment.

The third theory of interest has been published by Boissier et al.[19] They have considered the deeper question of how one can apply a pseudospin model to the CN^- rotational problem in the first place. They argued that in the frequency regime $\gamma\omega \ll 1$, where γ is the CN^- reorientation time, the CN^- ion finds itself in a "cage" which does not have cubic symmetry, but is distorted. In this region, and only in this region, one can develop probability functions for the CN^- orientation and can construct a theory in terms of pseudospin variables. They then wrote the free energy as a sum of the purely elastic energy, a term proportional to the pseudospin variable, λ, of the form

$$\frac{1}{2} J (T - T_0) \lambda^2 ,$$

where $J (T - T_0)$ represent the entropy and energy associated with the pseudospin variables, and a third term which contains a bilinear coupling of λ to elastic strain. In this area their theory is closely analogous to the theory of Rehwald et al. The expression for c_{44} which they obtain can be written as

$$c_{44} = c_{44}^o - \frac{b_{44}^2}{T - T_0} \qquad (8)$$

where b_{44}^2 contains the various coupling parameters and T_0 is the temperature at which a phase transition would occur under the influence of the pseudospin-pseudospin interaction alone. By fitting the data they found $T_0 = -86$ K, which means that this interaction alone could not produce the phase transition but that a coupling between pseudospin and strain is necessary for the phase transition to occur. Their result for c_{44} is obviously closely analogous to that of Rehwald et al., with suitable recombination

of parameters. In addition, the model of Boissier et al. offers a
deeper insight into the phase transition. It does not arise from
a dipole – dipole interaction, nor from a quadrupole – quadrupole
interaction. The lowest order term possible is an interaction between hexadecapoles.

Boissier et al. also offer a tentative explanation for our
pressure results in KCN. The value of c_{44} is only dependent on
the coefficients of the second order term, λ^2, in the free energy,
while the transition temperature, T_c, depends very critically on
the higher order coefficients. Our data could be explained by a
large pressure dependence of the higher order coefficients, thus
giving a pressure dependent T_c, but a very small pressure dependence
of the second order coefficient. However, the theory does not give
a reason why these coefficients should have different pressure dependences. The theory of Boissier et al. also has the advantage
that it has been applied to other results for KCN, such as Brillouin
linewidths and Raman intensities. However, as with the theory of
Rehwald et al., it must be modified to account for the (111) preferred ordering direction at pressures above 0.4 kbar.

None of these theories has yet yielded published results for
NaCN. This material presents a more interesting problem as the
elastic constants show frequency dispersion, c_{44} varies linearly
with temperature, and both T_c and T_o are pressure dependent. Pressure dependence of both T_c and T_o could be understood within the
context of the theory of Boissier et al. by simply having both
the second order coefficient of λ and the higher order coefficients
be pressure dependent. Furthermore, in order to reproduce the
linear temperature dependence of c_{44}, T_o', in Eq. (7) or T_o in Eq.
(8) must be a large negative number, implying that the pseudospin-
pseudospin interaction alone could even less cause the ordering.

We have given an explanation for the dispersion in NaCN[4] based
on comparison of the CN$^-$ reorientational time relative to the
period of ultrasonic or Brillouin elastic waves. Based on our
arguments one would say that NaCN is not in the region $\gamma\omega \ll 1$ for
Brillouin waves, where the importance of this parameter was discussed above for the theory of Boissier et al. It would be interesting to see experiments performed on samples of different origin
to see whether our experimental results for the dispersion of the
elastic constants is confirmed, as this point would bear on all of
the theories discussed above.

REFERENCES

1. C.W.F.P. Pistorius, Progress in Solid State Chemistry 11, 1
    (1976).

2. W. Dultz, Habilitationsschrift, Universität Regensburg (1976) unpublished.

3. S. Haussühl, Solid State Comm. 13, 147 (1973).

4. W.F. Love, H.D. Hochheimer, N.W. Anderson, R.N. Work, and C.T. Walker, Sol. State Comm. 23, 365 (1977).

5. W. Krasser, U. Buchenau, and S. Haussühl, Sol. State Comm. 18, 287 (1976).

6. S.K. Satija and C.H. Wang, J. Chem. Phys. 66, 2221 (1977).

7. M. Julian and F. Lúty, International Conference on Low Lying Lattice Vibrational Modes and Their Relationship to Superconductivity and Ferroelectricity, San Juan, Puerto Rico, December 1975, unpublished.

8. This cell was designed by H.D. Hochheimer and built in the laboratory of Prof. W. Gebhardt, FB Physik, Universität Regensburg, West-Germany.

9. H.D. Hochheimer, W.F. Love, and C.T. Walker, Phys. Rev. Lett. 38, 832 (1977).

10. F. Kohlrausch, Praktische Physik 1, (Teubner Verlagsgesellschaft, Stuttgart, 1953).

11. P.S. Peercy, G.A. Samara, and B. Morosin, J. Phys. Chem. Solids 36, 1123 (1975).

12. A.A. Maradudin and A.E. Fein, Phys. Rev. 128, 2589 (1962).

13. R.A. Cowley, Phil. Mag. 11, 673 (1965), Adv. Phys. 12, 421 (1963).

14. J.M. Rowe, J.J. Rush, N. Vagelatos, D.L. Price, D.G. Hinks, and S. Susman, J. Chem. Phys. 62, 4551 (1975).

15. R.T. Shuey and H.U. Beyeler, J. Appl. Math. Phys. (ZAMP) 19, 278 (1968).

16. K.H. Michel and J. Naudts, Bull. Am. Phys. Soc. 22, 463 (1977).

17. W. Rehwald, J.R. Sandercock, and M. Rossinelli, to be published.

18. W. Dultz, H. Krause, and J. Ploner, to be published in J. Chem. Phys.

19. M. Boissier, R. Vacher, D. Fontaine, and R.M. Pick, to be published.

ANHARMONICITY IN A15 SUPERCONDUCTORS:

A THERMAL EXPANSION APPROACH

T.F. Smith and T.R. Finlayson

Department of Physics, Monash University

Clayton, Victoria 3168, Australia

ABSTRACT

The highly anharmonic behaviour for the A15 compounds V_3Si and Nb_3Sn has been the subject of considerable theoretical and experimental study. The majority of theoretical treatments place the emphasis on the electronic contribution to the free energy.

Thermal expansion measurements provide a direct measure of anharmonicity and measurements have been made for a number of A15 compounds. The expansion behaviour for V-Si and Nb-Sn compounds, close to the stoichiometric A_3B composition, and V_3Ge is found to be highly anomalous. In the case of the V-Si compounds, it is argued from an analysis of the expansion behaviour in the superconducting state that the lattice is responsible for the anharmonic properties. The anomalous expansion in the Nb-Sn compounds is intimately associated with the occurrence of the cubic-tetragonal structural distortion. No specific conclusions have been reached concerning the origin of the anomalous expansion for V_3Ge.

INTRODUCTION

The A15 structure compounds represent one of the most extensively investigated groups of materials[1,2]. The high superconducting transition temperatures which are found for a number of these compounds have lead to the study of a wide variety of their physical properties including elastic, structural, lattice dynamics, heat capacity, electrical resistivity, Knight shift, thermo-power and magnetic susceptibility. Attention has been specifically focussed on V_3Si and Nb_3Sn, two of the high T_c compounds for which the

physical properties are often highly anomalous. Yet, in spite of
the wealth of information which has been collected, no clear understanding of the behaviour of these compounds has been formulated.

The usual approximation when describing the thermodynamic properties of a solid is to assume that the free energy may be separated
into its individual lattice and electronic components, namely[3]

$$F = F_\ell + F_e \qquad\qquad\qquad (1).$$

The most comprehensive theoretical treatments of the A15 compounds
have placed their emphasis upon the electronic contribution to the
free energy[2]. Such models are usually based upon the Weger-Labbé-
Friedel[4] 1-dimensional model for the electron density of states,
with the Fermi level situated close to the band edge singularity.
An alternative electronic model, first proposed by Gor'kov[5] and
recently elaborated by Bhatt and McMillan[6], attributes the anomalous behaviour to the formation of a charge density wave. While
such electronic models have often provided a convincing description
of the temperature dependence of the elastic and electronic properties, they are by no means complete.

It is a common feature of the electronic models that all anharmonic contributions to the free energy are associated with F_e
and it is assumed that F_ℓ may be treated within an harmonic approximation. In contrast to this approach, Testardi[7] has suggested a
model invoking a highly anharmonic potential, specifically designed
to provide a simple physical description of the anomalous elastic
properties of Nb_3Sn and V_3Si. This lattice model has recently been
applied[8] to a description of the anomalous temperature dependence
of the resistivity found in V_3Si.

Other recent theoretical developments have lead to a consideration of the possible role of lattice defects upon the macroscopic
properties of the crystal. Phillips[9] has suggested a model in which
the ordering of the local strain fields, produced by lattice defects,
is the origin of the stabilisation of the lattice instability,
which is driven by the electron-phonon coupling. A model calculation by Chui[10] of the phonon dispersion in an intrinsically unstable
two-dimensional lattice, stabilised by relaxation about a periodic
array of vacancies, leads to dispersion curves which are similar
in appearance to those observed for Nb_3Sn.

The thermal expansion is related to the free energy F by the
identity, $\beta = - K_T (\partial^2 F / \partial V \partial T)_{V,T}$ where K_T is the isothermal compressibility. Thus, expansion measurements provide a direct measure
of the volume dependence of the free energy. Under the assumption
(1), the expansion may be regarded to be composed of separable
lattice and electronic contributions. Such an approximation is
implicit in the normal practice of separating the low temperature

heat capacity into lattice, C_ℓ, and electronic, C_e, contributions

$$C = C_e + C_\ell$$
$$= aT + bT^3 \qquad\qquad (2)$$

in the low temperature limit where the Debye approximation for C_ℓ is valid. A similar separation into β_ℓ and β_e may also be made for the thermal expansion.

In discussing the relationship between the thermal expansion and the volume dependence of the free energy it is convenient to consider the dimensionless Grüneisen parameter which is defined as

$$\gamma = \frac{V}{T} \frac{(\partial^2 F/\partial V \partial T)_{T,V}}{(\partial^2 F/\partial T^2)_V} = \frac{V\beta}{C_V K_T} \qquad\qquad (3).$$

Substituting the individual lattice and electronic heat capacity and thermal expansion contributions into (3) gives the corresponding lattice γ_ℓ and electronic $\gamma_e = \frac{\partial \ln N(0)}{\partial \ln V}$ Grüneisen parameters respectively. γ_ℓ represents the weighted average $\frac{\Sigma C_i \gamma_i}{\Sigma C_i}$ of the individual mode gammas $\gamma_i = \frac{-\partial \ln \omega_i}{\partial \ln_V}$ where C_i is the heat capacity contribution for the mode frequency ω_i. $N(0)$ is the electronic density of states.

While comprehensive measurements have been made[1,2] of the heat capacity for the A15 compounds there has not been, to our knowledge, a systematic study of the low temperature thermal expansion of these compounds[11].

We are presently engaged in such a study and expansion measurements have been made between 1.5 and 300 K for a number of A15 compounds. Part of this work has been published in two earlier reports and reference should be made to these for experimental details. The present paper will be restricted to a summary of the results obtained and a discussion of their interpretation.

RESULTS

The highly anomalous variations of the linear expansion coefficient with temperature for V-Si and Nb-Sn compounds which have already been reported[12,13], are reproduced in figures 1 and 2.

The V-Si compounds in the composition range 25 to 26% Si exhibit negative expansion coefficients below a temperature between $\sim$ 25 and $\sim$ 60K, depending upon composition. The value of α, which

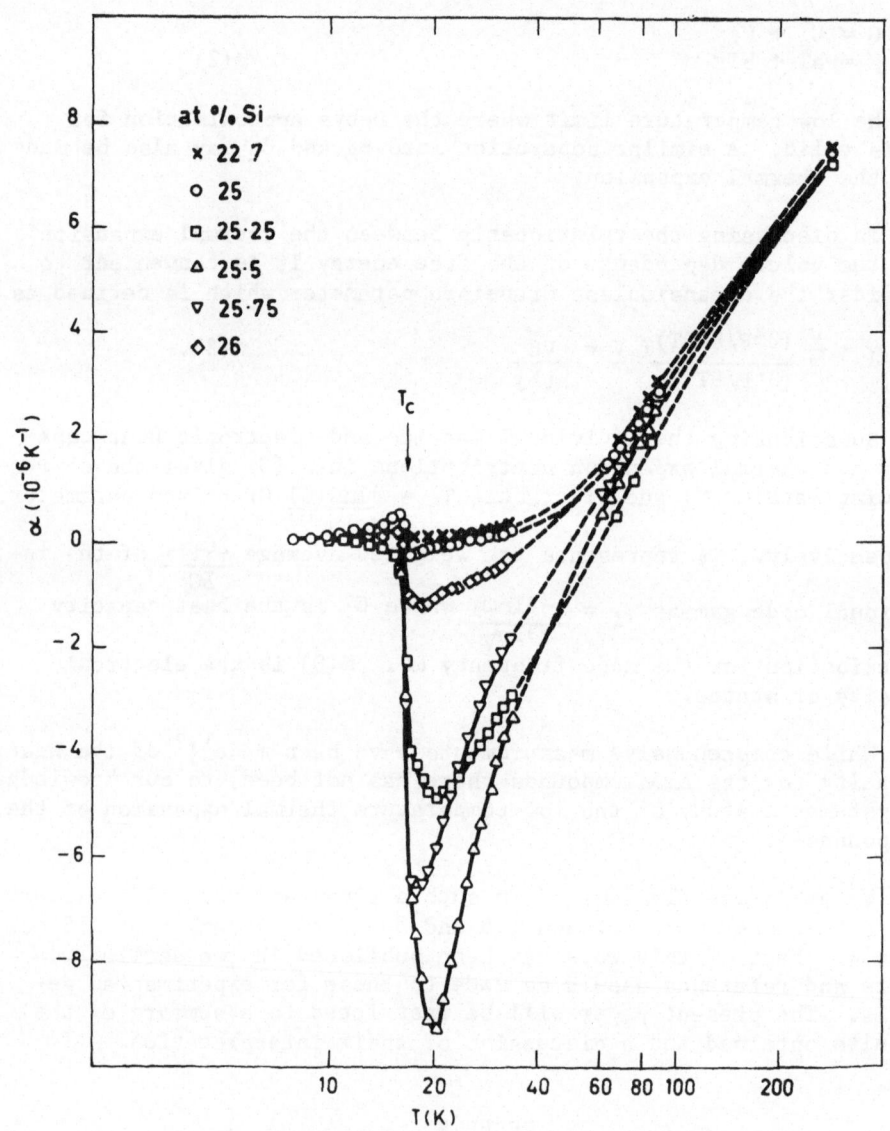

Figure 1.   The variation of the linear thermal expansion coefficient with temperature for heat treated V-Si compounds. The solid lines have been drawn smoothly
through the datum points and the broken lines serve to
bridge the interval between the temperature ranges over
which the data have been taken.[12]

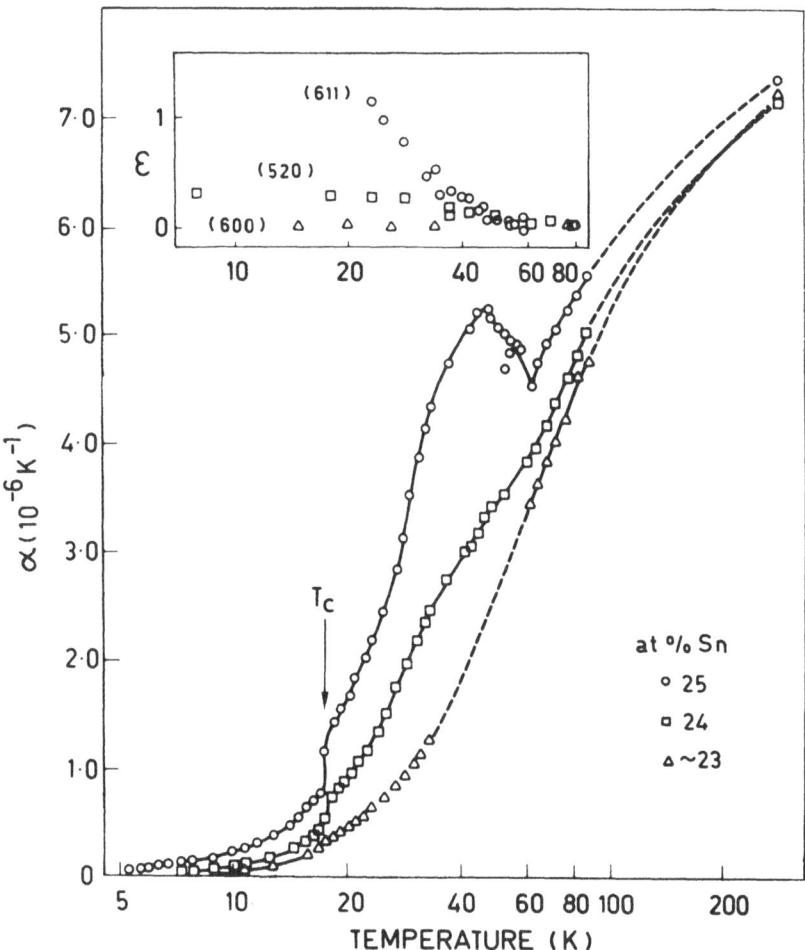

Figure 2.   The variation of the linear thermal expansion coefficient with temperature for Nb-Sn compounds.   The
insert shows the variation in X-ray diffraction line
widths as a function of temperature for the indicated
reflections.[13]

is strongly dependent upon both composition and heat treatment, passes through a maximum (negative) value close to 20K. The data shown in figure 1 are for samples which have been heat treated at 1200°C. An abrupt increase in α associated with the onset of super-conductivity occurs at a common temperature of $\sim$ 16.5K.

The possibility that the anomalous expansion is related to the cubic tetragonal structural transformation, which often occurs for V_3Si samples close to 20K, was discounted by an X-ray examination of powders made by crusting material taken from the ingots that the expansion samples were cut from. Only the 25.25 at% Si compound showed evidence of the line broadening indicative of a lattice distortion below 20K.

Although Nb_3Sn and V_3Si display a close similarity in their elastic properties, they show quite radically different expansion behaviours. The measurements for Nb_3Sn presented in figure 2 show a positive contribution to the thermal expansion coefficient, which X-ray measurements conclusively link to the occurrence of the structural distortion as evidenced from the increase in line widths, as a function of temperature, shown in the insert.

Figure 3 summarises unpublished measurements of the linear ex-pansion coefficients for Nb_3Al, Nb_3Ge, V_3Au, V_3Sn, $V_{74}Ga_{26}$ and V_3Ge. With the notable exception of V_3Ge, the expansion behaviour for these compounds follows a normal temperature dependence. In the case of the V_3Ge the expansion coefficient changes sign below 40K and is negative. This anomalous behaviour was observed in two independent-ly prepared samples and a more detailed plot below 36K of the data is shown in figure 4.

For those compounds where the expansion behaviour is normal a separation of the expansion coefficient into its electronic and lattice contributions has been made by plotting α/T against T^2 above T_c and fitting the data to the linear relationship

$$\alpha/T = A + BT^2 \tag{4}$$

where A and B are the electronic and lattice coefficients respective-ly. This procedure is demonstrated in figure 5 for Nb_3Al. While figure 5 illustrates the general character of a typical plot it is unusual in two respects; the high temperature to which the Debye approximation appears to hold and the remarkably close agreement between the lattice terms derived from above and well below T_c. The same situation has been found with the heat capacity[14].

Values for A and B are given in table 1. By combining these with the corresponding coefficients for the electronic and lattice con-tributions to the heat capacity, values for the electronic, γ_e,

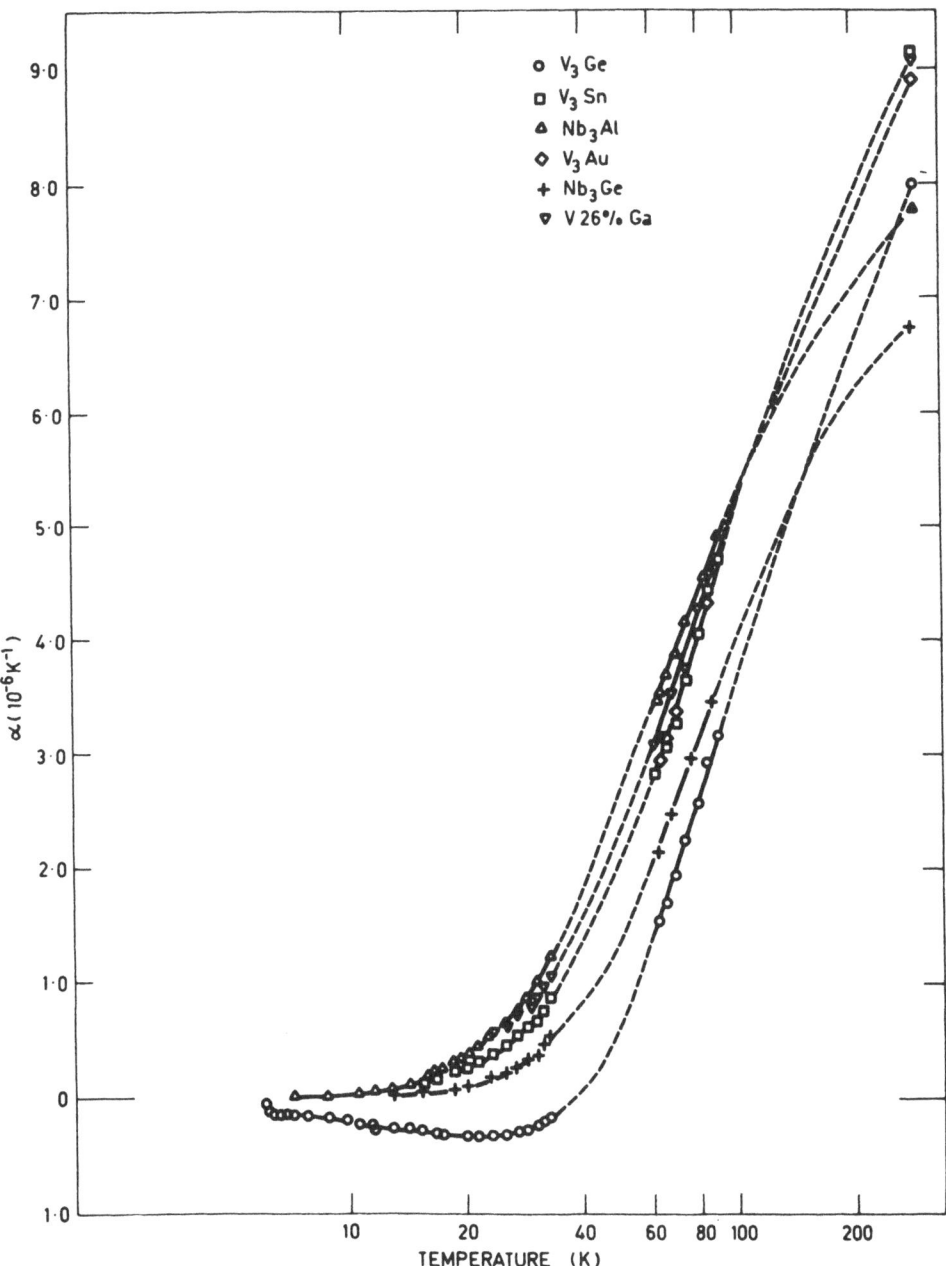

Figure 3. The variation of the linear thermal expansion coefficient for several A15 compounds.

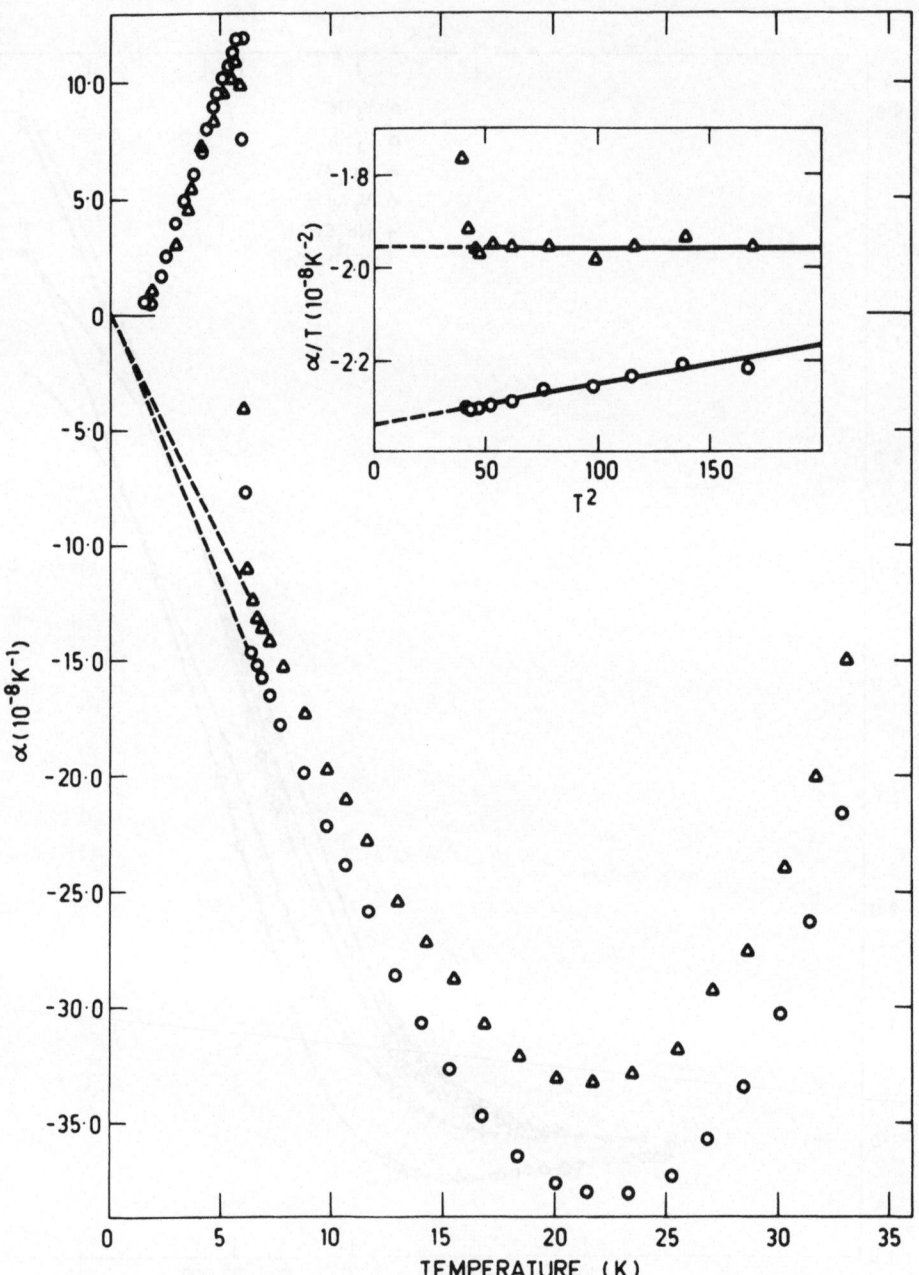

Figure 4.   The variation of the linear thermal expansion co-
            efficient below 35K for two samples of V_3Ge.   The
            broken lines are linear extrapolations to absolute
            zero.   The insert shows the plot of α/T against T^2
            from T_c to 12.5K.

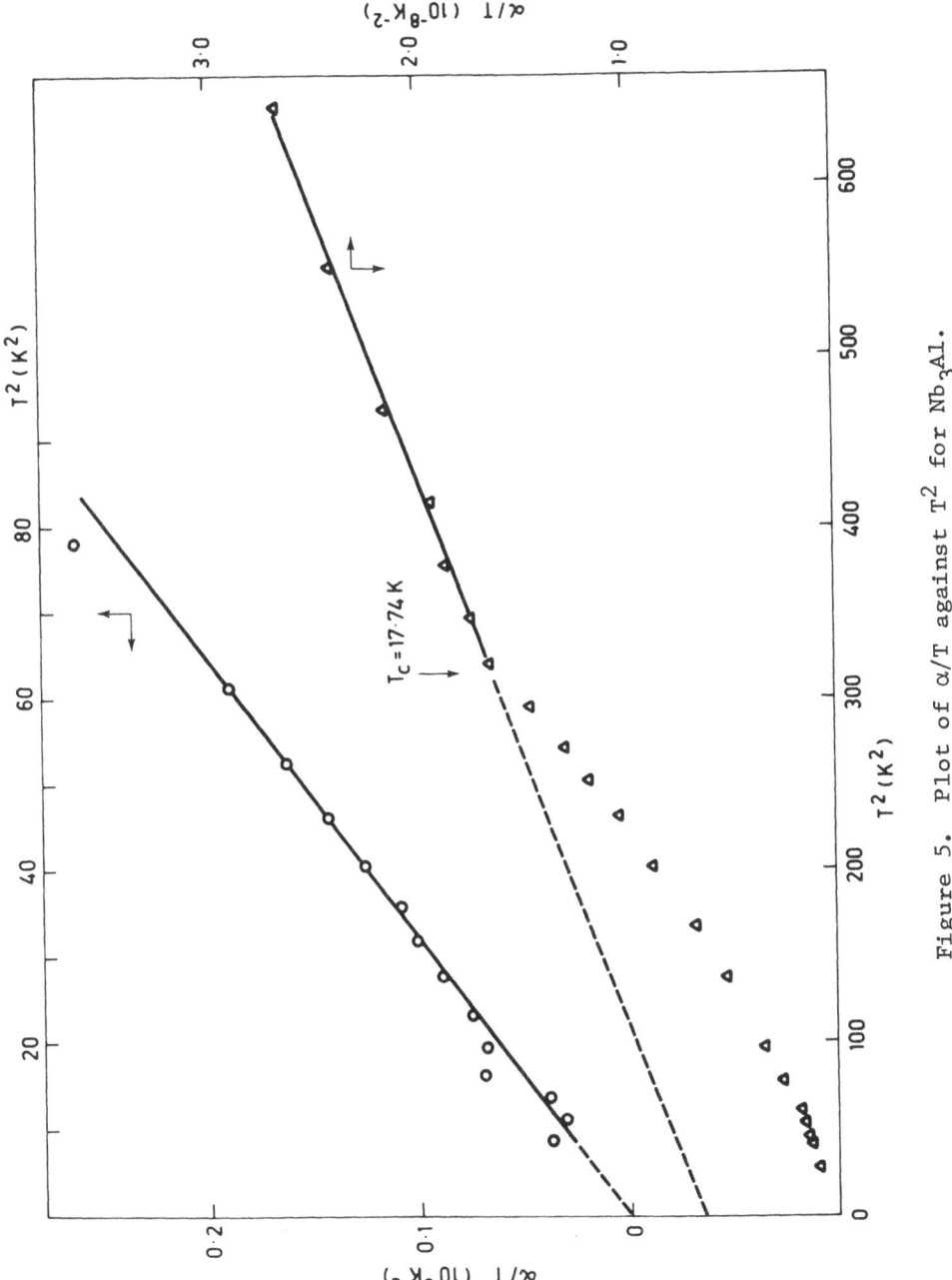

Figure 5. Plot of α/T against T^2 for Nb_3Al.

Table 1. Thermal expansion and Grüneisen parameters.

	Sample	A $(10^{-8}K^{-2})$	B $(10^{-10}K^{-4})$	γ_e	γ_o
Nb_3Al	T·> 17.4K	0.65±0.05	0.32±0.02	4.4±1.5	1.9±0.2
	T < 7.8K		0.31±0.01		
Nb_3Ge		0.15±0.01	0.082±0.004	2.3±0.2	2.2±0.1
V_3Sn		0.15±0.01	0.29±0.01	1.1±0.1	2.9±0.3
V_3Au	a	0.27±0.01	0.17±0.01	1.42±0.05	2.3±0.5
	b	0.32±0.01	0.16±0.01	1.34±0.04	
$V_{74}Ga_{26}$		1.32±0.03	0.19±0.01	2.9±0.5	1.7±0.2

a "as cast" $T_c < 1.2K$

b Following heat treatment at 1000°C and 600°C $T_c \sim 2.2K$

and zero temperature limit lattice, γ_o, Grüneisen parameters have
been calculated from the relationship

$$\gamma = \frac{3\alpha V\ B_S}{C_P} \qquad\qquad (5)$$

where V is the molar volume and B_S the adiabatic bulk modulus.
These values for γ_o and γ_e are also listed in table 1. As the necessary elastic data are not available for calculating B_S for V_3Sn,
V_3Au, $V_{74}Ga_{26}$, Nb_3Al and Nb_3Ge we have used the 4K values for
V_3Ge $(1.767 \times 10^{11}Pa)$[15] for the vanadium compounds and that for
Nb_3Sn at 35K $(1.65 \times 10^{11}Pa)$[16] for the niobium compounds. Using
the heat capacity data for V_3Sn published by Knapp et al.[17] and
Spitzli[18] and data, which we have taken for our Nb_3Al, V_3Ga and
V_3Ge expansion samples, the lattice Grüneisen parameters γ_ℓ have
been calculated as a function of temperature. These are shown in
figure 6. Heat capacity measurements on the thermal expansion
samples are still in progress and ultimately we intend to present
a detailed account of these and the thermal expansion measurements.
In the case of the V_3Si, V_3Ge and Nb_3Sn we have no means of making

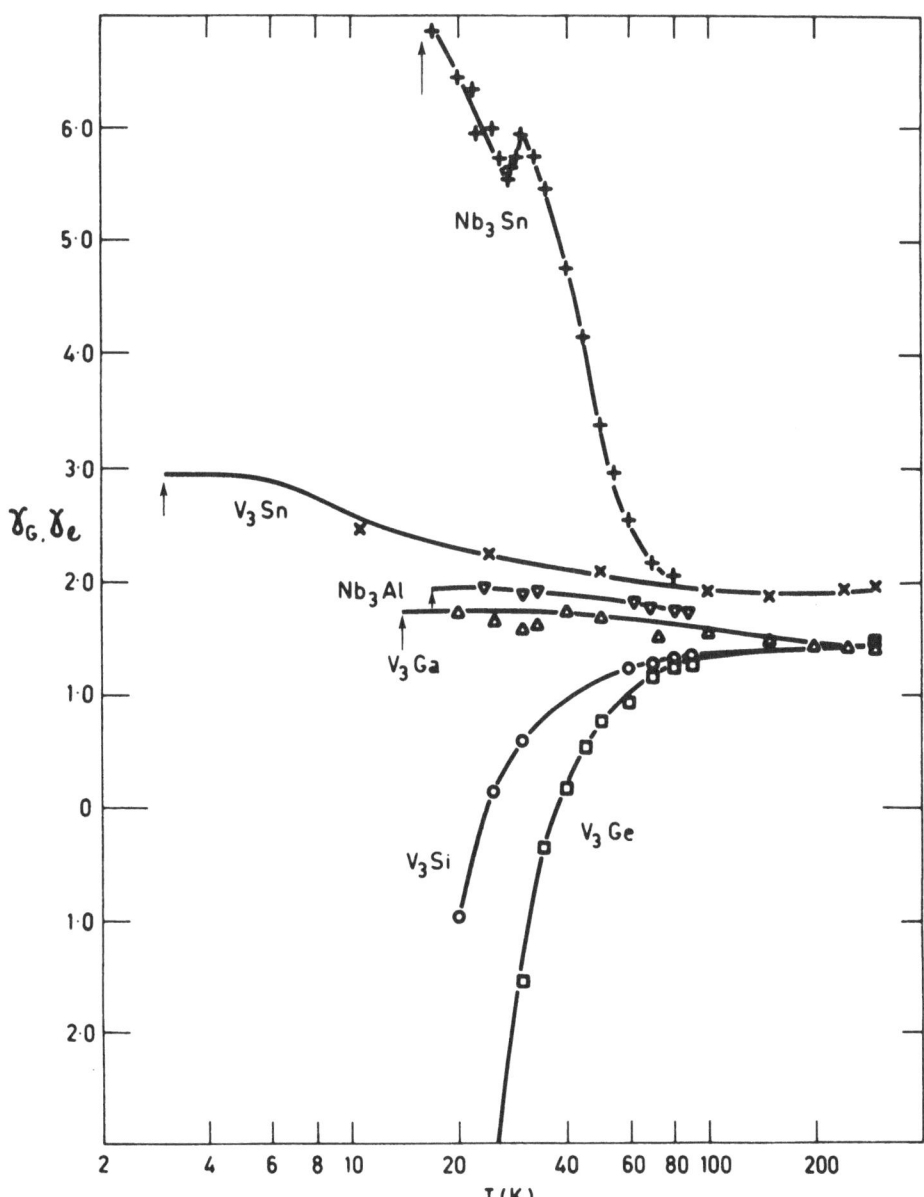

Figure 6. The variation of γ_G(V_3Si, V_3Ge, Nb$_3$Sn) and γ_ℓ(V_3Sn, V_{74}Ga$_{26}$, Nb$_3$Al) with temperature. The vertical arrows indicate T_c.

a reliable separation of the anomalous α into its electronic and lattice components and so plot the total Grüneisen parameter γ_G in figure 6 for these compounds.

DISCUSSION

This series of thermal expansion measurements were conceived with the expressed purpose of investigating the origin of the anharmonic behaviour in the A15 superconductors and evaluating to what extent it is responsible for their high T_c values. We have previously discussed [12,13] the possibility that the anomalous γ_G values for the V-Si and Nb-Sn compounds are manifestations of a highly anharmonic potential with a dominant contribution arising from a large mode gamma associated with a soft mode. It was argued that the sign of γ_G, and thus the corresponding volume dependence for both series of compounds, was not consistent with the sign of the pressure dependence for the transformation temperature.

The basis of this argument resided with the assumption that the negative γ_G for V_3Si implies a promotion of the lattice instability with decreasing volume, which is contrary to the observed[19] decrease of the transformation temperature with pressure. A similar argument was also applied to the Nb_3Sn, where the positive γ_G was considered to be inconsistent with the displacement of T_m to higher values under pressure[20]. Furthermore, even though it is not entirely clear to what extent mode softening influences T_c[21], it is reasonable to expect that a substantial displacement in soft mode frequency with volume would be apparent in the magnitude of dT_c/dP. Yet, for neither the V-Si nor the Nb-Sn compounds is there any correlation between the magnitude of dT_c/dP and γ_G.

As a consequence of these considerations and in view of the convincing descriptions which have been based upon electronic models of the normal state properties, it was suggested that the anamalous expansion behaviour is electronic in origin, and possibly associated with interband charge transfer. Such a model has been considered by Barsch and Rogowski[22] as an explanation for the pressure dependence of the elastic constants. Since it has been estimated by Labbé, Barisic and Friedel[23] that for both V_3Si and Nb_3Sn, T_c will be relatively insensitive to quite large changes in the d-band occupation, the absence of any correlation between dT_c/dP and γ_G could be accounted for.

After further consideration we now wish to argue against an electronic model, at least for the V-Si alloys, and advocate a lattice model to describe the thermal expansion. As it was commented earlier[12], the inability to decide conclusively between the lattice and electronic models stems from the inability to separate the expansion into its lattice and electronic components. Yet, it

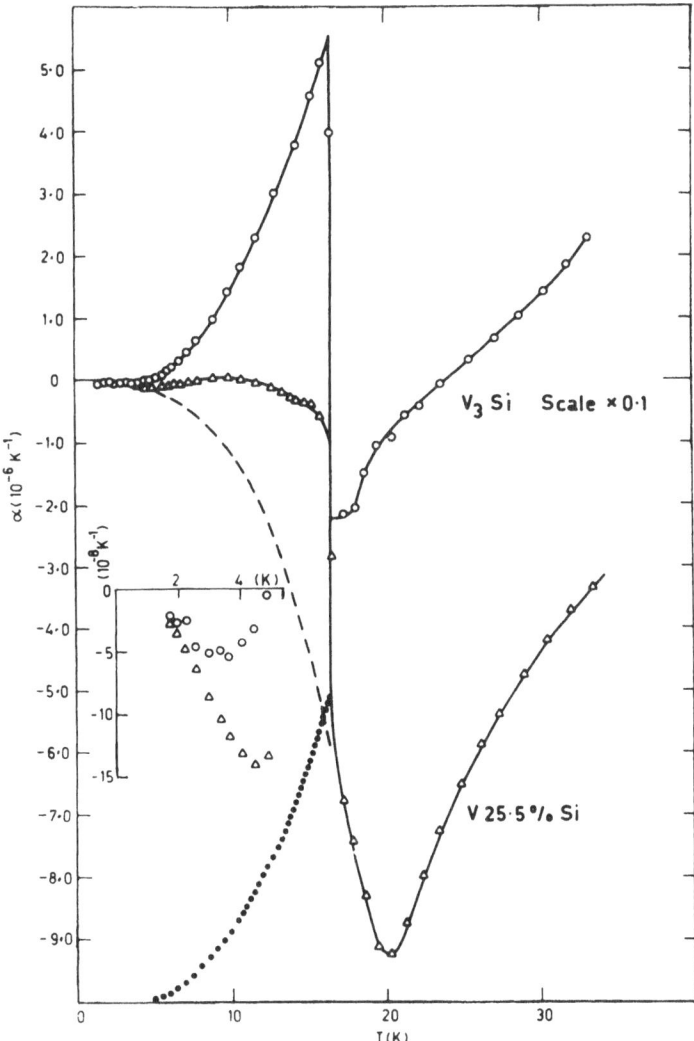

Figure 7. The variation of the linear thermal expansion coefficient with temperature below 35K for V_3Si and V25.5%Si.
The broken line represents a plausible extrapolation of
the normal state expansion, given by $-0.125 \times 10^{-8}T^3$,
to the data below 3K. The difference between this extrapolation and the smooth line drawn through the
measured values for the 25.5%Si is shown as the dotted
line with the temperature axis taken as the zero base
line on the vertical scale. This is taken to represent
the electronic contribution, α_{es}, in the superconducting
state. The insert shows the expansion behaviour below
4K.

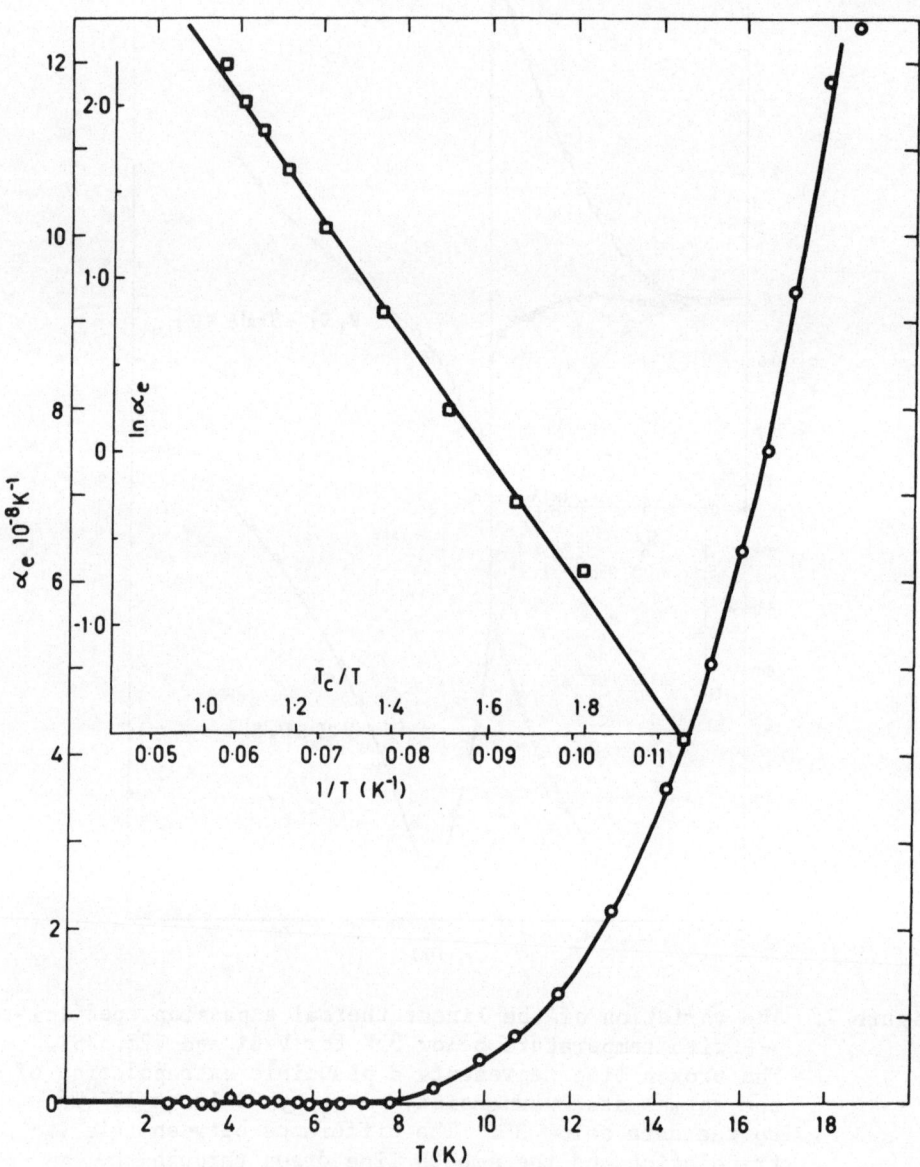

Figure 8.   The variation of α_{es} with temperature for Nb_3Al.   Upper
plot, $\ln\alpha_{es}$ as a function of $1/T$.

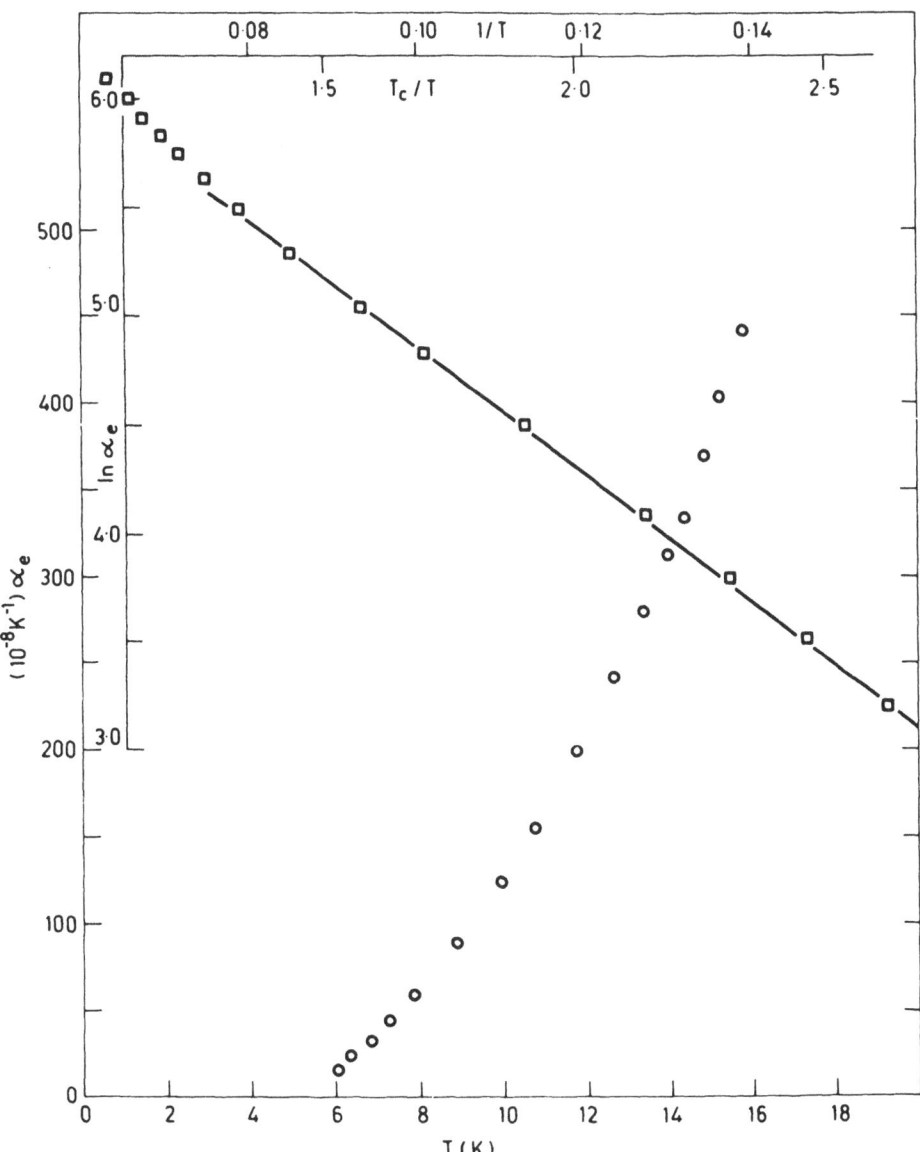

Figure 9. The variation of α_{es} with temperature for V25.5%Si.
Upper plot, ln α_{es} as a function of 1/T.

was noted that the negative expansion behaviour which is found for
the V-Si compounds at temperatures well below T_c is indicative of
a negative lattice contribution (see insert figure 7). Furthermore, in considering the broad nature of the peak in α at T_c for
the Si rich samples (figure 7 illustrates the relative shapes of
the α peak for V_3Si and V 25.5% Si) it was shown that this could
be explained by assuming that a strong negative lattice contribution (broken line) was superimposed upon the electronic contribution (dotted line). What was not emphasized was the extent to
which this analysis may be regarded as providing the required separation of the expansion into its electronic and lattice components
and establishing the latter as the source of the anomalous behaviour.

Since electron pairs have zero entropy, electronic contributions
to the thermodynamic properties rapidly drop to zero in the superconducting state as the number of unpaired electrons decreases approximately exponentially with temperature. The exponential temperature dependence of the electronic heat capacity C_{es} is probably the best known example of this situation[24]. However, a similar decline in the electronic contribution to α must also occur.
The actual form of the expression derived[25] within the BCS[24] formulation for the temperature dependence of α_{es} is somewhat more
complicated than that for C_{es} and cannot be represented by a simple
universal numerical approximation. Nevertheless, a close approximation to an exponential dependence of α_{es} upon temperature is generally expected and has been found[25]. By way of an illustration
for a "well behaved" superconductor, figure 8 presents the variation of α_{es} with temperature for Nb_3Al, derived from the data
shown in figure 4, after subtracting the lattice term represented
by $0.31 \times 10^{-10}T^3$. The insert shows the plot of $\ln\alpha_{es}$ against $1/T$
and clearly demonstrates a linear dependence for temperatures be-
tween T_c and $T_c/2$.

Returning to the V 25.5% Si, plots of the interpolated α_{es} as
a function of temperature, taken from figure 7, and $\ln\alpha_{es}$ as a
function of $1/T$, are shown in figure 9. Again, $\ln \alpha_{es}$ is linear
in $1/T$, in agreement with the expected behaviour. We regard this
as a very persuasive argument in support of our proposed separation of α into its lattice and electronic components in the superconducting state. We are unable to make any separation at temperatures above T_c, but it is implied from the superconducting state
analysis that the lattice term is the dominant contribution in the
normal state.

Having attributed the anomalous expansion to the lattice the
question now arises as to the nature of the lattice potential which
would result in the observed behaviour. Testardi's[7] picture of a
flat-bottomed potential offers an explanation of the temperature
dependence of the elastic behaviour, but without any specific in-

sight as to its origin and in particular, its volume dependence, it provides no insight into the thermal expansion.

Barron[26] has argued from a simple physical picture that an expansion of the lattice leads to an increase in the spring constant for transverse modes of vibration and consequently such modes will have negative Grüneisen parameters. It is interesting to recall that the lattice instability in V_3Si (and Nb_3Sn) is associated with the softening of the [110] transverse acoustic mode[16,27].

As noted above, the anomalous expansion behaviour for the Nb-Sn compounds is closely linked to the structural transformation. However, we are unable to offer any specific argument in favour of either an electronic or a lattice model. Of the other A15 compounds examined, only V_3Ge has been found to have an anomalous expansion (figure 4). While displaying a close overall resemblance to the form of the expansion for the V-Si compounds, there is quite a distinct difference in the power of the temperature dependence. Whereas the assumed lattice expansion for the V-Si varies as T^3, plotting α/T against T^2 reveals a linear temperature dependence of α between 6 and 13K for one V_3Ge sample and only a small additional T^3 term for the other (see insert figure 4). A linear temperature dependence is normally taken to be electronic in origin, but we hesitate to make any specific assignment at present. Following the same analysis adopted for the V-Si and plotting $\ln\alpha_{es}$ against $1/T$ fails to provide any grounds for a preference in attributing the anomalous expansion to the phonons or the electrons. Consequently, we plot the total Grüneisen parameter for V_3Ge in figure 6. γ_G decreases rapidly below 40K and at 10K is ~ -37.

The remaining Grüneisen parameters shown in Figure 6 are for the lattice. The only notable feature for those which have been determined above 20K is the slight decrease with increasing temperature, in contrast to the increase normally observed[28].

In attempting to understand the expansion behaviour described above we are faced with the dilemma so often encountered with A15 compounds; a diversity of behaviour which appears to defy interpretation within the framework of a single model. However, it is clear that there is a very significant degree of anharmonic behaviour which should not be ignored when constructing theoretical models. This is particularly important in the case of V_3Si where we would argue that the evidence suggests that the anharmonicity is associated with the lattice vibrations, which are usually treated as harmonic.

As a final comment we emphasize the lack of any correlation between the degree of anharmonicity and T_c. The expansion coefficients for the V-Si compounds vary by almost two orders of magnitude, yet all have the same T_c value. Below 40K the γ_G values for V_3Ge are

approximately an order of magnitude larger than those for V_3Si,
yet its T_c is 6.1K compared with the 16.8K for V_3Si. Transforming
Nb_3Sn and Nb_3Al have comparable T_c values but radically different
expansion behaviour. We also note that the expansion for the nontransforming Nb_3Sn sample is essentially the same as for the
Nb_3Al, with only a small reduction in T_c.

ACKNOWLEDGEMENTS

The expansion measurements were made while TFS was visiting
the National Measurement Laboratory, Sydney. We are indebted to
Dr. G.K. White for the use of his dilatometer and his instruction
in the art of expansion measurements. We also wish to thank
Dr. J.G. Collins for numerous discussions on lattice dynamics and
in particular his comments on the transverse vibration of a linear
chain of atoms.

We gratefully acknowledge the financial support of the
Australian Research Grants Committee.

REFERENCES

1. L.R. Testardi, Physical Acoustics, Vol. 10, ed. W.P. Mason and
    R.N. Thurston (Academic Press, N.Y.) 1973, pp193.
2. M. Weger and I.B. Goldberg, Solid State Phys., Vol. 28, ed.
    F. Seitz and D. Turnbull (Academic Press, N.Y.) 1973, pp1.
3. See for example D.C. Wallace, Thermodynamics of Crystals
    (John Wiley, 1972) p285.
4. M. Weger, Rev. Mod. Phys. 36, 175 (1964); J. Labbé and
    J. Friedel, J. Physique 27, 153 (1966); J. Labbé, Phys. Rev.
    158, 647 (1967); J. Labbé, Phys. Rev. 172, 451 (1968).
5. L.P. Gor'kov, Zh. Eksp. Teor. Fiz. Pis'ma Red. 17, 525 (1973)
    [JETP Lett. 17, 379 (1973)]; Zh. Eksp. Teor. Fiz. 65, 1658
    (1973)[Sov. Phys. JETP 38, 830 (1974)]; L.P. Gor'kov and
    O.N. Dorokhov, J. Low Temp. Phys. 22, 1 (1976; Zh. Eksp.
    Teor. Fiz. Pis'ma Red. 21, 656 (1975 [JETPlett. 21, 310 (1975)].
6. R.N. Bhatt and W.L. McMillan, Phys. Rev 14, 1007 (1976).
7. L.R. Testardi, Phys. Rev. B. 5, 4342 (1972).
8. P.B. Allen, J.C.K. Hui, W.E. Pickett, C.M. Varma and Z Fisk,
    Solid State Commun. 18, 1157 (1976).
9. J.C. Phillips, Solid State Commun. 18, 831 (1976); C.M. Varma,
    J.C. Phillips and S.T. Chui, Phys. Rev. Lett. 33, 1223 (1974).
10. S.T. Chui, Phys. Rev. B 11, 3457 (1975).
11. Testardi, ref. 7 reports expansion coefficients between 40 and
    300 K for V_3Si and V_3Ge. E. Fawcett, Phys. Rev. Lett. 26, 829
    (1971) has measured the thermal expansion for single crystal
    V_3Si at temperatures below 30K.

12. T.F. Smith, T.R. Finlayson and R.N. Shelton, J. Less-Comm. Metals 43, 21 (1975).
13. T.F. Smith, T.R. Finlayson and A. Taft, Commun. on Phys. 1, 167 (1976)
14. A. Junod, J.L. Staudenmann, J. Muller and P. Spitzli, J. Low Temp. Phys. 5, 25 (1971).
15. M. Rosen, H. Klimker and M. Weger, Phys. Rev. 184, 466 (1969).
16. K.R. Keller and J.J. Hanak, Phys. Rev. 154, 628 (1967).
17. G.S. Knapp, S.D. Bader, H.V. Culbert, F.Y. Fradin and T.E. Klippert, Phys. Rev. B. 11, 4331 (1975).
18. P. Spitzli, Phys. Kon. Mat. 13, 22 (1971).
19. C.W. Chu and L.R. Testardi, Phys. Rev. Lett. 32, 766 (1974).
20. C.W. Chu, Phys. Rev. Lett. 33, 1283 (1974).
21. P.B. Allen, Solid State Commun., 14, 937 (1974); P.B. Allen and R.C. Dynes, Phys. Rev. B. 12, 905 (1975).
22. G.R. Barsch and D.A. Rogowski, Mater. Res. Bull., 8, 1459 (1973).
23. J. Labbé, S. Barisic and J. Friedel, Phys. Rev. Lett., 19, 1039 (1967).
24. J. Bardeen, L.N. Cooper and J.R. Schrieffer, Phys. Rev. 108, 1175 (1957).
25. M.A. Simpson and T.F. Smith, to be published.
26. T.H.K. Barron, Ann. Phys. 1, 77 (1957).
27. L.R. Testardi and T.B. Bateman, Phys. Rev. 154, 402 (1967).
28. J.G. Collins and G.K. White, Progress in Low Temperature Physics IV, ed. C.J. Gorter (Amsterdam, North-Holland) pp450.

QUESTIONS AND COMMENTS

Unknown: If you take your log and semi-log plot and extract your energy gap, what energy gap would you have?

T.F. Smith: The actual theoretical expression does not give you an energy gap in a sense that you get something out from the heat capacity. We can actually calculate the superconducting state thermal expansion, and we've done this for niobium and several other materials where we've gotten quite extensive thermal expansion data. We've fitted the experimental data pretty well using values for the energy gaps you would expect from the heat capacity.

G.R. Barsch: I'm delighted to find further experimental and theoretical evidence for the large anharmonicity in A15 superconductors.

J. Wittig: If you would attribute the negative thermal ex-

pansion to electronic effects, wouldn't you get
stronger variations of the density of states at the
Fermi energy?

T.F. Smith: Yes, I might point out when I quoted Grüneisen
                  parameters of the order of 100 or so, it depends
                  on what you are going to put in for the heat capac-
                  ity in the calculation. I calculated total
                  Grüneisen parameters and then assumed that it's
                  dominated either by the lattice or by the electrons.
                  You would have the same sort of numbers for the
                  electronic Grüneisen parameters of the order -100
                  if the electronic contribution is dominant. That
                  might be okay, if you've got this very, very narrow
                  peak in the density of states, and if you believe
                  that sort of thing. You're shaking your head down
                  there, and I'm shaking my head to agree with you
                  because I think that even if you want to invoke
                  this sort of fine structure, it's going to get
                  washed out in these messy materials. So, again,
                  I think the numbers are unreasonable, but again,
                  I want to caution you that this sort of theory
                  is not set up to really handle numbers like -100
                  for a Grüneisen parameter. That's a quite big
                  Grüneisen parameter, which is a sort of fudge
                  factor to allow for the fact that we aren't treat-
                  ing the anharmonicity properly. We're feeding it
                  in as a correction to a harmonic model.

H.R. Ott: On the question of magnetostriction: From that
                  you could determine the volume effect on the
                  electronic part of the specific heat. It would
                  give you the electronic Grüneisen parameter. Then
                  you could say now that it would have to be in the
                  order of magnitude of 100 or 200 and this is rather
                  a large value.

G.A. Samara: You were saying that the crystal softens under
                  pressure. It wasn't clear to me why that was the
                  case.

T.F. Smith: If you've got a large negative Grüneisen parameter,
                  that means that as you decrease the volume, instead
                  of stiffening, the thing will soften.

G.R. Barsch: What precautions do you take to eliminate aniso-
                  tropic thermal expansion effects, so that what
                  you're presenting is really a volume effect?

T.F. Smith: Well first of all these are cubic materials.

G.R. Barsch: But how do you know that?

T.F. Smith: The A15 structure is cubic.

G.R. Barsch: Well, this is what everybody believes.

T.F. Smith: All right, then be like everybody else and believe
                    they're cubic.

G.R. Barsch: The anistropy might be small.

T.F. Smith: It's very difficult to argue against those sorts of
                    things, because we assume it's cubic when we can
                    only see cubic reflections in the x-ray patterns.
                    But, the Nb_3Sn material is sintered, so it's very
                    well randomized. The vanadium silicide has been
                    melted, crushed, remelted, and crushed about 5
                    or 6 times, and that's as mixed as we can make it.
                    So we can't argue against further orientation if
                    the thing transforms, but from our x-rays we don't
                    believe it transforms.

B.T. Matthias: I was surprised. Were the superconducting transi-
                    tion temperatures very different, that is, very
                    much higher or very much lower, in view of such
                    anomalous behavior?

T.F. Smith: Did you notice that the vertical rise in α, which
                    corresponds to the discontinuity at T_c, takes place
                    at exactly the same temperature for all of those
                    samples? So there's no correlation between a very
                    enormous thermal expansion behavior and the super-
                    conducting transition temperature, either the T_c
                    itself or dT_c/dP.

C.W. Chu: Did I understand you correctly when you said that
                    the lattice contribution is very important, you
                    only address yourself to the lattice transformation,
                    and you did not say too much about the superconduc-
                    tivity part?

T.F. Smith: All I'm saying is that the very large negative ex-
                    pansion behavior below T_c, we feel, is associated
                    with the lattice. It has to be, because the elec-
                    tronic part is decreasing to zero. By implication
                    we assume that it continues on through the minimum,
                    and that it's responsible for the very negative
                    thermal expansion.

C.W. Chu: It really implies that the soft phonons with $q = 0$

are not important for superconductivity.

T.F. Smith: Well, Phil Allen has argued that anharmonicity
                 isn't important either for superconductivity. I
                 still feel that if the phonon modes are moving a-
                 round that much, it should do something.

H.R. Ott: If indeed it is due to the lattice, then it is
                 really very strange, because then that is the only
                 superconducting material that has a negative
                 Grüneisen coefficient.

T.F. Smith: No. The tetrahedrally bonded compounds have nega-
                 tive Grüneisen parameters, for example, zinc
                 sulfide, cadmium telluride, and zinc selenide. And
                 there are also germanium and silicon. I haven't
                 brought this up, but there's a nice connection
                 between the two. I'm tempted to think in terms
                 of bonds in these materials rather than anything
                 else, which will no doubt delight Bernd.

B.T. Matthias: Absolutely.

HIGH PRESSURE STRUCTURAL AND LATTICE DYNAMICAL INVESTIGATIONS AT REDUCED TEMPERATURES

E.F. Skelton

Material Sciences Division
Naval Research Laboratory
Washington, D.C. 20375

and

I.L. Spain and F.J. Rachford

Department of Chemical Engineering
University of Maryland
College Park, Maryland 20742

ABSTRACT

Two diamond-anvil pressure cells designed for polycrystalline and single crystal x-ray studies at pressures up to 10 GPa and simultaneously at temperatures down to 2 K are discussed. Recent data concerning the low temperature phase diagram of Bi up to 8 GPa are reviewed along with the salient features of our x-ray detection systems --- including energy dispersive facilities. The objectives of our research program on high T_c superconductors are reviewed and some very recent data on V_3Si revealing an anomalous increase in the x-ray Bragg scattering on cooling are presented. A preliminary analysis of this anomaly suggests that the low temperature increase in the thermal vibrations may be attributed to the Si atoms and possibly represent a precursor to the martensitic transition.

INTRODUCTION

The concept of using single crystal diamonds in an opposed anvil geometry for high pressure x-ray diffraction measurements[1] and for infra-red studies[2] was introduced eighteen years ago.

During the almost two decades that have followed, numerous changes, improvements, and expanded applications of the diamond-anvil cell have taken place. Much of this pioneering work is reviewed by Hammon[3] and by Block and Piermarini.[4] We have recently reported the first coupling of a diamond-anvil pressure cell to a variable temperature cryogenic system for polycrystalline x-ray diffraction experiments[5]. In this paper we will firstly discuss some recent improvements in that facility which now permit both single crystal x-ray diffraction studies and simultaneous monitoring of changes in the magnetization and resistance of the sample, each as a function of both pressure and temperature. Secondly, an application of structural measurements at high pressure and low temperatures will be given concerning the phase diagram of Bi between 4 and 300 K, at pressures up to 80 GPa.

Finally the details of some ongoing research concerning the structural and dynamical properties of the high temperature superconductor V_3Si and other related compounds at high pressure will be discussed. Some very recent, temperature dependent x-ray data related to the lattice vibrational amplitudes at ambient pressure are presented. Energy dispersive diffraction techniques have been used to measure the integrated intensities of the (h00)-spectrum (h = 2,4...24) over the temperature range from 7 to 294 K. These data are reviewed in terms of the Debye-Waller factors for the V and Si atoms.

APPARATUS

As noted above, details of a variable temperature polycrystalline pressure cell have been recently described in Ref. 5. A development of this cell has now been constructed and tested, allowing single crystal x-ray diffraction measurements to be made between 2 and 300 K. The upper pressure limit has been set by the availability of pressurizing fluids. The 4:1::methanol:ethanol mixture tested by Piermarini et al.[6] allows very nearly hydrostatic pressure conditions to be maintained to $\geq$ 10 GPa at room temperature. Subsequent freezing of this fluid during slow cooling of the cell should not produce large non-hydrostatic stresses. Our observation of single crystal patterns from V_3Si at low temperature and high pressure support this view, but it is intended to check this in greater detail using the broadening of the ruby R-peaks as a diagnostic.

As shown in Fig. 1, the previously used hemi-cylindrical tungsten-carbide diamond-seat rockers are replaced by sintered boron-carbide (B_4C). Actually the B_4C-seats are shaped in the form of right conical frusta which are inserted, on one side, into a slidable plate, and, on the other side, into a hemispherical rocker (Fig. 1). Details of the diamond alignment procedure with the plate

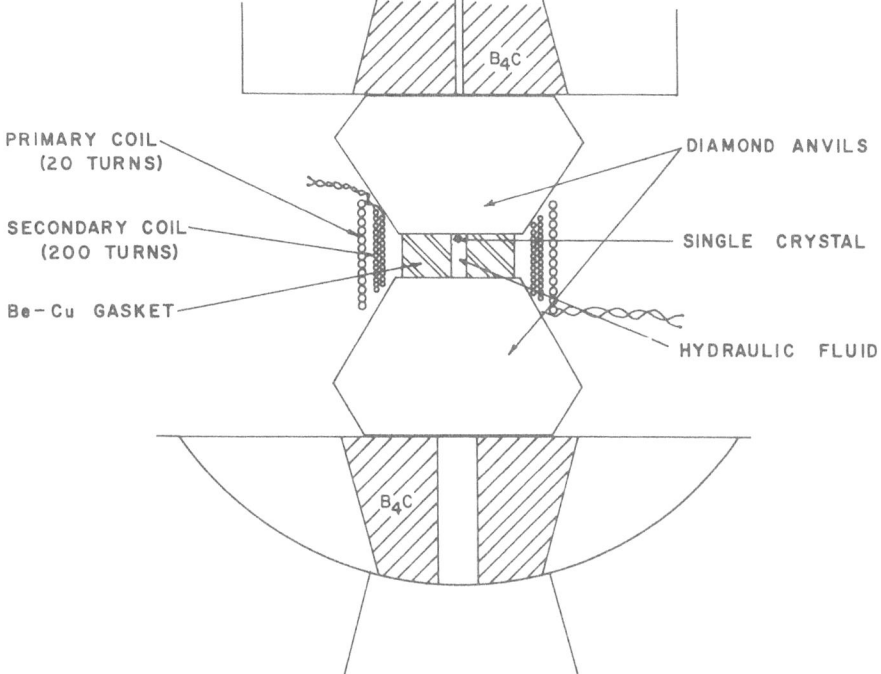

Figure 1. Schematic drawing of the interior of the single crystal diamond cell.

Figure 2. Pressure cell encased in Be cup and coupled to both the x-ray diffractometer and the Helitran refrigerator; the Si(Li) detector is shown in the left of the picture.

and rocker assembly are similar to those discussed by Barnett et al.[7]

As in our previous assembly,[5] the pressure cell is interfaced
to a biplanar x-ray diffractometer and coupled to a Heli-tran
cryogenic refrigerator[8] through an (x,y,z)-translation assembly
(Fig. 2). This coupling mechanism permits full lateral and
vertical adjustments of the cell.

The B_4C exit window seat provides a right conical vertex angle
of 70° which readily permits the recording of transmission Laue
photographs. Moreover, the entrance window is equipped with a 50°
angular slot so that x-ray oscillation photographs can also be recorded. Fig. 3 summarizes the x-ray measurement technique that
can be used to monitor and record the diffracted x-ray information,
viz., powder photographs, NaI(Tl) detector, and energy dispersive diffractometry.

As stressed in Ref. 9, the energy dispersive detector allows
x-ray diffraction information to be obtained in a very short space
of time, e.g., about 10 min. compared to standard 'wet-film' techniques requiring tens of hours. When a calibrant such as NaCl is
included with the sample, the pressure may also be estimated rapidly, although the precision with which the measurement can be made
is higher with more conventional methods (wet film or diffractometer scan). Uncertainties in the pressure are typically ±0.3
and ±0.1 GPa for energy dispersive and conventional techniques,
respectively. However, the energy dispersive technique is routine-
ly used for semi-quantitative scans and for determining the best
conditions under which conventional procedures are required; a
more complete description is given in Ref. 5.

NaCl has generally been used as a calibrant in our work in
conjunction with Decker's[10] equation of state calculations. Although Chhabildas and Ruoff have criticized this scale[11], it is
noted that pressures calculated using either Decker's equation or
that proposed by Chhabildas and Ruoff (Keane equation), differ by
less than 1% up to 10 GPa. Although a discussion will not be given
here, we also believe that Decker's equation is in error to a smaller extent above 10 GPa than that speculated by Chhabildas and
Ruoff.

Also represented schematically in Fig. 1 are the coils of an
inductive detector system. The interior coil, containing 200 turns,
is wound on a lucite form which is contoured to match the diamond
anvils; after winding and potting, the lucite form is dissolved in
acetone. This coil can be used as the inductive leg of a tunnel
diode oscillator circuit, similar to that discussed by Van Degrift.[12]
Resistive or superconductive transitions in the pressure cavity are
detected as frequency shifts in the tank circuit resonant frequency.

With a digital frequency meter and a digital-to-analogue converter, shifts in the resonant frequency can be detected to one part in 10^7.

Alternately, the interior coil system can also be used with a primary coil and a matched reference coil pair. The two halves of the primary are connected in series and the secondaries in series opposition. The real and imaginary components of the magnetic susceptibility of the sample can then be monitored with a mutual inductance bridge. The latter measurement gives information about the sample resistivity. Each of these circuits has been used to successfully measure the superconducting transition in a V_3Ga single crystal approximately 50 microns in diameter. Complete details of the electrical facility will be published elsewhere.[13]

In all the single crystal studies, the sample and pressure calibrant are contained between the diamond anvils by a hardened metal gasket. Be-Cu is frequently used for this purpose. Microphotographs of the gasket and coil assembly are shown in Fig. 4.

POLYCRYSTALLINE X-RAY DIFFRACTION STUDIES WITH Bi

To illustrate the application of polycrystalline x-ray diffraction studies at low temperatures, a brief discussion will be given of the Bi phase diagram. For convenience, a portion of the Bi phase diagram taken from the work of Homan et al[14,15] is reproduced in Fig. 5. Our measurements were made by incrementing the pressure at ambient temperature followed by changes in the temperature. This served to minimize the introduction of non-hydrostatic strains, but required that the cell be warmed before each pressure increment.

During cooling, the cavity pressure was observed to increase by about 10-15% between 294 and 4 K, due to differential thermal contraction of the cell. Accordingly, data were recorded at ambient temperatures followed by cryogenic cooling, as indicated in Fig. 5. Pressure was estimated at 4 K using the recent calculations of Decker[16] for the 0 K isotherm of NaCl. To our knowledge, these high pressure structural data are the first to be obtained on Bi below ambient temperature.

Figs. 6 and 7 illustrate energy dispersive data taken on polycrystalline Bi loaded into the cell with NaCl calibrant and the aforementioned alcohol mixture. Fig. 6 illustrates data taken at a pressure of about 0.1 GPa (A and A' in Fig. 5) and includes diffraction peaks from Bi and NaCl as well as fluorescence radiation excited in the sample and Mo characteristic radiation from the target. Usually the incident beam can be collimated to avoid the gasket, but in both Figs. 6 and 7, a strong diffraction peak from the gasket is present. Fig. 7 illustrates data at higher pressures

HIGH PRESSURE, VARIABLE TEMPERATURE
X-RAY FACILITY

• PRESSURE - VARIABLE UP TO 10 GPa (=100 k bar)
• TEMPERATURE - VARIABLE FROM 2 TO 300 K AND
    CONTROLLABLE WITHIN 10 mK

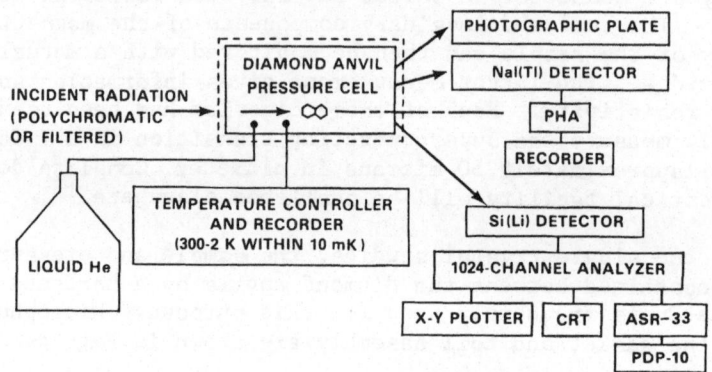

Figure 3.   Block diagram of the variable temperature x-ray de-
            tection facilities.

(B and B' in Fig. 5) corresponding to Bi-III. Extensive data have
been obtained on this phase at ambient temperature, most notably
by Brugger et al.[17,18] using neutron time-of-flight experiments,
although they could not index their pattern. The most important
feature of the present experiment is the absence of phase transitions between 294 and cryogenic temperatures.

    The existence of a phase transition at 4 GPa (Bi III-IV in Fig.
5) continues to be controversial. Since Bridgeman first observed
a 0.6% volume decrease at this pressure, there have been seven reported observations of a transition and six specific reports in
which a transition was sought but not observed. (See Ref. 14 and
19.) In two runs at ambient temperature and 3.7 and 4.4 GPa (point
C in Fig. 5), an additional line was observed at 2.67A not found in
Bi-III. However, it is not possible to make definitive statements
about the possibility of a transition before more complete results
and an indexing scheme for Bi-III are obtained.

    Two runs were made in the Bi-V region. The pattern obtained
suggests a larger unit cell or a lower symmetry than for Bi-III
because of the large number of low angle lines. By comparing data
at ambient (points D and E in Fig. 5) and low temperatures (points
D' and E' in Fig. 5), no change in structure could be observed.
This implies that the transition between Bi-V and -VIII observed
by Homan[14] using electrical resistance measurements is electronic
in origin. A more complete account of our measurements on Bi is
given in Ref. 5.

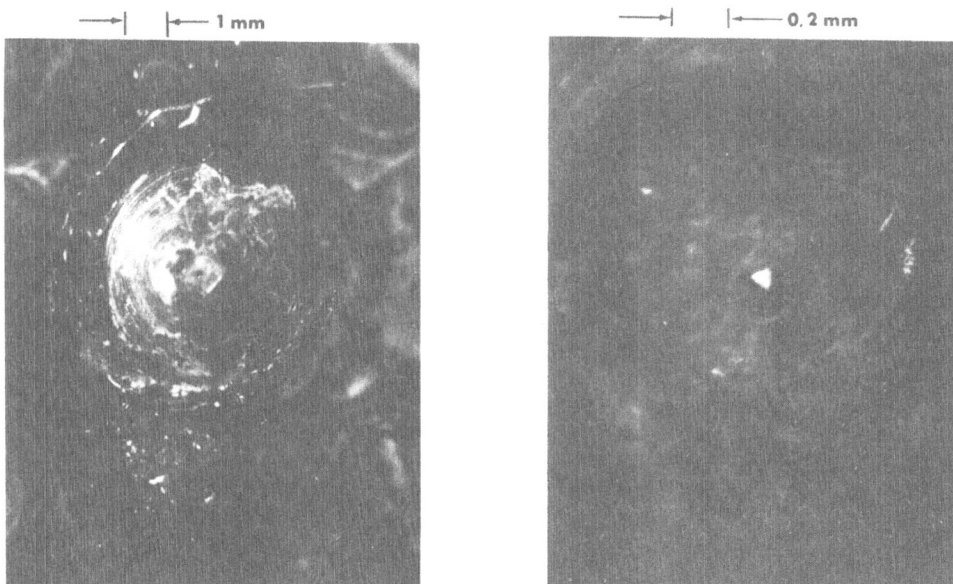

Figure 4. Left side: Induction coils mounted on lower diamond anvil (upper anvil removed) with Be-Cu gasket in position; Right side: enlargement of Be-Cu gasket showing the 50 micron V_3Ga single crystal.

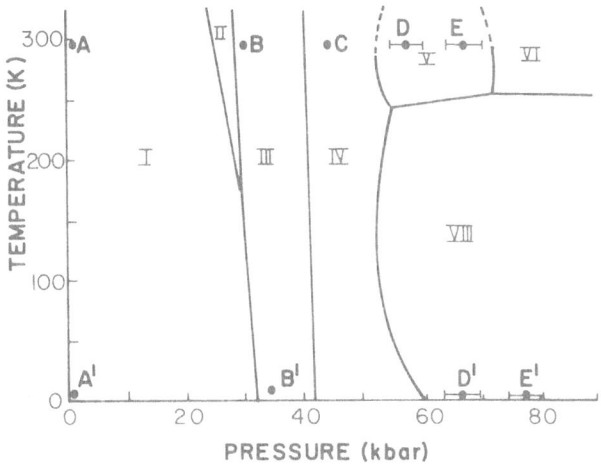

Figure 5. Partial phase diagram for Bi taken from Homan (Ref. 14).

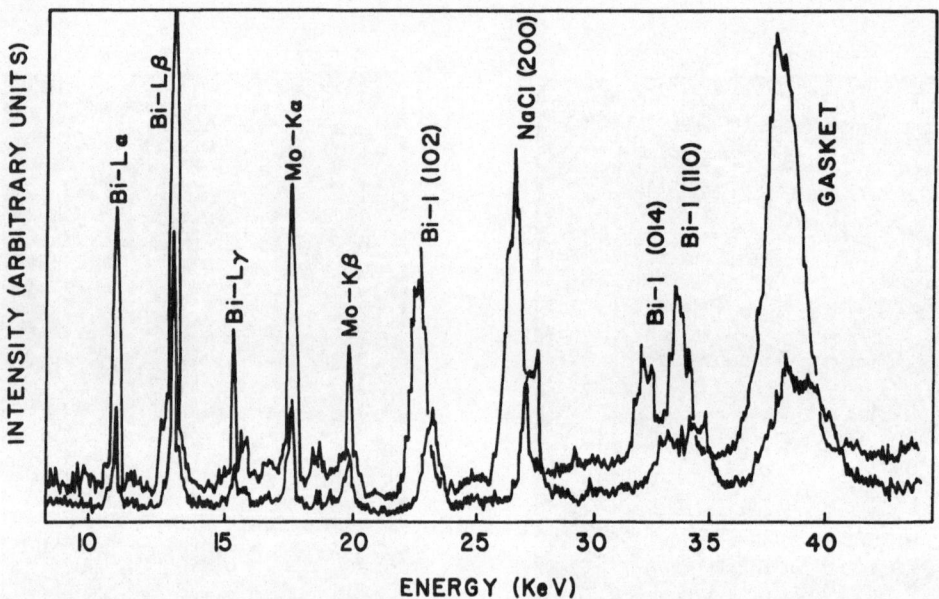

Figure 6.   Energy spectrum of Bi-I at P ~ 0.1 GPa;
             upper and lower curves recorded at 294 and 5 K, respec-
             tively.

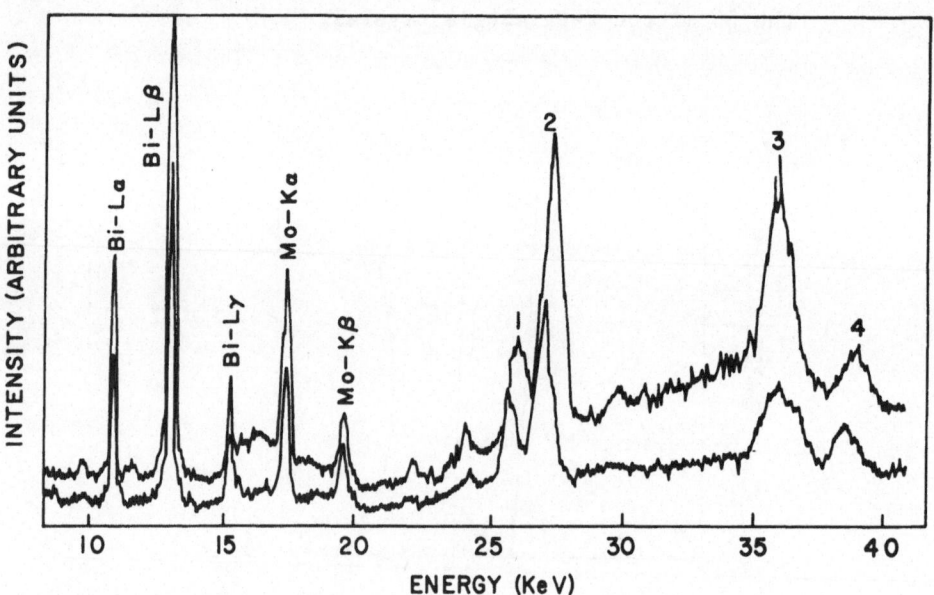

Figure 7.   Energy spectrum of Bi-III; the upper and lower curves
             were recorded at 3.4 GPa, 8K and 3.0 GPa 294 K,
             respectively.

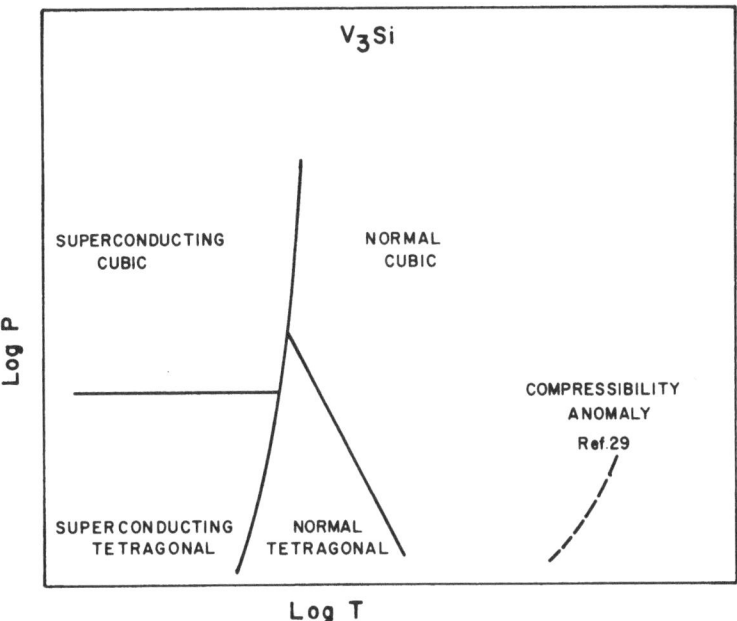

Figure 8. Partial phase diagram for V_3Si.

STRUCTURAL STUDIES ON V_3Si

Current studies are aimed at improving our understanding of the relationship between structural or lattice instabilities and superconductivity. It has been known for sometime that in highly perfect samples of V_3Si, the superconducting transition is preceded by a structural transition, often called a martensitic transition, from cubic to tetragonal symmetry.[20] This transition is accompanied by a softening of the shear acoustic mode propagating in the [110]-direction with polarization in the [$\bar{1}$10]- direction.[21,22] In this case the elastic modulus $[(C_{11}-C_{12})/2]$ vanishes at the transition.

Bhatt and McMillan[22,23] have proposed a phenomenological description of the transition based on Landau's theory.[24] According to this model, the slope of the superconducting phase line, dT_c/dP, should be positive for V_3Si, while the slope of the line dT_m/dP, separating cubic and tetragonal phases, should be negative. Allowing for a change in structure in the superconducting phase, and avoiding a quadruple point, the phase diagram should be similar to that shown in Fig. 8. Measurements of $T_c(P)$ have been made to 2.9 GPa[25,26] and $T_m(P)$ to 1.8 GPa,[27] in agreement with this model.*

----

* C.W. Chu reports at this Conference measurements of $T_c(P)$ and
(continued overleaf)

The single crystal x-ray diffraction cell was constructed specifically to explore the phase diagram of V_3Si and other high temperature superconductors in the low temperature region. We felt
it important to include the capability of making single crystal,
rather than polycrystalline, measurements following our experi-
ences with a similar transition in TeO_2.[28] By incorporating coils
around the diamond anvils it should be possible to obtain both
$T_c(P)$ and $T_m(P)$ to about 10 GPa with <u>in situ</u> structural measurements.

In addition to the low temperature transitions, a curious
anomaly has recently been found in the compressibility at room
temperature and about 0.9 GPa.[29] It has been proposed that this
phenomenon is related to an anomalously high concentration of
vacancies in the crystal.[30] Again it is of interest to study this
compressibility anomaly using single crystal rather than polycrystalline x-ray diffraction techniques, as in the earlier study.[29]

These and other studies are now being pursued and will be reported in due course. For the present, a more complete discussion
will be given of ongoing lattice dynamical studies on V_3Si.

THERMAL VIBRATION STUDIES IN V_3Si

The temperature dependence of the integrated x-ray intensities
of selected Bragg reflections in V_3Si is of great interest. The
measurements carried out here were prompted by calculations of the
mean-square thermal strain made several years ago by Testardi and
Bateman.[31] Basically they used the Debye model in conjunction with
their measured values of the temperature dependence of the elastic
constants to estimate the effect of the softening of the [110]-
transverse vibrational mode on the mean-sound velocity. Their calculations indicate that at about 40 K, there is an extremum, or
point of maximum stiffness of the lattice, to this particular vibrational mode. Therefore one might expect to see the influence
of such an extremum in any measurement involving this mode.

Bragg reflections from a crystal are, of course, attenuated by
atomic thermal vibrations; therefore such measurements provide information regarding the vibrational properties of the lattice. Although x-ray measurements cannot be used easily to probe a specific vibrational mode, they are biased in the direction of the diffraction vector, <u>i.e.</u>, the Debye-Waller factor involves the projection of the thermal ellipsoid onto the diffraction vector. Thus

---

$T_m(P)$ to the region where the curves intersected. However, it
is still of interest to carry out structural measurements in this
region.

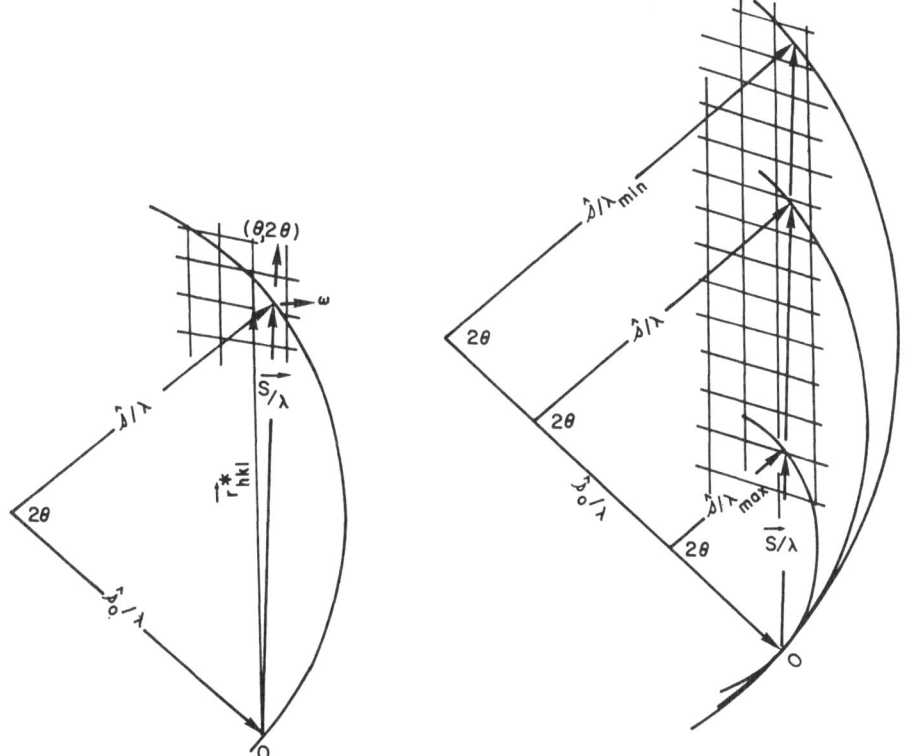

Figure 9. Left side: Section through Ewald sphere of reflection
           depicting a conventional diffraction event; Right side:
           Schematic representation of the intersection between
           the reciprocal lattice and the continuum of Ewald
           spheres encountered with polychromatic radiation.

atomic vibrations in the [110]-direction would be expected to contribute most strongly to the (h,h,0) Bragg reflections. In our
initial experiment, however, because of the external morphology
of the crystal, we chose to measure the (h,0,0) spectrum.

EXPERIMENTAL PROCEDURE

The technique used for these measurements was introduced by
one of us (E.F.S.) last year for rapid measurement of the temperature dependence of Debye–Waller factors.[32] The procedure is to
first irradiate the sample with characteristic "monochromatic"
radiation and orient it so that a particular reciprocal lattice
vector is coincident with the diffraction vector. The sample is
then illuminated with polychromatic radiation and multiorder
Bragg peaks belonging to the same class as the original diffrac-

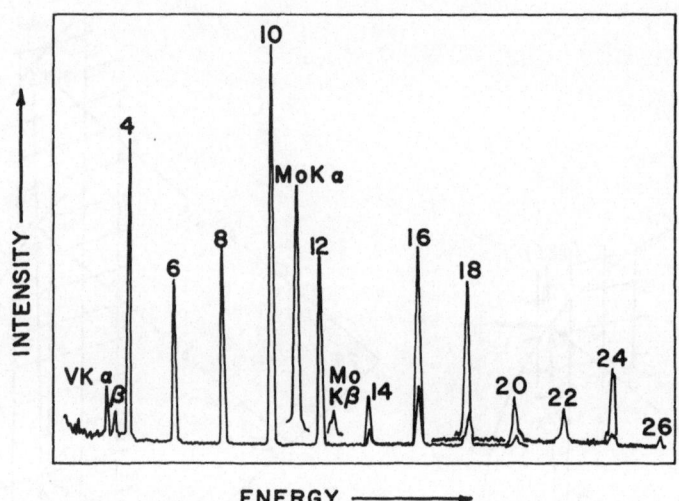

Figure 10. Bragg energy spectra of the (h,0,0) reflections of
            V$_3$Si recorded at 294 K (lower curve) and 7 K(upper
            curve); several different amplification factors are
            used in plotting the data.

tion vector are excited simultaneously. This situation is depicted schematically in Fig. 9. On the lefthand side of the figure the usual relationship between the reciprocal lattice and the
Ewald sphere of reflection (drawn with a radius of 1/λ) is shown.
When a reciprocal lattice point intersects the Ewald sphere, the
Bragg-von Laue conditions are fulfilled and a diffraction event
takes place. The diagram on the right-hand side corresponds to
the situation when polychromatic radiation is used, i.e., a continuum of Ewald spheres is considered, the largest being associated
with the energy of the most energetic photon in the incident beam
and the smallest being determined by the low-energy cut-off of
the detection system. Similarly, a continuum of diffraction vectors
is present ranging from $(\underline{S}/\lambda_{max})$ to $(\underline{S}/\lambda_{min})$. Thus when a reciprocal lattice vector, $\underline{r}^{*}_{hk\ell}$, is coincident with $\underline{S}$, then all orders of
that reflection, for which $\underline{S}/\lambda_{max} \leq r^{*}_{hk\ell} \leq \underline{S}/\lambda_{min}$ will be simultaneously excited. If the scattered radiation is measured with an energy
sensitive detector spanning the energy range of interest, then all
the excited reflections can be recorded simultaneously. V$_3$Si (h,0,
0) spectra recorded in this manner are shown in Fig. 10; the even-
order peaks for h = 4,6,8,...26 are recorded (odd-order (h,0,0)
peaks are forbidden by symmetry).

    The two spectra shown in Fig. 10 were recorded at different
temperatures, the upper curve at 7 K and the lower curve at 294 K.
The effect of temperature on the intensity of the Bragg reflections

is much more demonstrative for the higher order peaks, indeed the (22,0,0) and (26,0,0) cannot be "seen" in the spectrum at 294 K. In addition to the Bragg peaks, the Mo-Kα and Kβ characteristic radiation is recorded at 17.5 and 19.6 keV, respectively; the radiation used in the experiment was generated in a Mo x-ray tube operated at 48 kV and 14 ma. Fluorescence radiation is also excited in the sample: the V-Kα and Kβ peaks at 4.9 and 5.4 keV, respectively, are seen in Fig. 10, but the Si K-lines (Kα = 1.7 and Kβ = 1.8 keV) are below the low energy cut-off of the detector and consequently are not recorded.

The crystal used for the measurements was supplied by Dr. L. R. Testardi of Bell Laboratories; it is in the approximate form of a right circular cylinder of about 1/4 inch diameter and with a cylindrical axis approximately parallel to a [100]-axis. The crystal was mounted in a large Cu block of the Heli-tran refrigerator[8] and was aligned on the diffractometer so that the [100]-reciprocal lattice vector was parallel to the diffraction vector. Once properly aligned, the $(\theta, 2\theta)$-motion on the diffractometer permits adjustment of the energy at which the various (h,0,0)-reflections will occur. The 2θ-angle was set at the highest possible value which would allow a Bragg peak to straddle the Mo-Kα and Kβ peaks, viz., 112.85° .

The experimental procedure was to maintain a constant scattering geometry and vary the sample temperature. In operating the Heli-tran refrigerator, the temperature is reduced slightly below the desired set point by controlling the cryogen flow rate; the sample temperature is then electronically elevated and controlled. Uncertainties in the temperature measurement and control are believed to be well within 1%. Once thermal equilibrium was established at each temperature, the spectrum was recorded with a Li drifted Si detector for a preset period of time -- usually 20 min. The spectra were analyzed and stored in a 1024-channel analyzer which was set to span 41 eV per channel. The energy resolution of the Si(Li) detector was measured to be 176±6 eV based on the 5.994 keV Mn Kα line obtained from the decay of a Fe[55] source with activity in the low microcurie range. The energy calibration of the detector was based on the Mo and V Kα and Kβ peaks, in addition to the V_3Si diffraction lines using the calibration procedure discussed elsewhere.[33]

DATA REDUCTION AND ANALYSIS

The recorded spectra were outputted on an ASR - 33 teletypewriter and processed on a TI-ASC computer. The data reduction was based largely on a routine written by Larson and Repace[34] for analysis of spectral data collected with a solid state detector. This routine determines the location of each peak, applies a linear background correction, and then evaluates the integrated area.

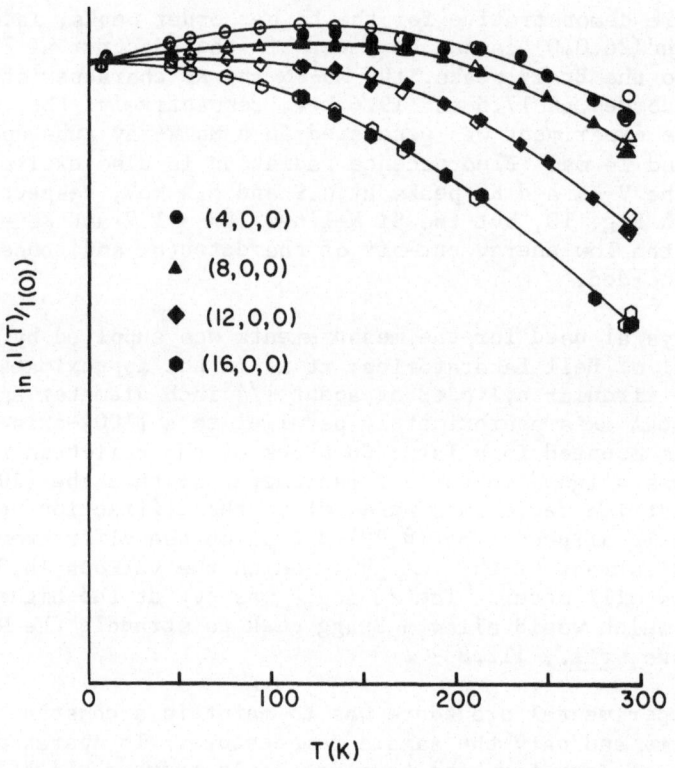

Figure 11.   Temperature dependence of the ratio of the background
             corrected integrated intensities of the (4,0,0),
             (8,0,0), (12,0,0), and (16,0,0) reflections.

     The ratios of the (h,0,0) peak areas (referenced to the lowest
recorded temperature) for h = 4, 8, 12, and 16 and h = 6 and 10 are
plotted in Figs. 11 and 12, respectively; the open and solid symbols represent two separate measurements. Although higher order reflections were recorded, the uncertainty in their integrated intensities at elevated temperatures was too large to permit meaningful
comparison. A striking feature of each set of curves is the apparent increase in the integrated intensity of the lower order reflections on warming from cryogenic temperatures. This anomalous
intensity increase appears to peak in the 100 to 150 K range.

     Our analysis of these results is, at present, rather specu-
lative and we stress that additional experiments to assess these
effects in greater detail are currently underway. Moreover, no
correction has yet been applied for thermal diffuse scattering
(TDS) contributions which will peak under the Bragg peaks and
hence contribute to the measured intensity. In addition to having

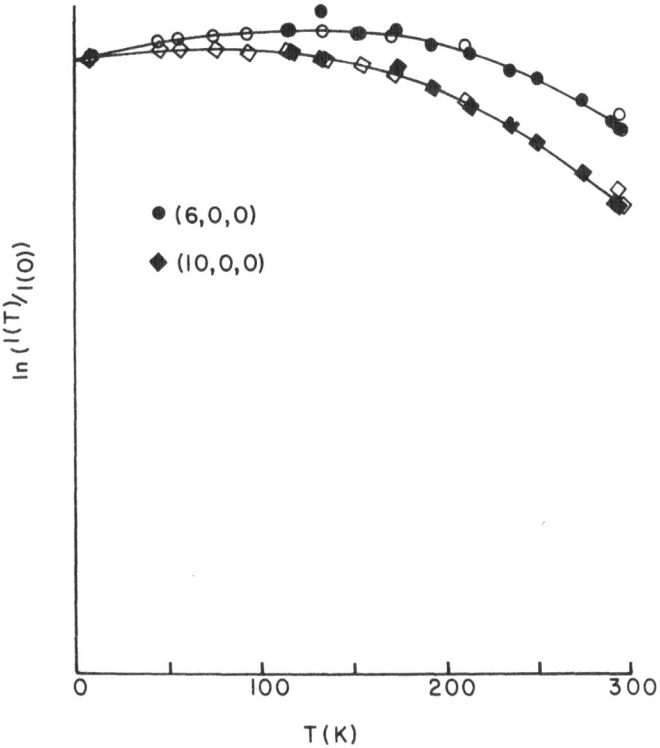

Figure 12. Temperature dependence of the ratio of the background
           corrected integrated intensities of the (6,0,0) and
           (10,0,0) reflections.

a temperature dependence inverse to the Bragg scattering, TDS
contributions to the measured intensity can be significant and,
more importantly, they can have a strong reflection dependence.
For example, Skelton et al.[35] found that integrated intensity
measurements of the (2,0,0) and (4,0,0) reflections of HgTe at 296 K
contained 0.02% and 0.12% TDS contribution, respectively, but the
contribution to the measured (16,0,0) peak under the same scattering
conditions was greater than 40%. Although detailed calculations
have not yet been carried out for V_3Si, we do not anticipate a
large TDS correction in this instance; using the expressions of
Skelton and Katz[36] and the elastic moduli of Testardi and Bateman[31],
the TDS correction for a V_3Si peak recorded under the same conditions as the HgTe-(16,0,0) reflection would only be about 5%.

    Assuming further that the quasi-harmonic approximation is
applicable in this case, the following interpretation is offered:
the integrated intensity from an extended face of a mosaic crystal such as the one used here can be expressed as follows:

$$I(T) = C\ P(\theta)\ \Omega(E)\ |F_{hk\ell}(T)|^2 \tag{1}$$

where C is a constant, $P(\theta)$ is a known function of the Bragg diffraction angle, $\Omega(E)$ is a function of the photon energy, and
$F_{hk\ell}(T)$ is the modulus of the structure factor . The function
$\Omega(E)$ depends on E through a number of terms whose values, in general, may be unknown, viz., the energy distribution of the incident
beam, the spectral response of the detection system, and the absorption and fluorescence properties of the sample.

The temperature dependence of the intensity is related to
the thermal vibrations of the crystalline lattice through the
Debye-Waller factor. The structure factor for the commonly known
A15 lattice (space group Pm3n), in which V_3Si crystallizes, can be
written, within the quasi-harmonic approximation, as follows:

$$F_{hk\ell}(T) = \sum_{j=1}^{8} f_j \exp\ [-2\pi i(hx_j+ky_j+\ell z_j)]\ e^{-q\langle u^2_{j,\underline{S}}\rangle} \tag{2}$$

where $q = 8(\pi\sin\theta/\lambda)^2 = 8[(E\sin\theta/(2\hbar c)]^2 \tag{3}$

and f_j is the j-th atomic scattering factor; (x_j,y_j,z_j) are the
fractional atomic coordinates; λ and E are the appropriate x-ray
wavelength and energy, respectively; the last term in eq. (2) is
the Debye-Waller factor; all other terms have their usual meaning.
The summation is carried out over the six V-atoms and the two Si-
atoms in the unit cell; the atoms are located in the special positions "a" and "c" (or "d") of the Pm3n space group.[37] In the
Debye-Waller factor, $\langle u^2_{j,S}\rangle$ is the average projection of the thermal displacement of the jth atom from its equilibrium position on-
to the diffraction vector, $\underline{S}$. The averaged thermal vibrations are
usually expressed in terms of the six anisotropic coefficients
of the thermal vibrational ellipsoid, however, the atomic site
symmetry imposes constraints on the number of independent ellipsoi-
dal coefficients. Based on the work of Peterse and Palm,[38] there
are 3 independent vibrational parameters for V_3Si: these are related to the following mean square displacements (MSD):
$\langle u^2_{V,\perp}\rangle$, $\langle u^2_{V,||}\rangle$, and $\langle u^2_{Si}\rangle$. The first two terms correspond to V
atomic vibrations normal to and parallel to, respectively, the [100]
direction; the site symmetry of the Si atoms is cubic and hence the
Si thermal vibrations are isotropic.

The (h,0,0)-reflections are themselves subdivided into two sub-
classes: h = 4n and h = 4n±2 (n = 1,2,3...). The analysis is simplified for the latter subclass because only two of the MSD's are
involved. (All three are involved for the 4n (h,0,0)-reflections.)
At this time, only the 4n±2 subgroup will be considered. The structure factor for this subgroup takes the following form:

$$F = 2[f_{Si} \, e^{-q \, <u_{Si}^2>} + f_V \, e^{-q \, <u_{V,\perp}^2>}] \tag{4}$$

and the temperature derivative is:

$$dF/dT = \{-2q \, f_{Si} \, e^{-q \, <u_{Si}^2>}[d<u_{Si}^2>/dT + R \, d<u_{V,\perp}^2>/dT]\} \tag{5}$$

where

$$R = (f_V/f_{Si}) \, \exp[-q(<u_{V,\perp}^2> - <u_{Si}^2>)]. \tag{6}$$

The maximum in the (6,0,0) curve in Fig. 12 implies that
$dF/dT = 0$ at the extremum. Since both the prefactor and the R-
term in eq. (5) are always positive, it follows that the temperature derivative of the two MSD's must have opposite signs. The
anomalous term, i.e., the negative derivative of the MSD, can
be attributed to the Si-atom assuming the R-factor to be thermally invariant based on the following argument: the R-factor has
a relatively strong reflection dependence because of the variations in the energy dependence of the atomic scattering factors.
The scattering factors computed by Doyle and Turner[39] were combined
with the published room temperature MSD's of Staudenmann et al.[40]
to determine the curve shown in Fig. 13. Since the magnitude of
R increases (over the range considered) for the higher order reflections and since the data suggest a monotonically decreasing
intensity ratio for the higher order reflections, we infer that
the V-atom contribution is more dominant for the higher order reflections where the usual negative slope of the intensity ratios
is seen. Based on these limited data and within the approximations
and assumptions cited above, we suggest that the MSD of the Si-
atom may decrease with increasing temperature up to about 100 to
150 K. Additional experimentation, however, is called for in
order to more precisely characterize the temperature dependence of
the MSD's in V_3Si and, as noted above, this work is currently underway.

The anomalous maxima reported here are similar to the observations of Kodess[41] and Geshko et al.[42] Kodess measured the x-ray
intensities of the (3,2,0), (3,$\overline{2,1}$), and (4,4,0) reflections at
80 and 300 K for both a transforming and a non-transforming V_3Si
crystal. The results were essentially the same for each crystal,
viz., I(80) > I(300) for the (3,2,0) and (3,2,1) reflections, but
I(80) < I(300) for the (4,4,0) reflection. Kodess interprets his
observations in terms of a sublattice distortion of the V-atoms
which accompanies the martensitic phase transition.

Geshko et al.[42] measured the temperature dependence of the
(4,4,0) reflection over the range from 80 to 300 K. They find
"an anomalous increase of relative intensities when the temperature is raised with a pronounced maximum at $T_i = 224$ K..." and

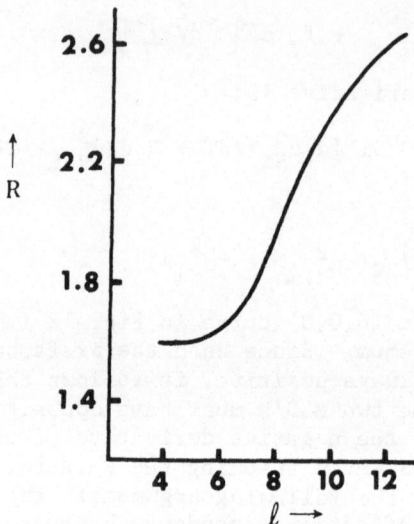

Figure 13.   Variation of the R-term (see text) with (h,0,0) reflection order (h = 4 through 12).

attribute this to "strong" anharmonic interactions. They further
suggest that T_i is the "temperature of maximum dynamic stability of
the V_3Si lattice."

Staudenmann[43] has carried out extensive single crystal measurements on V_3Si at 13.5 K and 80 K in a manner similar to his excellent 300 K work on the same crystal.[40] He infers from his data
"...that V_3Si is maybe preparing its martensitic transformation at
about 100 K." However, we are aware of the fact that the temperature dependence of the MSD measurements of Staudenmann[43] are not
as strong as that which we have observed. The origin of this discrepancy is presently being sought as this work continues.

Perhaps the single feature supported by all the recent low temperature x-ray studies on V_3Si is the existence of an apparent scattering anomaly above 80 K and below room temperature. The correlation between this phenomenon and the martensitic and superconducting
transitions, if such a correlation exists, is at present unknown.

                              ACKNOWLEDGEMENTS

    The support of the National Science Foundation for the Univer-
sity of Maryland personnel is gratefully acknowledged. We are also
indebted to Milton Shapiro of the University of Maryland for constructing the single crystal cell and to Dr. Testardi of Bell Laboratories for providing the V_3Si crystal used in this work.

REFERENCES

1. J.C. Jamieson, A.W. Lawson, and N.D. Nachtrieb, Rev. Sci. Instrum. 30 1016 (1959).
2. C.E. Weir, E.R. Lippincott, A. Van Valkenburg, and E.N. Bunting, J. Res. Natl. Bur. Stand. A 63, 55 (1959).
3. A.L. Hammon, Science 190, 967 (1975).
4. S. Block and G. Piermarini, Phys. Today 29, 44 (September 1976).
5. E.F. Skelton, I.L. Spain, S.C. Yu, C.Y. Liu, E.R. Carpenter, Jr., Rev. Sci. Instrum., 48, 879 (1977).
6. G.J. Piermarini, S. Block, and J.D. Barnett, J. Appl. Phys. 44, 5377 (1973).
7. J.D. Barnett, S. Block, and G.J. Piermarini, Rev. Sci. Instrum. 44, 1 (1973).
8. Model LT-3-110 manufactured by Air Products Inc., Allentown, PA 18105.
9. E.F. Skelton, C.Y. Liu, and I.L. Spain, "Simple Improvements to a Diamond Anvil High Pressure Cell for X-Ray Diffraction Studies" to be published in High Temp. - High Press.
10. D.L. Decker, J. Appl. Phys. 36, 157 (1965); 37, 5012 (1966); 42, 3239 (1971).
11. L.C. Chhabildas and A.L. Ruoff, J. Appl. Phys. 47, 4182 and 4867 (1976).
12. C.T. Van Degrift, Rev. Sci. Instrum. 46, 599 (1975).
13. F.J. Rachford, I.L. Spain, and E.F. Skelton - to be published.
14. C.G. Homan, J.Phys. Chem. Solids 36, 1249 (1975).
15. C.G. Homan, T.E. Davidson, and D.P. Kendall, Appl. Phys. Lett. 26, 615 (1975).
16. D.L. Decker, private communication, 1976.
17. R.M. Brugger, R.B. Bennion, and T.G. Worlton, Phys. Lett. A 24, 714 (1967).
18. R.M. Bruger, T.B. Bennion, T.G. Worlton, and W.R. Meyers, Trans. Am. Crystallogr. Assoc. 5, 141 (1969).
19. See J.F. Cannon, J. Phys. Chem. Ref. Data 3, 781 (1974) and references therein.
20. B.W. Batterman and C.S. Barrett, Phys. Rev. Lett. 13, 390 (1964).
21. L.R. Testardi, T.B. Bateman, W.A. Reid, and V.G. Chirba, Phys. Rev. Lett. 15, 250 (1965).
22. L.R. Testardi, in Physical Acoustics ed. by W.P. Mason and R.N. Thurston (Academic, New York, 1973),Vol. 10.
23. R.N. Bhatt and W.L. McMillan Phys. Rev. 14, 1007 (1976).
24. L.D. Landau and E.M. Lifshitz, Statistical Physics (Pergamon, London, 1958), pp. 430 - 456.
25. S. Huang and C.W. Chu, Phys. Rev. 10, 4030 (1974).
26. R.N. Shelton and T.F. Smith, Mat. Res. Bull 10, 1013 (1975).
27. C.W. Chu and L.R. Testardi, Phys. Rev. Lett. 32, 766 (1974).
28. E.F. Skelton, J.L. Feldman, C.Y. Liu, and I.L. Spain, Phys. Rev. 13, 2605 (1976).

29. R.D. Blaugher, A. Taylor, and M. Ashkin, Phys. Rev. Lett. <u>33</u>, 292 (1974).

30. C.M. Varma, J.C. Phillips, and S.T. Chui, Phys. Rev. Lett. <u>33</u>, 1223 (1974).

31 L.R. Testardi and T.B. Bateman, Phys. Rev. <u>154</u>, 402 (1967).

32. E.F. Skelton, Acta. Cryst. <u>A32</u>, 467 (1967).

33. E.F. Skelton, J. Appl. Cryst. <u>10</u>, 123 (1977).

34. R.E. Larson and J.L. Repace, NRL Memo Report 2658, NRL Computer Bull. <u>34</u>, Naval Research Laboratory, Wash. D.C. 20375.

35. E.F. Skelton, P.L. Radoff, P. Bolsaitis, and A. Verbalis, Phys. Rev. <u>5</u>, 3008 (1972).

36. E.F. Skelton and J.L. Katz, Acta. Cryst. <u>A25</u>, 319 (1969).

37. <u>International Tables for X-Ray Crystallography</u>, Vol. 1, SG Nr. 223 (Birmingham; Kynoch Press.)

38. W.J. A. M. Peterse and J.H. Palm, Acta Cryst. <u>20</u>, 147 (1966).

39. P.A. Doyle and P.S. Turner, Acta Cryst. <u>A24</u>, 390 (1968).

40. J. -L. Staudenmann, P. Coppens, and J. Muller, Solid State Comm. <u>19</u>, 29 (1976).

41. B.N. Kodess, Soviet Phys. Solid State <u>16</u>, 782 (1974).

42. E.I. Geshko, V.B. Lototskii, and V.P. Mikhal'chenko Ukr. Fiz. Zh. <u>21</u>, 186 (1976).

43. J. -L Staudenmann, ACA Program and Abstracts <u>5</u>, 82 (1977); private communication (1977).

QUESTIONS AND COMMENTS

J. Wittig: You showed us the phase diagram of bismuth including the phase IV? Maybe I'm not informed but I thought this phase died 20 years ago.

E.F. Skelton: There are 7 papers in the literature reporting a transition from phase III to phase IV. There are 6 papers in literature reporting a specific attempt to see this transition and a failure to detect it. So there are as many having seen it as having not seen it. We simply added to the confusion, I think, in that the patterns that we have of phase III differ from the one pattern we have probably of phase IV, in the sense that there is one additional line. This is certainly not strong evidence for the phase transition. I think this is still very much an open question.

T. Suski: In the place of such a measurement of structure like this one, you should be very cautious about conclusions about static atom displacements and about Debye–Waller factors. You should have as much information as possible to avoid possible mistakes connected with correlations between shift of atoms and their Debye–Waller factors.

E.F. Skelton: Yes, that is why I should stress that the analysis was carried within the limitations of the quasi-harmonic theory which perhaps under normal circumstances are not as severe as they appear here.

A.R. Moodenbaugh: What are the pressure limitations, when going low in temperature, before you crack the anvils?

E.K. Skelton: The system was designed for operation at 100 kbar. We have another one that is beefed up considerably in size, and it should go up to 300–400 kbar. Ian Spain designed it. But the one that I presented the data for, and the one that was in operation, should go up to 100 kbar. It's been up to about 80 kbar.

C.G. Homan: Can you determine if a volume change has occurred in the transition from phase III and IV from your data?

E.F. Skelton: It seems to be highly sample dependent due to the fact that there are so many papers where people specifically tried to see it.

C.G. Homan: Many of the people looking for the III–IV transition using differential thermal analysis techniques have not been successful. However, if one uses the largest reported volume change for this transition, the calculated latent heat may be in the background for DTA measurements.

J.B. Clark: Is there any sluggishness on the bismuth transition under discussion? One would expect some sluggishness at such low temperatures.

E.F. Skelton: Clarke (Homan) can answer about the sluggishness. Our data are recorded over an extended period of time. The procedure is as follows: the pressure is set, and after a reasonable wait of several hours we get it photographed. The photograph runs for some time. In terms of the energy disper-

sive technique, I don't think that the measurement is sensitive enough to see any small changes
in the pattern. Clarke (Homan) measures the
resistivity effectively on-line.

R. Viswanathan: Can you find out the static displacement from the
                   equilibrium positions?

E.F. Skelton: Yes, I can tell you the work of Staudenmann.
                   He measured several hundred reflections at 78 K,
                   13.5 K and room temperature. He has, I guess
                   most importantly at 13 K, attempted to refine
                   his data on the basis of a cubic structure, a
                   tetragonal structure, and an orthorhombic struc-
                   ture. The conclusion is that the data fit all
                   three of them equally well. That is, based on
                   his data, you can't really distinguish any change
                   in the atomic positions of the V_3Si sample. I
                   would say, the structure that has been proposed
                   for the Nb_3Sn, I think, seems to fit their data,
                   as they look at the R-factor. It's essentially
                   the same as the standard A15 structure, within
                   statistical uncertainties.

HIGH TEMPERATURE SUPERCONDUCTIVITY AND INSTABILITIES*

C.W. Chu

Department of Physics, Cleveland State University
Cleveland, Ohio 44115

and

Department of Physics[+], University of Houston
Houston, Texas 77004

ABSTRACT

It is known that high temperature superconductors are unstable. Different models have been proposed to explain the high temperature superconductivity and instabilities. High pressure experiments in a hydrostatic environment testing these models are reviewed and discussed. New data on single crystalline V_3Si and polycrystalline $Nb_3Sn_{1-x}Y_x$(Y = Al and Sb) are also included. Although no unique picture emerges at the present time, existing high pressure results tend to lend support to the electron model emphasizing the roles of electrons. Further experiments are proposed and planned for better understanding of the origin of high temperature superconductivity and instabilities, and their correlation.

INTRODUCTION

Immediately after the discoveries of the high temperature and high field superconductors V_3Si by Hardy and Hulm[1] and Nb_3Sn by Matthias[2], special effort has been made on the direct search for superconductivity of higher temperature in compounds similar to

---

*Research supported in part by National Science Foundation Grant No. DMR 73-02660
†Present address.

V_3Si and Nb_3Sn with A15 structure. Although high temperature superconductivity (e.g. > 12K) has been found in systems with other
structures, the current record high temperature superconductor
Nb_3Ge[3] still belongs to the A15 class. In an indirect but parallel
approach, extensive study has also been carried out on the understanding of the properties of the A15 material. It was found that
almost all A15 compounds with high superconducting transition temperature T_c behave anomalously compared with their low T_c counterparts, or the "normal" metals even in the non-superconducting
state[4,5]. The anomalies include, for instance, the large Sommer-
feld coefficient of the electronic specific heat, the strong temperature dependences of the magnetic susceptibility and the Knight
shift, the large negative curvature in the temperature dependence
of the resistivity, and etc. Through x-ray diffraction, a cubic-
to-tetragonal structural transformation was observed in some of the
V_3Si and Nb_3Sn samples. For V_3Si[6], the transformation occurs at a
temperature T_L ~21K with $c/a > 1$, while for Nb_3Sn[7], at T_L ~43K with
$c/a < 1$. Acoustic studies revealed mode softening on cooling in both
transforming and nontransforming V_3Si[8] and Nb_3Sn[9] at temperature far
above T_L. Phonon dispersion curves obtained from n-diffraction
measurements[10] demonstrated that this mode softening extended toward the Brillouin zone boundary. Furthermore, severe strain[11]
and radiation damage[12] were found to have large detrimental effect
on T_c, suggesting that the high temperature superconducting phase
in this class of material was metastable. More experiments[13] were
recently done and showed that the so called anomalies observed in
high T_c A15 compounds were so common in all high temperature superconductors that their absence became really abnormal.

In 1952, Frölich[14] already pointed out that the crystal lattice
of a high temperature superconductor was intrinsically unstable,
because of its strong electron-phonon interaction. Then Cochran[15]
and Anderson[16] formulated the soft mode theories which related the
structural transformation at T_L with phonons whose frequencies diminish as the transformation was approached from above. The BCS
theory[17] and the McMillan's strong coupling theory[18] together with
the occurrence of the aforementioned anomalies in high T_c compounds
suggest that high temperature superconductivity and instabilities
are closely related and all the observed anomalies are due to instabilities of one kind or the other.

Different models have been proposed to explain the observed
anomalies in high T_c superconductors in general and in those with
A15 structure in particular. In the past several years, many high
pressure experiments have been done on the T_L and T_c of unstable
superconductors. With some emphasis on the A15 compounds, we shall
discuss the results of these experiments and compare them with the
predictions of different models. It should be noted that both T_c
and T_L of unstable superconductors are extremely sensitive to pres-

sure inhomogeneity. In this paper, we concern ourselves only with experiments performed under hydrostatic pressure.

THE MODELS

In spite of the great many differences in details, all models proposed to explain the high temperature superconductivity and the different anomalies can be summarized into three, depending on their emphases: the electron model, the soft-mode model and the defect model. For the convenience of later discussion, we shall briefly describe the models and the roles of hydrostatic pressure in these models.

Electron Model

The main feature of the model is the assumption of the existence of density of states peaks of several meV wide near the Fermi level. Justification of these electron band fine structures is made[19] based on the linear chain arrangement of the transition metal atoms in the A15 compounds. The high T_c of the A15 compounds is then attributed to the high density of states[19,20], the unusual temperature dependences of the magnetic susceptibility[21], the Knight shift[22], the resistivity[23] and the elastic modulus[24] to the close proximity of the Fermi level to the narrow peak of the electron density of states, and the structural transformation[25] to the lower electron free energy state resulting from the removal of the 3-fold degeneracy due to the transformation. The high temperature superconductivity and instabilities are considered to be two aspects of one phenomenon, namely, the instability of the electron energy spectrum. According to this model, the application of hydrostatic pressure will increase both the interchain and intrachain couplings[26,27,28]. This can induce an interband charge transfer, broaden the density of states peak, and vary the electron-phonon interaction. T_c and T_L are changed under pressure as a result[29]. Unfortunately, the model allows too many adjustable parameters in order to fit the experimental data. In addition, Mattheiss' recent band structure calculation[30] does not provide evidence for the one- or quasi-one-dimensional characteristic, based on which the existence of fine structures in the density of states is originally assumed. However, it is recently suggested[31] that these fine structures can exist in a three-dimensional system. Later studies also seem to start to establish or identify the source of structural instabilities - electronic in nature, which will be discussed in the following paper[32].

Soft Mode Model

In this model[33], it is assumed that high T_c A15 structures in its strain-free-state is extremely unstable and large anharmonicity exists near the bottom of the lattice potential well at low strain-level. The flatness of the bottom of the lattice potential well automatically ensures the lattice softening at low temperature. The high T_c is then ascribed to the soft phonon modes, making use of the strong coupling theory for superconductivity[18] where details of the electron band structure play only a minor role, and the negative curvature[34] of the resistivity - temperature curve to the soft lattice enhanced electron-phonon scattering. It is further proposed[4,33] that the pressure dependence of the structural instabilities dictates the pressure dependence of superconductivity. Since instability is maximum at T_L, T_c is therefore enhanced (suppressed) as T_L is driven by pressure towards (away from) T_c. Unfortunately difficulties arise for any serious attempt to evaluate the model quantitatively. The model remains at a phenomenological stage.

Defect Model

It[35] focuses only on the relief mechanism for instabilities by assuming that all high T_c compounds are intrinsically unstable and thus extremely susceptible to defect formation. However, according to this model, it is the presence of defects that prevents any catastrophic transformation leading to a large T_c-reduction from happening. The lattice softness is preserved for the case of low defect concentration but removed for high defect concentration. The mode softening[36], the unusual temperature dependence of resistivity[35], the large and almost universal effect of radiation damage on T_c[35] and some other observations[35] have been attributed to the condensation of defects with decrease of temperature or the ordering of defects at low temperature. Pressure is suggested[36] to increase or decrease the defect concentration depending on the nature of the defects in the compound and hence to vary T_c and T_L accordingly. The model seems to suffer from the same weakness as the soft mode model, being phenomenological and qualitative.

RESULTS AND DISCUSSION

In spite of the early recognition of the coexistence of high temperature superconductivity and structural instabilities in A15 compounds, direct high pressure study on the correlation of these two phenomena was hampered by experimental difficulties due to the minute effect associated with the structural transformation in these compounds. The improvement in experimental techniques and the discovery of more unstable superconducting systems, many of which exhibit large anomaly in some easily measurable quantities like resistivity at the structural transformation, make it possible to investigate directly high temperature superconductivity and structural instabilities simultaneously under pressure. Almost all experiments done in this regard involve the determination of the hydrostatic effects on T_c and T_L. The systems studied consist of both the three-and two-dimensional unstable superconductors with different crystal structures. They are V_3Si[37], Nb_3Sn[29], V-Ru

near equi-atomic compounds[38], HfV_2[39], ZrV_2[39], La_3S_4[40], La_3Se_4[40],

$Cu_xMo_3S_4$[41], $2H-NbSe_2$[42,43], $2H-TaS_2$[44], $2H-TaSe_2$[45], $4Hb-TaS_2$[46]. The pressure coefficients of T_c and T_L, together with their T_c, T_L and crystal structures are summarized in Table I. Although the T_c's of some of these unstable systems are low compared with the current record of 23K, they may be considered high within their own crystal structure classes.

In the present paper, we shall concentrate only on the two cases where T_c's are the highest and the crystal structure is shared by most high temperature superconductors, i.e. V_3Si and Nb_3Sn. Results on other systems will also be commented on when appropriate. For simplicity of presentation, we shall discuss the results first in terms of the electron and soft mode models, and then the defect model.

A) In Terms of the Electron and Soft Mode Models

By analyzing the data of elastic moduli and the specific heats, a strong quadratic strain dependent T_c in V_3Si was predicted[47]. The prediction seemed to be consistent with the alloy results[48] at atmospheric pressure, which showed near-parabolic volume dependence for a number of A15 materials with maxima in T_c at V_3Si, V_3Ga, Nb_3Al and Nb_3Sn. However, it was found[49] that T_c was enhanced rapidly by pressure up to 24 kbar. To reconcile this, it was proposed[46] that the high temperature phase might not be ideally cubic due to the large anharmonicity and that hydrostatic pressure might have enhanced lattice instabilities. This proposition would move V_3Si away from the maximum in the T_c-volume curve and allow a positive pressure effect on its T_c in accordance with the soft mode model. In fact, pressure induced softening in the shear modulus C_s was observed at low temperature between 77 and 100 K, i.e. $\partial C_s/\partial p < 0$, in both transforming[50] and non-transforming[51] single crystalline V_3Si samples. Since the structural transformation is the result of an extreme C_s-softening, a pressure induced C_s -softening implies a pressure enhanced T_L in V_3Si[50]. Later direct measurements on T_L of V_3Si[37] up to 18 kbar show that hydrostatic pressure suppresses T_L instead. By extrapolation[37], a critical pressure of ~24 kbar was obtained for a complete suppression of the structural transformation in V_3Si down to its superconducting state. This means that hydrostatic pressure actually stabilizes the lattices, and suppresses the instabilities, implying a positive $\partial c_s/\partial p$ at temperature close to T_c. Indeed, a pressure induced c_s-stiffening below 50K was observed in the transforming sample following a softening between 50 and 100 K.

Table I. T_c, $\partial T_c/\partial p$, T_L, $\partial T_L/\partial p$ and the crystal structure of unstable superconductors. References are given in the text. For
compounds marked by "*", nonlinear pressure effect on T_c is observed and the $\partial T_c/\partial p$ shown is the low pressure value. $2H-TaSe_2$
undergoes an incommensurate-commensurate transition at 92.5K with
a negative pressure coefficient of 2.7×10^{-3} kbar.

Compounds	T_c (K)	$\partial T_c/\partial p$ (10^{-5} kbar^{-1})	T_L (K)	$\partial T_L/\partial p$ (10^{-4} kbar^{-1})	Structure
V_3Si	16.5	+3.65	21.5	-1.5	A15
Nb_3Sn	17.8	-1.40	43.2	+3.3	A15
$V_{0.54}Ru_{0.46}$	4.92	+0.91	45	-3.2	B2
HfV_2	8.9	+6.5	128	-8	C15
ZrV_2	7.9	0	124	-1	C15
La_3S_4*	8.1	+11	86	+13	$D7_3$
La_3Se_4*	7.6	+7	65	+38	$D7_3$
$CuMo_3S_4$*	10.5	+20	255	+85	Chevrel Phase
$2H-NbSe_2$	7.3	+4.95	31.3	-3.3	Hex layered
$2H-TaSe_2$	0.14	+1.3	122	+3.5	Hex layered
$2H-TaS_2$	0.49	+9.3	76	-2.2	Hex layered
$4Hb-TaS_2$	2.5	+3.5	22	-6.6	Hex layered

On the basis of the electron model, T_c and T_L depend on the band
structure and the position of the Fermi level, or the electron population in the d-subband, Q. The Q-dependences[25] of T_c and T_L have
been calculated for V_3Si and Nb_3Sn. They are schematically represented in Fig. 1 for V_3Si. Both T_c and T_L are non-monotonically increasing functions of Q and peak at Q_{cm} and Q_{Lm} respectively with $Q_{cm} > Q_{Lm}$.
The d-electron population of V_3Si is Q_0, lying between Q_{Lm} and Q_{cm}.
The application of hydrostatic pressure will increase the interchain coupling and can lead to a redistribution of charges between
bands. The observed pressure enhanced T_c and pressure suppressed T_L
are thus accounted for in terms of a pressure enhanced Q_0[29,37]. To
explain the unusual temperature dependence of $\partial c_s/\partial p$ of the transforming V_3Si sample[27,50], pressure enhanced Q_0 was deduced[27,28].
For Nb_3Sn, because of the less localized d-orbitals, Q_{Lm} is larger
than Q_{cm}. This difference is reflected, at least in part, in the
difference of the c/a-ratios in the low temperature phase of the
crystals, i.e. c/a > 1 for V_3Si and c/a < 1 for Nb_3Sn. With
$Q_{cm} < Q_0 < Q_{Lm}$ for Nb_3Sn, a pressure enhanced T_L and a pressure
suppressed T_c are therefore expected. As shown in Fig. 2, this is

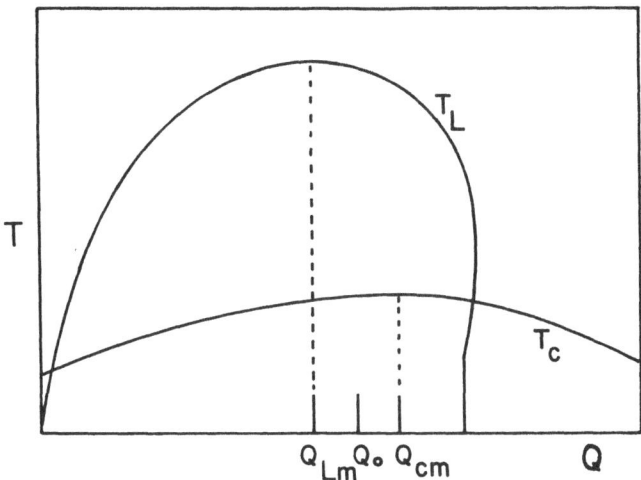

Figure 1.   Schematic Q-dependence of T_c and T_L of V_3Si.

indeed what has been observed[29]. It should be noted that the pressure effects on T_c and T_L are completely determined by the relative values of Q_o, Q_{Lm}, Q_{cm} and $\partial Q_o/\partial p$, according to the electron model. This means that no definite sign-correlation between $\partial T_c/\partial p$ and $\partial T_L/\partial p$ is expected[52].

The high pressure results on T_c and T_L of V_3Si and Nb_3Sn up to 19 kbar have been suggested[53] to be also consistent with predictions of the soft mode model, which are summarized in Fig. 3. Based on this model, pressure should always have opposite effects on T_c and T_L, and a maximum T_c in the T_c-P curve should be expected at $T_c \sim T_L$. Although $\partial T_c/\partial p$ and $\partial T_L/\partial p$ have opposite signs for many of the unstable superconductors, exceptions are recently observed as shown in Table I.

In an indirect search for the maximum T_c in the T_c-P curve, T_c's of both the transforming and nontransforming V_3Si single crystals were measured up to 20 kbar[54]. $\partial T_c/\partial p$ was found to be always positive, but about 30% smaller for the nontransforming samples. The results suggest the absence of a maximum T_c and the presence of a break in slope in the T_c-P curve at $T_c \sim T_L$ contradicting the predictions of the soft mode model. The observations can be attributed to the difference in the electron energy spectra associated with the distorted and undistorted phases. Later, T_c of a polycrystalline V_3Si sample was measured up to ~29 kbar[55] and failed to exhibit a maximum T_c except a decrease in slope beyond ~16 kbar. The general behavior of T_c is consistent with previous indirect

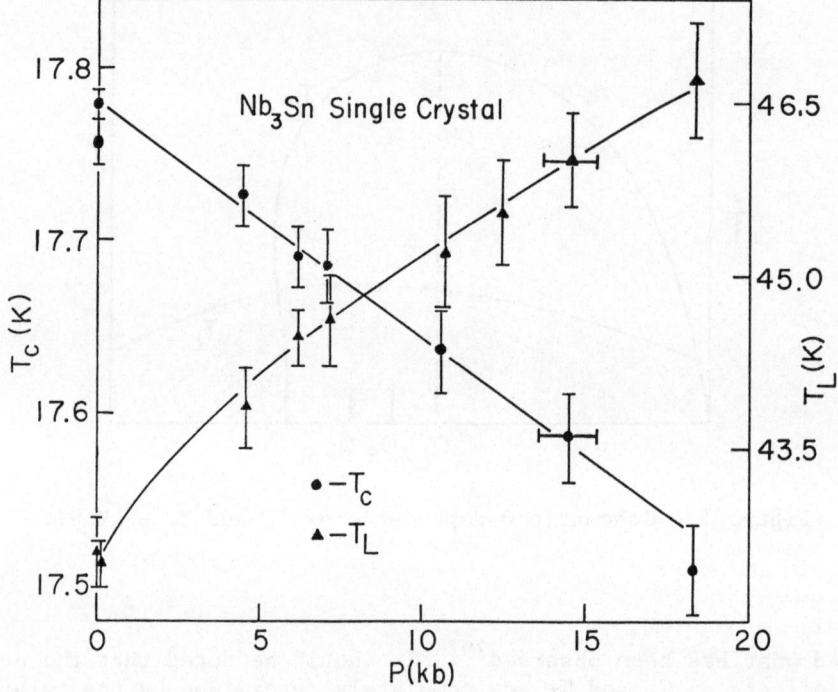

Figure 2.   Pressure dependences of T_c and T_L of Nb_3Sn simple crystal
            sample.   (after Ref. 29)

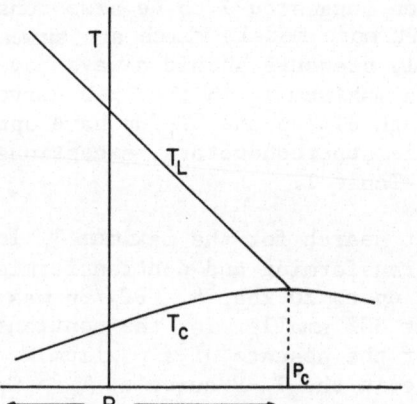

Figure 3.   The predicted pressure dependences of T_c and T_L on the
            basis of the soft mode model.   (after Ref. 33)

measurements[54]. However, the break in slope at ~16 kbar was attributed[55] to the possible pressure induced soft-stiff transition[56]
observed at room temperature, because of the lack of T_L-P data on
this sample and the fact that ~16 kbar was considerably lower than
the extrapolated critical pressure of ~24 kbar[37] for the complete
stabilization of the cubic phase. Recently, simultaneous measurements on T_c and T_L of a V_3Si single crystal have been extended to
~28 kbar by us[57]. The results are shown in Fig. 4. T_c increases with
pressure in a similar fashion as previously observed[55] except the
break in slope occurs at a slightly higher pressure ~19 kbar.
T_L decreases with pressure. It becomes increasingly difficult to
detect the structural transition above ~17 kbar, since the
anomalies in the specific heat and the temperature derivative of
resistivity at T_L diminish with increased pressure. However,
extrapolation gives a critical pressure of ~20 kbar for the
stabilization of the cubic phase in the superconducting state, suggesting that the change of slope in the T_c-P curve at ~19 kbar may
be related to the complete suppression of the structural transformation. This absence of maximum T_c and the presence of a break in
the slope of the T_c-P curve at $T_L \sim T_c$ have also been observed in
the V-Ru[38], 2H-NbSe$_2$[42] and 4Hb-TaS$_2$[46] compounds.

The $Nb_3Sn_{1-x}Y_x$(Y = Al and Sb) compounds were studied by us
under pressure in order to further test the electronic model. Both
the T_c and T_L have been previously investigated at atmospheric
pressure. A replacement of several percents of Sn by Al[58] is found
to enhance the T_c but suppress the structural transformation. However, a substitution of Sn by Sb[59] up to ~15% is observed to suppress
the T_c while leaving the T_L almost unchanged. The system is particularly interesting for the high pressure study because of its
reversal in sign[58,59] of the strain accompanying the structural
transformation. For example, the replacement of small amounts of
Sn by Sb results in a c/a ratio greater than unity similar to V_3Si,
while for pure Nb_3Sn and $Nb_3Sn_{1-x}Al_x$, c/a<1. The difference in the
relative values of Q_{cm} and Q_{Lm} between V_3Si and Nb_3Sn has been sus-
pected to be associated with the difference in c/a-ratios between
the two compounds. A simple minded picture based on the electron
model[25] would predict[29] a negative $\partial T_c/\partial p$ for $Nb_3Sn_{1-x}Al_x$ and a positive $\partial T_c/\partial p$ for $Nb_3Sn_{1-x}Sb_x$. Unfortunately, negative $\partial T_c/\partial p$ was observed for both systems, as shown in Fig. 5. Some of the possibili-
ties are: 1) whether Q_{cm} is smaller or greater than Q_{Lm} depends on
factors other than the c/a-ratio, 2) since the Sb-substitution
changes not only Q_0 but also c/a, the relative position of Q_{cm},
Q_{Lm} and Q_0 can be so displaced that the observed results are not
unreasonable and 3) the electron model is not valid. We are in
the process of determining the pressure effect on T_L of the compounds.

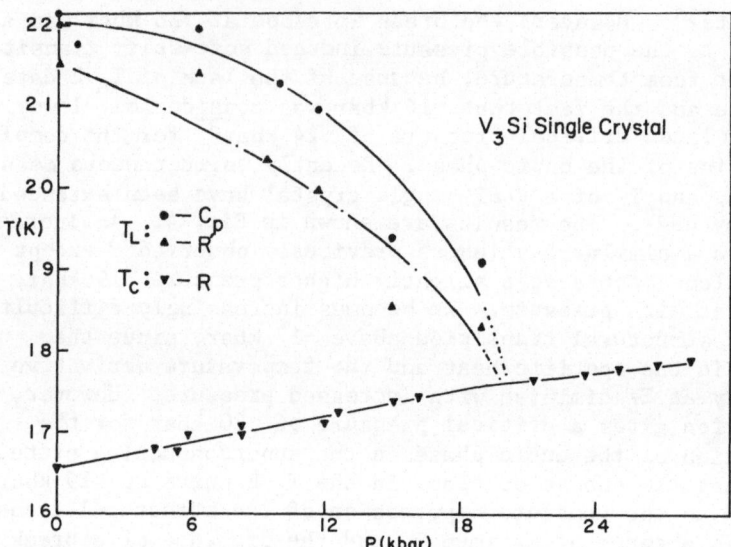

Figure 4.   Pressure dependences of T_c and T_L of a transforming V_3Si
            single crystal sample. (Present work). T_c was de-
            termined by resistivity (R) and T_L by calorimetry (C_p),
            and by resistivity slope (R').

B)   In Terms of the Defect Model

The defect model[35,36] has been used to explain qualitatively
many observations on unstable superconducting compounds. However,
only a few where high pressure can provide a test will be considered.

A soft-stiff transition[56] was found in V_3Si at room temperature
and ~12 kbar by X-ray measurements. Later static compressibility
measurements[60] confirmed the observation. In addition, the acous-
tically determined compressibility is considerably smaller than the
low pressure value determined by X-ray but equal to the high pressure
value[56], suggesting a mechanism with relaxation time longer than
~10^{-6} sec was involved. This was attributed to the possible removal
of defects in V_3Si under pressure[36]. It was further proposed[36] that
the difference in the pressure response of T_c and T_L between V_3Si and
Nb_3Sn rested on the difference in the pressure effect on the defect
concentrations between the two compounds. In other words[36], pressure
would squeeze out the defects in V_3Si where vacancies were assumed
to exist but would create defects in Nb_3Sn where anti-site defects
were supposed to prevail. Such pressure induced changes in the defect concentration would have resulted in a decrease in the resistivity for V_3Si and an increase for Nb_3Sn at room temperature under
pressure. However, no drastic resistivity reduction >0.3% correspond-

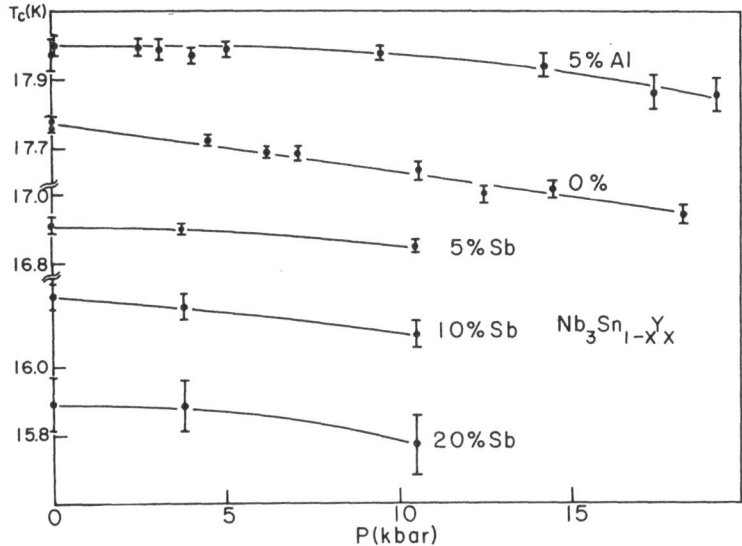

Figure 5. Pressure dependence of T_c of $Nb_3Sn_{1-x}Y_x$ with Y = Al and Sb.

ing to the squeezing out of defects, was detected by us in V_3Si up to ~22 kbar. For Nb_3Sn, on the contrary, a large decrease in resistivity ~10% was observed by us at ~22 kbar, in direct contradiction with predictions of the defect model. The positron lifetime in V_3Si[61] was found to have only one component and to be pressure independent up to 20 kbar at room temperature. This tends to suggest[61] that the postulated vacancies in V_3Si do not exist or do not depend on pressure. It is also possible[61] that vacancies are present but that they do not produce a sufficiently deep potential well to trap a positron.

Recent studies on Nb-Ge films[3,62] with different chemical compositions, prepared under different conditions and subject to different radiation does, reveals simple universal T_c-lattice parameter (a_0)[63] (or more exactly the reduction in T_c - the increase in a_0) and T_c-resistance ratio (R(300K)/R(25K), or R/R)[62] correlations shown in Fig. 6 and 7: T_c increases as a_0 decreases while R/R increases. These correlations also hold for a number of other high T_c bulk superconducting systems[63,64]. Since the R/R and a_0 are closely related to the defect concentration, it was then proposed[65] that the variation in T_c could be described by a single defect parameter and that the removal of defects would enhance the T_c. High pressure experiments will help to determine if the defects influence T_c through the defect induced volume expansion, and if defects are the only important parameter affecting T_c. According

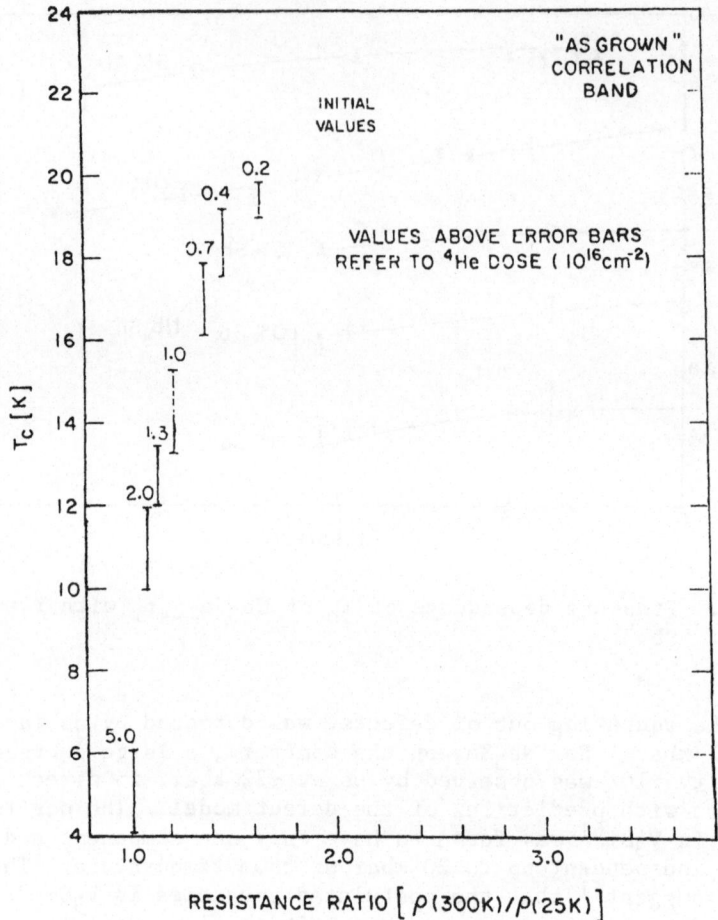

Figure 6. $\Delta T_c - \Delta a_o/a_o$ (Ref. 63).

to the T_c-R/R and $\Delta T_c - \Delta a_o/a_o$ correlations, any positive answers
to the above questions would imply a rapid T_c rise (drop) for the
low T_c compounds and a small T_c increase (decrease) for the high
T_c compounds under pressure, together with an increase (decrease)
in R/R. However, T_c's of the sputtered Nb-Ge films with T_c ranging
from 7 to 22K were found[66] to decrease always with increased pressure, as shown in Fig. 8. The rate of T_c-suppression increases
with T_c and no R/R-change greater than 0.2% was observed. The results suggest[66] that 1) defects affect T_c not through the volume
expansion generated by defects, 2) factor other than defect is important, and 3) pressure does not seem to influence the defect concentration strongly.

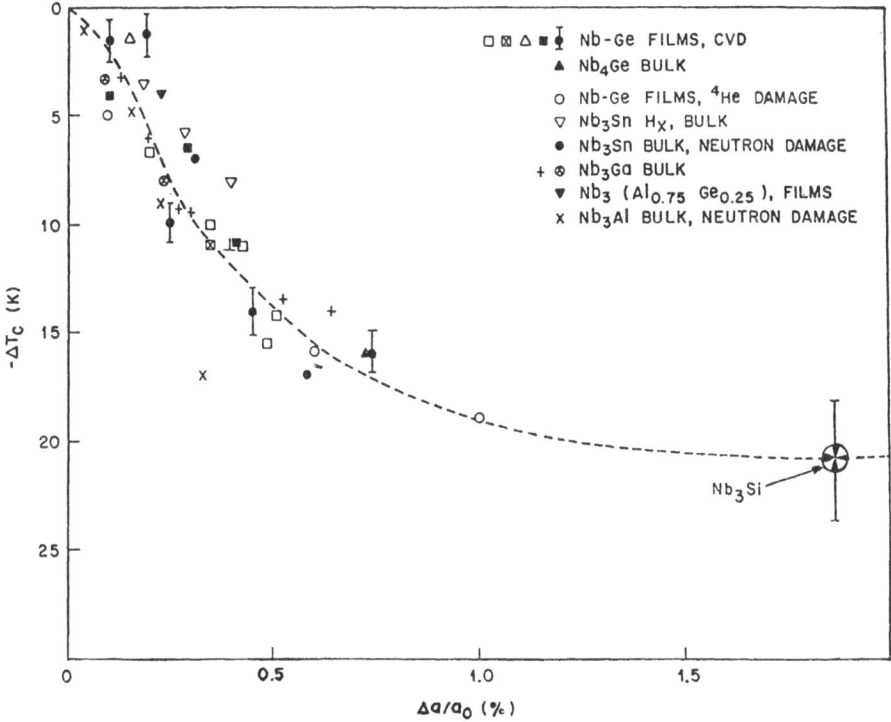

Figure 7. T_c - R/R correlation. (Ref. 3)

CONCLUSION

As shown in the previous discussion, results of high pressure experiments seem to favor the electron model, suggesting that high temperature superconductivity and instabilities are just two aspects of one phenomenon; namely, the instabilities in the electron energy spectrum. However, exceptions do exist. For instance, both the high T_c compounds of Nb$_3$Al[67] and Nb$_3$Ge[68] have low density of states, the mode softening in Nb$_3$Sn occurs also away from the Brillouin zone center[10], and etc. It is an experimental fact that soft phonon modes (including the long wavelength ones) exist in all high T_c material investigated[10,69] and that defects drastically suppress the T_c. Since it is the general belief that the

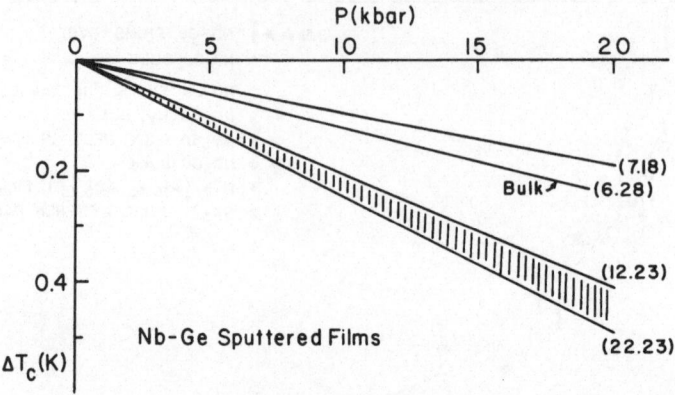

Figure 8.   Pressure dependence of T_c for Nb-Ge sputtered films.
The number inside the parenthesis stands for the T_c
at one atmospheric pressure.   (Ref. 66).

properties of matter must ultimately be predictable from the electron energy spectral of the matter, it becomes necessary that in
any future attempt to formulate any more realistic electron model,
the important roles of anharmonicity and defects should be taken
into consideration. A close examination of the three models described earlier indicates that the main difference between them
lies only in the different stages the assumptions are made and
the different specific roles of electrons, soft modes, or defects
the models emphasize. High pressure research on the problem of
possible identification of the structural transformation in unstable high T_c superconductors with the charge-density-waves
states[43,52,70,71,72] will provide additional insights about the
roles of electrons in these compounds.

The existence of high T_c in compounds not with A15 structure
and the absence of high T_c in some A15 compounds suggest that the
linear chain arrangement by the transition metal atoms in A15
structures is neither necessary nor sufficient for high temperature
superconductivity. Furthermore, defects and nonhydrostatic pressure both have large detrimental effects on T_c of unstable superconductors in contrast to the small effect due to hydrostatic compression. This leads us to the proposition that local symmetry
of the crystal lattice must play a crucial role in the unstable
high T_c systems. Systematic study of these compounds under uniaxial
stress will definitely shed light on the origin of high temperature
superconductivity and lattice instabilities, and their correlation.
It will also provide a direct basis for comparison of the hydrostatic pressure effect on T_c directly determined from experiments
with that thermodynamically deduced from acoustic and specific
heat data. At the present time, experiments of this kind are at
best sparse.

So far all high pressure studies on the problems of high temperature superconductivity and instabilities have been restricted to the pressure dependences of the superconducting and the structural transitions. It is known that the superconducting transition is affected more by phonons of short wavelengths whereas the structural transformation is a result of the extreme softening of the phonons of long wavelengths. In addition, both high pressure and alloy studies on unstable high T_c systems show that the structural transformation itself does not have too large an effect on T_c. Therefore a careful examination of the phonon dispersion relations of the unstable high T_c-systems into the short wavelength region by neutron diffraction under pressure will be very fruitful. With the recent successful development of the high pressure n-scattering technique, it is time for such an interesting study.

REFERENCES

1. G.E. Hardy and J.K. Hulm, Phys. Rev. 89, 884 (1958).
2. B.T. Matthias, T.H. Geballe, S. Geller and E. Corenzwit, Phys. Rev. 95, 1453 (1954).
3. L.R. Testardi, J.H. Wernick and W.A. Royer, Solid State Comm. 15, 1 (1974); L.R. Testardi, R.L. Meek, J.M. Poate, W.A. Royer, A.R. Storm and J.H. Wernick, Phys. Rev. B11, 4304 (1975).
4. For a review see L.R. Testardi, Physical Acoustics 10, 193 (1973).
5. For a review see M. Weger and I.B. Goldberg, Solid State Phys. 28, 1(1973).
6. B.W. Batterman and C.S. Barrett, Phys. Rev. Lett. 13, 390 (1964); Phys. Rev. 149, 296(1966).
7. R. Mailfert, B.W. Batterman and J.J. Hanak, Phys. Lett. A24, 315 (1967).
8. L.R. Testardi and T.B. Bateman, Phy. Rev. 154, 402(1967).
9. K.R. Keller and J.J. Hanak, Phys. Rev. 154, 628(1967).
10. G. Shirane, J.D. Axe and R.J. Birgeneau, Solid State Comm. 9, 397 (1971); G. Shirane and J.D. Axe, Phys. Rev. Lett. 27, 1803 (1971); J.D. Axe and G. Shirane, Phys. Rev. B8, 1965(1973).
11. B.T. Matthias, E. Corenzwit, A.S. Cooper and L.D. Longinotti, Proc. Nat. Acad. Sci. (USA) 68, 56(1971).
12. A.R. Sweedler, D.E. Cox, S. Moehlecke, L.R. Newkirk and F.A. Valencia, J. Low Temp. Phys. 24, 645(1976) and references therein.
13. See information included in Table I.
14. H. Fröhlich, Proc. R. Soc. A215, 291(1952).
15. W. Cochran, Phys. Rev. Lett. 3, 412(1959).
16. P.W. Anderson, Fiz. Dielectrikov, AN SSR, Moscow, 1960.
17. J. Bardeen, L.N. Cooper and J.R. Schrieffer, Phys. Rev. 108, 1175 (1957).
18. W.L. McMillan, Phys. Rev. 167, 331 (1968).

19. M. Weger, Rev. Mod. Phys. $\underline{36}$, 175 (1964); J. Labbé and J. Fridel
     J. Physique $\underline{27}$, 153, 303, 708(1966)·

20. J. Labbé, S. Barisić and J. Friedel, Phys. Rev. Lett. $\underline{19}$, 1039
     (1967).

21. J. Labbé, Phys. Rev. $\underline{158}$, 647(1967).

22. J. Labbé, Phys. Rev. $\underline{158}$, 655(1967).

23. R.W. Cohen, G.D. Cody and J.J. Halloran, Phys. Rev. Lett. $\underline{19}$,
     840 (1967). The analytical form of the electron density of
     states used is different from that in Ref. 19.

24. S. Barisić and J. Labbé, J. Phys. Chem. Solids $\underline{28}$, 2477(1967).

25. J. Labbé, Phys. Rev. $\underline{172}$, 451(1968).

26. S. Barisić, Phys. Lett. $\underline{26}$, 829(1971).

27. G.R. Barsch and D.A. Rogowski, Mat. Res. Bull. $\underline{8}$, 1459(1973).

28. C.S. Ting and A.K. Ganguly, Phys. Rev. B$\underline{9}$, 2781 (1974).

29. C.W. Chu, Phys. Rev. Lett. $\underline{33}$, 1283(1974).

30. L.F. Mattheiss, Phys. Rev. B$\underline{12}$, 2161(1975).

31. L.P. Gorkov and D.N. Dorokhov, J. Low Temp. Phys. $\underline{22}$, 1(1976)
     and C.S. Ting, private communication.

32. C.S. Ting, the following paper.

33. L.R. Testardi, Phys. Rev. B$\underline{5}$, 4342(1972).

34. P.B. Allen, J.C.K. Hui, W.E. Pickett, C.M. Varma and Z.Fisk,
     Solid State Comm. $\underline{18}$, 1157 (1976).

35. J.C. Phillips, Solid State Comm. $\underline{18}$, 831(1976) and references
     therein.

36. C.M. Varma, J.C. Phillips and S.T. Chui, Phys. Rev. Lett. $\underline{33}$,
     1233 (1974); S.T. Chui, Phys. Rev. B$\underline{11}$, 831(1976).

37. C.W. Chu and L.R. Testardi, Phys. Rev. Lett. $\underline{32}$, 766(1974).

38. C.W. Chu, S. Huang, T.F. Smith and E. Corenzwit, Phys. Rev.
     B11, 1866(1975).

39. T.F. Smith, R.N. Shelton and A.C. Lawson, J. Phys. F: Metal
     Phys. $\underline{3}$, 2157(1973).

40. R.N. Shelton, A.R. Moodenbaugh, P.D. Dernier and B.T. Matthias,
     Mat. Res. Bull. $\underline{10}$,111 (1975).

41. R.N. Shelton, Ph.D. Thesis, University of California, San Diego,
     1975.

42. C. Berthier, P. Molinie and D. Jerome, Solid State Comm. $\underline{18}$,
     1393 (1976).

43. C.W. Chu, V. Diatschenko, C.Y. Huang and F.J. DiSalvo, Phys.
     Rev. B$\underline{15}$, 1340(1977).

44. R. Delaplace, P. Molinie and D.Jerome, J. Physique Lett. $\underline{37}$,
     L.13(1976).

45. C.W. Chu, L.R. Testardi, F.J. DiSalvo and D.E. Moncton, Phys.
     Rev. B$\underline{14}$, 464 (1976).

46. R.H. Friend, D. Jerome, R. Frindt, A.J. Grant, and A.D. Yoffe,
     preprint.

47. L.R. Testardi, J.E. Kunzler, H.J. Levinstein and J.H. Wernick,
     Solid State Comm. $\underline{8}$, 907(1970).

48. See references in Ref. 33.

49. H. Neubauer, Z. Phys. $\underline{226}$, 211(1969); T.F. Smith, Phys. Rev.
     Lett. $\underline{25}$, 1483(1970).

50. P.R. Carcia, G.R. Barsch and L.R. Testardi, Phys. Rev. Lett. 27, 944(1971).
51. R.E. Larsen and A.L. Ruoff, J. Appl. Phys. 44, 1021(1973).
52. For the transition metal dichalcogenides, Friedel(J. Friedel, J. Physique Lett. 36, L. 279(1975)) proposed that the structural transformation at the onset of the charge density wave state will decrease the density of states at the Fermi level due to the creation of an energy gap and accordingly reduce T_c. With the application of pressure, the decrease of the density of states associated with the structural transformation is reduced, and the density of states is thus enhanced. Therefore opposite pressure effects are expected on T_L and T_c. It is also suggested that the argument may be valid for A15 compounds.
53. L.R. Testardi, Rev. Mod. Phys. 47, 637(1975).
54. S. Huang and C.W. Chu, Phys. Rev. B10, 4030 (1974).
55. R.N. Shelton and T.F. Smith, Mat. Res. Bull. 10, 1013(1975).
56. R. D. Blaugher, A. Taylor and M. Ashkin, Phys. Rev. Lett. 33, 292(1974).
57. The sample investigated here was the same one used in Ref. 37 except it was exposed to air at ~550K for about 10 hours. This might account for the slightly different values of T_c, T_L, $\partial T_c/\partial p$ and $\partial T_L/\partial p$ obtained in the present study from those in Ref. 35, for $P \leq 18$ kbar.
58. L.J. Vieland and A.W. Wicklund, Phys. Lett. 34A, 43 (1971).
59. L.J. Vieland, J. Phys. Chem. Solids, 31, 1449(1970).
60. L.R. Testardi and C.W. Chu, Phys. Rev. B15, 146(1977).
61. P. Sen, J.D. McGervey, C. Knox and C.W. Chu, preprint.
62. J.M. Poate, L.R. Testardi, A.R. Storm and W.M. Augustyniak, Phys. Rev. Lett. 35, 1290 (1975).
63. J. Nooland and L.R. Testardi, Preprint.
64. L.R. Testardi, J.M. Poate and H.J. Levinstein, Phys. Rev. Lett. 37, 637(1976).
65. L.R. Testardi, APS Bull. 21, 220(1976).
66. C.W. Chu, L.R. Testardi and P.H. Schmidt, to appear in Solid State Comm. (1977).
67. R.H. Willens, T.H. Geballe, A.C. Gossard, J.R. Maita, A. Menth, G.W. Hull, Jr. and R.R. Soden, Solid State Comm. 7, 837(1969).
68. J.M.E. Harper, T.H. Geballe, L.R. Newkirk and S.A. Valencia, J. Less. Com. Metals 43, 5(1975).
69. H.G. Smith, N. Wakabayashi and M. Mostoller, Superconductivity in d- and f-Band Metals, ed. by D.H. Douglass (Plenum, 1976), P. 223.
70. R.N. Bhatt and W.L. McMillan, Phys. Rev. B14, 1007(1976) and references therein.
71. P.H. Schmidt, E.G. Spencer, D.C. Joy and J.M. Rowell, Superconductivity in d-and f-Band Metals, ed. by D.H. Douglass (Plenum, 1976) p. 431.
72. C.W. Chu, C.Y. Huang, P.H. Schmidt and K. Sugawara, Superconductivity in d-and f-Band Metals, ed. by D.H. Douglass(Plenum, 1976) p.453.

QUESTIONS AND COMMENTS

T.F. Smith: You say it is now established that high T_c materials have anomalies associated with them. Just to throw in my own comment, how about Nb_3Al?

C.W. Chu: Yes, I didn't forget that case. That's why I said that when you don't find it, it's abnormal. But abnormality allows room for that.

T.F. Smith: I also understand that the Nb_3Ge deposited films that have 20 degrees superconducting transitions also don't have the structural transformation.

C.W. Chu: Well, in that case we have to be careful, because it's highly strained. Whether you can see the lattice transformation or not, it's not clear. But clearly the temperature dependence of resistance follows the same rule. It has a large negative curvature.

I.L. Spain: There's one point remaining about the phase diagram of V_3Si which I think is interesting, and I'm glad you raised this question. If you take a pressure below that triple point in pressure, then lower the temperature, the sequence of phase is normal cubic, normal tetragonal and superconducting tetragonal. If pressure is raised above the triple point as you drop the temperature down, the sequence is normal cubic straight to superconducting cubic. Now, if you get into the superconducting phase and raise pressure at constant temperature, you will go superconducting tetragonal to superconducting cubic. This means there is a quadruple point, not a triple point, which is forbidden thermodynamically. So, there's something more to the phase diagram than you indicated.

C.W. Chu: I see. There should be a **phase boundary** differentiating the cubic from the tetragonal superconducting phases.

I.L. Spain: Yes. In addition there should be two triple points.

C.W. Chu: Yes. But the problem is that we could not establish anything near the critical pressure because of the small size of the signal. Whether it is one quadruple point or two triple points is not yet known. I would like to add that I think it will be very interesting for any neutron study under pressure to find out the dispersion curve for this sample, and also I believe a uniaxial stress study on this system will be very interesting.

I.L. Spain: I think an x-ray structural study in the super-conducting phase would be very interesting.

J.S. Schilling: In your measurements of $T_c(P)$ on transforming V_3Si you find a kink at about 20 kbar. Do you ex-pect if the crystal does not transform that you'd also find a kink? Did you measure this?

C.W. Chu: If the kink is clearly related to the transforma-tion, we should see it. That is an interesting question. I have not yet measured the nontrans-forming samples to this high pressure. I remember Bob (Shelton) and Fred (Smith) measured four samples to 28 kbar. Did you (Smith) claim them to trans-form?

T.F. Smith: There was a problem. We've never known quite what we have in our samples, from the point of view of transformation. We were not able to detect the transformation in the same way that you can. Bob Shelton did some x-ray measurements, I recall.

C.W. Chu: At atmospheric pressure.

T.F. Smith: And we found the transformation.

C.W. Chu: We do have some samples which do not transform, and we're going to bring them up to that high pres-sure.

D. Jérome: I'm pleased to see the opposite effects of pres-sure on T_c and T_L. The correlation between the T_c and T_L phase transition is obvious. I want to make a comment on the layer structures, because the physics is just the same. I want to show you in the first slide the data on $2H-NbSe_2$ which are already old (C. Berthier, P. Molimé and D. Jérome, Solid State Comm. 18, 1393, (1976), see slide 1). The pressure suppresses T_L while enhances T_c sim-ilar to V_3Si. T_c continues to increase with pres-sure when T_L is completely suppressed to the super-conducting state, although at a smaller rate. I have another similar example, i.e. $4H_b-TaS_2$ as shown in slide 2 (see the article by R.H. Friend, D. Jérome, R.F. Frindt, A.J. Grant and A.D. Yoffe, J. Phys. C. 10, 1013 (1977)). A simple model

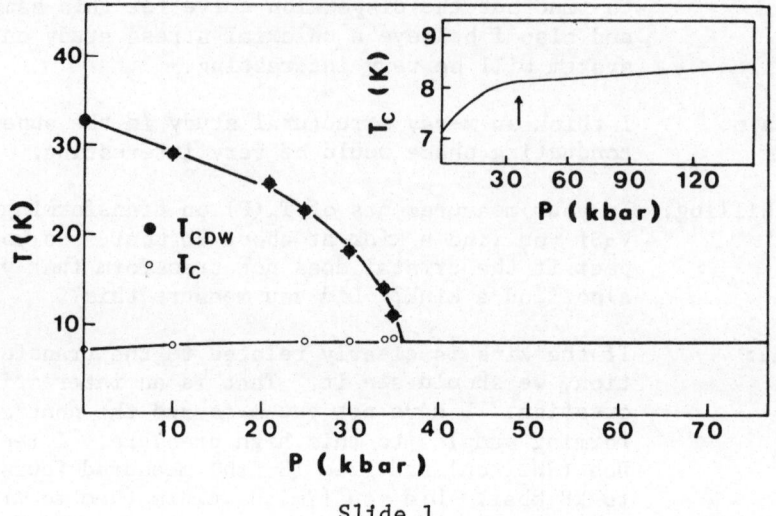

Slide 1

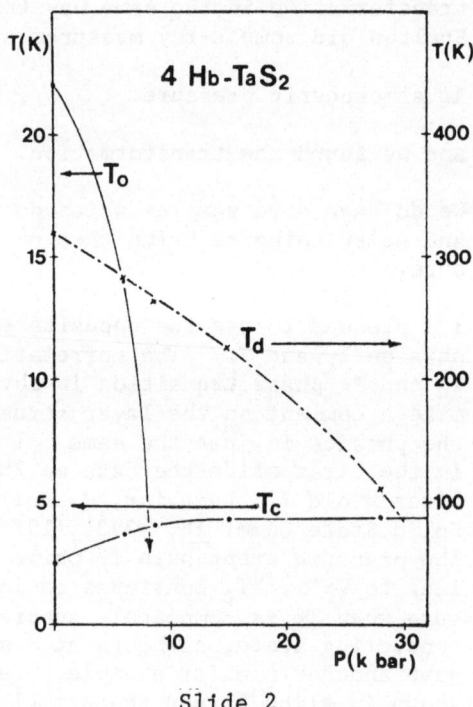

Slide 2

due to Friedel can explain all these. According to this model, the formation of the CDW's would create a gap and would result in a decrease of density of states. The pressure suppresses the CDW state and thus reduces the reduction in the density of states due to the CDW-gap. Therefore, T_c is enhanced by pressure.

C.W. Chu: Let me respond to that comment. Actually we observed that behavior in V-Ru back in 1972. I would also like to point out that there are exceptions. For instance, La_3S_4 and La_3Se_4 were observed by Sheldon and Matthias' group to have the same pressure effect on their T_c and T_L. This is also true for the $2H-TaSe_2$ samples, although our data disagree with Jérome's data. I believe the pressure effects on T_c and T_L are not always in the opposite directions.

D. Jérome: According to the data of Smith et al., it seems to be very difficult to know what the actual pressure dependence of the dT_c/dP is for $2H-TaSe_2$.

T.F. Smith: I wouldn't disagree with that at all. All we were saying was that we picked up an awfully small signal, which may or may not have been part of the transition. We were careful to state that we regarded the increase of T_c under pressure to be a tentative result.

C.W. Chu: We don't agree not only in the pressure effect on the CDW formation temperature, but also in the pressure effect on the incommensurate to commensurate transition temperature, as far as the sign is concerned.

E.F. Shelton: I just have a comment to add to the variety of pressure effects on T_c for the system of $CuMo_6S_8$. You have a positive and then a negative value of dT_c/dP. At the same time, the lattice transformation increases for the entire pressure region.

B.T. Matthias: I would like to point out one comment by Lawson who tabulated all phase transitions with respect to all the high and low superconducting transitions. The conclusion was, it makes no rhyme nor reason. Sometimes they go up, sometimes they go down, and frequently they don't go anywhere!

T.F. Smith: I hate to interrupt this lovely pattern every-

body's drawing up. We've been looking at this
2H-TaS/Se problem. By studying the Hall coefficient, which has an anomaly associated with the lattice distortion, we conclude that in the middle
of that system the lattice distortion is missing,
but the pressure dependence of T_c is still positive.

C.W. Chu: I feel that, in spite of the fact that there's no
correlation of any kind between T_c and T_L, the results are consistent with the electron model. This
is far from being a definitive answer. There is no
doubt that more studies are needed.

THE ISOTHERMAL PRESSURE DERIVATIVES OF THE ELASTIC SHEAR MODULUS FOR V_3Si*

C.S. Ting

University of Houston

Houston, Texas 77004

ABSTRACT

The effects of a hydrostatic pressure on the shear modulus c_s and the structural transition temperature T_m of V_3Si are studied on the basis of the coupled chain model of Gorkov (or the model proposed recently by Lee, Birman and Williamson). Taking into account the charge transfer effect between different bands and choosing suitable values for the Fermi energies, our results can account for the temperature dependences of $(\partial c_s/\partial p)_T$ quantitatively for transforming V_3Si and satisfactorily for nontransforming V_3Si.

There have been a number of intensive experimental and theoretical investigations [1,2] on the understanding of low-temperature behaviors of certain intermetallic compounds with β-tungsten structure in the past decade. These compounds (V_3Si and Nb_3Sn) have one feature in their structural transformation. This feature is that the transformation does not take place in all samples of the material[1]. The shear modulus c_s differs slightly, but the pressure derivative $(\partial c_s/\partial p)_T$ shows remarkably different behavior for transforming and nontransforming V_3Si[3,4]. For a transforming sample[3] $(\partial c_s/\partial p)_T \simeq 1.0$

---

*Research supported partly by Research Initiation Grant, Office of Research, at the University of Houston.

is positive and small, while for a nontransforming sample[4] $(\partial c_s/\partial p)_T \simeq$ -4.5 is negative and large near the $T \simeq 21^\circ K$. The temperature dependence of $(\partial c_s/\partial p)_T$ for transforming V_3Si was studied theoretically by Ting and Ganguly[5] and also by Barsch and Rogowski[6] using the Weger-Labbé-Friedel (WLF) linear chain model. These authors also took into account the effect of charge transfer from s band to d band (or the charge transfer parameter[5] $\nu > 0$) when pressure is applied. Their results [5],[6] agree reasonably well with the measurements of Garcia and Barsch[3]. The temperature dependence of $(\partial c_s/\partial p)_T$ for nontransforming V_3Si[4] however, has not been successfully explained, although some reasons have been suggested[6],[7]. In this paper I wish to reexamine the fundamental difference between the transforming and the nontransforming V_3Si by calculating $(\partial c_s/\partial p)_T$. For V_3Si the results of the band structure calculation of Matheiss[8] indicate that the Fermi level crosses Γ, R and M points in the reciprocal lattice space of V_3Si. There is a possibility that high density of states peak of electrons may locate at one of these points. In order to explain those anomalous electronic and elastic properties of V_3Si and Nb_3Sn, one needs a fine structure curve to describe the high density of states peak near the Fermi level. Because the energy resolution in the current band structure calculations is not high enough (the uncertainties in energy scale >10meV) to give such information, one has to use the results from model calculations. Different models[9-12] predict different shapes for the density of states peak. My approach will be based upon the Gorkov's coupled chain model[9] (or the model recently proposed by Lee[10] et al.). They have studied the electronic spectra of A-15 crystal structure by using group theory at several symmetrical points in the reciprocal lattice space. They found that a high density of states peak can be located at the edge from the R to M point[9], or at the R point[10]. The detailed shape of the density of states peak depends on parameters involved in the carrier spectrum, and sometimes need extensive numerical calculations[10]. For simplicity, the density of states with logarithmic singularity of Gorkov[9] will be used in this paper. As will be shown below, the behavior of $(\partial c_s/\partial p)_T$ depends critically on the position of the Fermi level. By choosing suitable values for the Fermi energy, the temperature dependence of $(\partial c_s/\partial p)_T$ for both transforming and nontransforming V_3Si can be reasonably accounted for. At the end of this paper the difficulty associated with the WLF model[11] will be pointed out and a correlation between the present approach and the defect model[7],[13] will be discussed.

Starting from the free energy F which is a combination of lattice contribution F_0 and electrons near the R-point in the reciprocal lattice space, we have[9],[14]

$$F(\varepsilon) = -NT \sum_{i=1}^{3} \int_{-|E_\sigma|}^{\infty} d\varepsilon \, n(\varepsilon) \, \ell n[1 + \exp(-\frac{\varepsilon+h_i-\mu}{T})] + F_0, \qquad (1)$$

where μ is the Fermi energy measured from $\varepsilon=0$. $h_1 = d(\varepsilon_{xx}-\varepsilon_{yy})$,
h_2 and h_3 are obtained by permuting x, y and z. ε_{xx}, ε_{yy} and ε_{zz}
are diagonal strain tensor. d is the deformation potential.
$3N = 5.73 \times 10^{22}/m^3$ is the number of vandium atoms per unit volume,
the density of states per atom is represented by $3n(\varepsilon)$, and $n(\varepsilon)$ is
taken to be[9]

$$n(\varepsilon) = \nu(o) \ln \left|\frac{\tilde{T}}{\varepsilon}\right| \cdot \quad \cdot \tag{2}$$

Here $\nu(o)$ is a constant and $\tilde{T}$ measures the transfer integral between
the nearest vanadium atoms on two neighboring orthogonal chains[9].
If we expand the free energy F in power series of the strain
ε_{xx}, ε_{yy} and ε_{zz}, we have   (3)

$$F(\varepsilon) = F(o) + \tfrac{1}{2} c_{11}(\varepsilon_{xx}^2+\varepsilon_{yy}^2+\varepsilon_{zz}^2) + c_{12}(\varepsilon_{xx}\varepsilon_{yy}+\varepsilon_{yy}\varepsilon_{zz}+\varepsilon_{zz}\varepsilon_{xx})+\cdots.$$

the elastic shear modulus is defined as $c_s = c_{11} - c_{12}$ and c_s is
obtained from eqs. 1 and 3

$$c_s = 3N \, d^2 \int_{-|E_o|}^{\infty} d\varepsilon \, n(\varepsilon) \cdot \frac{\partial f(\varepsilon)}{\partial \varepsilon} + c_s^o \ . \tag{4}$$

In eq. (4), the first term is due to the contribution of electrons
and the second term c_s^o comes from the lattice part which is assumed
to be temperature independent. $f(\varepsilon) = [1 + \exp(\varepsilon-\mu)]^{-1}$ is the usual
Fermi distribution function, where the Fermi energy μ depends on
temperature T through the equation of charge conservation:

$$3 \int_{-|E_o|}^{\infty} d\varepsilon \, n(\varepsilon) \, f(\varepsilon) = Q. \tag{5}$$

Here Q is the number of d electrons per vanadium atom which contribute
to the carrier pocket at the R point. Under a hydrostatic pressure
P, the lattice constant $a = a(1-\varepsilon)$, $N=N(1 + 3\varepsilon)$ and $\varepsilon = \frac{1}{3} kP$, where

k is the compressibility: for V_3Si $k^{-1} = 1.77 \times 10^{12}$ dyne/cm^2.
Moreover, the values of the charge transfer integral $\tilde{T}$, the constant
$\nu(o)$ and the cut-off energy $|E_o|$ will also be changed by the hydrostatic pressure. According to our previous experience[5], these volume effects on $(\partial c_s/\partial p)_T$ are very small and almost temperature independent. The dominant contribution to $(\partial c_s/\partial p)_T$ comes essentially
from the shift of Fermi level under pressure[5], or the charge Q in
Eq. 5 changes from Q to $Q + \delta Q$, with $\delta Q = -3\nu\varepsilon$. Here ν is a tempera-
ture-dependent quantity and measures the strength of the charge trans-
fer between the electron pocket at point R (d band) and some other
band (s band). Using the same method as Barisic and Labbé[17], one can
show that ν is of the following form

$$\nu = \alpha^{ds}/3(\beta^s+\beta^d)$$

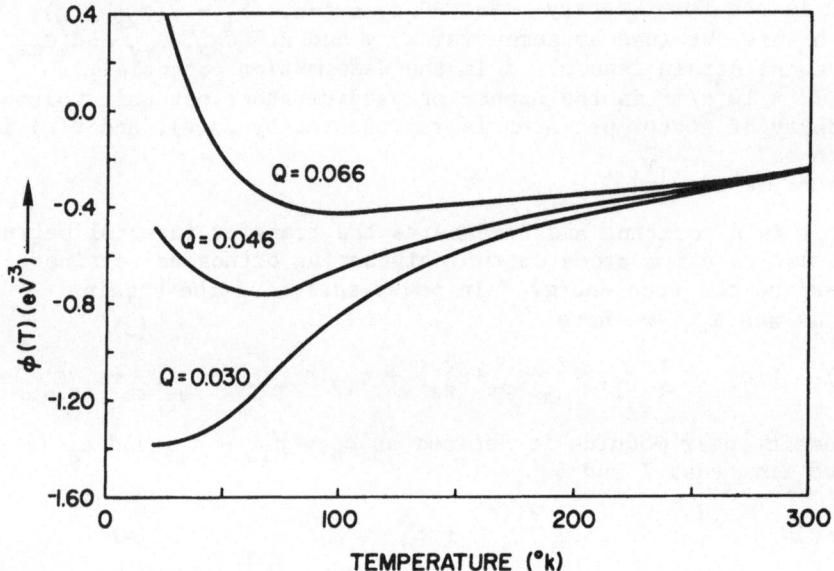

Figure 1. The function $\phi(T)$ which is defined in Eq. 7 is plotted
versus temperature for three different values of Q, the
number of d electrons per vanadium atom which contribute
to the carrier pocket at the R-point.

with

$$\frac{1}{\beta^s} = \int d\varepsilon \ n^s(\varepsilon) \ \frac{\partial f(\varepsilon)}{\partial \varepsilon} \ ,$$

$$\frac{1}{\beta^d} = 3 \int_{-|E_o|}^{\infty} d\varepsilon \ n(\varepsilon) \ \frac{\partial f(\varepsilon)}{\partial \varepsilon} \ ,$$

where $n^s(\varepsilon)$ and $3n(\varepsilon)$ are the density of states of the "s" band and
the "d" band. By neglecting the volume effects and taking into ac-
count only the charge transfer mechanism, we have

$$\left(\frac{\partial c_s}{\partial p} \right)_T \approx -3N \ kd^2\nu \ \beta^d \int_{-|E_o|}^{\infty} d\varepsilon \ n(\varepsilon) \ \frac{\partial^2 f(\varepsilon)}{\partial \varepsilon^2} + \frac{\partial c_s^o}{\partial p} \quad . \qquad (6)$$

In order to determine the parameters d, $|E_o|$, μ(or Q), $\tilde{T}$, ν(o), and
c_s^o, we fit c_s in eq. (4) with the measurements of Testardi and Bate-
man[15] for transforming V_3Si. We obtain d ≈ 2.06 eV $|E_o| = 80^oK$, $\mu=2^oK$

(Q = 0.054) $\tilde{T}$ = .862 eV. $3\nu(o)$ = 1.3/eV-atom and c_s^o = 10.75 x 10^{11} dynes/cm^2. These parameters, except $|E_o|$, are very close to those values suggested by Gorkov[9]. $-1/\beta^s$ is equal to the density of states of "s" electrons at the Fermi level. It is much smaller than the density of states for the "d" electron ($-1/\beta^d \simeq$ 9 eV^{-1} per vanadium atom at T = 0°K). Therefore a reasonable choice[18] for $-1/\beta^s \simeq$ 1 eV^{-1} per vanadium atom. We have calculated the quantity $\phi(T)$ which is defined by

$$\phi(T) = \frac{3}{d^2\alpha^{ds}} \left[\left(\frac{\partial c_s}{\partial p} \right)_T - \left(\frac{\partial c_s^o}{\partial p} \right)_T \right] \qquad (7)$$

as a function of temperature T for three different values of Q (or the Fermi energy μ) using the same set of parameters as given above. The results are given in Fig. 1 and show that the behavior of $\phi(T)$ depends critically on the value of Q. By choosing $(\partial c_s^o/\partial p)_T$ = 1.522 and the charge transfer parameter α^{ds} = 1.93 eV., the experimental behavior[3] of $(\partial c_s/\partial p)_T$ for transforming V$_3$Si can be accounted for very well. The result is shown by the solid line in Fig. 2. From the value of $(\partial c_s/\partial p)_T$ at the structural trans-formation temperature $T_m \simeq$ 21°K, we can determine the pressure de-pendence of T_m. Since T_m is determined from c_s = o, the Born's stability limit for the structural transition of a cubic crystal[16], and T_m (p) is defined as the temperature for $c_s + (\partial c_s/\partial p)_T\, p$ = o, we have

$$T_m(p) = T_m(o) + \alpha p \qquad (8)$$

$$\alpha = - [\partial c_s/\partial p/(\partial c_s/\partial T)_V]_T = T_m(o)$$

From the experimental data in Ref. 17, we have $(\partial c_s/\partial T)_V$ = 10^4 bar/°k, and from Fig. 2 we have $(\partial c_s/\partial p)_T \simeq$ 1.15. We therefore predict the optimal value of α for Garcia and Barsch's sample to be α = -1.15 x 10^{-4}°K/bar. This value for α is in qualitative agreement with Chu and Testardi's measurement on T_m under pressure[19]. They obtained α = (-1.5 ± 0.1) x 10^{-4}°K/bar. This difference between these two values might be due to different samples having been used for per-forming the experiments.

For a nontransforming sample, the role played by defects is very important. The effects of defects are expected to a) change the number of electrons and thus shifting the Fermi level μ, b) smear out the singularity in the density of states, c) alter the lattice properties or the values of c_s^o and $(\partial c_s^o/\partial p)_T$, and d) introduce randomness into the local free energy distribution and thus help to form microdomains in which $T_m(> T_c)$ might be different from domain to domain. An interpretarion by Varma et al.,[13] on the experiment of Blaugher, et al.,[19] indicates that the defects in V$_3$Si are primari-

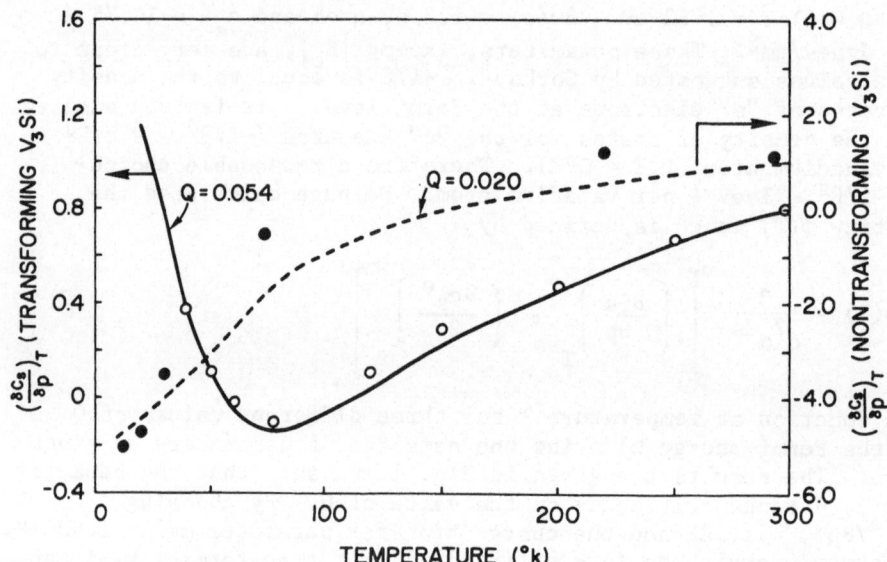

Figure 2. $(\partial c_s/\partial p)_T$ as a function of temperature for two different
values of Q. The solid line is for Q = 0.054, o denotes
the experimental data of Ref. 3 for transfroming V_3Si,
and they are plotted according to the scale of the left
ordinate. The dashed line is for Q = 0.020, • denotes
the experimental data of Ref. 4 for nontransforming
V_3Si, and they are plotted according to the scale of
the right ordinate.

ly vacancies. This means that the value of Q in nontransforming
V_3Si should be smaller than those of transforming ones. Because
we are not able to consider the effects of b), c) and d) in a
quantitative way, their contribution will be neglected. We only
examine the effect due to Fermi level shifting. By choosing Q =
0.020 (μ = -44°K) the experimental behavior[4] of $(\partial c_s/\partial p)_T$ for nontransforming V_3Si can be qualitatively accounted for. The result
is shown by the dashed line in Fig. 2. With this particular choice
of Q(or μ), we have checked the elastic shear modulus c_s obtained
from eq. (4) with the measurements[4] for nontransforming V_3Si. The
agreement is very close. We also wish to point out that the Fermi
level shift has little effect on the superconducting transition
temperature T_c as has been investigated by Gorkov and Dorokhov[20].
The behavior of $(\partial c_s/\partial p)_T$ and the value of α are strongly sample
dependent.

Finally we wish to discuss briefly why the transforming and the
nontransforming crystals behave so differently at low temperatures.
One of the primary causes is that the function $\phi(o)$ defined in
eq. (7) is proportional to the slope of the density of states at

the Fermi level

$$\phi(0) \propto - \left[\frac{d}{d\epsilon} n(\epsilon)\right]_{\epsilon=\mu}$$

using the density of states given by Eq. (2) $\mu = 2^{\circ}K, \phi(o) > o$, for transforming sample and $\mu = -44^{\circ}K$, $\phi(o) < o$ for nontransforming sample. In WLF model, however, the density of states there is of the form $n(\epsilon) \propto (\epsilon)^{-1/2}$ and $\epsilon > o$. No matter what the value of Q (or μ) is chosen, one always obtains $\phi(o) > o$, not $\phi(o) < o$. This is the reason why WLF model explains the behavior of $(\partial c_s/\partial p)_T$ for transforming but not for nontransforming V_3Si. In the above interpretation of experimental data, we have introduced a charge trans- "s" band to "d" band under pressure. There is also a possibility that the effect of charge transfer may come from defects. As the number of defects (vacancies of vanadium atoms) decreases under pressure[7], more electrons populate the vanadium chains. This interpretation is partly consistent with the defect model proposed by Varma et al.[13]

ACKNOWLEDGEMENTS

I would like to thank Professors J.L. Birman, S.J. Williamson and Dr. T.K. Lee for useful conversations. The graphs of the present work were drawn at the Naval Research Laboratory when the author visited there during the summer of 1977. The help from NRL is greatly appreciated.

REFERENCES

1. L.R. Testardi, Physical Acoustics, edited by W.P. Mason and R.N. Thurston (Academic, New York 1973).
2. M. Weger and I.B. Goldberg, Solid State Physics, edited by H. Ehrenreich, F. Seitz and D. Turnbull (Academic, New York, 1973) Vol. 28.
3. P.F. Garcia and G.R. Barsch, Phys. Status Solids B59, 595 (1973).
4. R.E. Larsen and A.L. Knoff, J. Appl. Phys. 44, 1021 (1973).
5. C.S. Ting and A.K. Ganguly, Phys. Rev. B9, 2781 (1974).
6. G.R. Barsch and D.A. Rogowski, Mater. Res. Bull. 8, 1459 (1973). In this work, the charge transfer parameter $\nu < o$ is assumed for nontransforming V_3Si.
7. J. Noolandi and C.M. Varma, Phys. Rev. B11, 4743 (1975).
8. L.F. Matteiss, Phys. Rev. B12, 2161 (1975
9. L.P. Gorkov, Zh Eksp. Teor. Pisma 30, 571 (1974); L.P. Gorkov and D.N. Dorokhov, J. Low Temp. Phys. 22, 1 (1976).
10. T.K. Lee, J.L. Birman and S.J. Williamson (to be published).
11. J. Labbe and J. Friedal, J.Phys. (Paris) 27, 153, 303, 708 (1966).
12. R.W. Cohen, G.D. Cody and J.J. Halloran, Phys. Rev. Lett. 19, 840 (1967).

13. C.M. Varma, J.C. Phillips and S.T. Chui, Phys. Rev. Lett. 33, 292 (1967).

14. Here we assume that the high density of states peak is located at or near the R-point in the reciprocal lattice, the cutoff energy E_0 in the lower bound of the integration has to be introduced.

15. L.T. Testardi and T.B. Bateman, Phys. Rev. 154, 402 (1967).

16. For V_3Si the structural transition is a weak first-order one, and the condition c_s = o at T_m in principle can no longer be applied to determine T_m. However, in the present model more than 95% of T_m comes from c_s = o for V_3Si. Therefore the use of Born's stability limit to determine T_m for V_3Si should be valid.

17. L.T. Testarid, J.E. Kunzler, H.J. Levinstein, J.P. Maita, and J.H. Weinick, Phys. Rev. B3, 107 (1971).

18. C.W. Chu and L.R. Testardi, Phys. Rev. Lett. 32, 766 (1974).

19. R.D. Blaugher, A. Taylor and M. Ashkin, Phys. Rev. Lett. 33, 292 (1967).

20. L.P. Gorkov and O.N. Dorokhov, Solid State Commun. 19, 1107 (1976).

QUESTIONS AND COMMENTS

T.F. Smith: If you take your electronic model (you may have gathered from my comments earlier on this morning, I don't altogether agree with the idea of electronic models), you should be able to take the appropriate derivatives and derive the thermal expansion.

C.S. Ting: It's possible, but I have not done that yet.

G.R. Barsch: I wonder if you have the results of the Gorkov model. Because I have a feeling that band structure calculations could give fine structures required in your calculations, because both Klein and Matheiss' band structure calculations agree quite well. However, they both show that the tabulated high density of states are near the Γ point.

C.S. Ting: Both Klein and Matheiss' band structure calculations indicate that high density of states appears at the Γ point and possibly at the M point in the reciprocal lattice of V_3Si. According to Gorkov's group theoretical approach, that high density of states comes from the edge of R to M, partly because the symmetry at R points involves six-fold

degeneracy. The argument for this discrepancy
may be that the band structure calculation is not
accurate enough to predict what Gorkov predicts.
For example the effect due to the Fermi liquid
theory is not included in band structure calculations. But nevertheless, I think my calculation
really doesn't depend on a particular model. All
we need is just a density of states which has a
sharp peak and has finite slopes on both sides
of the peak, independent of where it is located,
although I used the results of Gorkov's model.

M. Weger: COMMENT ABOUT THE A-15'S
When we talk about the linear chain model for material like V_3Si, Nb_3Sn, etc. we have to distinguish between the version originally proposed in
1963 and further developed in Jerusalem, and the
Labbé model. The Labbé model ignores the overlap
between wavefunctions belonging to different
chains. These overlaps, however, are large (for
all bands except the σ-band), and for the δ_2-band
(for example) exceed the overlap between neighbours
on the same chain (when the number of neighbours
is taken into account). Therefore electrons will
not generally move along the chains (except for the
σ-band). Under certain exceptional cases, however,
movement along the chains is possible. This is the
case when there is destructive interference, caused
by multiple scattering. A specific example is the
bonding states of the δ_2-band. This is illustrated
in Fig. 38 of the review paper, Weger & Goldberg,
Solid States Physics, 28, 1973, p. 98. The overlaps between a δ_2-function on a chain in the x-direction, and its two neighbours on a chain in the
y-direction, are very large, but of equal magnitude
and opposite sign for the two neighbours, and
therefore the scattering matrix element (or amplitude) cancels, and there is no scattering from the
chain in the y-direction to the one in the x-direction for $k_y = 0$. Thus, this property holds only
in part of k-space (in this case, the $k_y=0$ plane,
and its vicinity); still, the regions in k-space
where destructive interference plays an important
role, may be large and create large peaks in the
density of states. Such peaks are also seen in
APW calculations and the question is only whether
they are close to the Fermi level or not. According to the Jerusalem work (G. Barak and M. Weger,
J. Phys. C 7, 1117, 1974), the large crystal field
integrals raise the δ_2-band and cause this peak

in the density of states to be close to the Fermi
level. The crystal-field integrals are usually ig--
nored in APW calculations that make use of the muffin-tin potential (and thus assume spherical symmetry of a potential inside the muffin-tin spheres).
This approximation is justified when the local symmetry is cubic, but not in the A-15's, where the
local symmetry is very low ($\overline{4}2m$).

Thus, we see that the key property of the A-15's
structure is the destructive interference, rather
than chain structure.

The feature necessary for the destructive interference is a large unit cell, with a very high degree
of symmetry. In the A-15 structure, this is indeed
the case, since all 6 transition-metal atoms in the
unit cell are equivalent to each other by symmetry.

This destructive interference impedes the motion
of the electrons, and thus we have slowly-moving
wavepackets, which extend over many atoms. These
wavepackets are clearly very favorable for superconductivity.

If the electrons were localised by absence of large
matrix elements, as is the case for 4f electrons
in the rare earths, we would still get a high density of states, but magnetism instead of superconductivity. The extended nature of the slowly-
moving wavepackets of the A-15's ensures strong
electron-phonon coupling (since the matrix elements are large, so are their changes, caused by
the modulation of inter-atomic distances by the
phonons), while it is unfavorable for magnetism
(the large spacial extension makes U small).

As a counter-example to the A-15's, we may consider the σ-phase (Nb_2Ga, say, studied extensively
by Raub). Here the unit cell is large too, but
the transition metal atoms are not equivalent by
symmetry, but form 5 inequivalent groups. Thus
the material has to some extent an "amorphous"
nature, even in the crystalline state. Here the
conditions are unfavorable for destructive interference, and T_c is indeed low.

A "critical" experiment bearing on these ideas,
is the application of uniaxial stress. Such a
stress in the [100] direction, suppresses T_c con-

siderably (M. Weger, B.G. Silbernagel, and E.S. Greiner, Phys. Rev. Lett. 13, 521, 1964). Such stress does not destroy the chain structure and by the Labbé model it should actually make T_c go up (since T_c is not quite at the maximum of his T_c vs. band occupation curve, and this feature is essential to explain the increase of T_c with hydrostatic pressure, found by Smith, using the Labbé model). This experiment shows clearly that the chain structure is not sufficient for high T_c, but the cubic symmetry (ensuring the equivalence of the 6 atoms per unit cell) is essential. Thus, while some features of the Labbe model are correct (namely, that peaks in the density of states, associated with degenerate, itinerant electron states, cause elastic anomalies and a displacive transition; that the δ-band may contribute to the peaks in the density of states at the Fermi level), the neglect of interchain coupling is not justified, and any theory must take into account the phenomenon of destructive interference.

R.A. Hein:

I understand your model probably applies only to B atoms, being non-transition metal atoms, but what would transition metal B atoms (assuming B is a transition metal) do to this destructive interference right at the d-wave functions of the B atoms?

M. Weger:

For instance, you take V_3Pt instead of V_3Ge. It would change things, and you cannot say without a calculation, what this does. When you have a large unit cell you have a lot of Van Hove singularities. Now our argument here is that these Van Hove singularities are big because of the high symmetry. Pt shifts the Van Hove singularities but exactly where I don't know because no calculation has been done up to now. The picture I just described would be destroyed. That's obvious, but we may have other peaks being brought up.

LaAg UNDER HYDROSTATIC PRESSURE: SUPERCONDUCTIVITY AND PHASE TRANSFORMATION

J.S. Schilling and S. Methfessel*

Institüt für Experimentalphysik IV, Universität Bochum,

463 Bochum, Germany

and

R.N. Shelton

Institute for Pure and Applied Science, University of

California,

San Diego, La Jolla, Calif. 92093, USA

Measurements of the electrical resistivity of the superconducting (T_c = 1.062K) CsCl-compound LaAg from 1-300K reveal that the application of hydrostatic pressures greater than about 4 kbar induces a cubic-to-tetragonal lattice transformation at a temperature T_M. The variation of T_M with pressure, dT_M/dP = + 20K/kbar, is exceptionally rapid. The superconducting transition temperature T_c has a reversible oscillatory pressure dependence, increasing initially. There is no obvious correlation between the pressure dependence of T_c and T_M.

*Research supported in part by the Deutsche Forschungsgemeinschaft

Research sponsored by the Air Force Office of Scientific Research, Air Force Systems Command, USAF, under AFOSR Contract Number AFOSR/F – 49620-77-C-0009.

INTRODUCTION

It has been apparent for some time that systems with unusually high values of the superconducting transition temperature T_c usually display anomalous normal state properties. One such property is the presence of elastic instabilities which, in some systems, result in the onset of a lattice distortion at a temperature T_M;[1-5] another is a large negative curvature in the temperature dependence of the resistivity.[6] Variation of T_c and these anomalous properties on a single sample by the application of high pressure is a particularly clean technique for ascertaining their interrelationships, avoiding the chemical complications of alloying. Based on studies of the A-15 compounds and others, it was for a time generally believed that the pressure coefficients of T_c and T_M were opposite in sign. That this result, however, is not generally applicable was shown by Shelton et al.[2] in measurements on the Th_3P_4 - superconductors La_3S_4, La_3Se_4, and La_3Te_4, where T_c and T_M are found to have a parallel pressure dependence, and in the ternary superconductor $CuMo_3S_4$, where dT_M/dP remains positive while dT_c/dP changes sign between 0 and 15 kbar.[5] The negative resistance curvature is shown by Moodenbaugh and Fisk[7] to increase significantly with pressure in pure Ba or Y, two elements which only become superconducting under pressure and show $dT_c/dP>0$.

LaAg is a superconductor with T_c = 1.077K as determined by specific heat measurements,[8] in good agreement with the present value T_c = 1.062K, but about 0.1K higher than reported previously by Chao et al.[9] The resistivity for temperatures below 300K exhibits a slight negative curvature with no evidence of a phase transformation above T_c,[9,10] in agreement with X-ray measurements which show that the CsCl-structure is stable down to at least 10K.[10] Under pressure Chao et al.[9] find that T_c increased by about 0.2K at 10 kbar; however, no phase transition is found in a $\Delta V/V_o$ measurement to 35 kbar. A first order cubic-to-tetragonal lattice transformation is reported[10] in the alloy series $LaAg_{1-x}In_x$ for $x \geq 0.05$. T_M increases initially rapidly with x. The concentration dependence of the phase transformation has been calculated by Ihrig et al.[11] using an extension of the band Jahn-Teller theory originally applied by Labbé and Friedel[12] to the A-15 compounds. Other CsCl-compounds exhibiting similar structural transformations are LaCd,[10,13] LaCu,[14] YCu,[10,13] AuZn,[15] and $V_{.54}Ru_{.46}$,[4] whereas LaZn,[14] YAg,[11] YCd,[14] and YZn[10] remain cubic. It is interesting to note that all the transforming compounds are superconducting, whereas of the nontransforming compounds only LaZn is superconducting. In AuZn[15] and $V_{.54}Ru_{.46}$,[4] T_M is found to decrease with increasing pressure, whereas T_c increases initially, bending over at pressures above about 10 kbar.

In the present work the influence of hydrostatic pressure on superconductivity and phase instabilities in LaAg is studied. Ap-

plication of pressure is found to induce a lattice transformation
in LaAg, where T_M _increases_ very rapidly with pressure, in contrast to the results for AuZn[15] and V.54 Ru.46,[4] even though T_c (P)
for LaAg is qualitatively similar.

EXPERIMENT

The polycrystalline LaAg alloy was prepared from high purity
La(99.99%) and Ag (99.999%) by melting in an argon arc furnace
at 250 Torr. The sample was kept molten for several minutes, then
turned over and remelted four times. After an homogenizing anneal
in high vacuum at 800 C for three days, X-ray analysis at room
temperature revealed only reflexes of the cubic CsCl-type structure
with lattice constant 3.814 Å, in good agreement with previous
investigations.[10]

The sample for the high pressure measurements was spark cut
under mineral oil from the master ingot. The resulting piece
was a rectangular parallelepiped with a maximum dimension of approximately 4 mm. Four electrical leads of platinum wire (0.075
mm diam.) were attached to the sample by spot welding under a
coating of vaseline to prevent oxidation. No deterioration of the
sample was observed after immersion in the pressure field. Hydrostatic pressures were applied at room temperature using a one-to-
one mixture of n-pentane and isoamyl alcohol as the pressure transmitting medium.[16] The desired pressure was then sustained inside
a capsule by a clamp technique similar to that employed by Chester
and Jones.[17] A slight modification of the pressure cell permits
the exit of the four leads necessary for the resistivity measurements.

The applied pressure was determined at low temperature with
a superconducting lead manometer.[18] A pressure loss of 1-2 kbar
occurs on cooling from room temperature; therefore, the pressure
at the temperature of the lattice distortion (T_M = 70-190K) may
be slightly higher than that measured at 7K. Due to the fact that
the majority of the pressure drop occurs when the pressure medium
freezes,[19] this difference is not expected to be significant.

The high pressure technique for the quasihydrostatic measurements has been described in detail in an earlier publication.[20]

RESULTS AND DISCUSSION

Phase Transition

The results of the electrical resistivity measurements on
LaAg under hydrostatic pressure are shown in Fig. 1. The data
were taken in the order 0, 2.25, 6.31, 12.37, 9.62, and 4.20 kbar.[21]

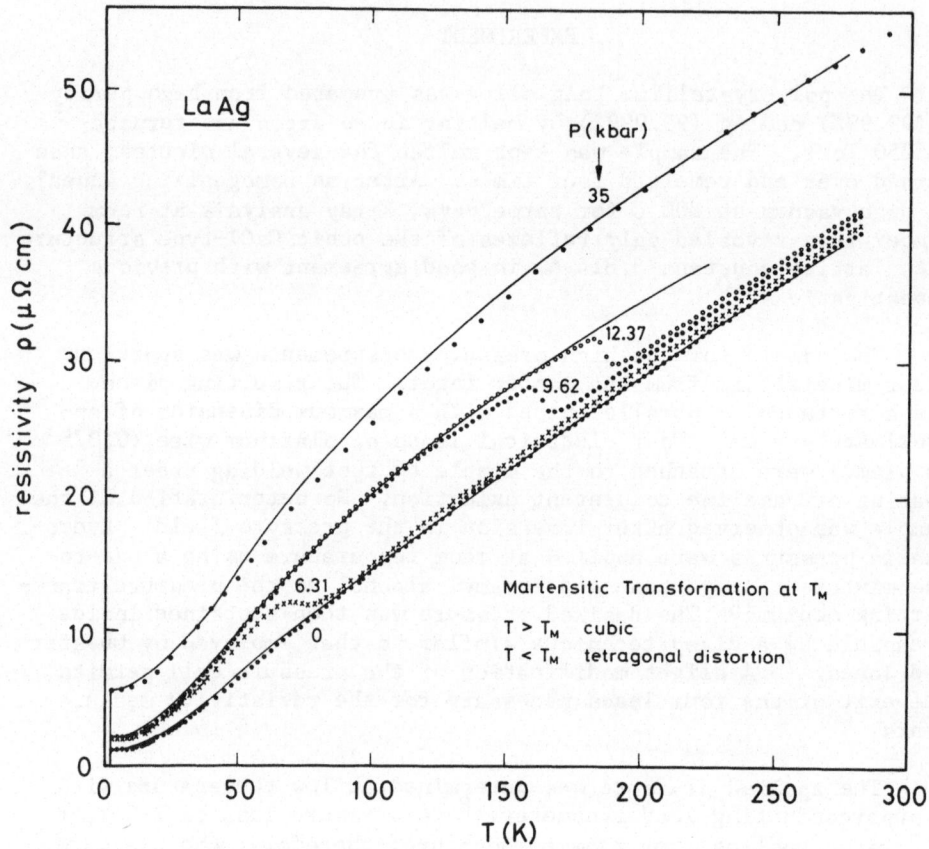

Figure 1.   Electrical resistivity of LaAg versus temperature at
            different pressures. Data taken upon cooling. Points
            and circles are data, solid lines are fits to data using
            theory of Cohen et al.[27] The jump in resistivity marks
            the lattice transformation at 73K for 6.31 kbar, 163K
            for 9.62 kbar, and 186K for 12.37 kbar. For the 35 kbar
            run, which was measured in a quasihydrostatic pressure
            cell, T_M must lie well above 300K. Due to sample de-
            formation in the quasihydrostatic cell, the 35 kbar
            data contain about $3\mu\Omega cm$ more defect-scattering resis-
            tivity than the hydrostatic data.

The data at 2.25 and 4.20 kbar differ only slightly from those at
P=0 and are not shown. For $P \leq$ 4.2 kbar no lattice transformation
is observed. The onset of a lattice distortion at $P \simeq$ 6.31 kbar
is signaled by a clear jump in the resistivity at $T_M \simeq$ 73K. The
transformation would appear to be first order, as was found for the
$LaAg_{1-x}In_x$ alloys.[10] The tetragonally distorted phase exists for
$T<T_M$, and the cubic phase for $T>T_M$. The increase in T_M with pressure is extremely rapid, $dT_M/dP \simeq +20K/kbar$: for comparison, Nb_3Sn
$+0.2K/kbar$, V_3Si $-0.15K/kbar$, La_3S_4 $+1.5K/kbar$, and $CuMo_3S_4$
$+3.0K/kbar$. These results on LaAg, obtained under hydrostatic
conditions, are in reasonable agreement with preliminary measurements in a quasihydrostatic pressure cell where at 16 kbar the
transformation temperature T_M has already reached 300K, shifting
to even higher temperatures as the pressure is increased further.

The cubic-tetragonal lattice transformation in LaAg has been
observed by Balster in X-ray measurements at room temperature under
quasihydrostatic pressure.[22] The results are shown in Fig. 2. At
300K, the cubic lattice distorts tetragonally at a critical pressure P_M which lies between 16 and 25 kbar, in agreement with the
resistivity results. The relative tetragonal distortion $(\frac{c-a}{a})$ increases from + 2.3% at 25 kbar to + 7.2% at 100 kbar. The compressibility at room temperature decreases from 0.13%/kbar for
$P<P_M$ to about 0.08%/kbar for $P>P_M$, with a volume collapse with increasing pressure of 1.85% at the phase transition. The observed
compressibility K agrees reasonably well with the piston-cylinder
measurements of Chao[9] $(k \simeq 0.1\%/kbar)$ who, however, did not observe
a volume collapse at T_M.

The occurrence of a cubic-tetragonal distortion is believed
to be favored in those systems where the Fermi level E_f lies in a
region of rapidly varying density of states $N(E)$. In such a situation, a tetragonal distortion of the lattice, and the ensuing
change of width and shifting of the structure in the density of
states, can lead to a significant repopulation of the various energy bands. When this repopulation reduces the total free energy
of the system, the tetragonal distortion can occur spontaneously.
Ihrig et al[11] estimated the concentration dependence of T_M in
$LaAg_{1-x}In_x$ by assuming a rectangular density of states whose width
changes when the lattice distorts tetragonally and thereby removes
the twofold degeneracy of the e_g-bands near the Γ-point at k= 0.
Ghatak et al[23] have recently derived an analytical expression for
T_M in a general treatment which takes into account both the width
change and the splitting of the e_g-band density of states by axial
crystal fields. The observed concentration dependence of T_M can be
qualitatively accounted for in these models by assuming that for
LaAg, E_f lies at or just below a sharp rise in $N(E)$. APW band
structure calculations by Hasegawa et al.[24] support this assumption.
In this picture E_f shifts up into the density of states peak as
Ag is replaced by In, precipitating the lattice distortion. Appli-

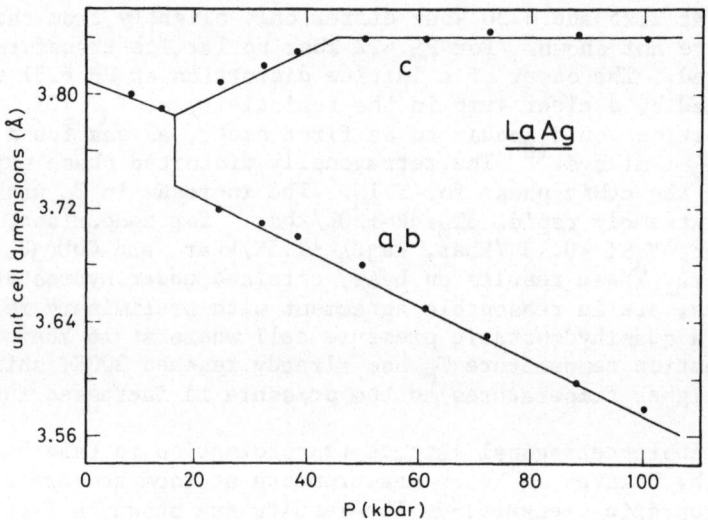

Figure 2. Lattice parameter versus quasihydrostatic pressure for
LaAg measured at T = 300K.

cation of pressure would presumably have a similar effect, leading
to the rapid increase of T_M observed in the present measurements.
It should be pointed out, however, that recent self-consistent
band structure calculations for LaAg by Tannous et al.[25] place E_f
well within the d-band density of states peak, a result which is
difficult to reconcile using the above models with the observed
phase diagram of $LaAg_{1-x}In_x$. Specific heat measurements[8] give
$N(E_f)=4.17$ states/eV-atom, in better agreement with the self-consistent calculation ($N(E_f)=3.84$ states/eV-atom) than the APW result ($N(E_f) = 2.44$ States/eV-atom). Kasuya has pointed out[14] that
peaks in density of states due to saddle points in the e_g-bands
for $k \neq 0$ would be split by a low symmetry distortion, for example by a pairing up of neighboring atoms in the tetragonal phase.
Evidence for such a sublattice distortion is indeed found in X-ray
studies.[10,13] A quantitative analysis of the present pressure experiments must await an accurate determination of the band structure
of LaAg as a function of lattice parameter. The rapid pressure
dependence of T_M should provide a critical measure for the correct-
ness of both band structure calculations and the various models of
the martensitic transformation.

The increase in the resistivity as the phase changes from cubic to tetragonal is opposite to what is observed in the A-15 compounds where lattice softening enhances the resistivity on the high temperature side of the transition. The reason for this resistivity increase is not clear but it is perhaps[14] a consequence of a gap appearing in the s-d Fermi surface due to the observed sublattice distortion for $T<T_M$.[10,13]

A further feature of the data in Fig. 1 is that the negative curvature in the resistivity is markedly larger in the tetragonal ($T<T_M$) than in the cubic ($T>T_M$) phase, whereas the pressure dependence of the curvature is small in either phase. These and other features of the data can be conveniently parameterized by an empirical formula first used successfully by Woodward and Cody[26] to fit data on Nb_3Sn films:

$$\rho(T)=\rho_o + \rho_1 + \rho_2\exp(-T_e/T) \qquad (I)$$

The values of the four parameters for the best fits are given in Table 1. Note that an increase in negative curvature is accompanied by a decrease in the value of T_e.

According to the theory of Cohen et al.[27] a negative resistance curvature is expected for those systems where the Fermi level at T = 0K is within the d-band and is separated by a small energy kT_0 from a sharp break in the density of states. The solid lines in Fig. 1 are fits using this theory with α =0.6[24] and Θ= 125K[8] for all data, where Θ is the Debye temperature and α is the ratio of the d-band to s-band density of states in the above model.[27] We find $T_0 \geq$ 1000K for the best fit to the P = 0 data and for both P = 12.37 and 35 kbar, T_0 = 325K, giving the stronger negative curvature characteristic of the tetragonal phase. It has, however, recently been pointed out by Allen[28] that the above theory is inaccurate because it neglects the energy dependence of the relaxation time $\tau(E)$. A more accurate calculation of the resistivity of Nb does give a negative curvature, which is, however, smaller than that observed in experiment. Fisk and Webb[29] have recently argued that the resistivity should bend over and saturate at a value ($\rho\approx150\mu\Omega cm$) where the electron mean free path becomes comparable to the interatomic spacing. It would seem unlikely that such a saturation effect could be important in the present studies where the resistivity of LaAg at room temperature ($\sim40\mu\Omega cm$) is well below the saturation value. Also, it would be difficult to explain the sudden large change in resistivity curvature on either side of the phase transition at T_M on the basis of such a saturation model.

It has recently been pointed out by Webb et al[30] that the high-T_c compounds Nb_3Sn, Nb_3Al, and Nb_3Ge all have both resistivi-

Table 1.   Parameters for fits of LaAg data to Eq. 1 and expression
for Cohen et al.[27]

P	T-range	ρ_o	ρ_1	ρ_2	T_e	T_o
(kbar)	(K)	($\mu\Omega$cm)	($\mu\Omega$cm)	($\mu\Omega$cm)	(K)	(K)
0	$\leq$300	1.28	0.0823	24.95	160	$\geq$1000
12.37	$\leq$180	2.2	0.0606	32.64	99.6	325
35	$\leq$300	5.8	0.0843	36.64	115	325

ties and lattice specific heats C_L which vary closely as T^2 in the
temperature range T$\leq$40K whereas for the isostructural low – T_c
compound Nb_3Sb, $\rho \propto T^{3.6}$.   It is interesting to note that the resistivity of LaAg at different pressures follows rather closely a
$\rho = A + BT^2$ law for T$\leq$40K in both cubic and tetragonal phases.   The
values of the parameters A and B are presented in Table 2 and compared to the results of Webb at al[30] on the A-15 compounds.   Note
that for LaAg both A and B increase reversibly with pressure.
Specific heat measurements[8] on LaAg at low temperatures show that,
however, C_L does not vary as T^2, in contrast to the above results
for the A-15 compounds.

Superconductivity

In the present experiments the superconducting transition temperature was determined by a mutual induction method and found to
be T_c = 1.062K at atmospheric pressure, in good agreement with the
specific heat result[8] T_c = 1.077K, but lying somewhat higher than
Chao's value[9] T_c = 0.94K.   Values of T_c determined resistively occur
5-10 mK higher in temperature.   The dependence of T_c(P) on hydrostatic pressure is shown in Fig. 3 and is seen to have an oscillatory character, going reversibly through two maxima.   The width of
the superconducting transition varies reversibly under pressure and
is largest when $dT_c/dP > 0$.   The non-linear behavior of T_c versus
pressure is characteristic of other materials that undergo crystallographic phase transformations.[1-5,31] It is noteworthy that
the abrupt appearance of the tetragonal distortion at 6.2 kbar,
which precipitates clear changes both in the magnitude and nega-

Table 2. Parameters for fits of LaAg and A-15 compound data to
          $\rho = A + BT^2$ for $T \leq 40K$.

System	P (kbar)	A ($\mu\Omega$cm)	B ($10^{-3}\mu\Omega$cm)
LaAg	0	1.08	2.43
LaAg	2.25	1.25	2.32
LaAg	4.2	1.56	2.43
LaAg	6.31	1.81	3.27
LaAg	9.62	1.88	3.56
LaAg	12.37	1.95	3.55
LaAg	35	5.6	4.15
Nb_3Sn	0	3.90	7.58
Nb_3Al	0	50.49	2.25
Nb_3Ge	0	24.16	4.0

tive curvature of the resistivity and probably in the susceptibility,[11] has no marked effect on T_c. The initial volume dependence
of T_c between 0 and 2.25 kbar is quite large ($\sim + 0.35K$ per % volume decrease), which is nearly the same as for pure La,[32] as illustrated in Fig. 4. It might, therefore, be speculated that the
rapid motion of T_M to higher temperatures for pressures above
4 kbar is related to the bending over of $T_c(P)$ as seen in Figs.
3 and 4. However, clearly, the monotonic $T_M(P)$ is not closely
correlated with the oscillatory $T_c(P)$ over the whole pressure range
to 35 kbar. In fact, the correlation between $T_c(P)$ and $T_M(P)$,
which was thought to perhaps exist for two further CsCl-compounds
$AuZn$[15] and $V_{.54}Ru_{.46}$,[4] should also be called into question. Whereas these compounds possess a pressure dependence of T_c analogous to
that of LaAg (T_c rises initially then bends over), their lattice
transformation temperatures T_M decrease with increasing pressure,

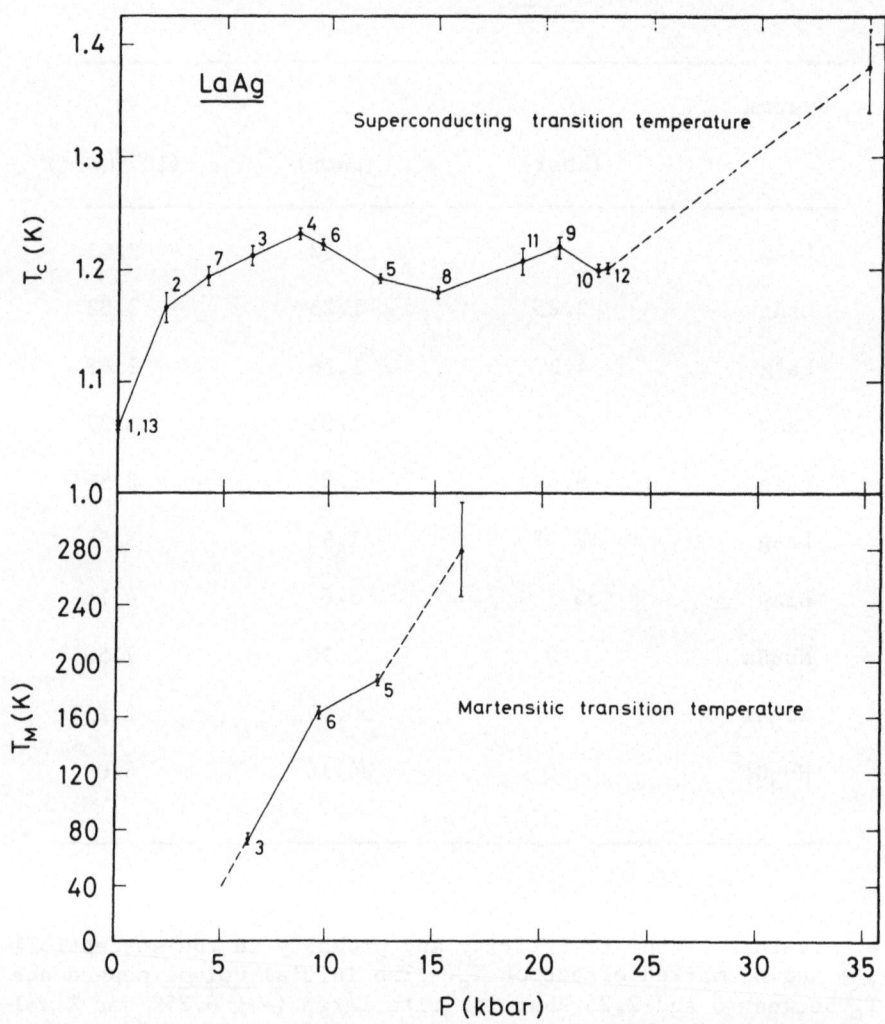

Figure 3.   Superconducting transition temperature T_c and Martensitic
             transition temperature T_M of LaAg versus hydrostatic
             pressure. Numbers near data points give order of meas-
             urement. "Error bars" give onset and termination tem-
             perature of transition. T_c value at 35 kbar and T_M
             value at 16 kbar were determined in a quasihydrostatic
             pressure cell. For P = 35 kbar, T_M lies well above
             300K.

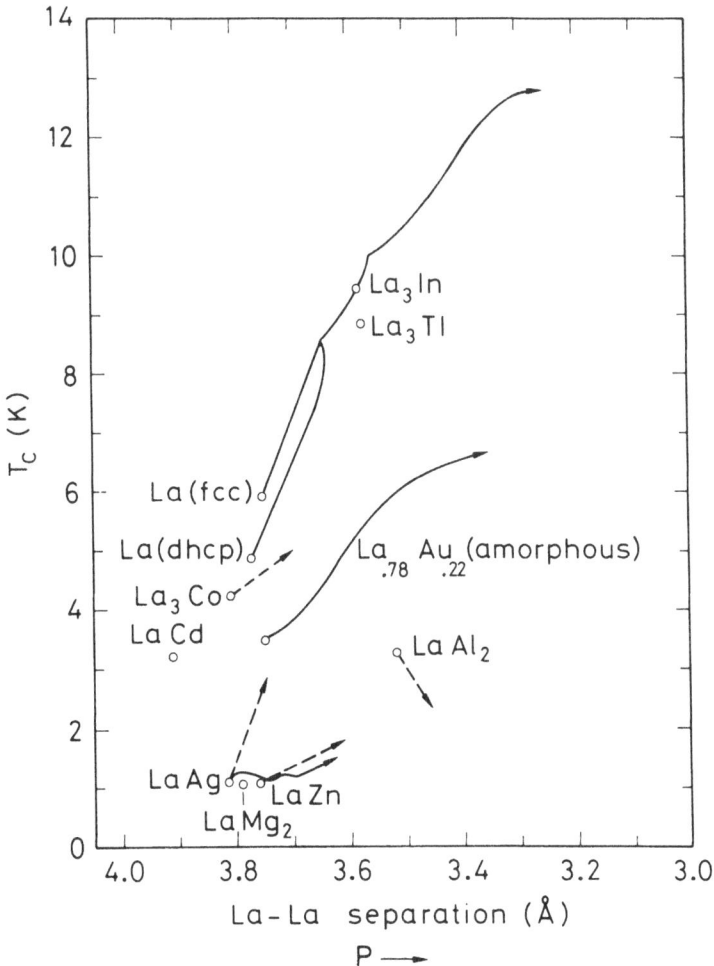

Figure 4.   Superconducting transition temperature T_c versus nearest neighbor La-La separation for various La compounds.[34] Dashed lines give initial slope.

in marked contrast to LaAg. In LaAg, therefore, and possibly in
AuZn and $V_{.54}Ru_{.46}$, there appears to be no obvious relation between phase instabilities and superconductivity. This result
supports the view expressed by Phillips[33] that a correlation between T_C and T_M only exists in materials with anomalously high
T_C-values.

In Fig. 4 the superconducting transition temperature T_C is
plotted versus the nearest neighbor La–La separation for a number
of La compounds.[34] None of the analogous compounds with Y,Lu, or
Sc are known to be superconducting. Also included are some recent data by Fasol et al[35] for amorphous $La_{.78}Au_{.22}$, when the compressibility and initial La–La separation are assumed to be the
same as for pure fcc La.[36] It can be seen that all compounds which
contain at least 50% La have T_C values which increase as pressure
reduces the La–La separation. A tendency for T_C for these compounds
to increase with decreasing La–La separation is also evident. However, clearly, other factors such as the structure and the individual constituents of the compound also play a role in determining
the T_C values.

ACKNOWLEDGEMENTS

One of the authors (J.S.S.) would like to express special
gratitude to H. Ihrig for suggesting high pressure measurements on
LaAg and for supplying him with a sample for the preliminary quasihydrostatic run. The authors acknowledge stimulating discussions
with T. Kasuya, D. Vigren and especially J. Kübler, who also critically read the original manuscript. The authors are grateful to
H. Balster for allowing them to show a graph of his unpublished
X-ray data on LaAg. Thanks are due H. Camen and P. Stauche for
preparation and analysis of the LaAg sample and to E. Havenstein
for her expert technical assistance. The interest and support
of B.T. Matthias is gratefully acknowledged by one of the
authors (R.N.S.).

REFERENCES

1. For a review of work on A-15 compounds see: L.R. Testardi,
    Rev. Mod. Phys. 47, 637 (1975); M.Weger and I.B. Goldberg,
    Solid State Phys. (1973), Vol. 28, ed. by H. Ehrenreich,
    F. Seitz and D. Turnbull.
2. R.N. Shelton, A.R. Moodenbaugh, P.D. Dernier, and B.T. Matthias,
    Mat. Res. Bull. 10, 1111 (1975).
3. T.F. Smith, R.N. Shelton, and A.C. Lawson, J. Phys. F: Metal
    Phys. 3, 2157 (1973).
4. C.W. Chu, S. Huang, T.F. Smith, and E. Corenzwit, Phys.
    Rev. B 11, 1866 (1975).

5. R.N. Shelton, D.C. Johnston, and J.J. Bugaj, Bull. Am. Phys.
    Soc. 22, 402 (1977).
6. Z. Fisk and A.C. Lawson, Solid State Commun. 13, 277 (1973);
    Z. Fisk, R. Viswanathan, and G.W. Webb, Solid State Commun. 15,
    1797 (1974).
7. A.R. Moodenbaugh and Z. Fisk, Phys. Lett. 43A, 479 (1973).
8. R.W. Hill, J. Cosier, and D.A. Hukin, J. Phys. F6, 1731 (1976).
    For $N(E_f)$ estimate, see also: R. Backhus (private communica-
    tion in same institute).
9. C.C. Chao, H.L. Luo, and T.F. Smith, J. Phys. Chem. Solids 27,
    1555 (1966).
10. H. Balster, H. Ihrig, A. Kockel, and S. Methfessel, Z. Physik
    B21, 241 (1975). For X-ray results on $LaAg_{1-x}In_x$, see also:
    H. Camen, Diplom-Thesis, Univ. Bochum, 1976 (unpublished).
11. H. Ihrig, D.T. Vigren, J. Kübler, and S. Methfessel, Phys.
    Rev. B8, 4525 (1973).
12. J. Labbé and J. Friedel, J. Phys. Radium 27, 153, 303 (1966).
13. H. Ihrig, thesis (University of Bochum, 1973) (unpublished).
14. T. Kasuya (private communication).
15. B.T. Matthias, E. Corenzwit, J.M. Vandenberg, H. Barz, M.B. Maple
    and R.N. Shelton, J. Less Common Metals 46, 339 (1976).
16. A. Jayaraman, A.R. Hutson, J.H. McFee, A.S. Coriell, and
    R.G. Maines, Rev. Sci. Inst. 38, 44 (1967).
17. P.E. Chester and G.O. Jones, Phil. Mag. 44, 1281 (1953).
18. T.F. Smith, C.W. Chu, and M.B. Maple, Cryogenics 9, 53 (1969).
19. N. Kawai and Sawaoka, Rev. Sci. Inst. 38, 1770 (1967).
20. J.S. Schilling, U.F. Klein, and W.B. Holzapfel, Rev. Sci. Instrum.
    45, 1353 (1974).
21 After the 0 and 6.31 kbar measurements it was necessary to recon-
    tact the sample. Near room temperature and for P<15 kbar we
    find $\rho(T,P) = 40 [1 + 0.0039P - 0.00338 (290-T)]$ μΩcm with P in
    kbar. The 35 kbar run was carried out in a quasihydrostatic
    pressure cell and was normalized to the hydrostatic data using
    the estimated pressure dependence of the resistivity at room
    temperature.
22. H. Balster (private communication in same institute).
23. S.K. Ghatak, D.K. Ray, and C. Tannous (to be published in Phys.
    Rev. B).
24. A. Hasegawa, B. Bremicker, and J. Kübler, Z. Physik B22, 231
    (1975).
25. C. Tannous, D.K. Ray, and M. Belakhovsky, J. Phys. F6, 2091
    (1976).
26. D.W. Woodward and G.D. Cody, RCA Rev. 25, 392 (1964).
27. R.W. Cohen, G.D. Cody, J.J. Halloran, Phys. Rev. Lett. 19,
    840 (1967).
28. P.B. Allen (to be published).
29. Z. Fisk and G.W. Webb, Phys. Rev. Letters 36, 1084 (1976).
30. G.W. Webb, Z. Fisk, J.J. Engelhardt, and S.D. Bader, Phys.
    Rev. B15, 2624 (1977).

31. R.N. Shelton, A.C. Lawson, and D.C. Johnston, Mat. Res. Bull. 10, 297 (1975).
32. H. Balster and J. Wittig, J. Low Temp. Phys. 21, 377 (1975).
33. J.C. Phillips, Phys. Rev. Lett. 26, 543 (1971).
34. Values of $T_c(P)$ for fcc-La and dhcp-La are from Balster and Wittig (Ref. 32) and for La_3Co, $LaAl_2$, LaZn, and $LaMg_2$ from T.F. Smith and H.L. Luo, J. Phys. Chem. Solids 28, 569 (1967). $T_c(P)$ for LaAg is from present work, for amorphous $La_{.78}Au_{.22}$ from Fasol et al (Ref. 35), and for LaCd from A.M. Stewart (private communication). The compressibility K for fcc-La was taken from Syassen and Holzapfel (Ref. 36); for dhcp-La and amorphous $La_{.78}Au_{.22}$ the same K was assumed as for fcc-La. For LaZn, K was taken to be same as for LaAg (~0.13%/kbar) and for $LaAl_2$ and La_3Co it was assumed $K \simeq 0.10\%/kbar$.
35. G. Fasol, J.S. Schilling, and C.C. Tsuei (to be published).
36. K. Syassen and W. B. Holzapfel, Solid State Commun. 16, 533 (1975).

QUESTIONS AND COMMENTS

J.A. Woollam: If you plot your resistivity versus temperature data, do they fit any particular model, such as a saturation model?

J.S. Schilling: Yes, this will be in the paper to be published in the proceedings. We have fitted the data with various types of models. The one normally used for A15 compounds is the exponential term $e^{-T_0/T}$. Also we fitted with the Cohen theory and tested for a low temperature dependence of T^2 which also fits some of the A15 compounds. As far as I know there's as yet no reliable theory of the resistivity curvature in these compounds. It's a matter of debate at the present time.

J. Wittig: I recall that the transition at 0 kbar was at 60K or 70K: the cubic to tetragonal transition.

J.S. Schilling: Yes.

J. Wittig: What will be the transition pressure at zero temperature?

J.S. Schilling: Oh, it's about 4 kbar. We also made measurements at 2 and 4 kbar. There is no clear evidence in these measurements of the phase transition. They differ only slightly from the 0 kbar measurements.

But if you extrapolate T_M down to zero temperature, it crosses the pressure axis at about 4 or 5 kbar.

J. Wittig: Isn't there an anomaly in T_c around this pressure?

J.S. Schilling: Well, you could possibly claim to see an anomaly, if you remember the one slide of the T_c versus La–La separation: the initial slope of the lanthanum–silver data is essentially the same as for pure lanthanum. Now one could think, perhaps, that $T_c(P)$ starts off with this large slope and the phase transition moving up has something to do with $T_c(P)$ bending over, but this is very speculative.

J.E. Schirber: Are there any of the band structure calculations performed in the tetragonal phases?

J.S. Schilling: No, only on the cubic phase. The tetragonal distortions are very small, of the order of a percent or less, and band structure calculations aren't accurate enough. A different type of model has to be used to calculate the shifts of the peaks and the broadening due to the distortion.

PRESSURE DEPENDENCIES OF THE SUPERCONDUCTING AND MAGNETIC CRITICAL

TEMPERATURES OF TERNARY RHODIUM BORIDES*

R.N. Shelton and D.C. Johnston

Institute for Pure and Applied Physical Sciences

University of California

San Diego, LaJolla California 92093

ABSTRACT

The pressure dependencies of the superconducting and magnetic transition temperatures of ternary rhodium borides, MRh_4B_4, are reported for hydrostatic pressures up to 21 kbar. The magnetic ordering temperature is enhanced by pressure for every magnetic compound (M=Gd, Tb, Dy, Ho). In contrast, the superconducting transition temperature, T_c, may be either increased or depressed by pressure depending on the size of the third atom (M= Nd, Sm, Er, Tm, Lu, Th, Y). For the lutetium ternary an unusual nonlinear pressure dependence of T_c may indicate the presence of a pressure induced phase transformation.

INTRODUCTION

The existence of superconductivity and magnetic order has recently been reported for the class of ternary compounds MRh_4B_4 where M can be one of several rare earth elements or some transition elements.[1,2] For M = Nd, Sm, Er, Tm, Lu, Th or Y the compounds become superconducting at temperatures ranging from 2.5 K

---

*Research sponsored by the Air Force Office of Scientific Research, Air Force Systems Command, USAF, under AFOSR Contract No. AFOSR/ F-49620-77-C-0009.

to 11.8 K, whereas for M = Gd, Tb, Dy or Ho magnetic order is observed between 5.6 K and 12.0 K. For one compound, $ErRh_4B_4$, the superconductivity which occurs at T_c = 8.6 K vanishes at the lower temperature of 0.9 K at which magnetic ordering sets in.[3]

In order to compare and contrast the behavior under pressure of the magnetic and superconducting critical temperatures (T_M and T_c), we have completed a study of all ternary rhodium borides up to hydrostatic pressures of 21 kbar and have found a rich variety of pressure dependencies. Each magnetic ordering temperature shows a linear increase under pressure; however, the superconducting compounds have both positive and negative values of dT_c/dp, depending on the size of the rare earth atom. Additonally, the highest T_c ternary boride, $LuRh_4B_4$ displays a distinct nonlinear character in $T_c(p)$.

EXPERIMENTAL DETAILS

Except for $NdRh_4B_4$, the XRh_4B_4 samples studied in this investigation were synthesized from stoichiometric mixtures of the high purity elements by arc melting in a Zr-gettered argon atmosphere. The samples were then annealed at 800-1200°C for periods of from several days to several weeks. The compound with X = Nd could not be obtained by arc melting and annealing the composition $NdRh_4B_4$, but the tetragonal $NdRh_4B_4$ phase was obtained in an arc melted ingot with composition $NdRh_6B_6$, as previously reported.[1,2]

Hydrostatic pressures were generated at room temperature using a one-to-one mixture of isoamyl alcohol and n-pentane[4] to transmit the pressure. A "clamp" device similar to the one introduced by Chester and Jones[5] was utilized to maintain the necessary pressures at low temperatures. A typical series of data for each sample consisted of an initial determination of the zero pressure critical temperature followed by measurements at increasingly higher pressures until a maximum pressure of 20-25 kbar was attained. Immediately following this series, the pressure was released and the zero pressure value of the critical temperature redetermined. In no instance was any hysteresis observed. For $ThRh_4B_4$, $TmRh_4B_4$ and $LuRh_4B_4$, a high density of data points were taken to check for possible nonlinear behavior in $T_c(p)$. All transition temperatures were determined by monitoring the ac magnetic susceptibility.

RESULTS

The effect pressure on T_c for the superconducting XRh_4B_4 compounds is shown in figures 1 and 2. The data for both of the light rare earth members (X = Nd, Sm) are grouped with those for $ThRh_4B_4$

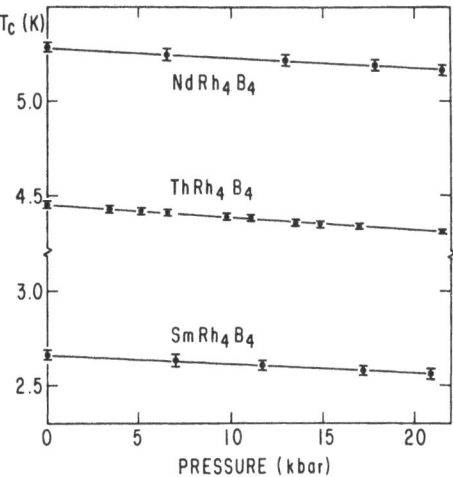

Figure 1. T_c versus hydrostatic pressure for three ternary rhodium borides with negative values of dT_c/dp. Vertical error bars indicate the width of the transition into the superconducting state.

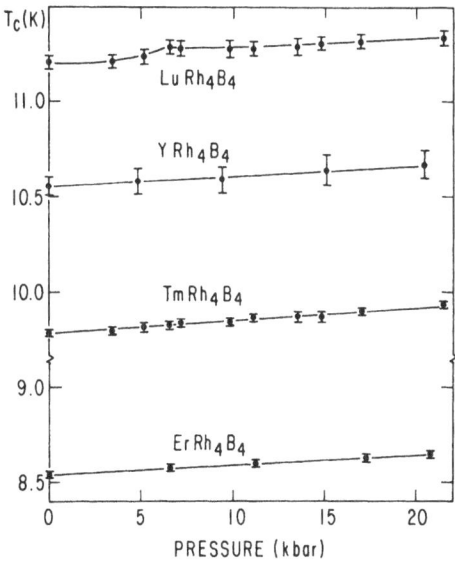

Figure 2. T_c versus hydrostatic pressure for four ternary rhodium borides with positive values of dT_c/dp. Vertical error bars indicate the width of the transition into the superconducting state.

in figure 1 and the transition temperatures of all three materials
are observed to decrease linearly with pressure. This behavior is
in contrast to that observed for YRh_4B_4 and the heavy rare earth
ternaries, shown in figure 2. The higher transition temperatures
of this latter group of compounds increase with pressure. For three
of the four samples shown in figure 2, namely YRh_4B_4, $ErRh_4B_4$ and
$TmRh_4B_4$, this enhancement is linear. The fourth compound, $LuRh_4B_4$,
displays a behavior unique to this set of materials. The initial
effect of pressure on T_c is minimal up to roughly 3.5 kbar. A
sharp rise in T_c then occurs in the pressure interval of 3.5 to 6.6
kbar with a rate of change of T_c approximately four times as large
as that observed for the other superconducting compounds. For pressures just above this region of rapid increase, the pressure dependence of T_c is zero. The transition temperature remains constant from 6.6 kbar to approximately 12 kbar, but beyond this pressure a linear increase in T_c is observed.

A nonlinear pressure dependence of T_c at these relatively low
pressures and under hydrostatic conditions has been related to the
presence of crystallographic instabilities in other high T_c superconductors.[6,7] This instability can be either temperature induced
as is the case for some Chevrel-phase ternary compounds[8,9] and many
binary superconductors,[6,10] or it may have its origin in the application of pressure as observed for the CsCl superconductor LaAg.[11]
Two additional factors may indicate the presence of an instability
in $LuRh_4B_4$. First, this compound has the highest T_c among this isostructural series and secondly, Lu is the smallest atom that will
stabilize this ternary phase.[1,2] Since lutetium is at the lower
size limit of stability, the crystal structure of $LuRh_4B_4$ might be
more susceptible to external pressure than other members of the
series. Because $LuRh_4B_4$ exhibits no lattice distortion above 4 K
at zero pressure,[12] experiments are planned to clarify the reasons
for the unique behavior of T_c under pressure for this compound.

The effect of pressure on the magnetic ordering temperatures
of the remaining four rare earth rhodium borides is shown in figure
3. Values taken for T_m correspond to the position of the cusp in
the ac susceptibility versus temperature curve. Surprisingly,
pressure enhances the magnetic ordering temperature for all four
compounds. This pressure dependence is similar to the effect of
pressure on T_c observed for the superconducting heavy rare earth
ternaries. In every instance T_m increases linearly; however,
these magnetic materials are more sensitive to pressure than their
isostructural superconducting counterparts. The values of dT_m/dp
calculated from the data in figure 3 are a factor of two to four
greater than the magnitudes of dT_c/dp calculated from figures 1
and 2.

Data for each sample except $LuRh_4B_4$ were fit by least squares
to obtain the pressure derivatives listed in Table I. A value char-

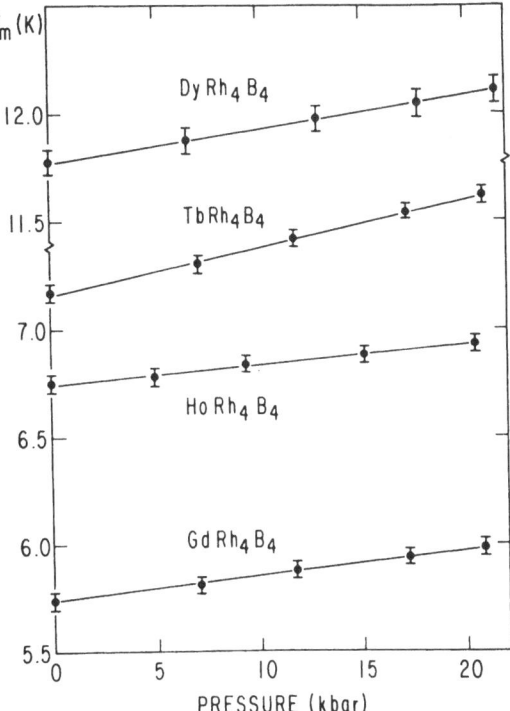

Figure 3.   The magnetic ordering temperature versus hydrostatic
            pressure for four rare earth rhodium borides. Vertical
            error bars indicate the uncertainty in determining T_m.

acteristic of the overall pressure effect from 0 to 21 kbar was
taken for $LuRh_4B_4$. These data clearly show the greater sensitivity
to pressure of the magnetic compounds and reemphasize that the only
negative pressure derivatives occur for the superconducting compounds containing the larger third atoms.

DISCUSSION

In order to search for any possible trends in T_c or dT_c/dp a-
mong the XRh_4B_4 compounds, these quantities are plotted against the
metallic radius of the third atom $R_{12}(X)$ in figure 4. The 12 coordination radii are based on the observed atomic volumes of the
metals.[13] In both the upper and lower graphs, lines have been
added to indicate the behavior for the rare earth ternary rhodium
borides. YRh_4B_4, containing the much lighter yttrium atoms, is
distinguished from the rare earth (RE) compounds by a higher T_c
and a much more positive value of dT_c/dp than would be predicted

TABLE 1

Pressure Dependencies of the Critical Temperatures, T_c and T_m, of Ternary Rhodium Borides

Compound	$T_c(0)$ (K)	dT_c/dp $(10^{-5}$ K/bar)
$ThRh_4B_4$	4.469– 4.441	−0.66
$NdRh_4B_4$	5.307– 5.259	−0.55
$SmRh_4B_4$	2.690– 2.637	−0.49
YRh_4B_4	10.610–10.502	0.56
$ErRh_4B_4$	8.554– 8.530	0.50
$TmRh_4B_4$	9.800– 9.770	0.69
$LuRh_4B_4$	11.240–11.175	0.71

Compound	$T_m(0)$ (K)	dT_m/dp $(10^{-5}$ K/bar)
$GdRh_4B_4$	5.74 ± 0.04	1.18
$TbRh_4B_4$	7.17 ± 0.04	2.19
$DyRh_4B_4$	11.78 ± 0.06	1.52
$HoRh_4B_4$	6.75 ± 0.04	0.88

from its radius. In contrast, $ThRh_4B_4$ with the heavier and tetravalent thorium atoms shows a slightly more negative pressure dependence of T_c than observed in the rare earth series. Focusing on the rare earth compounds, one observes that T_c rises at either end of the series while the magnetic ordering temperatures peak in the middle with $DyRh_4B_4$. The variation of the pressure dependence of T_c versus $R_{12}(RE)$ for the superconducting compounds is essentially linear across the rare earth series. This includes negative as well as positive values of dT_c/dp and contrasts with the magnetic compounds where dT_m/dp passes through a maximum at $TbRh_4B_4$.

Extrapolation of the linear relationship for dT_m/dp versus $R_{12}(X)$ among $TbRh_4B_4$, $DyRh_4B_4$ and $HoRh_4B_4$ to the compound $ErRh_4B_4$ yields a value of $(0.1 - 0.2) \times 10^{-5}$ K bar^{-1} for dT_m/dp of the latter. Since $T_m = 0.9$ K, this value of dT_m/dp is consistent with

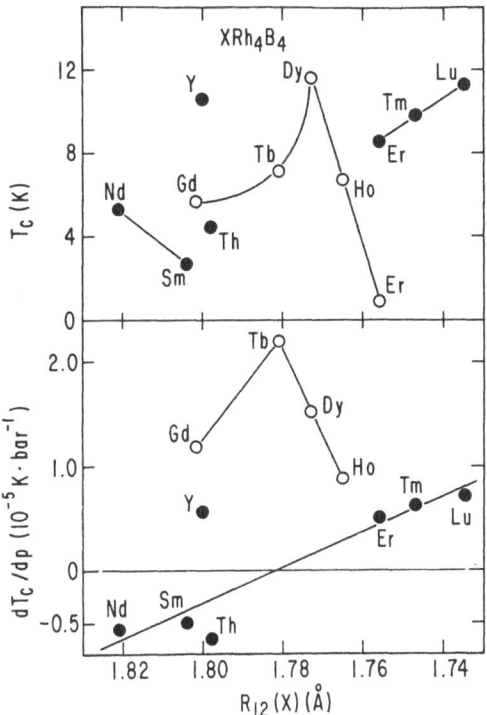

Figure 4. Superconducting (•) and magnetic (o) critical tempera-
           tures and their pressure derivatives for XRh_4B_4 compounds
           plotted versus the twelve-coordination metallic radius
           of the X atom. Lines are drawn to indicate the trends
           in these quantities as X varies across the rare earth
           series.

our not observing magnetic order in $ErRh_4B_4$ above our present low
temperature limit of 1.02 K for pressures up to 21 kbar. Measurements below 1 K are planned for $ErRh_4B_4$ to determine whether pressure will enhance T_m as well as the superconducting transition temperature for the same material.

     Some general conclusions with respect to the superconducting
members of the series XRh_4B_4 may be formulated. The pressure dependence of T_c is linear for every material studied with the exception of $LuRh_4B_4$. The anomalous behavior found for the latter
compound plus the fact that lutetium is the smallest X atom to
form this phase may be indicative of a pressure induced phase transformation. For the isoelectronic rare earth series there is a correlation between the atomic radius of the rare earth atom and the
effect of pressure on T_c: ternaries comprised of smaller rare earth

elements display positive values of dT_c/dp, whereas compounds containing larger rare earth atoms show a depression of T_c with pressure. This relationship between the volume occupied by the third element and the superconducting properties of the compound under pressure is an important result of this investigation.

ACKNOWLEDGMENTS

We are grateful to Prof. B.T. Matthias for numerous discussions concerning the nature of these compounds and to Dr. A.C. Lawson for permission to refer to his low temperature X-ray results prior to publication.

REFERENCES

1. B.T. Matthias, E. Corenzwit, J.M Vandenberg and H.E. Barz, Proc. Nat. Acad. Sci. USA 74, 1334 (1977).
2. J.M. Vandenberg and B.T. Matthias, Proc. Nat. Acad. Sci. USA 74, 1336 (1977).
3. W.A. Fertig, D.C. Johnston, L.E. DeLong, R.W. McCallum, M.B. Maple and B.T. Matthias, Phys. Rev. Lett. 38, 987 (1977).
4. A. Jayaraman, A.R. Hutson, J.H. McFee, A.S. Coriell and R.G. Maines, Rev. Sci. Inst. 38, 44 (1967)
5. P.E. Chester and G.O. Jones, Phil. Mag. 44, 1281 (1953).
6. R.N. Shelton, A.R. Moodenbaugh, P.D. Dernier and B.T. Matthias, Mat. Res. Bull. 10, 1111 (1975).
7. A.C. Lawson and R.N. Shelton, Ferroelectrics, to be published.
8. A.C. Lawson, Mat. Res. Bull. 7, 773 (1972).
9. A.C. Lawson and R.N. Shelton, Mat. Res. Bull. 12, 375 (1977).
10. R.N. Shelton and T.F. Smith, Mat. Res. Bull. 10, 1013 (1975).
11. J.S. Schilling, S. Methfessel and R.N. Shelton, paper presented at this conference.
12. A.C. Lawson, private communication.
13. W.H. Zachariasen, J. Inorg. Nucl. Chem. 35, 3487 (1973).

QUESTIONS AND COMMENTS

J.A. Woollam: How sensitive is T_c to stoichiometry?

R.N. Shelton: We found that these are essentially line compounds, that they form in the ratio 1 to 4 to 4. If you vary the relative amount of, let's say, erbium in erbium rhodium boron, you wind up getting different phases from the ternary phase diagram.

T.A. Kaplan: What happens to the superconductivity when you go to the magnetic ordering at low temperature?

R.N. Shelton: Superconductivity disappears, and you see a mag-
                 netic ordering that occurs at 0.9 K.

T.A. Kaplan: Ferromagnetic?

R.N. Shelton: I don't think we can say definitely the type of
                 magnetic ordering yet. A number of measurements
                 have been done. There was a Phys. Rev. Letter
                 ($\underline{38}$, 987 (1977)) that came out recently, reporting
                 resistivity versus temperature, susceptibility,
                 heat capacity. The susceptibility, for example,
                 shows a Curie Weiss law and shows essentially a
                 very small value for the Curie temperature: one
                 degree, plus or minus a degree. But some other
                 things indicate it might not be a ferromagnetic
                 ordering.

D.B. McWhan: The team of D.E. Moncton, E. Corenzwit, J. Eckert,
                 G. Shirane, W. Thomlinson and I have made elastic
                 neutron scattering measurements on $ErRh_4B_4$ at 1.4
                 and 0.07 K. We have shown that $ErRh_4B_4$ is a sim-
                 ple ferromagnet below 1 K and the data are consis-
                 tent with the spins being in the basal plane.

THE AuGa$_2$ DILEMMA - SUPERCONDUCTING VERSION

R.A. Hein and J.E. Cox
Naval Research Laboratory
Washington, D.C. 20375

R.W. McCallum*
Institute for Pure and Applied Physical Sciences
University of California
San Diego, La Jolla, California 92037

ABSTRACT

The magnetic response of two nominal AuGa$_2$ samples has been studied as functions of temperature, applied magnetic field and pressure. These samples display widely different superconducting transition temperatures, critical magnetic field curves and pressure effects. Variations of the initial slope, i.e., $(dH_c/dT)_{T_0}$, with pressure supports the concept of an electronic transition at about 0.55 GPa. Changes in the magnetic response as a function of pressure suggest that some sort of lattice transformation must also occur at this pressure.

I. INTRODUCTION

The normal state electronic properties of intermetallic compounds AuAl$_2$, AuGa$_2$ and AuIn$_2$ have been extensively studied[1]. Although these compounds all form in the cubic fluorite structure the magnetic properties of AuGa$_2$, i.e. bulk susceptibility, Ga spin-lattice relaxation time and the Ga Knight shift, are significantly different from those of AuAl$_2$ and AuIn$_2$. These facts caused Jaccarino, et al.[2] to point out the existence of the "AuGa$_2$ Dilemma."

*Research supported by the U.S. Energy Research and Development Administration under Contract No. ERDA E(04-3)-34PA227.

Schirber's study[3] of the Fermi surface of $AuGa_2$ as a function of pressure was interpreted as evidence for an isostructural Lifshitz "electron transition". In this transition, a second flat energy band which is located below the Fermi energy at zero pressure, approaches the Fermi level as one applies pressure (P) and actually passes above the Fermi level at P>0.6 GPa. On the basis of this and the results of some $Au_{1-x}Pd_xGa_2$ studies[4], Schirber predicted[3] and experimentally verified[5] that the superconducting transition temperature, T_o, of $AuGa_2$ does indeed increase rapidly for 0.55 GPa<P<0.60 GPa[$\Delta T_o \approx 0.8K$]. These data provide one with a basis for explaining the "$AuGa_2$ dilemma" but the follow-on work of Smith, et al.[6] presents us with a new "$AuGa_2$ dilemma - superconducting version." Smith, et al.[6] observed that at higher pressures, i.e. P>1.5 GPa, the ac magnetic susceptibility in zero applied dc magnetic field exhibited a differential paramagnetic effect[7] (DPE) and supercooling.[8] This latter phenomenon is usually restricted[9] to pure, strain-free samples, surfaces of which are defect-free, and usually amounts to only a few percent of H_c, the thermodynamic critical magnetic field of a type 1 superconductor. How can a "purely electronic transition" give rise to such an effect?

The fact that $AuGa_2$ exhibits well defined zero field transitions[5,6] and the magnitude of the changes in T_o with pressure suggested to us that one should be able to verify the predicted change in $N_{bs}(0)$, band structure electronic density of states evaluated at the Fermi level, from measurements of the initial slope of the critical magnetic field curves $(dH_c/dT)_{T_o}$. A study of the magnetic response in an applied field as functions of pressure and temperature also allows one to examine in detail the reported pressure induced supercooling and DPE reported by Smith, et al.[6] Such determinations are the objectives of this investigation.

II. EXPERIMENTAL DETAILS

A) Sample Preparation

Two samples of nominal $AuGa_2$ were used in this study. One was a polycrystalline sample prepared by Dr. Ch. J. Raub of the Institute for Noble Metals, Schwäbish-Gmund, Germany. It was prepared by flame melting in an evacuated quartz tube and annealed at 400 C for one week. Metallographs indicated a small amount of a second phase. The other sample was a single crystal prepared by Dr. R.J. Soulen, Jr. of the National Bureau of Standards, Gaithersburg, Maryland. It was also melted in a quartz tube but directionally cooled. ESCA and Auger spectroscopy revealed no free Ga in this sample.

B) Superconducting Measurements

AC mutual inductance techniques using lock-in amplifiers were used to measure the in-phase and out-of-phase components of the low frequency (17-77 Hz) ac magnetic susceptibility. In zero applied dc magnetic field such measurements are a measure of the initial magnetic susceptibility of the samples, i.e.,

$$\chi_{ac} = \left(\frac{dM}{dH}\right)_{H_o = 0} = \frac{\Delta M}{\Delta H}$$

where ΔH is the magnitude of the peak ac measuring field H_{ac}. In the presence of a longitudinal dc magnetic field H_o, where $H_o \gg H_{ac}$ such measurements yield data which are proportional to the differential magnetic susceptibility

$$\chi_{ac} = \left(\frac{dM}{dH}\right)_{H_o \neq 0} \qquad \text{where } dH = H_{ac}.$$

Thus one can study the detailed magnetic response of the sample as functions of applied magnetic field, H_o, temperature, T, and pressure, P.

C) Temperature Ranges

Several cryostats were used in this study. At La Jolla, temperatures down to 0.95K were produced by pumping on liquid ^{4}He, while temperatures down to 0.40K were produced by means of a ^{3}He cryostat. At NRL temperatures down to 1.25K were produced by pumping on ^{4}He. Temperatures were deduced from either vapor pressure measurements or from calibrated carbon or germanium resistor thermometers.

D) Pressure Measurements

Pressures were produced by means of piston and clamp pressure cells of the Chester type. Two kinds of pressure fluids were used; a 1:1 mixture of pentane and isoamyl alcohol; and a methanol-ethanol-glycerine mixture. For pressure determinations a Sn manometer was incorporated in the cell along with the sample under investigation. In the critical magnetic field studies, the Sn was removed and pressure was deduced from the measured T_o of the AuGa$_2$ sample itself.

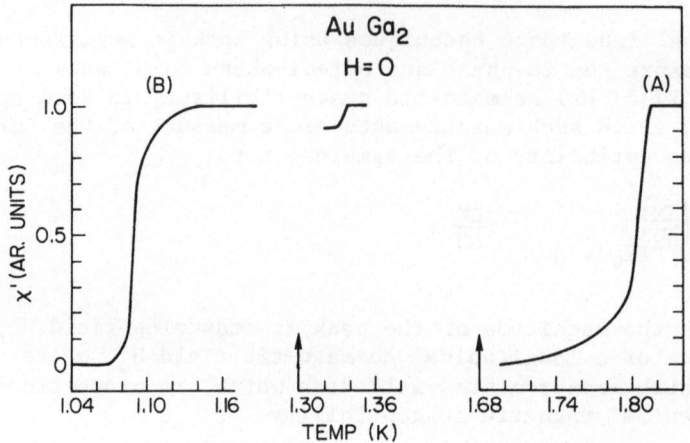

Figure 1. The in-phase component, χ', of the ac magnetic susceptibility of two "AuGa$_2$" samples as a function of temperature.
The values are given in normalized units were $\chi' = 1$ in
the normal state and $\chi' = 0$ in the superconducting state.
A denotes the polycrystalline and B denotes the single
crystal data. The small second "transition" common to
both samples is also shown.

E) Critical Field Measurements

These data were obtained by either varying the temperature while
the sample was subjected to a constant applied magnetic field (temperature sweep) or by varying the applied dc magnetic field at a
constant rate while the temperature was maintained (field sweep).
We define critical field values or critical temperatures in an
applied field by extrapolations of the descending portions of the
DPE regime (see Figure 4).

III. RESULTS AND DISCUSSION

A) Zero Pressure Data

1) <u>Transition Temperature, T_0</u>. The two nominal AuGa$_2$ samples
exhibit considerably different transition temperatures. Figure 1
contains plots of the in-phase, χ' component of the ac magnetic
susceptibility as a function of the temperature. For comparison pur-

poses we present the data in normalized form setting $\chi' = 1$ for the normal state and $\chi' = 0$ for the superconducting state. The poly-crystalline sample has a transition centered about 1.81K with a low-temperature "tail". A smaller second transition was observed at 1.34K. The single crystal data shows a transition centered about 1.09K, with a gradual tailing towards the higher temperatures. It, too, exhibited a small second transition at 1.34K. It should be noted that the extreme T_0 values span the range previously reported for the zero pressure values of AuGa$_2$ and $(Au_{1-x}Pd_x)Ga_2$ as well as the high pressure "phase" of AuGa$_2$. We tentatively attribute the small transition seen at 1.34K to the presence of AuGa as a second phase in both samples. The initial study of superconductivity in the AuGa system by Hamilton, et al.[10] reported that AuGa (orthorhombic B31 structure) was a superconductor at 1.2K and that an argon arc melted sample of AuGa$_2$ was not superconducting down to 0.34K. Subsequent work by Wernick, et al.[4], led to the discovery of super-conductivity in AuGa$_2$, with $1.05 < T_0 < 1.12$K. Their sample was prepared by induction melting in an argon atmosphere, followed by zone refining. While their stated belief was that AuGa$_2$ is a stoichio-metric line compound, we believe the difference in T_0 values observed for our two samples is due to the existence of the fluorite phase of the AuGa system, over a range of compositions, in keeping with the observations of Longo, et al.[11] These workers pointed out that single crystals of the "AuX$_2$" compounds of the fluorite structure possess the best resistivity ratios if one prepares them with a slight excess of the X constituent-in the case of AuGa$_2$, 0.3 to 0.4 excess wt% Ga. They cited the work of Straumanis and Chopra[12] as evidence that "AuAl$_2$" exists with Al concentrations ranging from 66.10 at % to 67.74 at %, and emphasized that at stoichiometry there are 0.076 empty Au sites per unit cell and 0.152 vacant Al sites. This question of deviations from stoichiometry in AuX$_2$ compounds was also raised, but not pursued, by Carter, et al.[13]. We believe a consistent picture with regard to all the superconducting studies on "AuGa$_2$" at zero pressure is obtained if one postulates that truly stoichiometric, or perhaps Ga deficient, "AuGa$_2$" is not supercon-ducting above 0.34K and that T_0 is an increasing function of the excess Ga in the compound. This implies that the electronic tran-sition discussed by Schirber, which was produced by the application of pressure, can also be induced by changes in the Ga content of the compound. A detailed investigation of composition effects is underway.[14]

2) <u>Critical Magnetic Field Data</u>. The purpose here is twofold: (1) to evaluate the initial slope of the critical magnetic field curves and to ascertain if these samples exhibit a DPE. The ob-servation of Smith, et al.[6] of a DPE in zero applied field at P>1.5 GPa means that in the zero applied field data there must have been an ambient field H_a present which has a component $|H_a| > |H_{ac}|$.

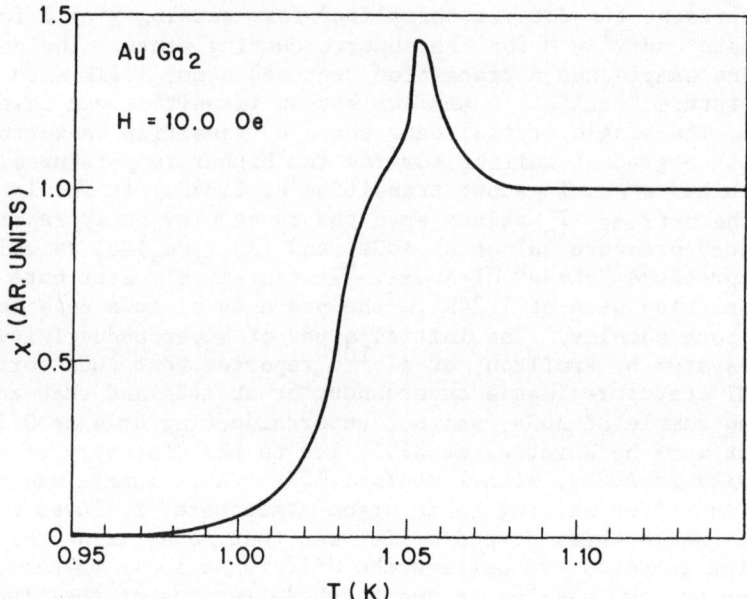

Figure 2. The in-phase component, χ', of the ac magnetic susceptibility of the single crystal "AuGa$_2$" sample as a function of temperature for an applied dc magnetic field of
10.0 Oe. In these normalized units $\chi' > 1$ corresponds to
positive values for $\left| dM/dH \right|_{H_0 \neq 0}$.

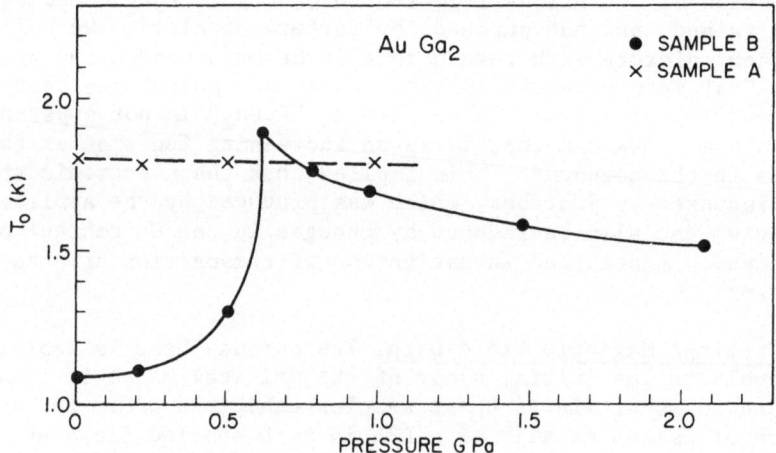

Figure 3. Zero applied field transition temperatures T_0 as a
function of pressure for the polycrystalline sample
(X) and single crystal sample (o).

Since H$_{ac}$ is usually of the order of 10^{-2}Oe one sees that the earth's
field can fulfill the above requirement. We stress this point because if one does not observe a DPE, it means that the intermediate
state portion of the sample magnetization curve is not reversible
and critical field values deduced from the χ_{ac} transitions may not
reflect properties of the bulk of the sample.

In Figure 2 we show the temperature dependence of the in-phase
component, χ', for an applied dc field of 10 Oe. These data are for
the single crystal and one sees the region of positive χ', (in our notation for $\chi' > 1$) i.e. a DPE. While the shape of the DPE is non-
ideal[7] it does indicate that one is probably observing "bulk" properties. The polycrystalline sample did not exhibit a DPE. From a
series of such "temperature sweeps" for different values of the
applied field one obtains initial slope values of - 125 Oe/deg for
a single crystal and -358 Oe/deg for the polycrystalline sample.

B) Pressure Effects

1) <u>Transition Temperature T$_o$</u>. Zero applied field transition
temperatures for the two AuGa$_2$ samples as functions of the applied
pressure[15] are presented in Figure 3. A comparison with the results of Schirber[5] and Smith, et al.[6] indicates that our single
crystal data is in excellent agreement with theirs. The nearly
pressure independent results (small decrease) for the polycrystalline sample is a new facet to the AuGa$_2$ story, but is suggestive
of the data of Smith, et al.[6] for 2.4 at % Pd in $(Au_{1-x}Pd_x)Ga_2$.
Because of these T$_o$ results we have concentrated our present study
on the single crystal sample.

2) <u>Critical Magnetic Fields.</u> Since the object of this study
was to detect changes with pressure in the initial slopes of the
critical magnetic field curves as well as to investigate the supercooling phenomenon reported by Smith, et al.[6], the data were taken
with just the AuGa$_2$ sample in the high pressure cell. Pressure
values quoted henceforth are deduced from the T$_o$ values of AuGa$_2$
and the data presented in Figure 3.

The region of interest corresponds to temperatures above 1.4K.
Thus, the pressure cell was immersed directly in the liquid helium
bath and data were obtained by holding the temperature constant and
changing the dc magnetic field at some fixed rate. Figure 4 presents some typical recorder traces of the in-phase components of
χ_{ac} as a function of magnetic field. To obtain values of H$_c$(T) we
extrapolate the descending high field portion of the DPE as shown.
From a series of pressure values, critical magnetic field data are
obtained as a function of pressure and are presented in Figure 5.
In Table I we present estimates of $(dH_c/dT)_{T_o}$.

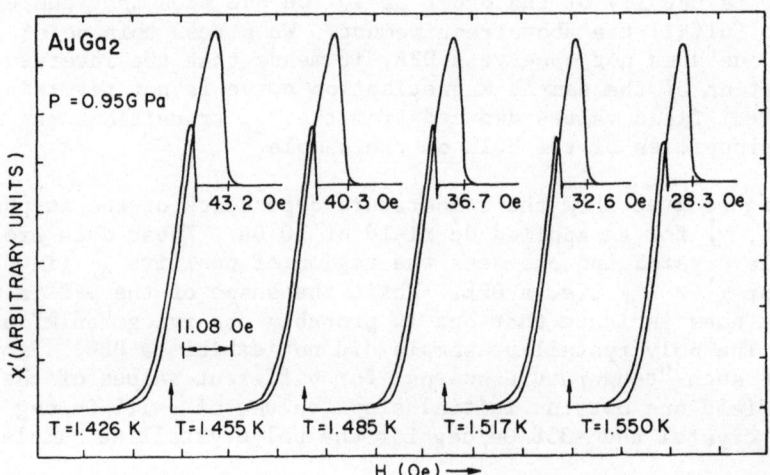

Figure 4.   XY recorder traces of χ' as a function of "swept" dc
            magnetic field for several values of the temperature.
            Note the offsets of the field axis.   The field scale
            is in millivolts with 10 mv = 11.08 Oe.

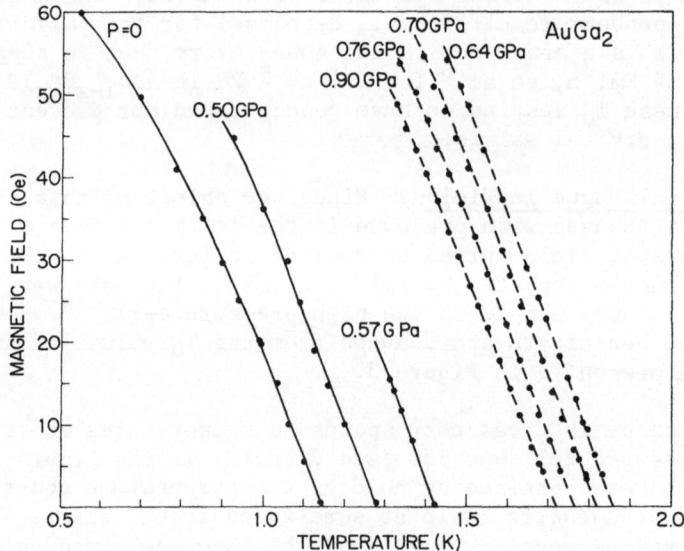

Figure 5.   Critical magnetic fields of the single crystal sample as
            a function of temperature for various values of the pres-
            sure.

Table I. Superconducting and Normal State Parameters of "AuGa$_2$".

T_o	$\left(\dfrac{dH}{dT}\right)_{T_o}$	V	γ^*	λ	$\gamma^*/1+\lambda$	P
(K)	(-0e/K)	(cm^3/mol)	(MJ/mol-K^2)		(MJ/mol-K^2)	(GPa)
1.14	125	33.4	6.93	0.442	4.87	0
1.28	130	33.2	7.45	0.452	5.13	0.50
1.42	142	33.2	8.88	0.462	6.08	0.57
1.72	126	33.0	7.00	0.483	4.72	0.90
1.76	134	33.0	7.82	0.486	5.27	0.76
1.82	136	33.1	8.19	0.490	5.50	0.70
1.86	144	33.1	9.05	0.493	6.06	0.64

Using McMillan's[16] formula for T_o with a Debye temperature of 196K[4] and assuming $\mu^* = 0.1$ we obtain the values of λ presented in Table I. Based on published values of the lattice constant (a_o = 6.075A) we calculate a molar volume, V, of 33.4 cm^3 and using Testardi's data[17] we correct the volume for the effects of pressure, Table I. Given the values of λ and volumes shown in Table I, we calculate a band structure density of states via the BCS expression[18]

$$\gamma^* = (V/18.0)\ (dH_c/dT)^2_{T_o}$$

and

$$N_{bs}(0) \approx \frac{\gamma^*}{1+\lambda} = \frac{V}{18.0(1+\lambda)}\left(\frac{dH_c}{dT}\right)^2_{T_o}$$

These values are tabulated in Table I and presented as a function of pressure in Figure 6. These data clearly support Schirber's contention of an electronic transition resulting in a sudden increase in the density of electronic states at the Fermi level. The percentage increase between P=0 and the peak T_o value (P≈0.6 GPa) is about 25%, very close to the 30% increase observed by Pd doping.[2] Such data suggest that the effects on the band structure produced by pressure or by Pd doping are closely related.

Our polycrystalline sample which exhibited a T_o = 1.8K at zero pressure, exhibited typical alloy behavior and its initial slope of -358 Oe/deg is not felt to be representative of the "bulk." Clearly an unreasonably large value of $\gamma*$ is obtained from such an initial slope.

3) <u>Supercooling.</u> One of the most intriguing aspects of the data of Smith, et al.[6] was the observation of supercooling, in zero applied field, for pressures in excess of 1.5 GPa. As mentioned in the introduction, one normally only sees "bulk" supercooling in pure stress free samples and it is associated with the absence of any nucleation sites in the sample.[9] Harris and Mapother[19] observed supercooling in very pure polycrystalline 0.080 in diameter Al wires; the supercooling amounted to 16% (assuming $H_{sc}/H_c = 0.84$) at P=0 and decreased to 5% at higher pressures; their maximum pressure was 0.05 GPa.

The mere appearance of supercooling at high pressures P>1.5GPa as reported by Smith, et al.[6] is perplexing, but the sharp "switching" from a non DPE, no supercooling response to a DPE, with supercooling response observed in this study (Figure 7) is very intriguing. The data in Figure 7 shows that 0.05 GPa "switches" the sample into a state with a new type of response. We have now narrowed the transition range of pressures to about 0.02 GPa.

It is clear from the data of Figure 3 that any inhomogeneity in the amount of stress induced in the sample, by any means, can lead to "phase" inhomogeneity in the 0.55 GPa to 0.6 GPa pressure range. This will produce a spread of T_0 values and result in a lack of a DPE. However, since we did not see supercooling at P=0, where presumably T_0 is not a sensitive function of stress, why did we not see supercooling at P≠0 along with the DPE? Thus we feel stress considerations per se cannot explain the sudden appearance of supercooling.

A second and equally puzzling aspect of the data is the magnitude of this "apparent" supercooling in Figure 4. Here we see $H_{sc}/H_c = 0.85$, a rather remarkable value for bulk samples in which no extraordinary care was taken with surface preparations. This aspect of the $AuGa_2$ problem is being investigated in detail.

IV. CONCLUSIONS

Critical magnetic field data yield information about the variation of the electronic density of states which support the pressure induced "electronic phase transition" in $AuGa_2$. The dependence of T_0 upon sample preparation suggests that $AuGa_2$ is not a line compound and that T_0 depends upon composition within the cubic fluorite phase field. Finally the observation of a large pressure

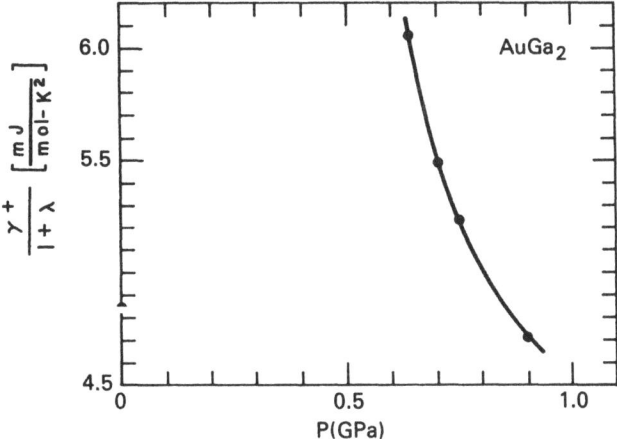

Figure 6. The quantity $\gamma*/(1+\lambda)$, which is proportional to the band structure density of electronic states at the Fermi level plotted as a function of pressure.

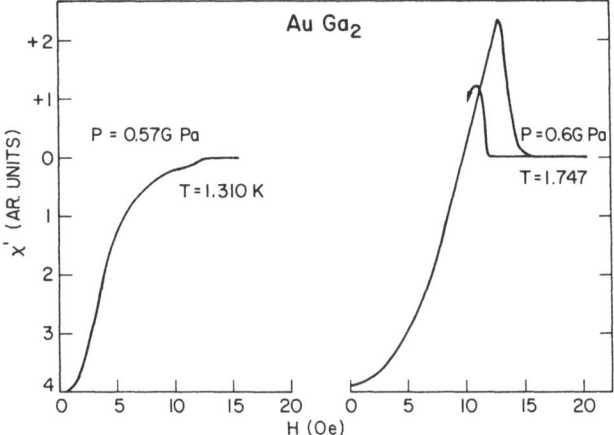

Figure 7. In-phase component χ' of the ac magnetic susceptibility as a function of magnetic field for two values of pressure corresponding to the quoted T_0 values. These data were selected at the given temperatures so as to produce nearly the same $H_c(T)$ values.

induced "apparent" supercooling suggests that there is either some subtle lattice transformation (role of defects?) occurring as a function of pressure or we are seeing some sort of "surface" super-conductivity being induced at high pressures. Since all measurements to date indicate the lack of any lattice transformation, one is faced with the problem of explaining how a purely "electronic transition" gives rise to ideal magnetic response. Thus the $AuGa_2$ dilemma continues. Recent work by Soulen et al.[20] includes magnetization curves and specific heat data for $AuGa_2$ which yield values for γ in agreement with the P=0 value presented in Table I.

ACKNOWLEDGMENTS

The authors are deeply indebted to Dr. Ch. J. Raub and Dr. R.J. Soulen, Jr. for their discussions and samples and to Dr. A.C. Lawson and Dr. F. Rachford for X-ray data on these samples. The cooperation of Professors B. Maple and B.T. Matthias is also acknowledged.

REFERENCES

1. H.T. Weaver, J.E. Schirber and A. Narath, Phys. Rev. B8, 5443, (1973) (References to earlier work are included).
2. V. Jaccarino, M. Weger, J.H. Wernick and A. Menth, Phys. Rev. Letts. 21, 1811 (1968).
3. J.E. Schirber, Phys. Rev. B6, 333 (1972).
4. J.H. Wernick, A. Menth, T.H. Geballe, G.Hull and J.P. Maita, J. Phys. Chem. Solids 30, 1949 (1969).
5. J.E. Schirber, Phys. Rev. Letts. 28, 1127 (1972).
6. T.F. Smith, R.N. Shelton, and J.E. Schirber, Phys. Rev. B8, 3479 (1973).
7. R.A. Hein and R.L. Falge, Jr., Phys. Rev. 123, 407 (1961).
8. Supercooling is most commonly observed by a hysteresis in the magnetization curve where the expulsion of flux, upon decreasing the magnetic field from above H_c, does not occur until $H_{sc} = H_c(1-x)$ where x is a few percent of unity.
9. See Schoenberg, D., "Superconductivity," Cambridge Univ. Press (1965).
10. D.C. Hamilton, Ch. J. Raub, B.T. Matthias, E. Corenzwit and G.W. Hull, Jr., J. Phys. Chem. Solids 26, 665 (1965).
11. J.T. Longo, P.A. Schroeder and D.J. Sellmyer, Phys. Rev. 182, 658 (1969).

12. M.E. Straumanis and K.S. Chopra, Z. Physik. Chem. Neue Folge 42, 344 (1964).
13. G.C. Carter, I.D. Weisman, L.M. Bennett and R.E. Watson, Phys. Rev. B5, 3621 (1972).
14. Ch. J. Raub, R. Soulen, R.A. Hein and J. Willis (to be published).
15. The authors are indebted to Dr. R. Shelton for his help in making these measurements.
16. W.L. McMillan, Phys. Rev. 169, 331 (1968).
17. L.R. Testardi, Phys. Rev. B1, 4851 (1970).
18. J. Bardeen, L.N. Cooper and J.R. Schrieffer, Phys. Rev. 108, 1175 (1957).
19. E.P. Harris and D.E. Mapother, Phys. Rev. 165, 522 (1968).
20. R.J. Soulen, D.B. Utton and J.H. Caldwell, Bull. A.P.S. 22, #3, 403 (1977).

QUESTIONS AND COMMENTS

T.F. Smith: I seem to recall, without remembering any of the details, that there's been some work on the anharmonicity of AuGa₂ from X-ray studies. Have you thought that it is caused by some mode-softening effect which is altering as a function of temperatures?

R.A. Hein: Yes, there is a report by Hollenberg and Batterman (Phys. Rev. B, 10, 2148 (1974)) detailing an X-ray study of the anharmonicity in AuGa₂. No, we have not considered this aspect in any detail.

SUPERCONDUCTIVITY OF BCC BARIUM UNDER PRESSURE

J. Wittig and C. Probst*

Institut für Festkörperforschung, Kernforschungsanlage Jülich

D-5170 Jülich, West-Germany

ABSTRACT

The bcc phase of barium is a superconductor. T_c increases steeply with pressure from 0.06 K at 37 kbar to 0.5 K at 48 kbar. A fit of McMillan's expression for T_c suggests a T_c already in the millikelvin range at 30 kbar. Phonon softening as inferred from the decreasing melting curve is insufficient to account for the rise of T_c with pressure. A He^3/He^4 dilution refrigerator is described which allows high pressure experiments at temperatures down to 0.05 K.

Three high pressure modifications of barium are known to be superconducting.[1,2] In the phases II and III, T_c increases from ≈1 K at 55 kbar to ≈ 3 K just before the transformation to the phase BaIV, which possesses a T_c of approximately 5.5 K.[2,3] The room-temperature resistivity of BaII and BaIII increases also considerably under pressure.[1] This is apparently another indication of the steady increase of the electron-phonon interaction with pressure.[4] No superconductivity has been so far observed in the low-pressure bcc phase (BaI). Interestingly enough, a strong increase of the room-temperature resistivity occurs also in BaI in the pre-transitional range of the transformation to BaII.[1] The lower

*New address: Zentralinstitut für Tieftemperaturforschung der Bayerischen Akademie der Wissenschaften, D-8046 Garching.

straight line in Fig. 1 represents our resistance data for one particular pressure cell. The data are normalized to R_{min}, the minimum
of R(P) occurring near 10 kbar. Another interesting property of Ba
is the decreasing melting line above 15 kbar and the low melting
temperatures in the phases II and III.[5,6] Relying on the Lindemann
formula, we would like to suggest that the decreasing melting line
signals a pressure-induced softening of the average phonon frequencies. Summarizing, the pressure range just below the BaI-II phase
boundary seemed the most probable region in which superconductivity
in BaI could be expected to occur.

We have searched for superconductivity in BaI under pressure.
As indicated by the arrow in Fig. 2, the sample remained normal
down to a lowest temperature of 0.05 K at 34 kbar. Between 37 and
48 kbar superconducting transitions were observed with T_c varying
over an order-of-magnitude between 0.06 and 0.5 K. The horizontal
bars show the pressure distribution as inferred from the width of
the superconducting transition of the Pb-manometer.[7] The numbers
indicate the sequential order of the experiments. It is seen that
T_c depends reversibly on pressure.

In the experiment No. 8 (which is missing from Fig. 2) the
pressure was increased to 51 kbar at room-temperature. The sample
transformed partially to BaII on cooling as seen from a hysteresis
loop in the R-T characteristic. This mixed-phase sample showed a
broad superconducting transition between 0.5 K and 0.9 K, the
latter being the T_c of BaII at the I-II equilibrium pressure. In
this experiment, the residual resistance was markedly higher (Fig.
1, upper curve). We attribute this to an additional scattering
from phase boundaries. The straight-line behavior at lower pressures therefore proves that the sample was single-phase BaI up to
48 kbar. The pressure distribution turned out to be more inhomo-
geneous upon release of pressure (Fig.2, data point labelled 9).
Hence the width of the Ba-transition is broader than in the other
experiments.

The pressure technique is essentially standard.[7] Fig. 3 shows
the compact mechanical press which has been developed for the
present experiment. Six screws generate a force of the order of
10 tons. The total force acting on the Bridgman anvils is measured
by strain gages mounted on the outer surface of the press body vis-
à-vis of the anvils. Each screw compresses a set of 7 disk springs.
They effectively reduce the spring constant of the whole system
thereby minimizing the changes of force due to differential contraction. The piston pusher is machined to a close fit in the
press body and centers the lower anvil with respect to the upper
one. Other features are readily clear from Fig. 3. A few details
have been omitted for clarity. Those include Speer carbon resis-

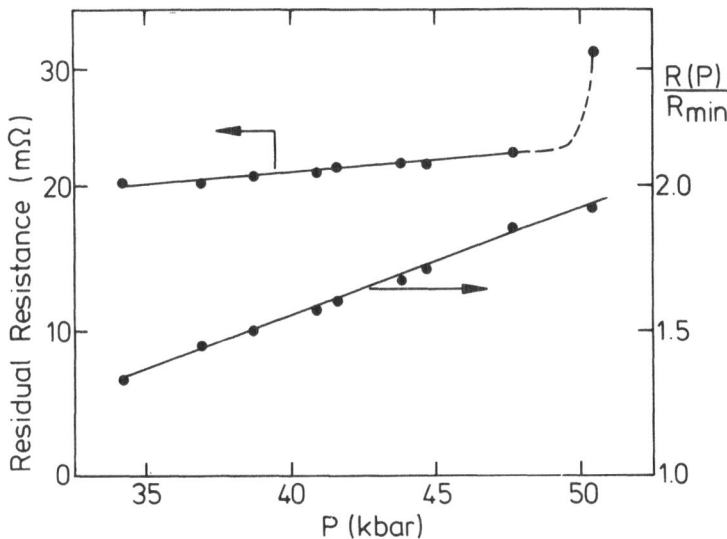

Figure 1:   Room-temperature resistance of bcc Ba vs pressure (lower
            line). The residual resistance (upper curve) shows a
            marked increase at the onset of the BaI → II transition.

tance thermometers attached to each binding ring and other resistance thermometers in various locations on the press body. The
piston pusher and, in turn, the lower anvil is thermally anchored
to the mixing chamber by a flexible connection made from high conductivity copper wire. The reason for this will be given below.
Many ventholes allow the thorough removal of the He-gas which circulates freely around the press during the initial cool-down. The
total mass is of the order of 6 kg. All moving parts are lubri-
cated with PTFE-spray and MoS_2 powder. As indicated in Fig. 3,
the press is bolted to the copper mixing chamber of a He^3/He^4 dilution refrigerator.

Fig. 4 shows the dilution refrigerator. Major parts are identi-
fied in the figure caption. One has access to the press after re-
moving the lower part of the vacuum mantle and the 3 radiation
shields. The 4.2 K - radiation shield is In-sealed to the bottom
of the 4.2 K - liquid He reservoir. The He^3 dilution circuit, in
particular the flow impedances and the heat exchanger, have been
designed following the prescriptions of Anderson.[8] No problems
have been encountered with the operation of the dilution process.

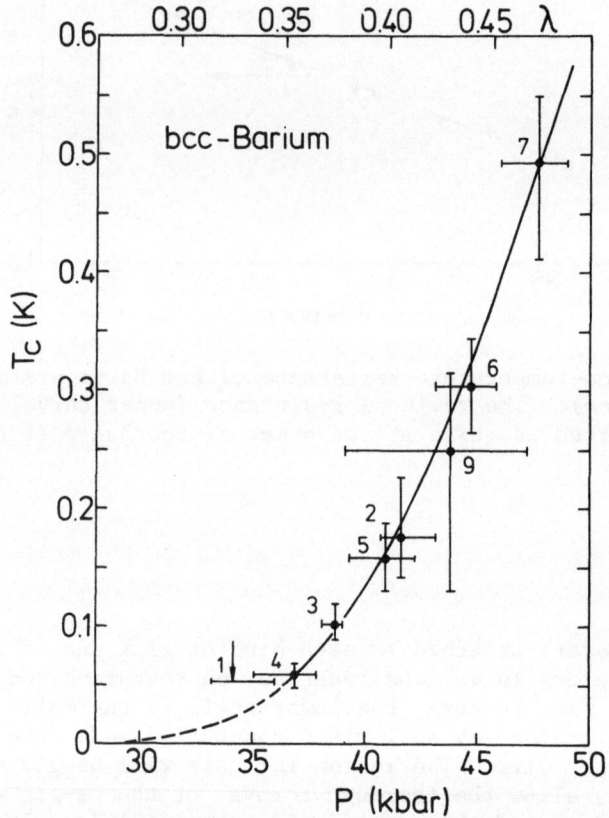

Figure 2: T_c vs pressure for one particular pressure cell. The
bars indicate transition width and pressure inhomogeneity.
Numbers denote sequence in which data were taken.   The
line is a fit of McMillan's formula for T_c as a function
of λ (top scale).

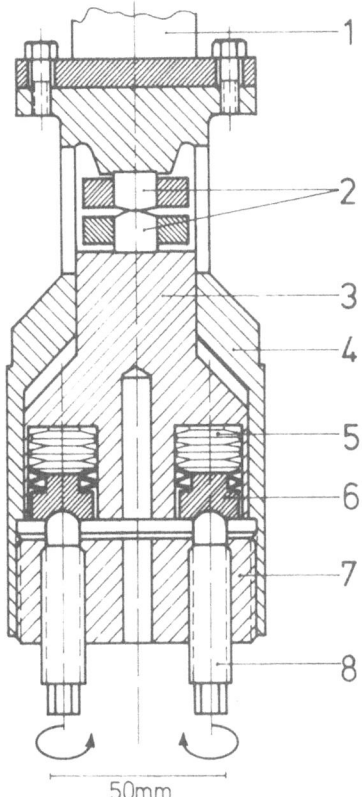

50mm

Figure 3: Mechanical 10 t-press for high-pressure low-temperature
experiments. 1 Mixing chamber of refrigerator shown in
Fig. 4. 2 High pressure anvils with binding rings.
3 Piston pusher. 4 Press body. 5 Set of disk
springs. 6 Hemispherical bearing. 7 Demountable end
piece. 8 Screw with hexagonal head. With the exception of the anvils, all parts are made from hardened 2%
BeCu alloy.

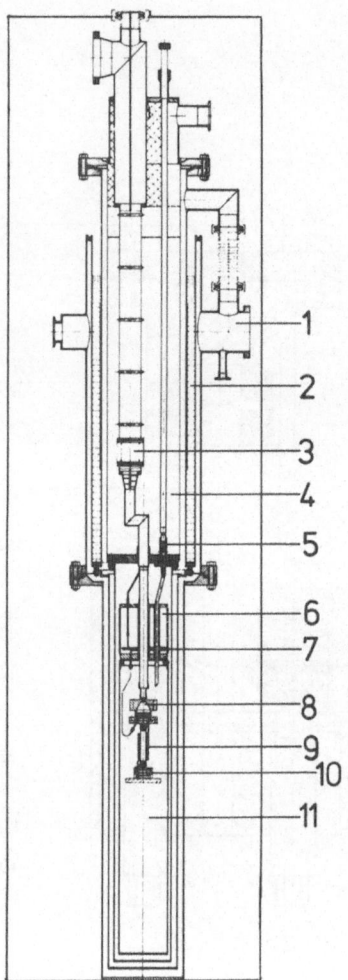

Figure 4: He^3/He^4 dilution refrigerator for high pressure experiments. 1 Main high vacuum connection. 2 Liquid
nitrogen open ring tank. 3 He^3 pumping line with
radiation baffle. 4 Liquid He^{4}reservoir. 5 Cold
valve connecting spaces 4 and 11, operated from top.
6 Pumped He^4 bath at 1 K. 7 He^3 condensor. 8 Still.
9 Concentric tube heat exchanger. 10 Copper mixing
chamber with flange. 11 Experimental working space
for press shown in Fig. 3. The cryostat is constructed
from austenitic stainless steel and OFHC copper. Not
shown are the filling and vent lines of the 1 K tank as
well as the pumping line for the vacuum space 11 and a
separate Mu-metal shield for screening the earth magnetic field.

A typical experiment proceeds as follows. After adjusting the pressure at room-temperature the cryostat is closed. Precooling to liquid nitrogen temperature conveniently takes place over night with liquid nitrogen in tank 2 and 4 (cf. Fig. 4). Space 11 contains He exchange gas and the cold valve 5 is closed. After removal of the liquid nitrogen from tank 4, liquid He[4] is transferred to tank 4. The cold gas is forced through the open valve 5 and circulates freely around the press for an efficient heat exchange. It leaves space 11 through a hole in the bottom of the 1 K - radiation shield and a vent tube (not shown in Fig. 4). If the press is at approximately 10 K, the cold valve is closed. The storage tank 4 is filled, while space 11 is pumped. Eight liters of liquid He are sufficient for cooling down the entire refrigerator and for at least 12 hours of continuous operation.

The pressure cell has been described previously[7]. The pressure is determined from the superconducting transition temperature of a Pb sample. We rely on a previous fixed point calibration.[7] Concerning the accuracy of this pressure scale we just mention that a Ba sample at a mean pressure of 51 kbar (see above) did not transform to BaII at room-temperature in agreement with the at present accepted transition pressure of 55 ± 1 kbar.[9] However, the transformation began on cooling below ≃ 200 K. This is in satisfying agreement with the small positive slope of the BaI-II phase boundary. Experimentally, the Pb-transition is measured by maintaining a small amount of exchange gas in space 11 and regulating the temperature of the press with a heater which is mounted on the mixing chamber.

After pumping high vacuum in space 11 by using a diffusion pump as well as a highly efficient internal charcoal absorption pump, the dilution refrigeration run is started. It takes approximately 2 hours to reach a lowest temperature of 60 mK. In a single shot (with the He[3] return stopped) a lowest temperature of 50 mK can be attained.

It has been found that the piston pusher (cf. Fig.3) is in fairly poor heat contact with the rest of the press (even under high load). It turned out that it becomes virtually thermally disconnected around 140 mK. We have remedied this problem by providing a direct connection to the mixing chamber made from high conductivity copper wires. Although this thermal link has not yet been optimized, the temperature of the upper and lower anvil now agree to within ≃ 5 mK at the lowest temperature in the steady state. The thermometers have been calibrated with CMN. We now return to a discussion of the T_c data for BaI.

For a tentative comparison of the results of Fig. 2 with theory[10] we assume that the electron-phonon coupling parameter λ increases linearly in the pressure range of interest: $\lambda = a + bP$. The linear rise of the room-temperature resistance (Fig. 1) may hint at such a simple relation, although it must be stressed that the actual connection of λ with transport properties is more complicated[11], [12]. The curve in Fig. 2 is a fit of McMillan's formula for $T_c = T_c(\lambda)$ through the data points 4 and 7, thereby fixing the parameters a and b. The Debye temperature was taken to be 110 K and the Coulomb-pseudopotential $\mu^* = 0.13$. Both quantities were assumed to be independent of pressure. The calculated T_c at 30 kbar is 2.4 mK.

An estimate shows that the anticipated decrease of the average phonon frequencies toward the BaI-II phase transformation is insufficient to explain the observed increase of T_c. Employing the Lindemann relation, the melting curve data lead to $d(\ln T_M) = d \ln <\omega^2>_M \simeq -0.07$ for our pressure interval between 37 and 48 kbar. In the absence of better knowledge, we take the freedom to set $<\omega^2>_M = <\omega^2>$ with $<\omega^2>$ having the usual meaning[10]. This effect would lead to an increase of λ by 0.03. However, λ changes from 0.35 to 0.47 (Fig. 2, upper scale). Thus, the electronic quantities, i.e. the density of states at the Fermi level and the electron-phonon matrix elements in the expression $\lambda = N(o) <I^2>/M<\omega^2>$ must appreciably increase with pressure.

From the present and previous data[2] it is now clear that the T_c of Ba can be raised by more than two orders of magnitude by applying pressure. Crystallographic phase transformations introduce only minor discontinuous changes of T_c whose directions however follow the general trend[2]. This simple behavior should have a rather general origin. Some final remarks therefore pertain to the possible overall electronic band structure of Ba under pressure. From their band structure calculations, Vasvari et al.[13] concluded that the 5d component in the conduction electron wave functions increases strongly under pressure. In other words, Ba is gradually turned into a 5d-transition metal, the d character resulting from s-d hybridization at the Fermi level. However, we think another effect is even more important. We would like to suggest that the unusually low melting temperature of Ba at high pressure[5], [6] points to the presence of considerable 4f character of the Bloch waves. This is an idea which has been previously discussed in the literature in connection with the low melting temperature of the neighboring element La [14-17] and, also in the particularly striking case of Ce at high pressure[17]. We regard it to be a rather well-founded concept now that unusually soft phonon modes are a characteristic property of f-hybridized metals[18] (although, admittedly, some gaps of knowledge do exist).

The spectroscopic properties of the neutral and singly-ionized Ba atom lend some support to the hypothesis that Ba metal has incipient rare earth character at high pressure. In the latter case of Ba^+, the binding energy of the outer electron in the excited 4f state shows a marked increase over the "hydrogen value" Ry $2^2/4^2 =$

27,432 cm^{-1} for the nuclear charge 2 and the principal quantum number 4. The reported value for Ba$^+$ is 32,428 cm^{-1} [19]. This has been interpreted[20] in terms of the 4f wave function being partially trapped by the inner valley of the double well potential which is due to the centrifugal barrier and is characteristic for f electrons. The corresponding value for the neutral Ba atom is 7,430 cm^{-1} [19]. This must be compared with the hydrogen value Ry/4^2 = 6,858 cm^{-1} and, in particular, with the observed binding energies[19] in the other earth alkaline metals (Ca: 7,135 cm^{-1}, Sr: 7,175 cm^{-1}). The effect still seems to exist although it is much more marginal than for Ba$^+$.

ACKNOWLEDGMENTS

    Discussions with G. Eilenberger, D. Rainer and H. Wühl are gratefully acknowledged. We would like to thank W. Bünten for valuable technical assistance.

REFERENCES

1. J. Wittig and B.T. Matthias, Phy. Rev. Lett. 22, 634 (1969).
2. A.R. Moodenbaugh and J. Wittig, J. Low Temp. Phys. 10, 203 (1973).
3. The onset of superconducting transitions has been repeatedly observed at 1.3 K and 55 kbar (unpublished results) pointing to a T$_c$ of approximately 1 K for BaII at this pressure.
4. A large rise of the room-temperature resistance occurs at the III-IV transformation [e.g. ref. 1] . It can however not be attributed to an increase of the electron-phonon interaction. It is caused by an anomalous increase of the residual resistivity in the phase BaIV (unpublished data).
5. A. Jayaraman, W. Klement, Jr., and G.C. Kennedy, Phys. Rev. Lett. 10, 387 (1963).
6. J.P. Bastide and C. Susse, High Temperatures-High Pressures 2, 237 (1970).
7. A. Eichler and J. Wittig, Z. Angew. Physik 25, 319 (1968).
8. A.C. Anderson, Rev. Sci. Instrum. 41, 1446 (1970).
9. D.L. Decker, W.A. Bassett, L. Merrill, H.T. Hall, and J.D. Barnett, J. Phys. Chem. Ref. Data 1, 773 (1972).
10. W.L. McMillan, Phys. Rev. 167, 10 (1968).
11. B. Chakraborty, W.E. Pickett, and P.B. Allen, Phys. Rev. B14, 3227 (1976).
12. B.L. Gyorffy, in Superconductivity in d- and f-Band Metals, edited by D.H. Douglass, Jr. (Plenum Press, N.Y., 1976), p. 29.
13. B. Vasvari, A.O.E. Animalu, and V. Heine, Phys. Rev. 154, 535 (1967).
14. D.C. Hamilton, Amer. J. Phys. 33, 637 (1965).
15. B.T. Matthias, W.H. Zachariasen, G.W. Webb, and J.J. Engelhardt,

Phys. Rev. Rev. Lett. $\underline{18}$, 781 (1967).
16. K.A. Gschneidner, Jr., J. Less-Common Metals $\underline{25}$, 405 (1971).
17. J. Wittig, Comments Solid State Physics VI, 13 (1974).
18. C. Probst and J. Wittig, in Handbook of the Physics and Chemistry of Rare Earths, edited by K.A. Gschneidner, Jr., and LeRoy Eyring (North Holland Publishing Co., Amsterdam) Chapter 10, in print.
19. C.E. Moore, Atomic Energy Levels (National Bureau of Standards, Circular No. 499, U.S. Government Printing Office, Washington, D.C., 1950), Vols. 2 and 3.
20. J.P. Connerade and M.W.D. Mansfield, Proc. R. Soc. Lond. A $\underline{346}$, 465 (1975) and references therein.

SUPERCONDUCTING BEHAVIOUR OF Nb_3Ge THIN FILMS UNDER HYDROSTATIC PRESSURE [+]

H.F. Braun [++], E.N. Haeussler, W.W. Sattler and E.J. Saur

Institute of Applied Physics, University of Giessen

Giessen, West Germany

INTRODUCTION

Type II superconductors are characterized by high upper criti-cal induction values and high current carrying capacities, which are caused by pinning of the flux-lines at structural inhomogenei-ties of the samples. Fietz and Webb[1] and recently Kramer [2] have shown that the dependence of the pinning force densities on tempera-ture and magnetic induction is governed by "scaling laws". This means that the pinning force densities due to a particular pinning mechanism follow a specific function of the reduced magnetic induc-tion the form of which is not changed with temperature. The mag-netic induction is reduced with respect to the measured upper crit-ical induction which will depend on temperature and the orientation of the sample surface relative to the magnetic field.

In this work for the first time investigations under hydro-static pressure at different temperatures shall be carried out in order to get informations on the behaviour of the superconducting properties under these conditions. As suitable materials cosput-tered and coevaporated thin films of Nb_3Ge shall be used.

[+]Work supported by the Deutsche Forschungsgemeinschaft

[++]Present address: Institute for Pure and Applied Physical Sciences, University of California at San Diego, La Jolla, California

EXPERIMENTAL

Preparation and Characterization of Samples

Nb$_3$Ge films have been deposited on sapphire substrates by simultaneous dc cosputtering in a pure argon atmosphere similar to the
method first reported by Gavaler[3] and by thermal multisource coevap-
oration similar to the method described by Tarutani et al.[4].

All measurements under pressure were done on one selected
Nb$_3$Ge cosputtered film of size 12 x 2,4 mm^2 and one Nb$_3$Ge coevaporated film of equal size. The thickness of the evaporated film was
224 nm and the thickness of the sputtered Nb$_3$Ge film was about
600 nm on the average with a gradient in thickness along the sample.

The crystal structure of the films was examined by X-ray diffractometer analysis. The selected samples prepared by either
method have A15-structure with a lattice parameter of 5.14 Å. In
these films no second phase could be detected within the limit of
the diffractometer ($\approx$ 5%).

The midpoint value of the transition curve was 21.2 K for the
cosputtered and 20.6 K for the coevaporated sample.

Pressure Technique

A clamp device was used for pressure generation. The pressure
applied at room temperature by a standard press can be locked, permitting the removal of the high pressure bomb from the press and its
transfer to a cryostate.

The pressure was contained in a cylinder fabricated from hardened CuBe alloy (2 weight % Be) with an o.d. of 35 mm and an i.d. of
about 7 mm. The moveable piston was made of a high quality hardened
tool steel. The sample was mounted on a plug made of CuBe, through
which the electrical leads were fed in. The pressure bomb was filled
with a 1:1 mixture of n-pentane and iso-pentane as hydrostatic pressure transmitting medium at low temperatures. It can be assumed[5]
that at low temperatures the deviation from hydrostatic conditions
around the sample is only slight and does not influence the determination of transition temperature or critical current noticeably.

The pressure at low temperature was determined by the resistance of a manganin gauge mounted as a coil around the sample. This
pressure bomb allows to get pressures up to 16 kbar at liquid hydrogen temperature.

Measurements

The superconducting properties of the samples were measured in a liquid hydrogen bath cryostate. No special effort was made to cool the pressure bomb slowly, since no difference could be observed between transitions measured after cooling the sample to liquid nitrogen temperature over several hours or several minutes. Thus normal procedure adopted was to precool the clamp by direct immersion in liquid nitrogen and then transfer it to a Dewar which will be filled with liquid hydrogen. The total cooling time from room temperature to 20 K was usually about 30 minutes.

For temperatures above normal boiling point of hydrogen the cryostate could be pressurized up to 2 bar by heating the hydrogen bath. The liquid hydrogen is well mixed by the ascending boiling bubbles and thus temperature gradients are avoided.

Any temperature between 24 K and the triple point of hydrogen (14 K) can be adjusted by pumping the bath and setting the pressure value by a manostate. The temperature is determined from the hydrogen vapour pressure using the IPTS 1968 vapour pressure-temperature relation given by Durieux[6].

Absolute temperature determination is believed to be accurate to ± 0.1 K, the reproducibility of temperature is about 0.02 K.

Standard four probe resistance measurements have been used for the determination of transition temperatures, critical currents and upper critical inductions. The magnetic field is oriented either normal ($B_\perp$) or parallel ($B_{\parallel}$) to the film surface in a plane normal to the current direction.

Critical currents were measured at constant field and temperature by increasing the current until a voltage of 1 μV along the sample has developed.

A sample resistance of 1% of the normal value on the magnetic transition curve was defined as criterion for the upper critical induction.

RESULTS

Pressure Effect on Transition Temperature

For comparison of transition temperatures under pressure both samples were mounted on one plug and pressurized together in four steps up to 13.6 kbar. Fig. 1 shows the decrease of transition temperature with applied hydrostatic pressure for both samples. The results are two different pressure coefficients for cosputtered

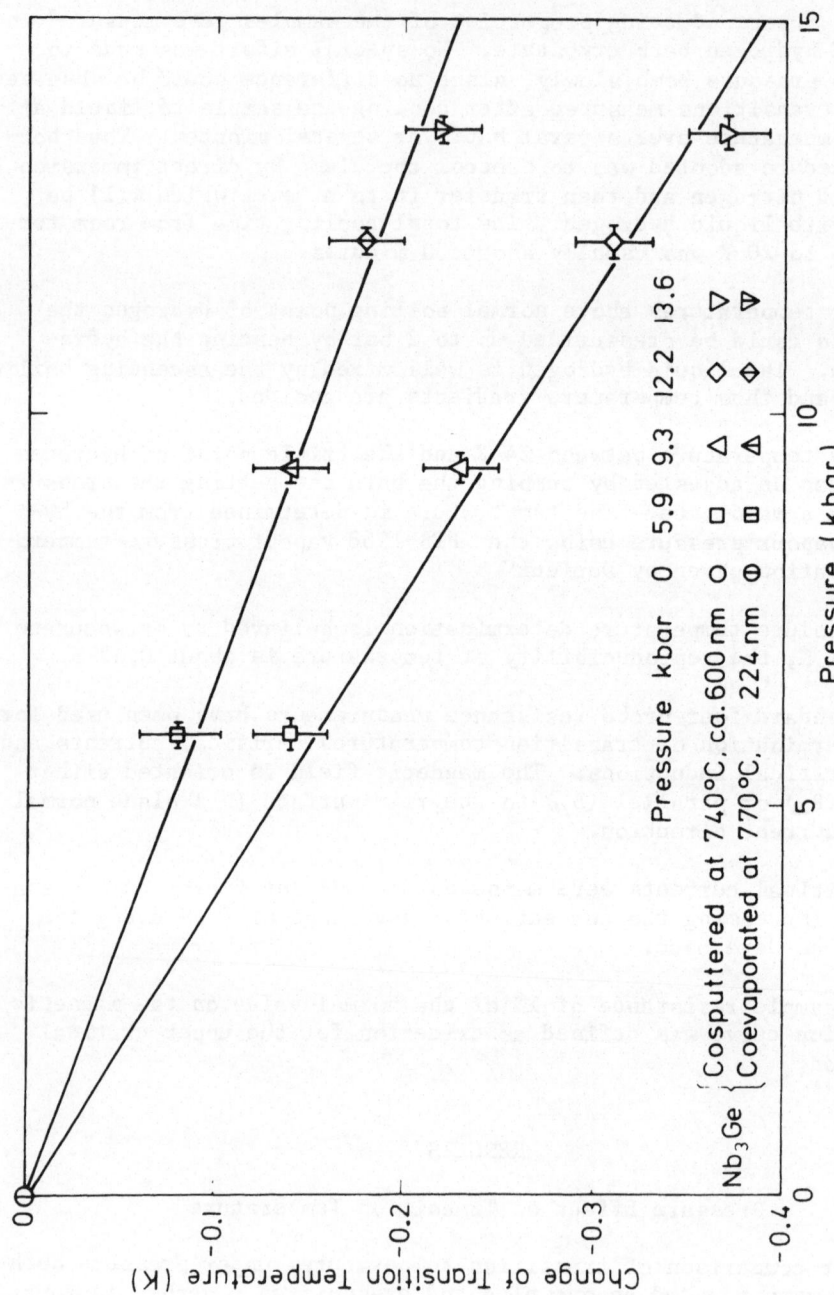

Figure 1:   Change of transition temperature with pressure for cosputtered and coevaporated Nb_3Ge.

$\left(\frac{dT_c}{dp} = -2.6 \cdot 10^5 \frac{K}{bar} \right)$ and coevaporated $\left(\frac{dT_c}{dp} = -1.5 \cdot 10^5 \frac{K}{bar} \right)$ films.

These different coefficients may be explained by the different microstructure of the films. Sputtered films show a columnar growth of crystallites normal to the substrate surface[7], whereas no special preferred orientation of the crystallites was observed with the evaporated films[8]. At the interface of sapphire substrate and film a certain inhomogeneity of pressure may arise which will influence the superconducting properties of both types of layers in different ways. In subsequent pressure cycles the same variations of the transition temperatures were reproduced. In order of magnitude the pressure coefficients agree with those of other A15 compounds.

Angular Dependence of Upper Critical Induction Under Pressure

The different shapes of the crystallites in the two types of thin Nb₃Ge films influence the orientation dependence of the critical current densities and upper critical fields. In the evaporated film (Fig. 2) the angular dependence of the upper critical induction is in fairly good accordance with the theory of Tilley and Ward[9]. The upper critical inductions of the sputtered film (Fig. 3) do not follow this law as can be expected from the microstructure of this sample[8].

Without changing the shape of the curves, pressure lowers the upper critical induction values for both samples. This decrease of the upper critical induction at constant temperature is about proportional to the increase of the applied pressure and therefore proportional to the degradation of the transition temperature under pressure.

Quenching Curves Under Pressure

Critical currents were measured at constant fields and temperatures with pressure as parameter in the two distinguished field orientations normal and parallel to the sample surface.

For the cosputtered sample the quenching curves up to 12.3 kbar are given in Fig. 4 at three different temperatures. In this sample the critical current is lower for parallel than for normal field orientation, whereas the opposite is true for results of the co-evaporated sample given in Fig. 5 up to 14.5 kbar at three different temperatures. In general the critical current decreases with increasing pressure in Nb₃Ge films. Again there is a stronger effect of pressure on the critical current in the sputtered Nb₃Ge film than in the evaporated one.

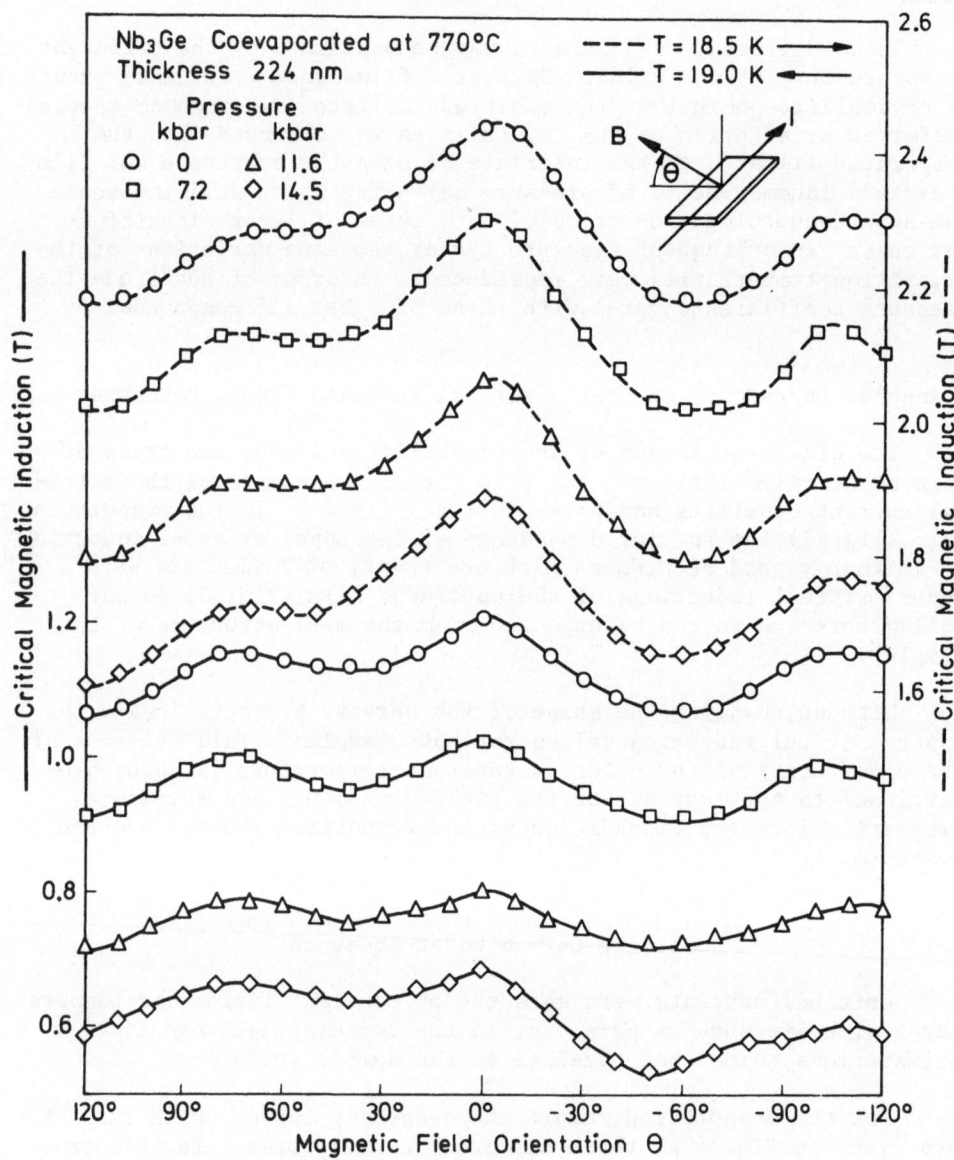

Figure 2:   Upper critical induction for coevaporated Nb$_3$Ge versus magnetic field orientation.

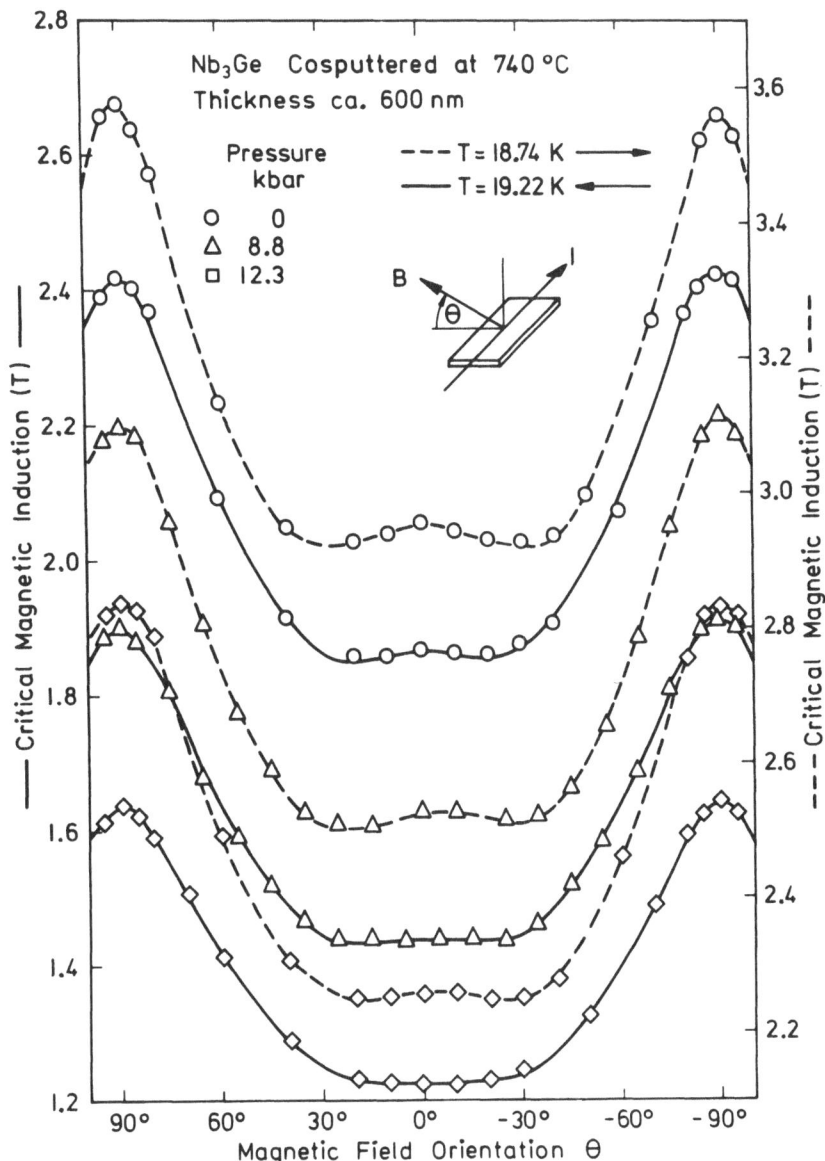

Figure 3: Upper critical induction for cosputtered Nb₃Ge versus magnetic field orientation.

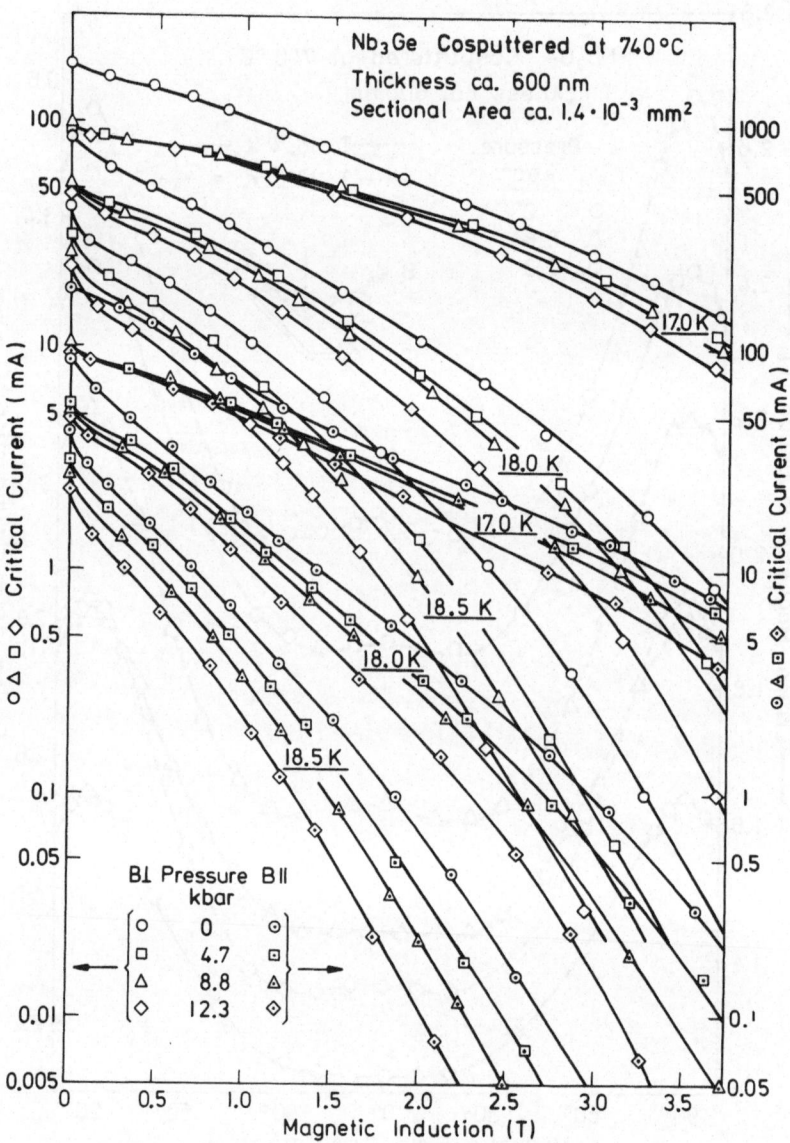

Figure 4:   Quenching curves of a cosputtered Nb₃ Ge film for paral-
            lel and normal field orientation under pressure.

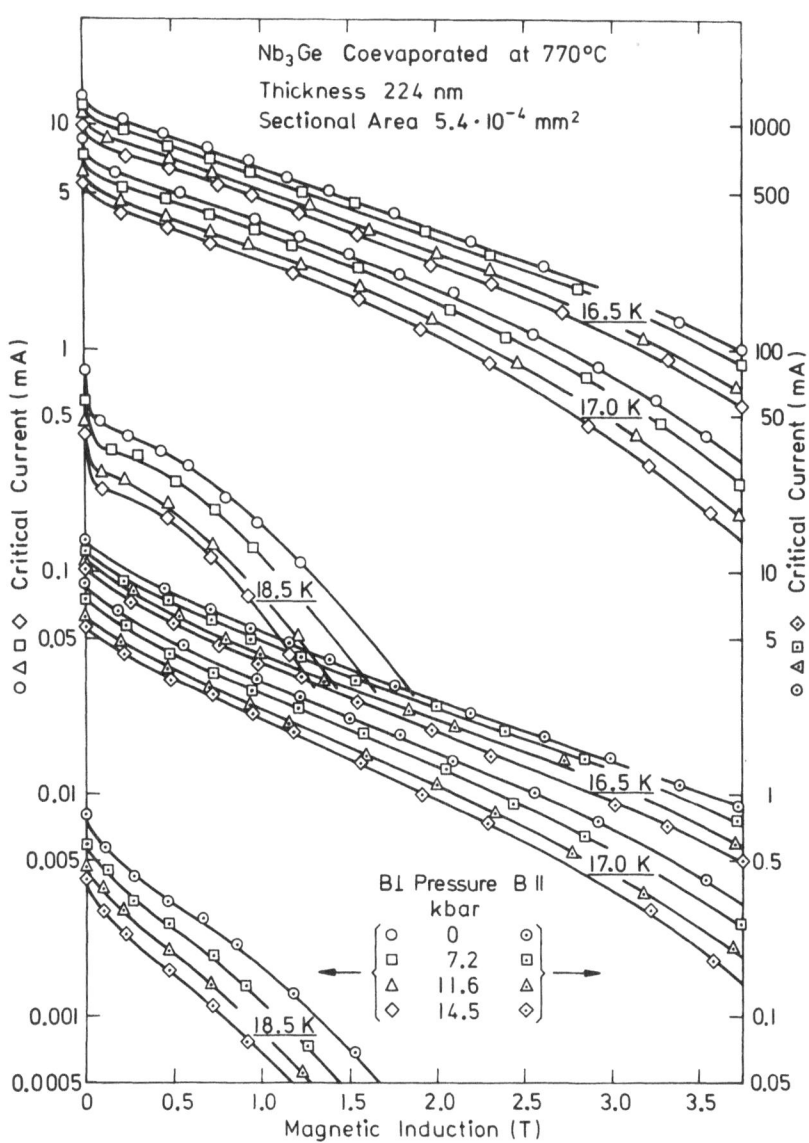

Figure 5: Quenching curves of a coevaporated Nb$_3$Ge film for parallel and normal field orientation under pressure.

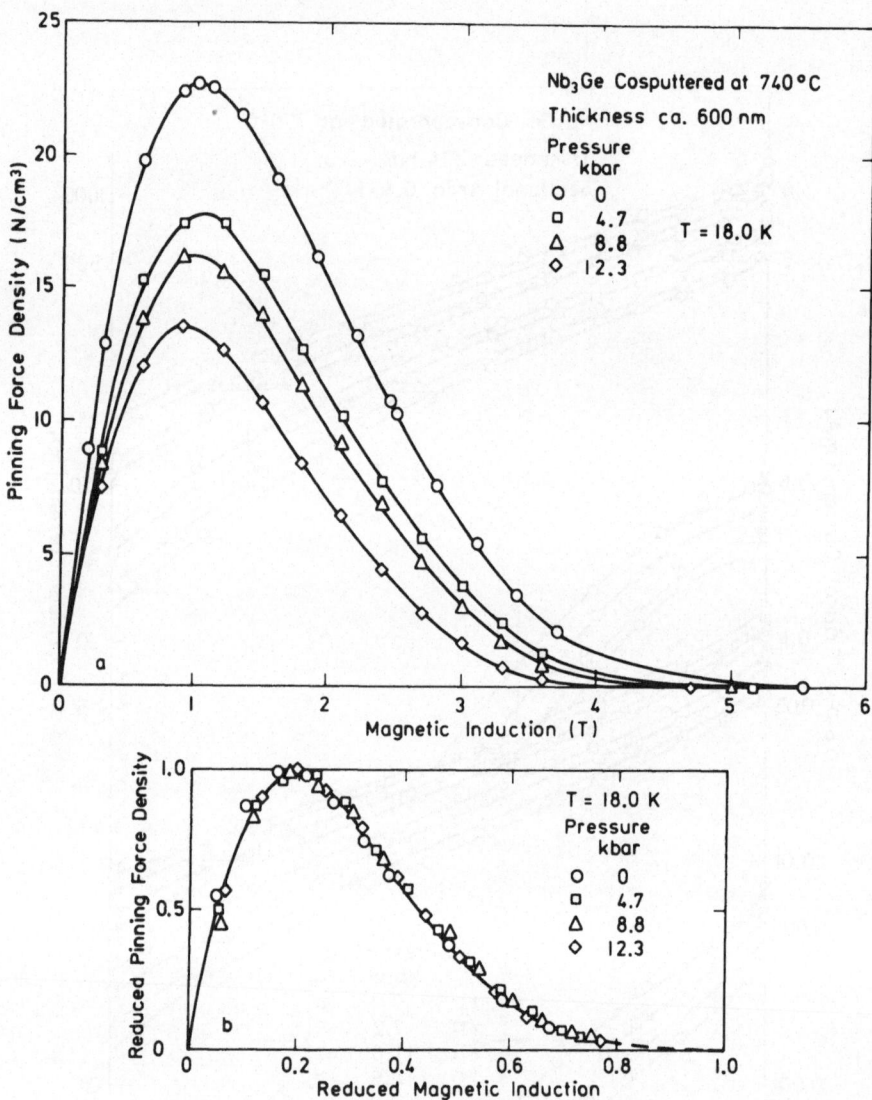

Figure 6. a. Pinning force density versus magnetic induction at
              normal field orientation and
           b. reduced pinning force density (F_p/F_{pmax}) versus
              reduced magnetic induction ($B_\perp/B_{c2}$), for cosputtered
              Nb_3Ge.

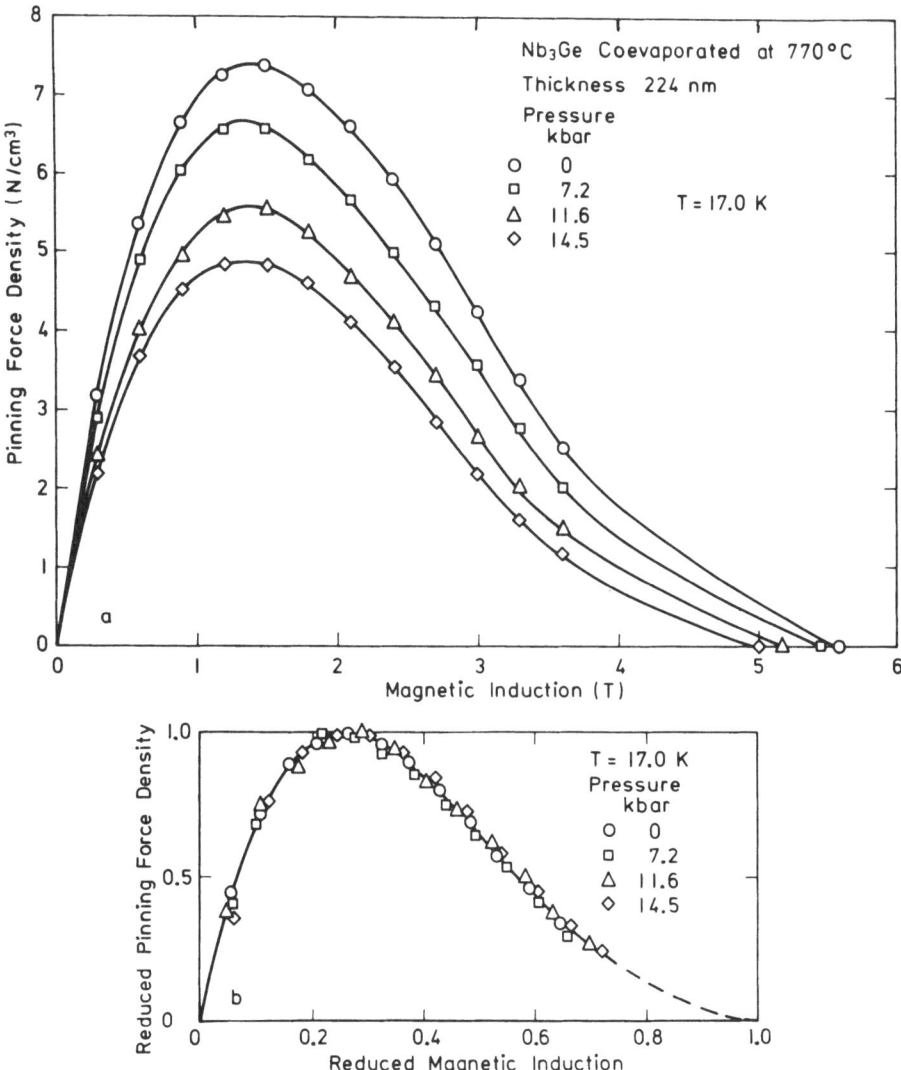

Figure 7.   a. Pinning force density versus magnetic induction at
               normal field orientation and
            b. reduced pinning force density (F_p/F_{pmax}) versus
               reduced magnetic induction ($B_\perp/B_{c2}$) for coevaporated
               Nb₃Ge.

Pinning Force Density and Scaling Laws

The pinning force density is given by the product of the critical current density and the magnetic induction. Fietz and Webb[1]
and Kramer[2] have given scaling laws for the temperature and field
dependence of the pinning forces. At constant temperature, the
pinning force density reduced to its maximum value is a function
of the reduced magnetic induction the form of which is independent
of temperature. The maximum value scales as a power of the upper
critical induction.

Braun and Winkelmann[10] have shown that for cosputtered Nb_3Ge
films the reduced pinning force density follows a function

$$f(b) = b^{1/2}(1-b)^2 \qquad\qquad (1)$$

for both normal and parallel field orientation at temperatures
between 19 K and 4.2 K where b means the reduced induction. In a
coevaporated Nb_3Ga film they found

$$f(b) = b(1-b)^2 \qquad\qquad (2)$$

for parallel field orientation.

The quenching curves at different pressures and constant temperature shall be evaluated in a similar way as for different temperatures. The pinning force density versus induction for the
sputtered film as calculated from the quenching curves of Fig. 4
is shown in Fig. 6. With increased pressures, the pinning force
density and the upper critical induction B_{c2} decrease and the maxima shift to lower inductions (Fig. 6a). In the reduced plot
(Fig. 6b.) all experimental points for normal field orientation
coincide with one single curve which is given by equation (1). The
results for different temperatures and for parallel field orientation fit the same curve. A corresponding behavior has been found
for the coevaporated Nb_3Ge film. The results are given in Figs.
7a and 7b. For parallel field orientation, equation (1) seems to
give a fairly good fit, whereas for normal orientation the experimental points are better fitted by equation (2). This is just opposite to the results on evaporated Nb_3Ga films mentioned above.

The experimental results confirm that the pinning force density
under hydrostatic pressure scales in the same way as under temperature. Thus the pinning mechanism which acts in the Nb_3Ge thin film
samples is not changed under pressure.

ACKNOWLEDGEMENTS

We wish to thank the Deutsche Forschungsgemeinschaft for
financial support of this work.

REFERENCES

1. W.A. Fietz and W.W. Webb, Phys. Rev. <u>178</u>, 657 (1969)
2. E.J. Kramer, J. Appl. Phys. <u>44</u>, 1360 (1973)
3. J.R. Gavaler, Appl. Phys. Letters <u>23</u>, 480 (1973)
4. Y. Tarutani, M. Kudo and S. Taguchi, Proc. ICEC 5, 477 (1975)
5. H. Neubauer, Z.f. Phys. <u>226</u>, 211 (1969)
6. M. Durieux, Progr. Low Temp. Phys., C.J. Gorter edit., <u>6</u>, 405 (1970)
7. H.F. Braun, E.J. Saur, Proc. ICEC 6, 411 (1976)
8. H.F. Braun, E.N. Haeussler, E.J. Saur, IEEE Transactions on Magnetics, <u>Mag. 13</u>, 327, (1977)
9. D. Tilley and R. Ward, J. Phys. <u>C3</u>, 2119 (1970)
10. H.F. Braun, M.A. Winkelmann, to be published

QUESTIONS AND COMMENTS

R.A. Hein: I missed what the current density was when you had a current of 100 mA.

E.J. Saur: You can calculate it because I gave the cross-section of the sample. One has to divide the current by this sectional area 1.4×10^{-3} mm^2. The critical current density will depend on the orientation of the sample with respect to the field.

J.A. Woollam: You showed two different scaling laws. What are the physical assumptions for these two equations?

E.J. Saur: It has to do with the orientation. Dew Hughs has newly given some more formulas, but that's for special arrangements of the pinning centers. The whole behavior depends on the arrangement of the pinning centers.

J.S. Schilling: What was the initial value of T_c you started out with?

E.J. Saur: About 21 K.

J.S. Schilling: Does pressure decrease this?

E.J. Saur: The temperatures decrease under the pressure.

J.S. Schilling: How much is this depression?

E.J. Saur: It is about the same as in the other A15 compounds. It's about 10^{-5} K/bar.

C.W. Chu: What's the physical explanation for decreasing
                   pinning force density under pressure?

E.J. Saur: That's a hard question. But you see, the behavior
                   of the change of pinning force density with pres-
                   sure is similar to that with temperature. It's
                   the same behavior, because the two formulas are
                   related for this change. But I can't tell you why.
                   We only conclude that the pinning mechanism de-
                   pends on temperature in the same way with and with-
                   out the pressure. This doesn't say anything about
                   the size, and doesn't explain the mechanism itself.

PRESSURE DEPENDENCE OF T_c IN THE SUPERCONDUCTING A15 COMPOUNDS $V_{75-x}TM_xGa_{25}$ (TM = Cr, Mn, Fe, and Co)

D.U. Gubser*, K. Girgis, and H.R. Ott

Eidgenössische Technische Hochschule

Zürich, Switzerland

ABSTRACT

The superconducting transition temperatures T_c and the pressure dependences of T_c have been determined as a function of x in the compounds $V_{75-x}TM_xGa_{25}$ where TM is the transition metals Cr, Mn, Fe, or Co. In these systems T_c is a strongly dependent, but a universal function of the electron per atom e/a ratio. Systematic trends in the relative pressure derivatives $(1/T_c)(\partial T_c/\partial p)$ have also been observed for these compounds, but the best correlation of the pressure data is with concentration x not with e/a.

INTRODUCTION

The interpretation of hydrostatic pressure (Volume) effects on the superconducting transition temperature T_c provides a useful test for the various theoretical models of superconductivity. For instance, the large depressions of T_c in nontransition metals is consistent with the dependence of the electron-phonon coupling parameter λ on the square of the average phonon frequency[1,2], while in transition metals the smaller (both positive and negative) pressure effects have been understood by the more complete formulation of λ which takes into account both phononic and electronic processes[2,3].

---

*Permanent address: Naval Research Laboratory, Washington D.C.

The effects of pressure on A15 compounds, however, are not easily understood. Because of the strong energy dependence of the
electronic density of states near the Fermi level, no simple generally accepted equation for T_c exists to help formulate the problem. Furthermore, no general trend in the existing pressure data
on T_c of binary A15 compounds has been observed which might suggest the importance of a specific parameter[4].

In an attempt to discover a systematic behaviour in the pressure derivative of T_c, the superconducting properties of the
$V_{75-x}TM_xGa_{25}$ (TM = Cr, Mn, Fe, and Co) A15 compounds have been
studied. In these compounds part of the vanadium atoms are systematically replaced by other transition metal atoms which increase the electron per atom e/a ratio. Previous work on the Cr,
Mn, and Fe compounds[5], as well as the Co studies reported here,
have shown that T_c is a universal function of e/a. Variations
in the pressure derivatives of T_c herein reported for all these
A15 compounds are functions of x and e/a.

SAMPLE PREPARATION

The samples were prepared by melting the constituent elements
in an argon atmosphere. The samples were subsequently wrapped in
tantalum foil and annealed under vacuum at 800°C for approximately
4 weeks. The final compositions of the samples were determined
from a microprobe analysis and the lattice parameter was measured
using X-ray techniques. Neutron diffraction studies were also
made to determine the distribution of TM atoms and the long range
order parameter. Complete details of the sample preparation and
metallographic studies will be reported elsewhere.

TRANSITION TEMPERATURES T_c

All superconducting transition temperatures reported here were
measured inductively with a standard ac mutual-inductance bridge.
The transition widths were typically 0.5 K wide with no apparent
systematics. T_c was always defined as the midpoint of the transition.

The superconducting transition temperatures for TM = Cr, Mn,
and Fe have been reported earlier[5] while those of Co are first
reported here. The results of these investigations are shown in
Figure 1 where T_c is plotted against concentration x (Fig. 1a) and
electron per atom e/a ratio (Fig. 1b). When plotted against x,
families of curves appear which are characterized by the different
transition metal atoms; however, all the data coincide to a universal curve when T_c is plotted against the e/a ratio. Over the

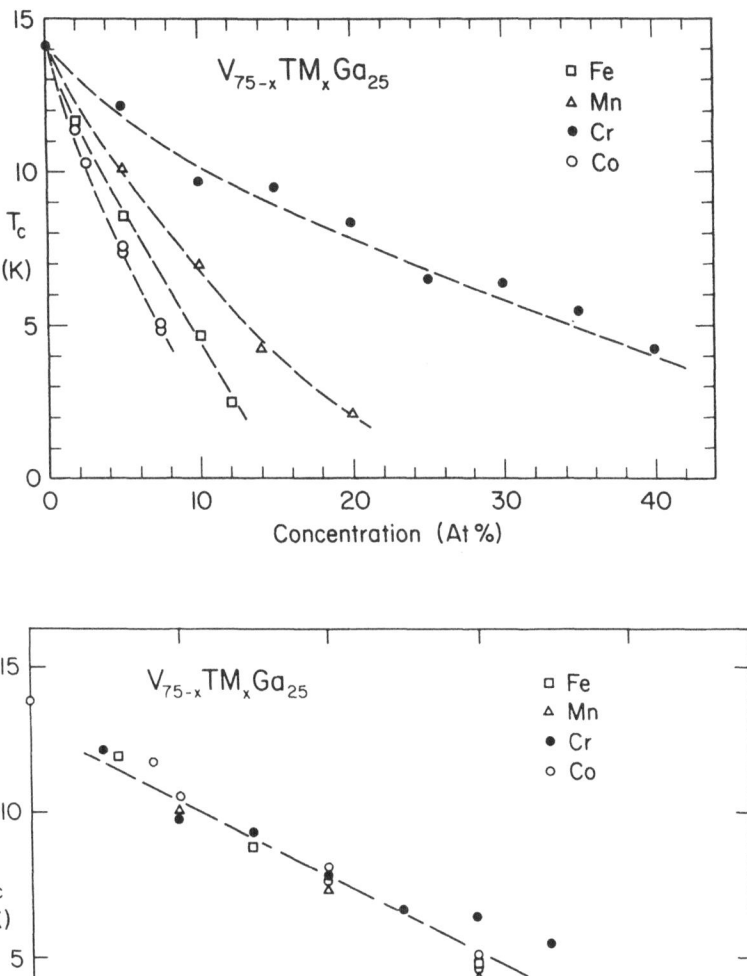

Figure 1.   Change in T_c for $V_{75-x}TM_xGa_{25}$ compounds as a function of concentration x (1a) and electron/atom e/a ratio (1b)

majority of this range, T_c is linearly related to e/a through the relation

$$T_c = 13 - 27(e/a - 4.5) \qquad\qquad (1)$$

as shown in Figure 1b. At e/a values near 4.5 (x = 0), thermal ordering becomes important and the T_c values begin to deviate upward from the straight line[6,7]. The highest concentration Cr data also show a deviation from the straight line for an unknown reason.

PRESSURE EFFECTS $(1/T_c)(\partial T_c/\partial p)$

The apparatus for measuring the effects of pressure on T_c was a standard piston-cylinder, clamp apparatus whereby hydrostatic pressures to approximately 20 kbars could be applied to the sample. The pressure cylinder was hardened BeCu having a 5mm bore while the piston material was WC. Pressure was applied at room temperature and retained by a lock-nut mechanism. The sample, along with an In pressure calibrant and the pressure transmitting fluid (glycerine), were all contained inside a nylon cell. Pressures were determined at low temperatures by simultaneously measuring the T_c of the In calibrant whose $T_c(p)$ curve is well documented[8]. T_c was again measured inductively while slowly sweeping the temperature which was determined with a germanium resistance thermometer.

The effect of pressure on A15 compounds is not large, in fact, the total shift for the compounds reported here is less than the width of the transition temperature itself. Nevertheless, the shapes of the transitions proved invariant under pressure and the midpoint was reproducibly defined to about 20 to 30 mK. Since the total shifts were from 0.05 K to approximately 0.25 K at 20 kbars, such an accuracy is necessary to obtain reliable $\partial T_c/\partial p$ values. Figure 2 shows the actual data for the $V_{65}Mn_{10}Ga_{25}$ sample as a function of pressure. From data such as these, values for T_c and $\partial T_c/\partial p$ were obtained.

Figure 3 shows the results of the pressure measurements where the ratio $(1/T_c)(\partial T_c/\partial p)$ is plotted against the parameters x (Fig. 3a) and e/a (Fig. 3b). This ratio is related to the volume derivative of T_c through the relation

$$\partial \ln T_c/\partial \ln V = - (1/\kappa T_c)(\partial T_c/\partial p) \qquad\qquad (2)$$

where κ is the compressibility of the sample. For purposes of comparing data from a variety of samples, this volume derivative is more meaningful than the pressure derivative alone. The value of $(1/T_c)(\partial T_c/\partial p)$ for binary $V_{75}Ga_{25}$ determined in this study $(0.8 \pm 0.1 \times 10^{-6}$ bar$^{-1})$ agrees quite well with the value determined previously by Smith[4] $(0.75 \pm 0.03 \times 10^{-6}$ bar$^{-1})$. The most

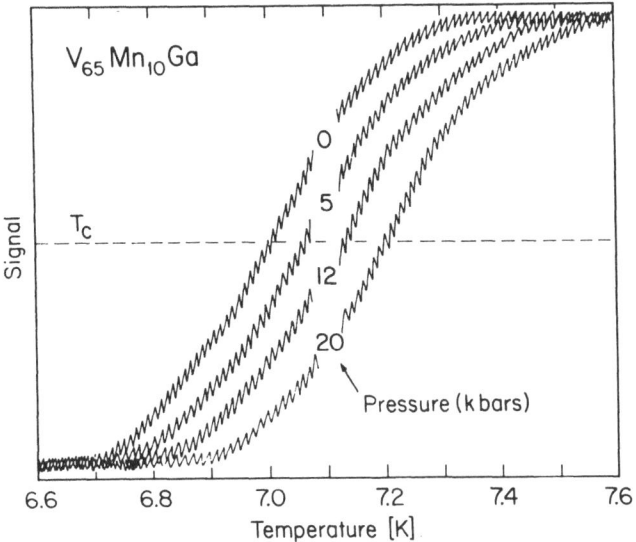

Figure 2. Superconducting transitions for $V_{65}Mn_{10}Ga_{25}$ showing the T_c definition and the effects of pressure. Numbers on the individual curves refer to pressure in kbars.

prominent feature of these plots is that the ratio $(1/T_c)(\partial T_c/\partial p)$ increases significantly as one adds TM atoms, rising to a value that is approximately double that of $V_{75}Ga_{25}$ for the higher concentration alloys.

DISCUSSION

The superconducting transition temperature is generally given by a formula of the form

$$T_c = \theta \exp(-1/g) \qquad (3)$$

where θ is a characteristic phonon frequency and $g(\lambda, \mu^*)$ is a quantity relating to the overall strength of the interaction producing the superconducting state; λ is the attractive electron-phonon coupling parameter, and μ^* is the Coulomb repulsion parameter. Assuming $\partial \mu^*/\partial V = 0$, the volume derivative of (3) becomes

$$\partial \ln T_c/\partial \ln V = -(1/\kappa T_c)(\partial T_c/\partial p) = \partial \ln \theta/\partial \ln V + f(\lambda, \mu^*) \ln \frac{\theta}{T_c} (\partial \ln \lambda/\partial \ln V)$$

$$(4)$$

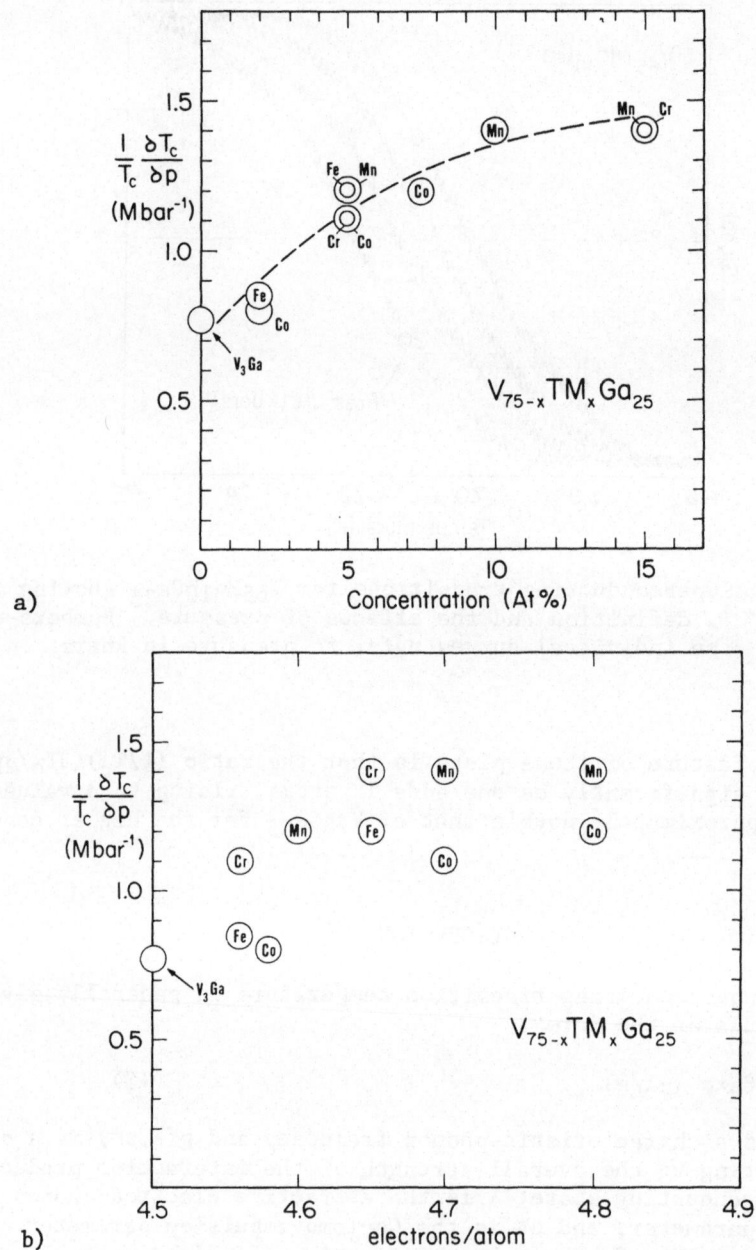

Figure 3.   Change in the pressure derivative $(1/T_c)(\partial T_c/\partial p)$ for
$V_{75-x}TM_xGa_{25}$ compounds as a function of concentration
x (3a) and electron/atom e/a ratio (3b).

where

$$f(\lambda, \mu^*) = (\lambda/g)(\partial g/\partial \lambda).$$

The function g is usually given in the form proposed by McMillan[9]

$$g = (\lambda - \mu^*(1 + 0.62\lambda))/(1.04(1 + \lambda)) \tag{5}$$

and the volume derivative of θ is the Grüneisen parameter, γ_G. The central problem in describing pressure effects in ordinary metals is therefore that of reconciling experimental and theoretical values of $\partial\ln\lambda/\partial\ln V$.

For A15 materials, the rapid energy variation of the electronic density of states near the Fermi level creates many additional complications. First, the functional form of g is unknown and second, both g and θ may depend explicitly on additional parameters such as the e/a ratio. The effects of e/a values on the pressure derivative $(1/T_c)(\partial T_c/\partial p)$, have not previously been studied and are not easily anticipated theoretically because one does not have a generally accepted T_c equation. Since the T_c behaviour of the $V_{75-x}TM_xGa_{25}$ compounds were well behaved and quite systematic in e/a, similar systematic trends were sought in the pressure data as reported here.

There are numerous theoretical models[10-13] for A15 superconductors, based on various assumptions concerning the form of the electronic density of states, however, the only model for which pressure effects have been theoretically calculated is the Labbé, Barisic, Friedel (LBF) model[10,14]. Under their assumptions, one finds

$$2kT_c = B^2\lambda^2L^2 \tag{6}$$

where B is a normalization constant proportional to J^{-1}, J being the d-band intrachain overlap integral; λ is the electron-phonon coupling parameter which is proportional to $q_o^2J^2$, q_o being the Slater coefficient describing the exponential decay of the wave function, and L is a smoothly varying function of $Q_d/B^2\lambda$ where Q_d is the number of d electrons. The volume dependent terms are those containing the overlap integral $J \propto$ exp. $(-q_o a)$ (a is the lattice parameter). Under the assumption of no interchain or interband charge transfer, i.e. $\partial Q_d/\partial V = 0$, the function L is volume independent since the product $B^2\lambda$ is not a function of J. The volume derivative of T_c is therefore a consequence of the product $B^2\lambda^2$ and is independent of Q_d. According to Barisic[14],

$$\partial\ln T_c/\partial\ln V = - aq_o \tag{7}$$

This result predicts pressure effects of the right order of magnitude for V based compounds, but as shown by Smith[4], it can not be generally applied to all A15 compounds.

The prediction of the LBF model that the pressure derivative should be independent of $Q_d(e/a)$ was tested in the alloys reported here. As seen in Figure 3, trends in $(1/T_c)(\partial T_c/\partial p)$ do exist, but the best correlation for the pressure data is with concentration x rather than with e/a as in the T_c data. This fact suggests that what is important for explaining the observed trends in the pressure data is simply the number of times one physically disrupts the V chains with another atom and not the e/a value. Since the cohesion energy of the A15 structure is closely related to the formation of d-electron bonds between neighboring V atoms along the chain, the disruption of these bonds may affect the cohesion energy of the structure and would be reflected as a change, proportional to the number of impurity atoms, in the isothermal compressibility. If this were the case, then the volume derivative of T_c which involves κ as seen in eq. (2) may in fact be constant across these A15 compounds as predicted by the LBF model.

Unfortunately, no compressibility data exist for these compounds or, in fact, for most of the individual A15 binary systems. In a recent study, however, Skelton et al.[15] showed that stoichiometry can have a large effect in the compressibility of an A15 material. They found that in $Nb_{75}Ga_{25}$ the compressibility is a factor of 3 smaller than in off stoichiometric $Nb_{82}Ga_{18}$; hence, there is supporting evidence that compressibility changes of a size necessary to explain the upward systematics in these $V_{75-x}TM_xGa$ systems may indeed be reasonable (κ-changes by factor of 2).

CONCLUSION

Studies of T_c and $(1/T_c)(\partial T_c/\partial p)$ in the $V_{75-x}TM_xGa_{25}$ systems have revealed a systematic behaviour of T_c on e/a and a systematic behaviour of $(1/T_c)(\partial T_c/\partial p)$ on concentration x. These data suggest that although the e/a ratio is important for determining T_c values it is not as important for influencing the volume effects in these compounds. Changes in the compressibilities for these materials due to deviations from the ideal A_3B composition may explain the increase in $(1/T_c)(\partial T_c/\partial p)$ values.

ACKNOWLEDGEMENTS

The authors are grateful to Prof. J.L. Olsen for his continuing interest and support during this work. Financial support from the Schweizerische Nationalfonds zur Förderung der wissenschaftlichen Forschung is gratefully acknowledged.

REFERENCES

1. H.R. Ott and R.S. Sorbello, J. Low Temp. Phys. 14, 73 (1974).
2. J.W. Garland and K.H. Bennemann, in "Superconductivity in d-and f-Band Metals", AIP Conf. Proc. No. 4, D.H. Douglas, ed. pg. 255 (New York, 1972).
3. A. Birnboim, Phys. Rev. B 14, 2857 (1976).
4. T.F. Smith, J. Low Temp. Phys. 6, 171, (1972).
5. K. Girgis, W. Odoni, and H.R. Ott, J. Less Comm. Met, 46, 175, (1976).
6. R. Flukiger, J.L. Staudenmann, A. Treyvaud, and P. Fischer, Proc. LT14, Vol. 2, pg. 1, North Holland and American Elsevier, New York, (1975); and R. Flukiger, J.L. Staudenmann, and P. Fischer, J. Less Comm. Met. 50, 253 (1976).
7. B.N. Das, J.E. Cox, R.W. Huber, and R.A. Meussner, Met. Trans. A 8A, No. 4, 541 (1977).
8. L.D. Jennings and C.A. Swenson, Phys. Rev. 112, 31 (1958). T.F. Smith and C.W. Chu, Phys. Rev. 159, 353 (1967).
9. W.L. McMillan, Phys. Rev. 167, 331 (1968).
10. J. Labbé, S. Barisic, and J. Friedel, Phys. Rev. Letters, 19, 1039 (1967).
11. R.W. Cohen, G.D. Cody, and L.J. Vieland, Proc. 3rd Mat. Res. Sym. on Electronic Density of States, National Bureau of Standards, special publication 323 (1969).
12. L.P. Gor'kov and O.N. Dorokhov, Solid State Comm. 19, 1107 (1976).
13. P. Horsch and H. Rietschel, Z. für Physik B 27, 153 (1977).
14. S. Barisic, Phys. Letters, 34A, 188 (1971).
15. E.F. Skelton, D.U. Gubser, S.C. Yu, I.L. Spain, R.M. Waterstrat, A.R. Sweedler, L.R. Newkirk, and S.A. Valencia (to be published).

QUESTIONS AND COMMENTS

B.T. Matthias: If that is true, that the e/a determines the transition to such a beautiful line, why didn't you go to the other side?

H.R. Ott: We did, experiments are going on right this week.

B.T. Matthias: Did the temperature go up?

H.R. Ott: No.

B.T. Matthias: Is this the end of the straight line?

H.R. Ott: It looks like that. It almost peaks at the binary line. We also would like to see what the pressure dependence does, and whether it goes down.

T.F. Smith: You are on the small lattice parameter side of
                   V_3Ga. Therefore, your increase of T_c under pres-
                   sure is in disagreement with the correlation be-
                   tween T_c and the parameter.

H.R. Ott: Right.

R.A. Hein: I'm curious about your last statement. How did
                   you know where the impurity atom was going? You
                   said it was going onto the A site, let's accept
                   that. What I'm curious about is since we know
                   that when V_3Ga deviates from stoichiometry, the
                   long range order and transition temperature de-
                   crease; how do you know that you're keeping the
                   same amount of crystallographic order?

H.R. Ott: We know that this order is probably about the same
                   in all the samples, because at the end when we
                   go to the binary, the best straight line doesn't
                   fit. If you go close to or directly to the pure
                   V_3Ga, T_c lies too high above this straight line.
                   I don't know whether you saw that. So, it seems
                   that the disorder effect is more or less the same
                   in all the compounds.

R.A. Hein: Let me ask you another question. Knowing that the
                   impurity atom is a defect, one could calculate the
                   decrease in long range order due to, say, adding
                   2 percent of Fe or Mn. Did you compare that with
                   the decrease in long range order due to defects
                   introduced thermally?

H.R. Ott: No, we haven't actually looked into that.

NEW HYPOTHESIS CONCERNING THE COMMENCEMENT OF THE RARE EARTH SERIES AT HIGH PRESSURE

J. Wittig

Institut für Festkörperforschung, Kernforschungsanlage Jülich

D-5170 Jülich, West-Germany

ABSTRACT

It is suggested that the elements Ce, La, Ba and Cs represent one common family of metals with strongly 4f-hybridized Bloch waves at high pressure. There is evidence that f-hybridized metals possess soft phonon spectra in general and rather large overall electron phonon coupling.

In 1965, Hamilton published an interesting paper in which he came to the conclusion that lanthanum may not be a plain sd-band metal.[1] Instead, he suggested that La might be a real rare earth metal with the 4f level being involved in the electronic band structure, however, without the usual local-moment magnetism. He first drew attention to the fact that some physical properties of La (melting temperature, superconducting T_c, crystal structure, ...) deviate from what one should expect, if these quantities would vary smoothly and systematically in the periodic table. As indicated below, Hamilton viewed La as having a 4f virtual bound state on each atomic site. Unfortunately, the paper has not found much recognition in the literature. We therefore repeat an important argument which comes from the physics of the atomic spectra of La.

The ground-state configuration of a free La atom is $6s^2 5d^1$. Naturally, various excited states of La contain a 4f electron. Since the orbit of a 4f electron (with principal quantum number 4) should be rather large in comparison to the core electrons, one

expects that the 4f electron moves essentially in a Coulomb potential. Its binding energy (with respect to the ionization thres-
hold) should be therefore the "hydrogen value": Ry $1/4^2$ = 6,858
cm^{-1}. However, spectroscopic data completely disagree with this
expectation. For instance, there is an excited configuration containing a 4f electron which has the large binding energy of 21,418
cm^{-1}.[2] This huge binding energy has been interpreted in the following way. The inner well of the double well potential, which is
responsible for the binding of the atomic 4f electrons in the elements beyond La, must have already in La an appreciable depth.
The 4f wave function has therefore a finite probability inside the
core at a distance from the nucleus which is of the order of 0.4 Å.
This is just a variant of the square-well textbook problem in quantum mechanics. The inner well is not yet deep enough to bind one
electron; it just causes a scattering resonance. The possible consequences in metal physics have been clearly forseen by Hamilton:
one may expect 4f-hybridized conduction electron wave functions at
the Fermi surface of La. On the other hand, if La atoms are di-
lutely dissolved in a non-f-band host, we may expect a scattering
contribution from a 4f virtual bound state.

At present, most band structure theorists (and experimentalists as well) decline the possibility of 4f-band electrons in La
and Ce. Therefore, we see the purpose of this talk in again[3-6]
discussing phenomenologically some anomalous properties of La and
Ce at high pressure. On the phenomenological level, fairly definite
conclusions can be drawn for the electronic band structure of these
elements. We then briefly turn to the elements Ba and Cs. Indeed,
there are some indications that these metals are also 4f-band metals
at high pressure. However, much more experimental information about
the alkali and alkaline earth metals in general is needed before
definite conclusions can be drawn. In this context, we should consider the rumor that the energy of the 4f states always <u>increases</u>
under pressure with respect to the Fermi level.[7-10] According to
our hypothesis, it may be just the converse in Ba and Cs. Therefore we would like to point out that the postulated upward movement of the 4f level is obviously an ad hoc-assumption in connection with the properties of Ce which has, to our knowledge, not
yet found any direct physical proof. In summary, we think it is
premature to ask why the 4f levels may come down under pressure to
the Fermi surface in Ba and Cs. It is perhaps worth mentioning
that Fermi has thought about this possibility in the case of Cs as
early as in 1950.[11]

As noted by Hamilton, the melting temperature of La is
anomalously low.[1] Fig. 1 shows the melting temperatures of the d-

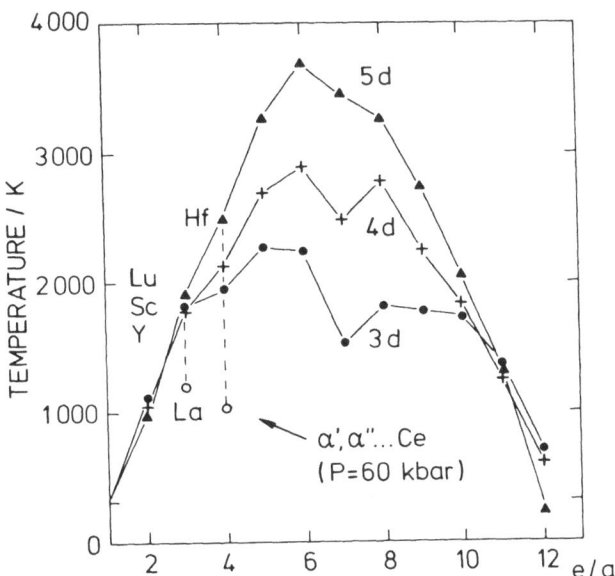

Figure 1:   Melting points of the d-transition metals.   Adjoining
            elements are included.   The melting point of La is low
            in comparison to the other trivalent metals.   The melting
            point of Ce is catastrophically low.[3,4,6]

transition period metals and the adjoining elements.   It is seen
that the melting point of La is roughly 700 K lower than for lutetium which is the corresponding member of the 5d period and thus
the proper element for a comparison.   La has a larger metallic
radius than Lu.   Thus for a fair comparison of the melting points,
one should compress La, so that its metallic radius equals that of
Lu.   After this correction,[6] the melting point turns out to be only
≈400 K lower than for Lu.   One may hence think that the melting
point depression of La is a debatable effect.   Below we discuss the
properties of Ce.   The discussion will appreciably strengthen our
faith that the low melting point of La is nevertheless a significant effect.

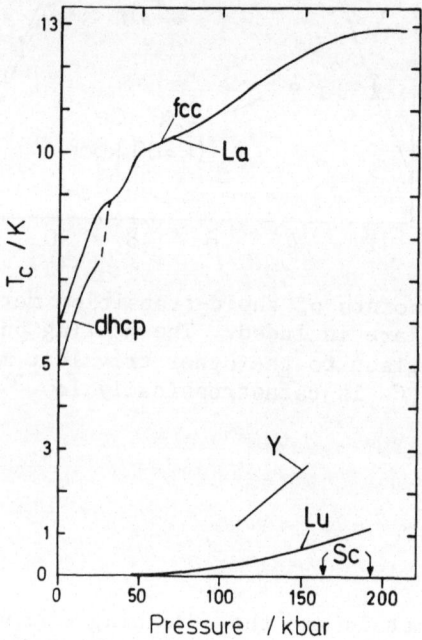

Figure 2: Pressure dependence of T_c for La, Y and Lu.[6]

Fig. 2 shows the superconducting transition temperatures as a function of pressure for La and Lu. It is seen that the T_c of La is an order-of-magnitude higher than for Lu at the highest pressure at which both elements have been so far studied. The Lindemann-formula ($T_M \propto \langle\omega^2\rangle_M$) suggests a particularly soft phonon spectrum for La, i.e. a rather low value of $\langle\omega^2\rangle_M$ for the average phonon frequencies. Because of the lack of better knowledge, we take the freedom and set $\langle\omega^2\rangle_M \approx \langle\omega^2\rangle$ where $\langle\omega^2\rangle$ has the usual meaning.[12] The well-known expression for the electron-phonon coupling parameter $\lambda = N(0)\langle I^2\rangle/M\langle\omega^2\rangle$ now teaches us that we can hope for a high overall electron-phonon coupling strength λ, as is in fact observed. I think the presence of soft phonons is one reason for the relatively high T_c of La and strongly points to the fact that La is not a plain sd metal rather than a sd 4f-band metal.

Up to the present time, the 4f electrons are generally ignored by the majority of the band structure theorists.[13] Breaking with this traditional approach, Glötzel and Fritsche[14] have recently carried out band structure calculations for fcc-La using the cellular method (Fig. 3). The upper band structure is the result for zero pressure; the lower one belongs to a pressure of 120 kbar. The analysis of the occupied band states in terms of angular momenta revealed f-admixtures of $\simeq 0.3$ electron.[14] A self-consistent calculation did not lead to major changes of the result in Fig. 3.[15] Finally, we would like to draw attention to the interesting topological changes of the Fermi surface which occur as a function of pressure in the Σ-direction. According to Lifshitz,[16] a subtle "electronic" phase transition may occur if the Fermi energy is shifted across a Van Hove singularity. It is quite possible that the sharp kink in the T_c-P dependence of La at 53 kbar (cf. Fig. 2) is related to such an isostructural low temperature lattice instability.[17] Similar sharp kinks have been observed among the pure elements so far only in the 5f-band metal uranium.[18] Clearly, f bands (or f-hybridized bands) will have in general relatively flat dispersion relations at the Fermi surface. One should therefore expect that Van Hove singularities are stronger in f-band metals than in the normal spd metals.

Another group of theorists[19] has recently shown that the most important anomaly of La, its double-hcp crystal structure, is not at all related to the anticipated 4f-band character. From their calculations, the d-hcp structure is stable at normal pressure, because La has 0.6 d electrons more than Lu. Within this theory, Lu should transform to the Sm-type structure at very high pressure due to an increase of the number of d electrons. In this connection, they quote data[20] for a hcp→Sm-type transformation. However, the existing experimental evidence for such a transition is, in our view, to put it mildly, meager. In fact, we would expect that Lu never transforms into the Sm-type or d-hcp structure,

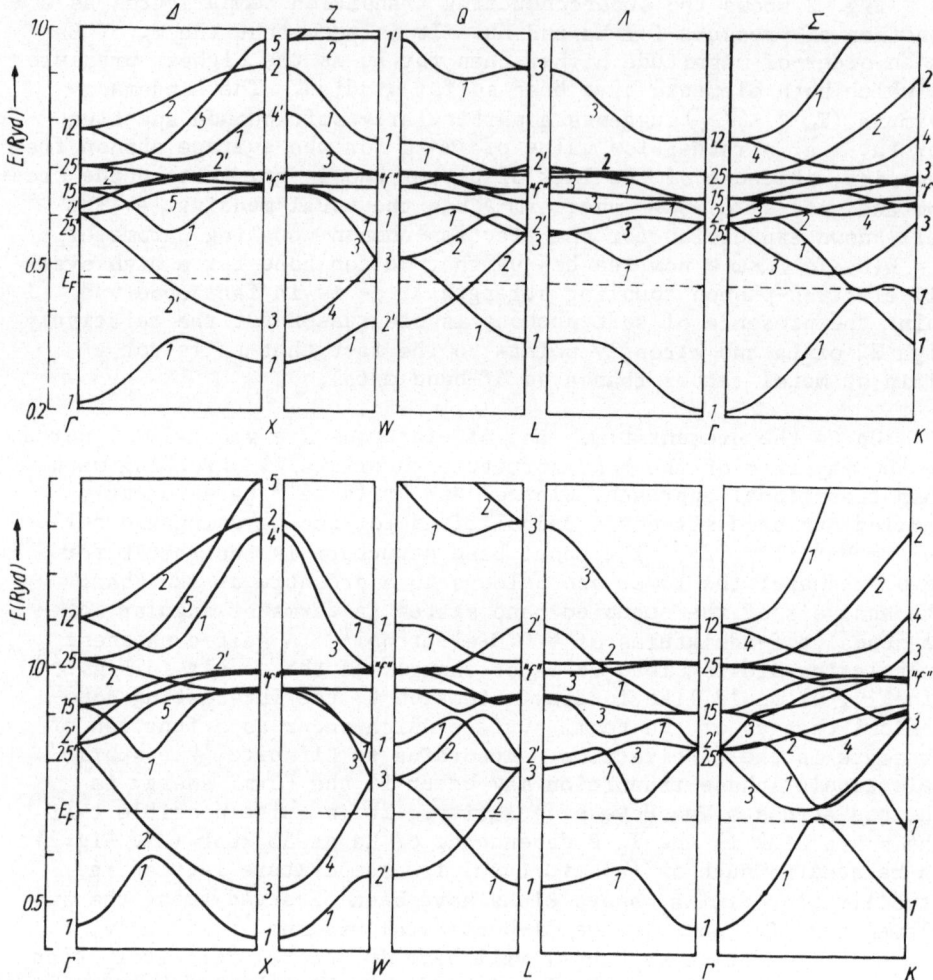

Figure 3: Band structure of fcc-La at normal pressure (top) and at 120 kbar (bottom). After Glötzel and Fritsche.[14]

since the occurrence of these structures seems to be related to the presence of 4f character at the Fermi surface.[6] This suggests an important experiment: one should pressurize Lu up to the highest possible pressures ($\approx$400 kbar) in order to determine the crystal

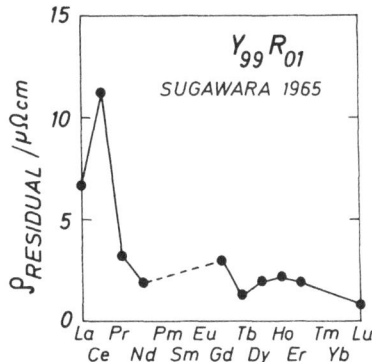

Figure 4: Residual resistivity caused by 1 at % rare earth metal dissolved in yttrium.

structure with certainty. In view of the discrepancies on the theoretical side, a phenomenological approach[6] seems at present about the best that can be done for getting a hold of the situation.

We believe there is one experiment in which the 4f-scattering resonance on the La atom has been seen.[3] Sugawara has studied the residual resistivity of yttrium, doped with 1 at % rare earth metals. The results are shown in Fig. 4. Randomly distributed Ce impurities are the strongest scatterers for the conduction electrons. We will not go into this interesting detail here. The weakest scatterers are trivalent Lu-impurities. The reason is probably that the 5d-wave function of the Lu-atom matches fairly well with the 4d-wave functions of the Y-host. From this reasoning, it is all the more surprising that La atoms are much stronger scatterers than Lu atoms. It was therefore concluded[3,6] that this experiment may have sensibly checked the additional scattering due to the 4f-virtual bound state. These experiments should be repeated, since much purer rare earth metals have now become available. This will help to reduce the considerable experimental uncertainty of the data points which is not shown in Fig. 4.

In summary, we think there is considerable evidence now, especially from the macroscopic properties of La metal, that it has 4f-
hybridized band states. This conclusion[6] will be strongly corroborated from the following discussion of the properties of Ce at
high pressure.

The physical properties of cerium metal undergo unusual
changes under pressure.[6,10] For instance, a minimum occurs in the
melting curve (Fig. 5). This has been attributed to a general
phonon softening in the P-T surroundings of the critical point in
which the γ/ α- phase line terminates.[17] In the α', α'' ... phase
range, Ce is a fairly "good" superconductor with a $T_c \simeq 1.9$ K at
$\simeq 40$ kbar, whereas in the α phase it is superconducting in the ten-
millikelvin range[6,21] (Fig. 6). The entire loss of magnetism
under pressure has been attributed either to the promotion of the
4f electron to the sd-band or, alternatively, to a delocalization
of the 4f electron into a 4f-band electron.[6,10] We think we can
prove quite convincingly that the first explanation is not correct, at least in so far it requires the radical depopulation of
the 4f band in α', α'' ... Ce.[6] The phase α', α'' ... Ce should
have the same $(6s5d)^4$-band configuration as Hf in the radical promotional model. In other words, it should be a sd-transition
metal with very similar properties as Hf.[3,4] This is, however, not
the case. In Fig. 7 we have listed the observed range for the
T_c's of α', α'' ... Ce in the column e/a = 4, together with Zr and
Hf. It is seen that the T_c's of Ce are an order-of-magnitude higher
than the T_c of Hf. This is a strong deviation from the regular
pattern of the superconducting transition temperatures as evidenced
from Fig. 7. It has hence been concluded that the relatively high
T_c of Ce in the α', α'' ... phase range points to the fact that it
is not a pure sd-transition metal rather than a 6s5d4f-band metal
with a share of nonmagnetic 4f electrons.[3,4,6]

As seen from Fig. 1, the melting temperature of α', α'' ...
Ce is 1500 K lower than for Hf. Certainly, a correction must again
be applied for the different metallic radii. Taking this into account,[6] the melting temperature turns out to be at least 1000 K
too low in comparison to Hf. The most reasonable conclusion therefore is that Ce is not a sd-band metal at high pressure rather than
a sd4f-band metal. We believe that the relatively high superconducting T_c in the α', α'' ... phase range is a consequence of the
soft phonon spectrum as discussed above for La.

An interesting point is the comparatively low T_c in the α
phase (Fig. 6). In this phase, Ce has a fairly high density of
states as inferred from the electronic specific heat coefficient γ.

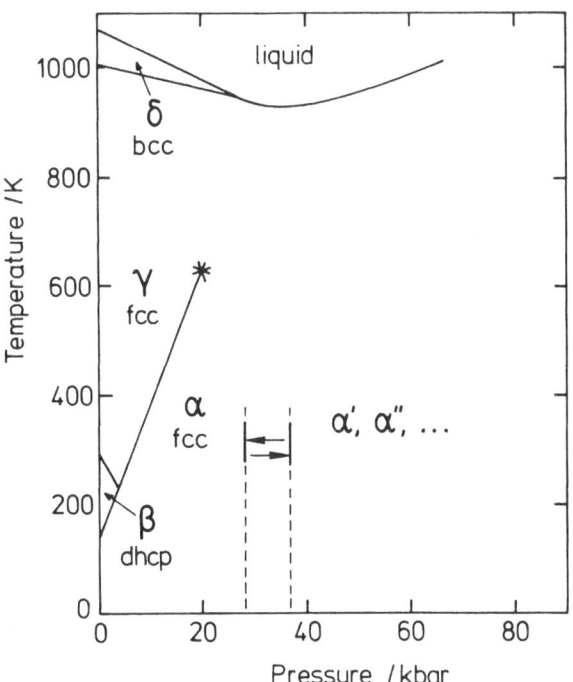

Figure 5: Phase diagram of Ce[6] The symbol α', α''... denotes the
existence of phase mixtures. Note that the $\alpha \rightarrow \alpha'$,
α''... phase boundary occurs at much lower pressure
than reported in the literature.[10]

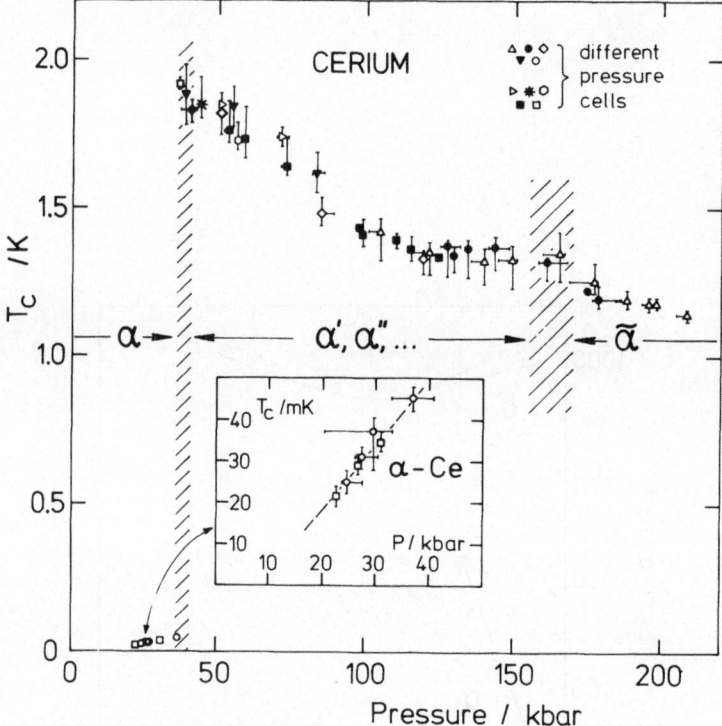

Figure 6: Pressure dependence of T_c for Ce.[6,21]

The reported value at normal pressure is 12.8 mJ/g-at. K^2.[10] As
judged from the melting temperature, the average phonon frequencies
$\langle\omega^2\rangle$ should be similarly low as for the α', α''... phase. Both
these are favourable conditions for the overall electron-phonon
coupling parameter $\lambda = N(0)\langle I^2\rangle/M\langle\omega^2\rangle$ being similarly large as for
α', α'' ...Ce. It is therefore quite conceivable that the Coulomb
repulsion is particularly strong in α-Ce and is responsible for the
low T_c. In conclusion, we think there can be no doubt at present
that Ce is a 4f-band metal at all pressures. Nothing can be said
about the total number of 4f electrons or how this number changes
with pressure. However, the melting point catastrophe of Fig.1

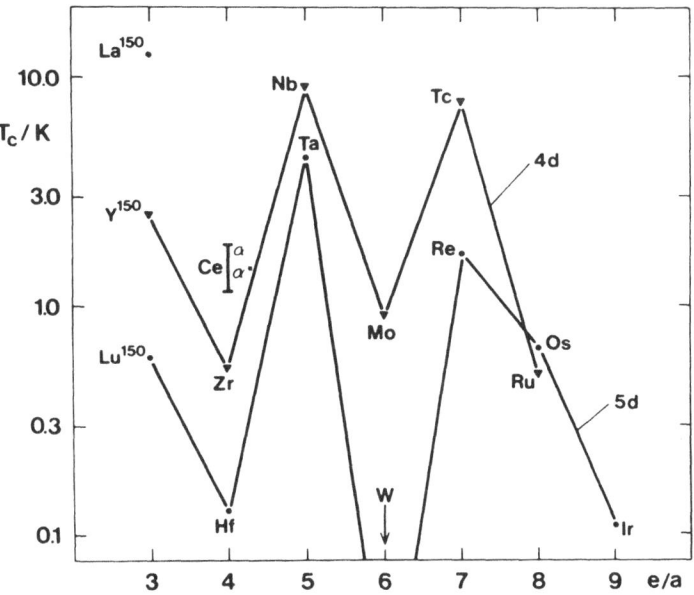

Figure 7: T_c as a function of the electron per atom ratio for the 4d and 5d transition metals. Also included are the T_c's of La, Y and Lu at a pressure of 150 kbar and the range of the T_c's for α', α'' ... Ce.

suggests that the 4f-band occupation is not just small. In our whole discussion we avoided the term "valence change"[10], since we think it obscures the situation. It looks as if Ce has always 4 conduction electrons. From the relatively small variation of the melting temperature up to 70 kbar one may infer that the 4f component in the wave functions does not change fundamentally.

A couple of years ago I have pointed out that La's two left-hand neighbours in the periodic system, barium and cesium, possess anomalously low melting temperatures under pressure.[3] On the other side, these metals are fairly "good" superconductors at high pressure.[3] We believe both properties are related to each other by an unusually soft phonon spectrum, as in La and Ce. We have

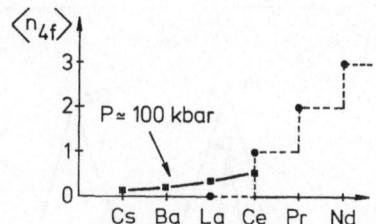

Figure 8:   Hypothetical occupation number of the 4f shell for the
            metals at the beginning of the 4f group at high pressure.
            The 4f group commences very gradually with increasing
            atomic number. In the traditional view (circles) the
            4f group is thought to begin abruptly at the element Ce
            (at normal temperature and pressure).

therefore (and for some other reasons) suggested that Cs and Ba
may be turned into 4f-band metals under high pressure. In other
words, Cs and Ba may be pressure-induced nonmagnetic rare earth
metals.[3,5] This is illustrated in Fig. 8. The squares indicate,
qualitatively, the anticipated occupation of the 4f shell with increasing atomic number at, say, 100 kbar. The new idea is that the
rare earth series commences gradually in the metals featuring altogether four nonmagnetic 4f-band metals at pressure. This is in
contrast to the conventional picture of an abrupt onset of the rare
earth series with, essentially, an integer 4f-shell count for Ce,
Pr, Nd, ... (dashed curve) and nil for La and even for Ce at high
pressure.

     What should be the next steps? Clearly, a systematic investi-
gation should be carried out of all the alkali and earth alkaline
metals for superconductivity up to the highest possible pressures.
Such investigations are now under way and new data for Ba are presented at this conference. The investigation will clarify whether
Cs and Ba are, due to a 4f-band admixture, both distinguished by a
particularly high T_C from the rest of those elements which, naturally, cannot have 4f-band character at the Fermi surface. It
would be equally important to have melting-curve data for these
elements up to the highest possible pressures in order to establish
with certainty that Cs and Ba are the only elements of these groups
which possess anomalously low melting points under high pressure.
Finally, we would like to draw attention to an interesting paper by
Matthias and co-workers in which they gave a general survey of melting points in the periodic system.[22] They noted that particularly

the 5f-group metals possess extraordinarily low melting temperatures. In closing we would like to ask the theorists: why do f-band electrons reduce so effectively the mean phonon frequencies?

REFERENCES

1. D.C. Hamilton, Amer. J. Phys. $\underline{33}$, 637 (1965).
2. C.E. Moore, Atomic Energy Levels (National Bureau of Standards, Circular No. 499, U.S. Government Printing Office, Washington D.C., 1950), Vols. 2 and 3.
3. J. Wittig, in Festkörperprobleme, edited by H.-J. Queisser (Vieweg, Braunschweig, Germany, 1973), Vol. 13, p. 365.
4. J. Wittig, Comments Solid State Physics VI, 13 (1974).
5. J. Wittig, in Proceedings of the 12th Rare Earth Research Conference, Vail, Colorado 1976, edited by C.E. Lundin, Vol. 2, p.873.
6. C. Probst and J. Wittig, in Handbook of the Physics and Chemistry of Rare Earths, edited by K.A. Gschneidner, Jr., and L. Eyring (North-Holland Publishing Co., Amsterdam) Chapter 10, in print.
7. C.F. Ratto, B. Coqblin, and E. Galleani d'Agliano, Advan. Phys. $\underline{18}$, 489 (1969).
8. B. Coqblin, J. Phys. Suppl. $\underline{32}$, C1-599 (1971).
9. L.L. Hirst, J. Phys. Chem. Solids $\underline{35}$, 1285 (1974)
10. D.C. Koskenmaki and K.A. Gschneidner, Jr., in Handbook on the Physics and Chemistry of Rare Earths, edited by K.A. Gschneidner, Jr. and LeRoy Eyring (North-Holland Publishing Co., Amsterdam) Chapter 4, in print.
11. R. Sternheimer, Phys. Rev. $\underline{78}$, 235 (1950).
12. W.L. McMillan, Phys. Rev. $\underline{167}$, 10 (1968).
13. A.J. Freeman, in Magnetic Properties of Rare Earth Metals, edited by R.J. Elliott (Plenum Press, London and New York, 1972) p. 245.
14. D. Glötzel and L. Fritsche, phys. stat. sol. (b) $\underline{79}$, 85 (1977).
15. D. Glötzel, private communication.
16. I.M. Lifschitz, Sov. Phys. JETP $\underline{11}$, 1130 (1960).
17. H. Balster and J. Wittig, J. Low Temp. Phys. $\underline{21}$, 377 (1975).
18. T.F. Smith and E.S. Fisher, J. Low Temp. Phys. $\underline{12}$, 631 (1973).
19. J.C. Duthie and D.G. Pettifor, Phys. Rev. Lett. $\underline{38}$, 564 (1977).
20. L. Liu, J. Phys. Chem. Solids $\underline{36}$, 31 (1975).
21. C. Probst and J. Wittig, in Proceedings of the 14th Int. Conf. on Low Temp. Physics, Otaniemi, Finland, 1975, edited by M. Krusius and M. Vuorio (North-Holland Publishing Co., Amsterdam) Vol. 5, p. 453.
22. B.T. Matthias, W.H. Zachariasen, G.W. Webb and J.J. Engelhardt, Phys. Rev. Lett. $\underline{18}$, 781 (1967).

QUESTIONS AND COMMENTS

B.T. Matthias: I'm sorry, but there must be a few historical corrections. The first one is that Landau and
Lifshitz, who noted in their book on Statistical Mechanics that the position of lutetium and
lanthanum should be interchanged. They were right,
but then who cares? That was true. However, the
first one to find out the 4f character of lanthanum was neither David (Hamilton) nor I, but was, of
all people, Kondo. In his "Progress in Theoretical
Physics" of 1963, which was a very interesting
paper, he finds out that maybe there is some 4f
character in lanthanum. Hume-Rothery did not consider 10 electrons for platinum and palladium because he defined them as zero. So the Hume-Rothery parameter goes only to 7 and then just to zero,
for the eighth group, that is to say, from iron to
platinum. The first person to assign these electrons to these atoms was a man by the name of
Schubert, but he didn't do it a very good way, so
it never became popular. But this block from iron
to platinum must be defined as having zero contributions. Otherwise, you would never get the
three phases.

J. Wittig: It is true that Kondo (Progr. Theor. Phys. (Kyoto)
$\underline{29}$, 1 (1963)) first speculated on the contribution
of the 4f states in lanthanum. But if you compare
and weigh these papers, you see that he merely
speculated and that Hamilton really had the physical idea.

J.S. Schilling: Do you say that lanthanum at 100 kbar has more
4f- character than at zero pressure?

J. Wittig: I can only say what the calculations bring out.
The calculations by Glötzel and Fritsche (Ref. 14)
show that the f-character doesn't strongly change,
so that the fraction of f-character was, I think,
0.3 electrons per atom at normal pressure and
something like 0.4 at 120 kbar. The main effect
seems to be the band broadening. This was also
evident from the band structure (Fig. 3).

I.L. Spain: One of the things that worries me is your use of
the residual resistivity plot. Normally residual

resistivities are associated with both defects and impurities. There would have to be some pretty strong justification as to why one can make comparison of the sort you make in that part of the plot. Would you like to comment on that?

J. Wittig: I mean, if you dissolve copper in silver, you have a certain amount of scattering by the different scattering centers, and the residual resistivity is not only caused by physical defects, but also by chemical defects. We are here looking at the chemical component. There was an additional background scattering just because the rare earth impurities were very impure at that time. This has been subtracted in this plot. This may be a slightly questionable procedure.

B.T. Matthias: One more comment, Zachariasen measured the number of f-electrons in lanthanum. That can be done without any great problem, as far as he's concerned, by determining the size of the radius, and he came to the conclusion that the concentration of 4f-electrons in lanthanum can under no circumstances exceed 0.05 of the net f-electron per atom.

J. Wittig: Landau and Lifshitz (Quantum Mechanics, Vol. 3, Pergamon Press, 1959, P. 245) criticized the chemists' listing Lutetium as the last member of the rare earth group. Instead, they place Lutetium together with Lanthanum in the 5d period below Yttrium. Our paper shows that there is new evidence from high pressure metal physics that Lu is indeed not a rare earth element. In addition, however, there are strong grounds now to believe that La metal belongs to the 4f series proper, rather than to the 5d series.

PRESSURE-INDUCED COVALENT-METALLIC TRANSITIONS

S. Minomura

Institute for Solid State Physics, University of Tokyo

Tokyo 106, Japan

ABSTRACT

This review will be concerned with those aspects of the covalent-metallic transitions in crystalline and amorphous semiconductors under pressure that provide information on lattice and electron instabilities. The emphasis will be on the difference of transition processes between crystalline and amorphous semiconductors. The new characteristics of structures and electron states for the high-pressure modifications of Si, Ge, III-V compounds, Se, Te and chalcogenides will be reported. The lattice instabilities will be evidenced by the softening of Raman active modes with pressure. The current status of theoretical attempts to describe the covalent-metallic transitions under pressure will be presented.

INTRODUCTION

Crystalline semiconductors are defined by periodicity or long-range order. Amorphous semiconductors lack periodicity, but the atoms may have some local organization or short-range order. The structural characterization of amorphous semiconductors remains controversial and the most models of prototype structures are described as random networks or microcrystallites. The random network models [1-3] involve the deviations and variations in topology of local atomic environments which are fitted to the radial distribution functions, the bond-angle and dihedral angle distributions, the ring statistics, the densities and the strain energies. The microcrystallite models [4-6] propose the existence of randomly oriented microcrystals which are connected by disordered boundaries.

Covalent crystalline and amorphous semiconductors are classified with common properties which are characterized by the nature of chemical bonds. The first part of this review will describe the covalent-metallic transitions in tetrahedrally bonded semiconductors of Si, and Ge and some III-V compounds under pressure. The second part will deal with the transitions in lone pair semiconductors of Se, Te and some chalcogenides under pressure.

TETRAHEDRALLY BONDED SEMICONDUCTORS

The quantum dielectric theory developed by Phillips and Van Vechten[7],[8] for tetrahedrally bonded crystalline semiconductors has refined the homopolar covalent contribution E_h and the heteropolar ionic contribution C to the energy gap E_g between bonding and antibonding (sp^3) hybridized orbitals, as given by

$$E_g^2 = E_h^2 + C^2 . \qquad (1)$$

The values of E_g are derived from the low-frequency electronic dielectric constant or the square of the optical index of refraction and the plasma frequency of all the valence electrons. In the elemental crystals, C = 0 by definition so $E_g = E_h$. In the diatomic crystals of formula $A^N B^{8-N}$, C is defined as the electronegativity difference and the ionicity factor f_i is given by

$$f_i = C^2/E_g^2 . \qquad (2)$$

The covalent radii for all elements of the same row are assumed equal. The changes in covalent radii for the transformations from fourfold to sixfold coordination are given by the hypothesis, $r_6 = 1.0925 \, r_4$. This hypothesis has been fitted to the empirical change in volume for the transformation from α-Sn to β-Sn, $V_{\alpha\beta} = -0.209 \, V_\alpha$. The pure covalent-metallic transition energies for the elemental and diatomic crystals give a good fit to a curve as a function of pressure.

Tetrahedrally bonded crystalline semiconductors under pressure show the covalent-metallic transitions accompanied by discontinuous changes in electrical resistivity and structure.[9-12] Crystalline Si, Ge, GaSb and InSb with lower ionicity factors (0 <f_i <0.321) transform to a β-Sn structure, while crystalline InAs, InP and II-VI compounds with higher ionicity factors (0.357 <f_i < 0.717) transform to a NaCl structure. The critical ionicity factor of $f_i = 0.785$ separates the tetrahedrally bonded crystals from the NaCl-type crystals at normal temperature and pressure.

Tetrahedrally bonded amorphous semiconductors under pressure show new aspects of the covalent-metallic transitions which differ

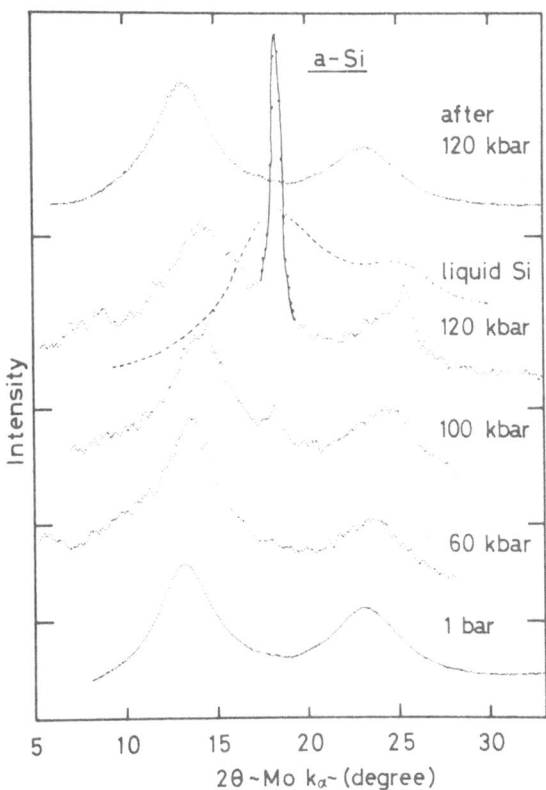

Figure 1a. The X-ray diffraction patterns for amorphous Si under
pressure: 1 bar - 120 kbar.

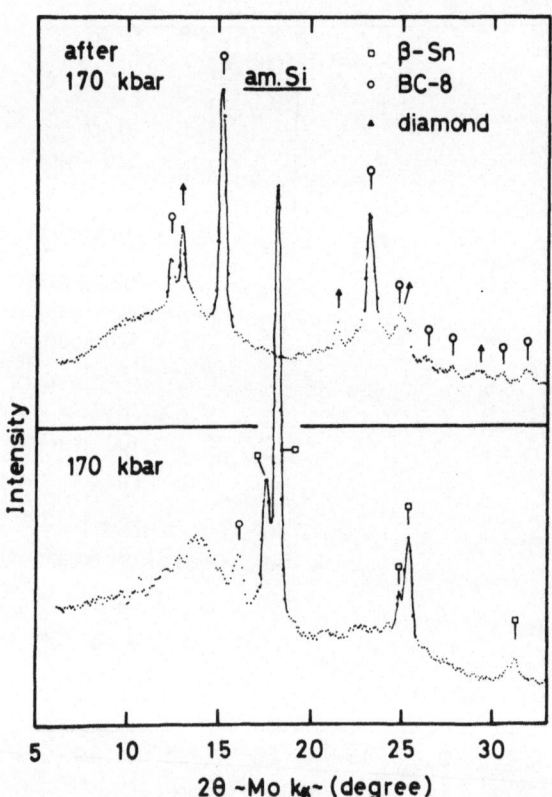

Figure 1 (b).   The X-ray diffraction patterns for amorphous Si under
pressure: 170 kbar.

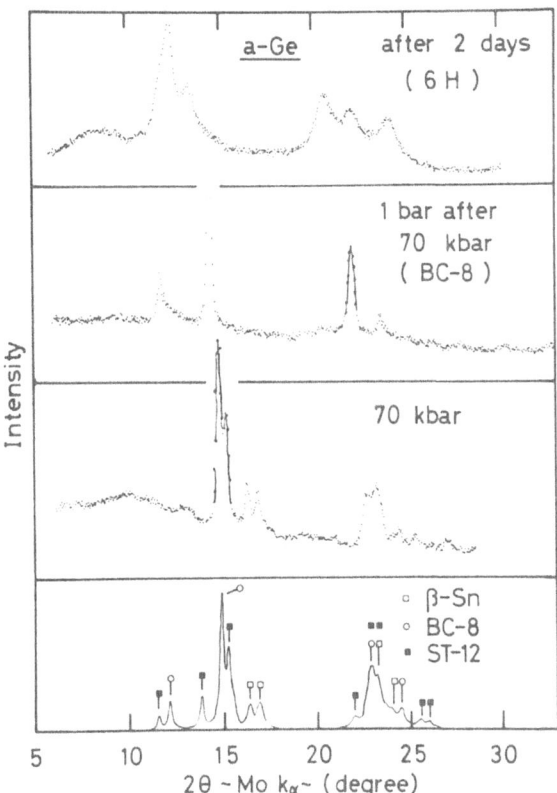

Figure 2(a). The X-ray diffraction patterns for amorphous Ge under pressure: 70 kbar.

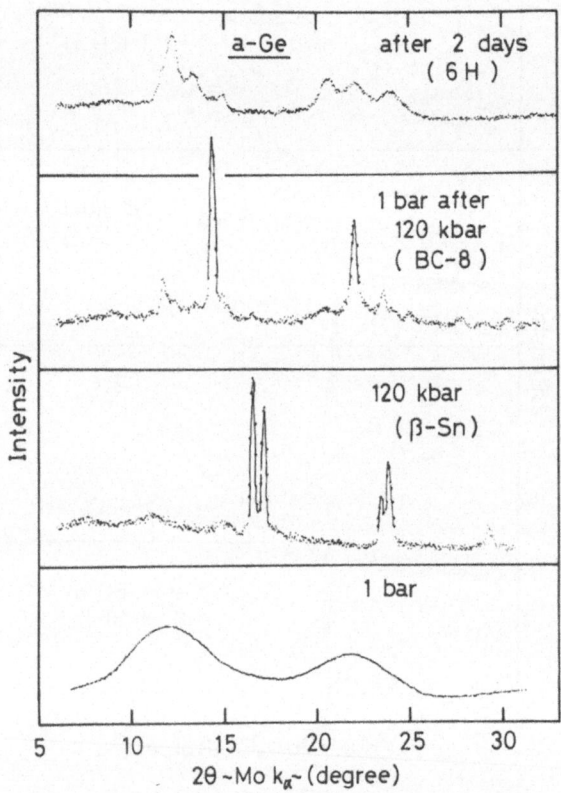

Figure 2(b).   The X-ray diffraction patterns for amorphous Ge under
              pressure: 1 bar - 120 kbar.

from those of the crystals.[13-15] The transition pressures of amorphous semiconductors are much lower. The associated changes in electrical resistivity are characterized by a sharp drop at the transition pressure for amorphous Si and Ge, a continuous decrease in the wide range of pressure for amorphous GaSb and GaAs, and a sluggish decrease at the critical pressure for amorphous InSb and InAs.

Amorphous Si under pressure shows a transition to an intermediate metallic modification at about 100 kbar with a further transition to a distorted β-Sn structure at about 150 kbar as in Fig. 1. The intermediate metallic modification of Si is evidenced by the broad X-ray patterns which correspond to the (101) and (211) spacings of the β-Sn structure. After releasing pressure, this modification transforms to the amorphous state. Amorphous Ge under pressure shows a series of transitions to an intermediate metallic modification with a distorted β-Sn structure and an intermediate semiconducting modification with a body centered cubic (BC-8) or diamond (FC-2) structure at about 60 kbar, with a further transition to a distorted β-Sn structure at about 100 kbar as in Fig. 2. After releasing pressure, a hexagonal modification with a 6H structure is recovered from a cubic anvil apparatus, while a cubic (FC-2) or tetragonal (ST-12) modification is recovered from an opposed anvil apparatus. Amorphous GaSb under pressure transforms to the metallic modification which remains amorphous at about 25 kbar as in Fig. 3. With increasing pressure, the diffuse haloes for the metallic modification at higher angles continuously increase, while those for the tetrahedrally bonded material at lower angles decrease. After releasing pressure, the metallic modification of GaSb transforms to the terahedrally bonded material. Amorphous GaAs under pressure appears to be an orthorhombic structure at about 90 kbar. Amorphous InSb under pressure shows a transition to an intermediate metallic modification with a NaCl structure below 15 kbar, with a further transition to a β-Sn structure at about 30 kbar as in Fig. 4. Amorphous InAs under pressure also transforms to the NaCl structure. After releasing pressure, the NaCl-type modification of InSb and InAs can be retained at atmospheric pressure.

The surprising result is that the NaCl-type modification of InSb can be prepared either by recovering from the high-pressure experiments or by tetrode sputtering with anode voltage above 60 V at argon pressure of 10^{-4} Torr onto glass substrate from the crystalline target. The amorphous semiconductor of InSb can be prepared by tetrode sputtering with anode voltage below 60 V. The structures of sputtered films are not affected by the deposition rate, argon pressure, substrate and thickness, but they reflect the electron states in the flying ions. The sputtered films of InSb with the NaCl structure are in the polycrystalline state. The lattice constant of 6.12 Å and the density of 6.92 g/cm^3 are equal to those of the high-pressure modification. It shows an exothermal

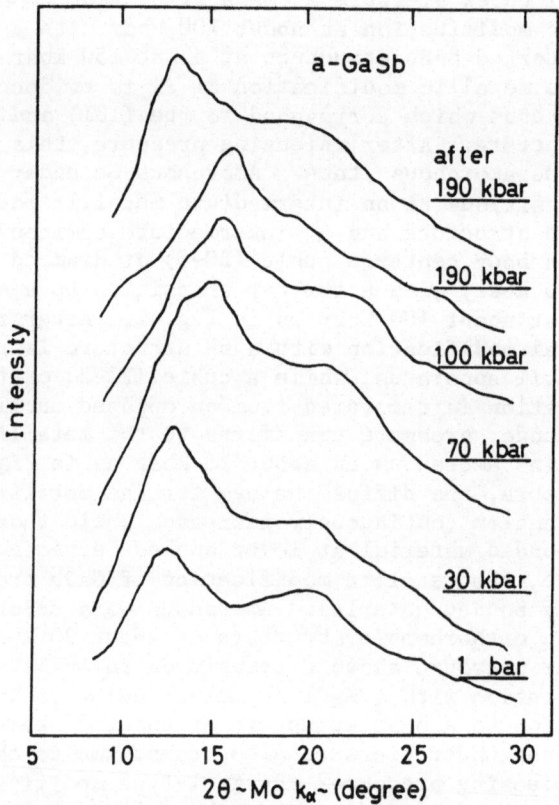

Figure 3(a). The X-ray difraction patterns for amorphous GaSb under
pressure: the first and second diffuse haloes.

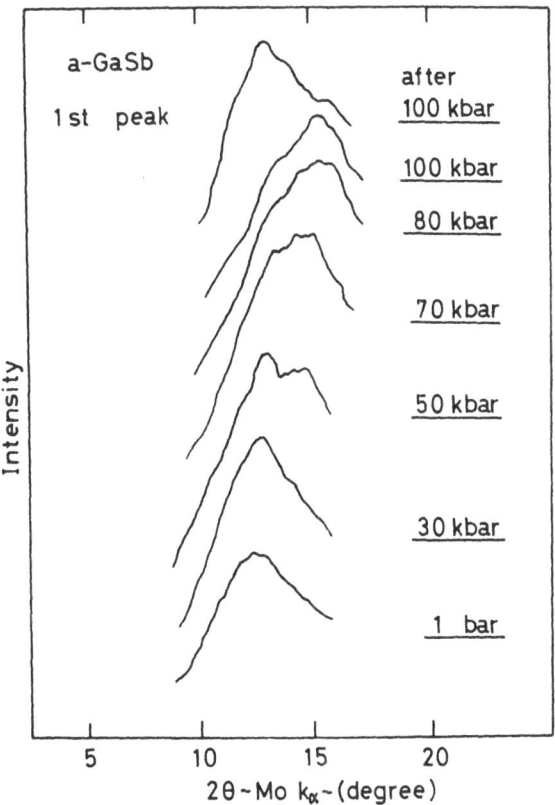

Figure 3(b).   The X-ray diffraction patterns for amorphous GaSb under pressure: the first diffuse haloes.

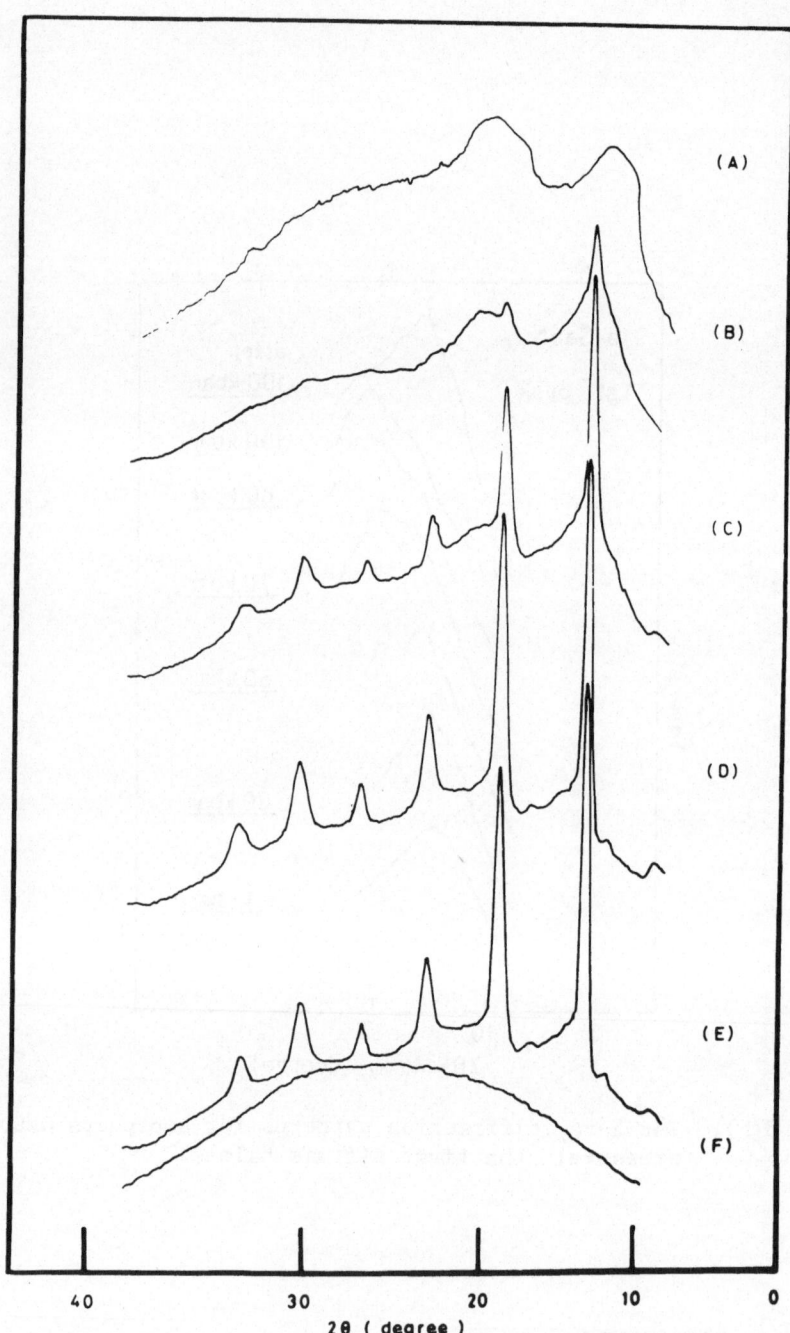

Figure 4(a).   The X-ray diffraction patterns for amorphous InSb under
               pressure: (A), 1bar; (B), 4kbar; (C), 14 kbar;
               (D), 20 kbar; (E), after 20 kbar; (F), diamond anvil
               background.

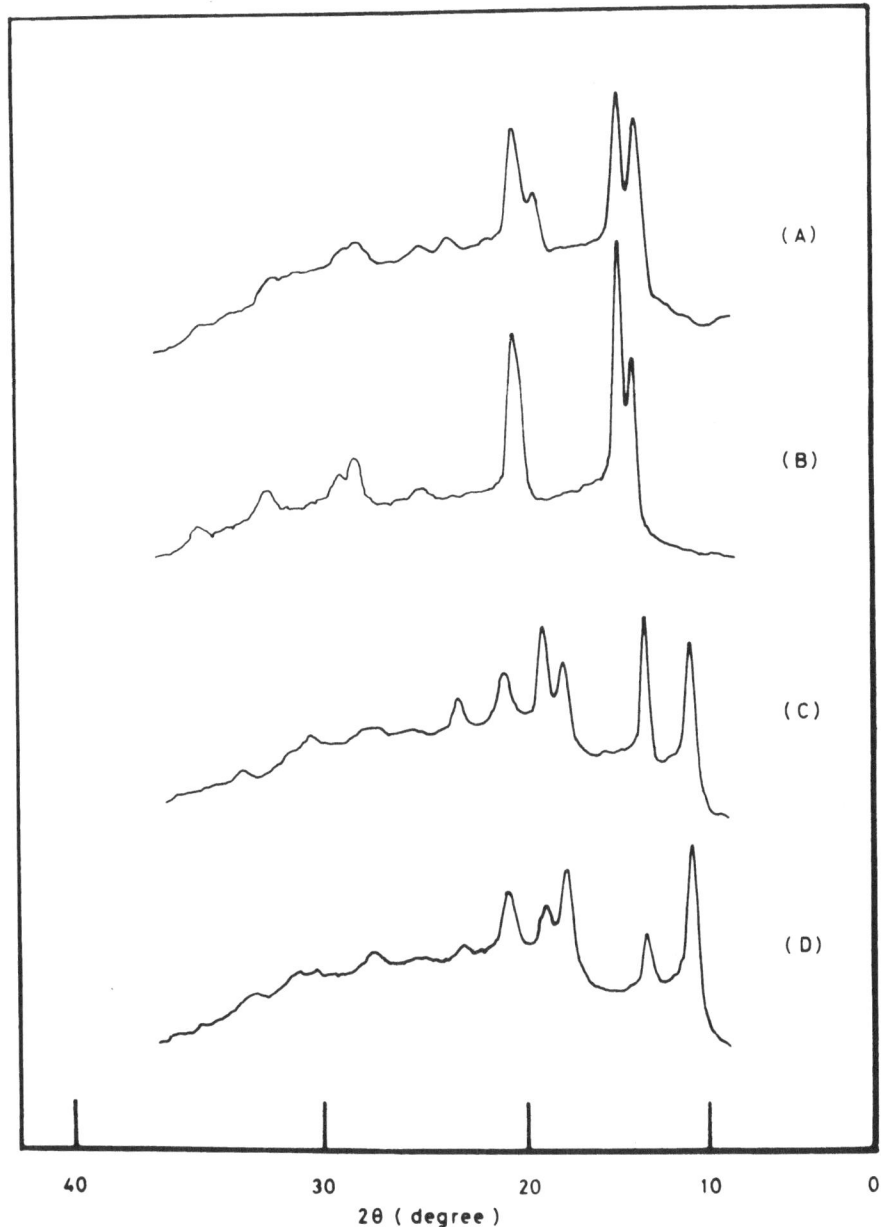

Figure 4(b): The X-ray diffraction patterns for amorphous InSb under
pressure: (A), a modified bct structure at 28 kbar;
(B), a distorted β-Sn structure at 30 kbar; (C), on
releasing pressure; (D), a mixture of NaCl- and
ZnS-type structure after 30 kbar.

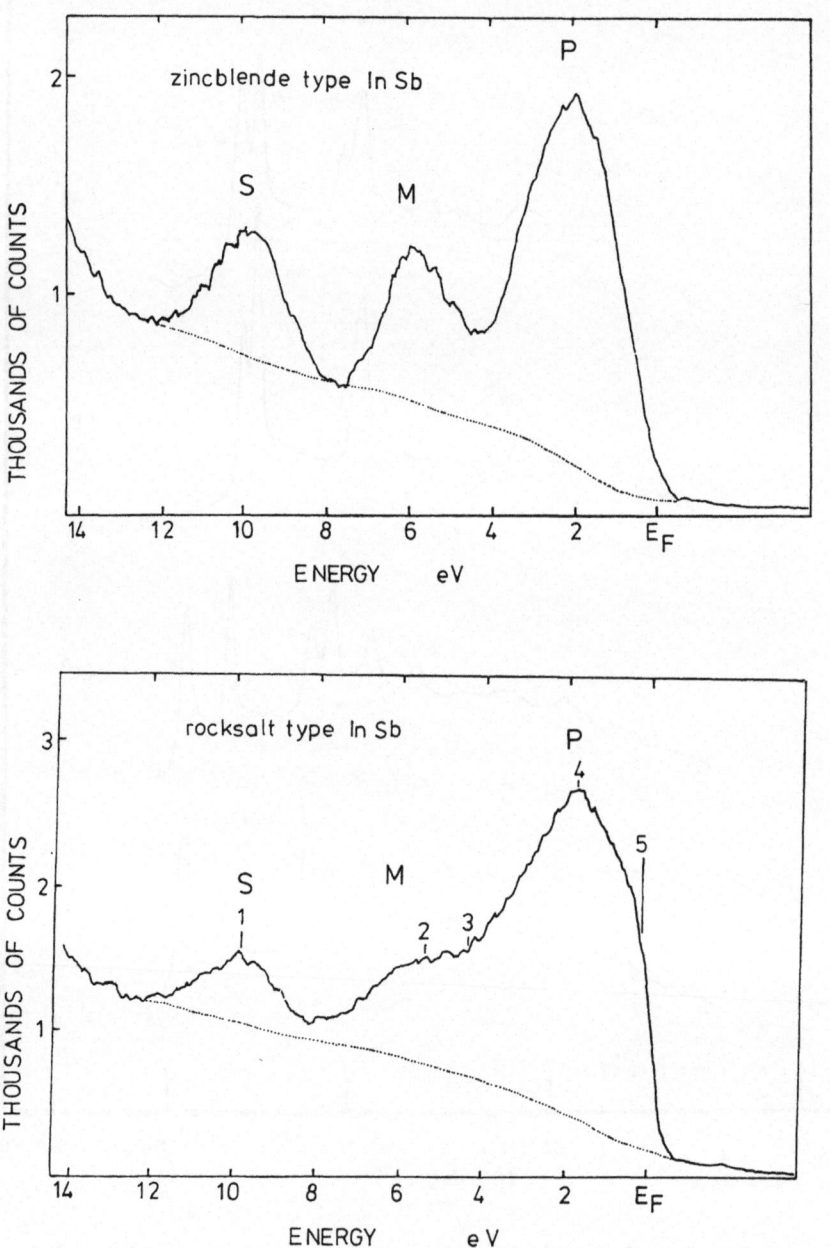

Figure 5.   The valence band X-ray photoemission spectra for ZnS-
           and NaCl-type crystals of InSb.

transformation to a ZnS structure. The interesting difference between the ZnS- and NaCl-type crystals of InSb is found in the valence band X-ray photoemission spectra as in Fig. 5. In the ZnS-type crystal, the valence band spectra consist of three regions arising predominantly from p-like bonding states (P), s- and p-like hybridized states (M), and s-like localized states (S).[16] In the NaCl-type crystal, the densities of states increase in the P region whereas they decrease in the S region. The spectra of the P and M regions are shifted to higher energies. The upper edge at the Fermi level rises more sharply. The spectra of the S region preserve the separation. These features suggest that the nature of chemical bonds in the NaCl-type modification of InSb appears to be a mixture of covalent and ionic as well as metallic character. This modification at 4.2 K shows the electrical resistivity of 9.5×10^{-5} Ωcm and the negative magnetoresistance in the order of 10^{-4}. The Hall coefficient is negative and decreases with increasing magnetic field: -14.5×10^{-4} cm^3/C at the initial field, -3.3×10^{-4} at 20 kG and -1.0×10^{-4} at 45 kG. The characteristics of electrical transport properties at low temperature indicate that the NaCl-type modification is not a semimetal, but it is an uncompensated metal with a negative direct energy gap between valence and conduction bands.

Weinstein and Piermarini[17] have measured the frequency shift of the first and second order Raman active phonons in crystalline Si and GaP under pressure up to 135 kbar. The first order Raman spectra for the TO and LO modes at the zone center are shifted to higher frequencies, while the second order Raman spectra for the TA mode at the zone boundary are shifted to lower frequencies. The softening of the TA mode is explained by the weakening of the effective ion-bond-charge coupling and the associated noncentral forces which help to stabilize the structure against the shear distortions. It appears that the covalent-metallic transitions in tetrahedrally bonded crystalline and amorphous semiconductors under pressure are connected with the shear instabilities. The deviations of bond angles and dihedral angles in the amorphous semiconductors cause the lower critical pressure. The difference of the pure covalent-metallic transition energies between crystalline and amorphous semiconductors is approximately equal to the heat of crystallization.

The superconducting modifications have been found in crystalline and amorphous Si, Ge[13,18] and InSb,[14,19-21] and crystalline GaSb[22] and AlSb.[23] The transition temperature T_c and the critical magnetic field H_c are given in Table 1. The modifications of Si and Ge show the negative pressure dependence of transition temperature, $dT_c/dP = -2.1 \sim -2.5 \times 10^{-2}$ K/kbar. The four modifications of InSb show the transition temperatures varying from 2.1 to 4.3 K. The crystalline modifications behave much like the amorphous modifications. These modifications appear to be a hard superconductor with high critical field.

Table I. The superconductivity of high-pressure modifications.

Material	P kb	T K	Structure	T_c K	dT_c/dP 10^{-2} K/kb	H_c kG	Ref.
c-Si	150		β-Sn	6.7			18
a-Si	100			7.6	-2.5	3.3	13
c-Ge	100		β-Sn	5.3	-2.1		18
a-Ge	60			5.5	-2.5	1.7	13
c-InSb	30		β-Sn	2.1 (1b)			19
	30		orthor.	3.4		1.3	20
	60	500		4.1			21
a-InSb	15		NaCl	3.4 (1b)		9.5	14
c-GaSb	67	473	β-Sn	4.2 (1b)		2.6	22
c-AlSb	104		β-Sn	2.8			23
c-Se (m)	100			7			33
a-Se	105			7			34
c-Te (t)	40		monoc.	3.3		0.25	35
a-As$_2$Te$_3$	100			4.4			26

LONE PAIR SEMICONDUCTORS

Lone pair crystalline and amorphous semiconductors of chalcogen elements (S, Se and Te) and chalcogenide compounds (As_2S_3, As_2Se_3 and As_2Te_3) are characterized by the uppermost valence band which is derived from the nonbonding orbitals. Trigonal Se and Te consist of infinite helical chains. Orthorhombic S and monoclinic Se are composed of eight-membered rings. Within the chains or rings each atom is tightly bonded to two neighbors with covalent character. The bonding between individual chains or rings is much weaker. Monoclinic As_2S_3 and As_2Se_3 are described as a puckered layer structure (orpiment structure), which consists of twelve-membered rings with threefold As and twofold S or Se. The bonding between layers is much weaker than the intralayer bonding. Monoclinic As_2Te_3 is built up of three-dimensional covalent networks with fourfold and sixfold As and threefold Te.

Trigonal, α-monoclinic and amorphous Se under pressure show a covalent-metallic transition accompanied by a discontinuous change in electrical resistivity at about 180, 100 and 105 kbar, respectively, as in Fig. 6.[24] The amorphous films of Se which are aged in the air for a period of month show a reverse transition accompanied by an upward drift of resistivity at about 105 kbar, with a further transition to the metallic state at about 180 kbar.

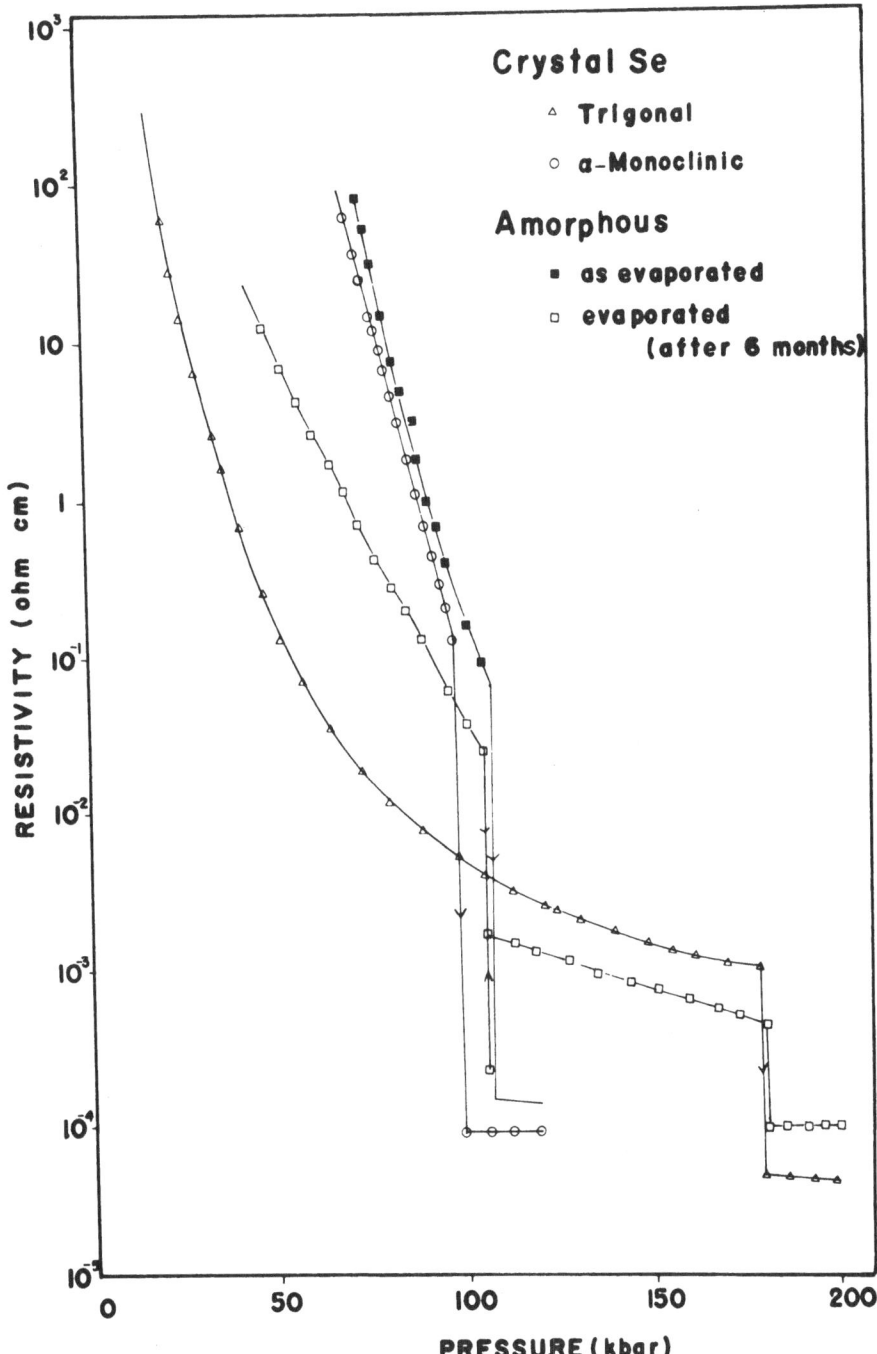

Figure 6.   The changes in electrical resistivity with pressure for
trigonal, α–monoclinic and amorphous Se.

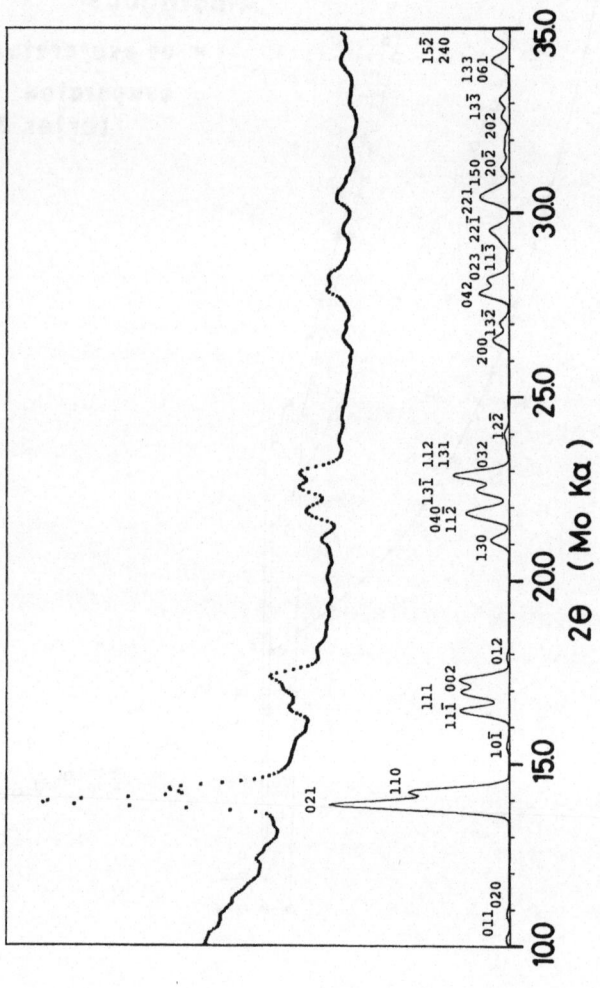

Figure 7.   The X-ray diffraction patterns for monoclinic Te at 45 kbar.

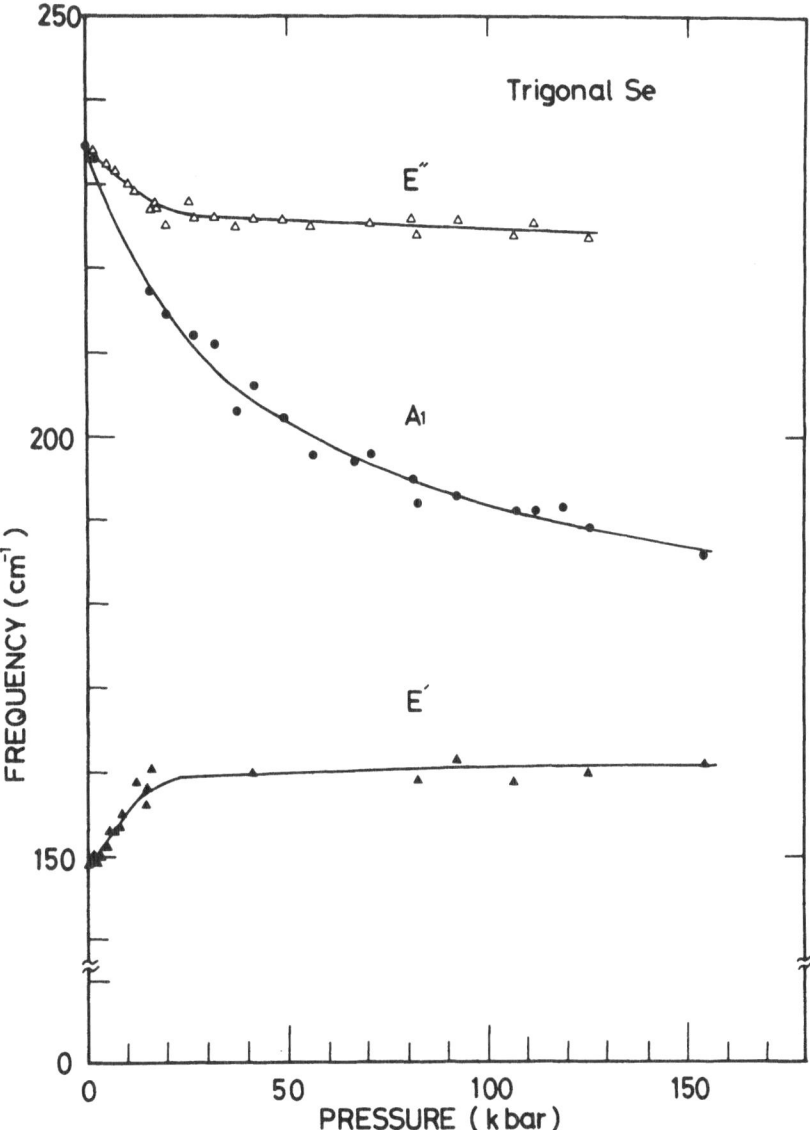

Figure 8. The frequency shift of the first order Raman active phonons for trigonal Se under pressure.

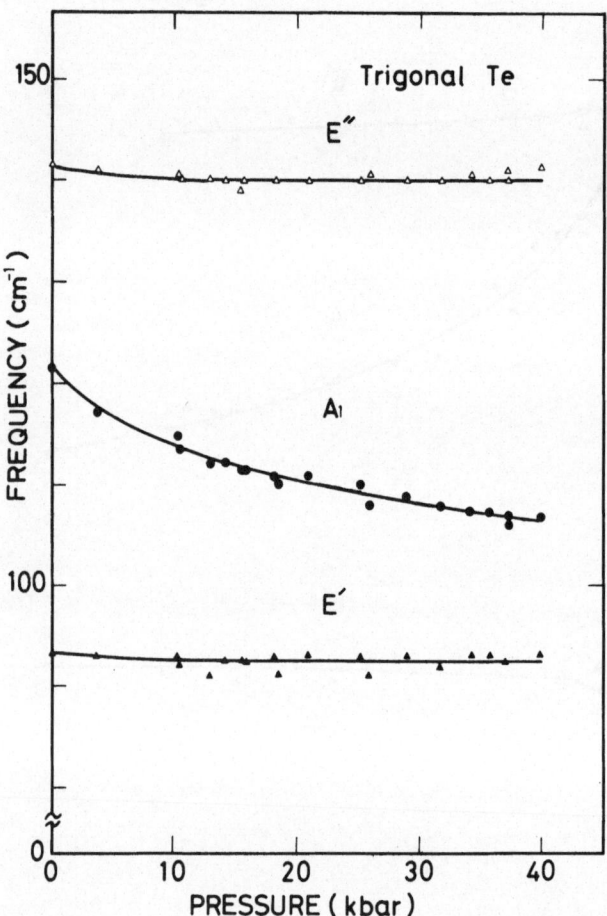

Figure 9.   The frequency shift of the first order Raman active
            phonons for trigonal Te under pressure.

Trigonal Te under pressure also shows the transition at about 40 kbar.[25] Amorphous As_2Te_3 and $Ge_{16}As_{35}Te_{28}S_{21}$ under pressure show the metallic behaviour accompanied by a continuous decrease in resistivity at about 100 and 150 kbar, respectively.[24,26]

The X-ray diffraction patterns taken for Te at 45 kbar are shown in Fig. 7.[27] These patterns are indexed as monoclinic, space group C_2^2 . The lattice parameters are a = 3.104 Å, b = 7.513 Å, c = 4.766 Å, and β = 92.709°. The calculated density is 7.66 g/cm^3, which is higher by a factor of about 1.226 than the density of normal phase. This structure is described as puckered layers. Each atom has four neighbors within the same layer and four next neighbors within the same double layers. Trigonal Se and Te under pressure show a rapid decrease in lattice parameter a or interchain atomic distance and a small increase in lattice parameter c or intrachain bond angle. The structural transformation occurs at the c/a ratio of about 1.42 both for trigonal Se and Te.[28,29] The structures of high-pressure phases of Se remain unsolved. However, the X-ray diffraction patterns taken for trigonal, α-monoclinic and amorphous Se under pressure appear to be similar to those of monoclinic Te. Jamieson and McWhan[28] have reported that Te under pressure show a further transition to β-Po structure.

Trigonal Se and Te under pressure show a nonlinear softening of the Raman active modes as in Figs. 8 and 9.[27] With increasing pressure, the phonon frequencies of the A_1 mode decrease non-linearly. The phonon frequencies of the two E modes are shifted by about 10 cm^{-1} to the opposite directions for Se up to 25 kbar, but they are shifted by a small amount within 3 cm^{-1} for Se above 25 kbar and Te up to 40 kbar. Richter et al.[30] have reported that all the Raman frequencies decrease linearly with increasing pressure up to 8 kbar. The lattice dynamics of trigonal Se and Te have been studied on the basis of the valence force field models.[31,32] The lattice energy is represented in terms of intrachain and interchain force constants which are determined by the observed six elastic constants. The softening of the Raman active modes is explained by the interference of interchain interactions with the intrachain bonding. The nonlinear softening suggests that the force constants and the geometrical coefficients vary non-linearly with pressure.

The superconducting modifications have been found in monoclinic and amorphous Se,[33,34] trigonal Te,[35] and amorphous As_2Te_3.[26] The values of T_c are given in Table 1. The two modifications of Se at about 105 kbar become superconducting at about 7 K. The modification of Te at about 56 kbar becomes superconducting at about 3.3 K with a critical magnetic field of about 0.25 kG. It appears to be a soft superconductor. The modification of As_2Te_3 at about 100 kbar becomes superconducting at about 4.4 K.

REFERENCES

1. D.E. Polk, J. Non-Cryst. Solids, 5, 365 (1971).
2. D. Turnbull and D.E. Polk, J. Non-Cryst. Solids, 8 - 10, 19
    (1972).
3. D. E. Polk and D.S. Boudreaux, Phys. Rev. Lett., 31, 92 (1973).
4. M.L. Rudee and A. Howie, Phil. Mag., 25, 1001 (1972).
5. A. Howie, O. Krivanek and H.L. Rudee, Phil. Mag., 27, 235
    (1973).
6. F.C. Weinstein and E.A. Davis, J. Non-Cryst. Solids, 13, 153
    (1973).
7. J.C. Phillips, Bonds and Bands in Semiconductors (Academic
    Press, New York and London, 1973).
8. J.A. Van Vechten, Phys. Rev. B, 7, 1479 (1973).
9. S. Minomura and H.G. Drickamer, J. Phys. Chem. Solids, 23 451
    (1962).
10. G.A. Samara and H.G. Drickamer, J. Phys. Chem. Solids, 23, 457
    (1962).
11. J.C. Jamieson, Science, 139, 762 & 845 (1963).
12. P.L. Smith and J.E. Martin, Phys. Lett., 6, 42 (1963) & 19,
    541 (1965).
13. O. Shimomura, S. Minomura, N. Sakai, K. Asaumi, K. Tamura,
    J. Fukushima and H. Endo, Phil. Mag., 29, 547 (1974).
14. O. Shimomura, K. Asaumi, N. Sakai and S. Minomura, Phil. Mag.,
    34, 839 (1976).
15. S. Minomura, O. Shimomura, K. Asaumi, H. Oyanagi and K. Takemura,
    Proceeding of the 7th International Conference on Amorphous
    and Liquid Semiconductors, ed. by W.E. Spear (Edinburgh, Scot-
    land, 1977).
16. L. Ley, R.A. Pollak, F.R. McFeely, S.P. Kowalczyk and
    D.A. Shirley, Phys. Rev. B, 9, 600 (1974).
17. B.A. Weinstein and G.J. Piermarini, Phys. Rev. B, 12, 1172
    (1975).
18. J. Wittig, Z. Phys., 195, 228 (1966).
19. B.R. Tittman, A.J. Darnell, H.E. Bommel and W.F. Libby, Phys.
    Rev. A, 135, 1460 (1964).
20. D.B. McWhan and M. Marezio, J. Chem. Phys., 45, 2508 (1966).
21. M.D. Banus and M.C. Lavine, J. Appl. Phys., 40, 409 (1969).
22. D.B. McWhan, G.W. Hull, T.R.R. McDonald and E. Gregory, Science,
    147, 1411 (1965).
23. J. Wittig, Science, 155, 685 (1967).
24. S. Minomura, K. Aoki, O. Shimomura and K. Tanaka, Electronic
    Phenomena in Non-Crystalline Semiconductors, ed. by B.T.
    Kolomiets (Academy of Sciences of USSR, Leningrad, 1976) p.289.
25. P.W. Bridgman, Proc. Am. Acad. Arts Sci., 81, 165 (1952).
26. N. Sakai and H. Fritzsche, Phys. Rev. B, 15, 973 (1977).
27. S. Minomura, K. Aoki, N. Koshizuka and T. Tsushima, Proceedings
    of the 6th AIRAPT International High Pressure Conference, ed.
    by K.D. Timmerhaus (Boulder, Colorado, 1977).
28. J.C. Jamieson and D.B. McWhan, J. Chem. Phys., 43, 1149 (1965).

29. D.R. McCann and L. Cartz, J. Chem. Phys., $\underline{56}$, 2552 (1972).
30. W. Richter, J.B. Renucci and M. Cardona, Phys. Status Solidi B, $\underline{56}$, 223 (1973).
31. R.M. Martin and G. Lucovsky, Phys. Rev. B, $\underline{13}$, 1383 (1976).
32. T. Nakayama and A. Odajima, J. Phys. Soc. Japan, $\underline{33}$, 12 (1972).
33. J. Wittig, J. Chem. Phys., $\underline{58}$, 2220 (1973).
34. A.R. Moodenbaugh, C.T. Wu and R. Viswanathan, Solid State Commun., $\underline{13}$, 1413 (1973).
35. B.T. Mattias and J.L. Olsen, Phys. Lett., $\underline{13}$, 202 (1964).

QUESTIONS AND COMMENTS

I.L. Spain: I would like to comment that we have measured the crystalline (cubic) form of GaSb transforming to the high pressure (tetragonal) form at 62+2 kbar.

E.F. Skelton: I noticed your lattice curves seem to have a lot of structure in them. Is there any possibility in analyzing them in terms of their radial distribution function?

S. Minomura: No, we have not. We are going to do it with a different and more precise technique. What you saw is only a photograph, not a precise data-plot.

NEW HYDROSTATIC PRESSURE RESULTS ON SULPHUR IMPURITY CENTER IN GaSb

L. Dmowski, S. Porowski

High Pressure Research Center, Polish Academy of Sciences

Warsaw

and

M. Baj

Institute of Experimental Physics, University of Warsaw

Warsaw

INTRODUCTION

The pressure measurements of the resistivity of GaSb samples doped with sulphur at temperature T = 300K reported by B.B. Kosicki (1) revealed the steep increase of the resistivity with pressure. This increase was thought to result from deionization of sulphur impurity level. From the slope of the experimental dependence $\rho(p)/\rho_0$ for pressures p>10kbar, B.B. Kosicki found the pressure coefficient of the sulphur donor level relative to L minimum $\partial(E_D-E_L)/\partial p = -6.4$ meV/kbar which in relation to Γ minimum gives the value $\partial(E_D-E_\Gamma)/\partial p = -16$ meV/kbar. Basing on the obtained value of the pressure coefficient B.B. Kosicki related the observed level to the X minimum of the conduction band.

On the other hand A.Ya.Vul´ et al. (2) presented the results of the resistivity measurements as a function of <100> and <111> uniaxial stress at various temperatures and basing on the piezo-resistance tensor symmetry, they related the observed sulphur donor level to the L minimum of the conduction band.

The following paper aims at explaining the inconsistencies of the above opinions. Our measurements were carried out at pres-

sures up to 14 kbar. The range of pressures used includes pressures p > 10 kbar for which the analysis presented by B.B. Kosicki
is surely correct as well as the range of lower pressures for which
the results of B.B. Kosicki and A.Ya, Vul' et al. are inconsistent.

EXPERIMENTAL RESULTS

Measurements of the Hall coefficient, R_H, and conductivity, σ,
as a function of pressure p, up to 14 kbar at fixed temperatures
T = 48.8K, 140K, 160K, 206K, 251K, 293K, and 350K have been made
on GaSb samples doped with sulphur. Complementary measurements
of R_H and σ as a function of temperature have been made within the
temperature range 4.2K - 458K at atmospheric pressure.

In the experiment standard DC method was used. The samples were
rectangular bars 0.7 x 1.4 x 8.0 mm. The ohmic contacts were soldered
with indium containing 1% tellurium or with tin in the case of measurements carried out at temperatures exceeding 300K. The temperature
was stabilized with the accuracy better than 0.5K. The samples were
mounted in the high-pressure Be-Cu bomb which was placed between the
electromagnet pole pieces. Helium gas was used as a pressure transmitting medium. The experimental bomb was connected to the helium
gas compressor by a flexible high pressure capillary tube.

Within the temperature range T$\geq$ 120K the temperature dependence
of the Hall coefficient is determined by change of the occupation
of the deep sulphur impurity level. At temperatures T<120K the
changes of the resistivity and the Hall coefficient with time are
observed. It means that the change of the occupation of the sulphur
impurity level takes place with some relaxation time. When the temperature is decreased further this process becomes so slow that the
occupation of the electron level at temperatures T<60K becomes metastable. According to the rate of cooling process different values
of the Hall coefficient and the resistivity i.e. different metastable occupations at T = 60K can be obtained. Similar effects were
observed for CdTe doped with Cl (3). Within the temperature range
4.2K<T<60K the temperature dependence of the hall coefficient reveals the existence of another impurity level much shallower than the
sulphur impurity level that is active at higher temperatures.
Figures 1 and 2 present the pressure dependence of the Hall coefficient for different temperatures. The increase of the Hall coefficient with pressure is determined by the increase of the occupation of the electron impurity level. Fig. 3 shows the pressure
dependence of the Hall mobility R_H x σ for three temperatures. These
dependences exhibit the change of contribution of Γ (high mobility
of electrons) and L (low mibility) minima to the conduction process
due to the change of their mutual positions.

The pressure dependence of the resistivity is due to the change

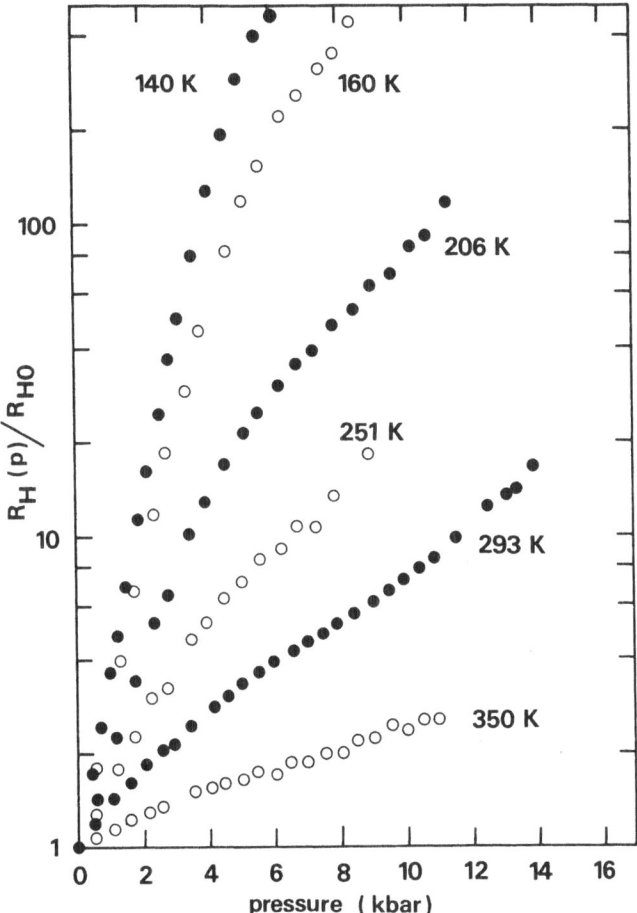

Figure 1.   Relative variation of the Hall coefficient with pressure
at different temperatures.

of the donor level position as well as the change of the band structure.   The pressure dependence of $R_H \times \sigma$ is mainly caused by the
change of the band structure with pressure.   Independent analysis
of the resistivity and $R_H \times \sigma$ as a function of pressure, enables
separation of the effects of pressure on the band structure and on
the donor level position.

ANALYSIS

Considering that electronic transport in GaSb samples takes
place in both the Γ and L minima, there are two contributions to the
conductivity, as well as to the Hall coefficient and the Hall mobility $R_H \times \sigma$:

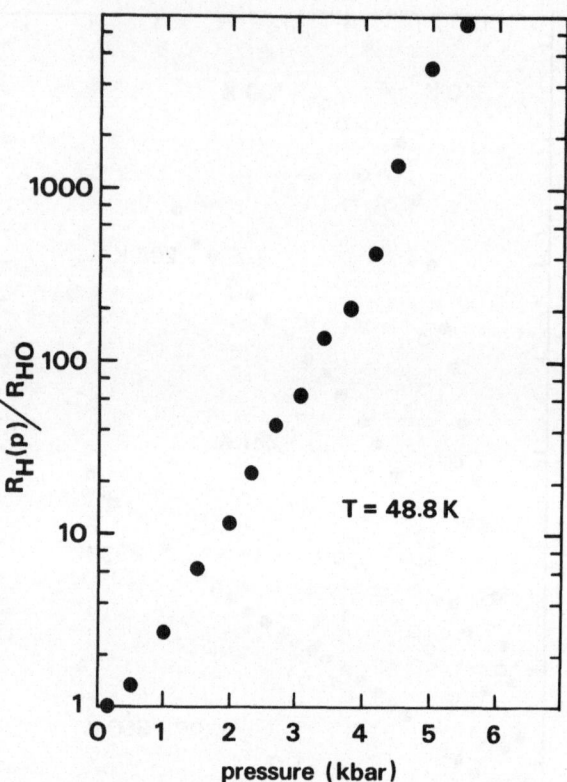

Figure 2.   Relative variation of the Hall coefficient with
            pressure at T = 48.8K.

$$\sigma = q \ (n_\Gamma b + n_L) \ \mu_L \tag{1}$$

$$R_H = \frac{1}{q} \ \frac{n_\Gamma b^2 + n_L}{(n_\Gamma b + n_L)^2} \neq \frac{1}{q \ (n_\Gamma + n_L)} \tag{2}$$

$$R_H \times \sigma = \frac{n_\Gamma b^2 + n_L}{n_\Gamma b + n_L} \mu_L \neq \mu_L \neq \mu_\Gamma \tag{3}$$

where, n_Γ and n_L are the electron concentrations in Γ and L
    minima respectively,
    μ_Γ and μ_L are the electron mobilities in Γ and L
    minima respectively; $b = \dfrac{\mu_\Gamma}{\mu_L}$

Using classical statistics and assuming the parabolic shape of the

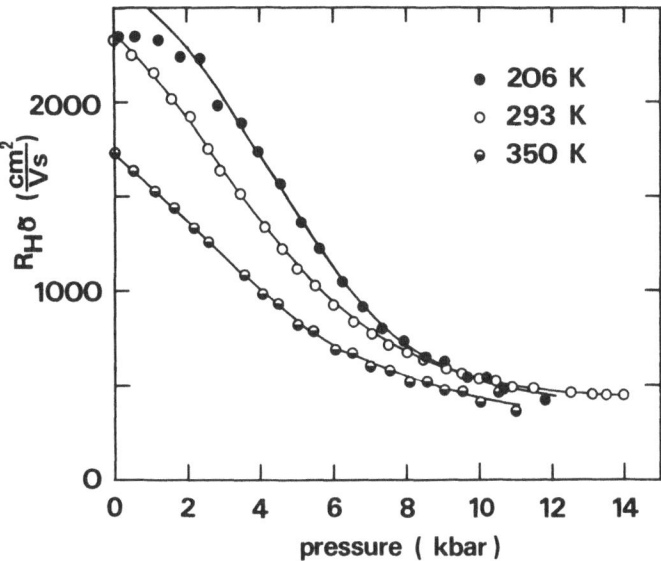

Figure 3. Typical pressure dependence of the Hall mobility $R_H \times \sigma$ at temperatures 206K, 293K, 350K. Smooth curves are the curves computed using the parameters presented in Table 1.

conduction bands the electron concentrations in Γ and L minima may be given as follows:

$$n_\Gamma = \frac{(2\pi m_\Gamma^* \, kT)^{3/2}}{4\pi^3 \, \hbar^3} \quad \exp \frac{E_F - E_\Gamma}{kT} \tag{4}$$

$$n_L = 4 \, \frac{(2\pi m_L^* \, kT)^{3/2}}{4 \, \pi^3 \, \hbar^3} \, \exp \frac{E_F - E_L}{kT} \tag{5}$$

Regarding this, the Hall mobility $R_H \times \sigma$ can be expressed by:

$$R_H \times \sigma = \frac{b^2 + 4 \left(\frac{m_L^*}{m_\Gamma^*}\right)^{3/2} \exp\left(-\frac{(E_L - E_\Gamma)(p)}{kT}\right)}{b + 4 \left(\frac{m_L^*}{m_\Gamma^*}\right)^{3/2} \exp\left(-\frac{(E_L - E_\Gamma)(p)}{kT}\right)} \tag{6}$$

Table 1.

T(K)	b	$\mu_L \left[\dfrac{cm^2}{V.s}\right]$	a	$\gamma \left[\dfrac{meV}{kbar}\right]$
350	8.3	310	4.7	−10.5
293	7.7	340	2.6	−10.5
251	7.6	360	1.5	−10.0
206	7.0	400	0.7	−9.5
160	7.4	340	0.15	−9.5

From the expression (6) it results that the observed pressure changes of the $R_H \times \sigma$ are mainly due to the pressure change of the band
structure:

$$\left(E_L - E_\Gamma\right)(p) = \left(E_L - E_\Gamma\right)_o + \frac{\partial\left(E_L - E_\Gamma\right)}{\partial p}\Delta p$$

The fitting of the experimental data to the theoretical dependences
$R_H \times \sigma$ has yielded the parameters which enable separation of the two
contributions to the pressure changes of the Hall coefficient. The
first contribution is due to the lowering of the donor level while
the second is due to the increasing participation of the L minimum
in the electronic transport. This fitting was made assuming that
μ_L and b do not depend on pressure. This assumption seems to be
reasonable because the energy gap in GaSb is much larger than its
change caused by application of pressure. From this fitting we
have obtained the following parameters:

$$b; \quad \mu_L; \quad a = 4 \left(\frac{m_L^*}{m_\Gamma^*}\right)^{3/2} \exp\left[-\frac{\left(E_L - E_\Gamma\right)_o}{kT}\right]; \quad \gamma = \frac{\partial\left(E_L - E_\Gamma\right)}{\partial p}$$

The values of these parameters for different temperatures are presented in Table 1. Smooth curves in Fig. 3 are the curves computed
using the obtained parameters.

On the basis of experimental dependences of the Hall coefficient
$R_H(p)$, using the obtained values of parameters and the electron

effective mass $\frac{m_\Gamma^*}{m_0} = 0.047$ (4,5) the Fermi level energy E_f as a function of pressure was calculated for all temperatures at which the measurements have been made. Assuming the energy of the observed sulphur impurity level $E_D = -70$ meV at room temperature and ambient pressure (1,2,6) the number of the sulphur impurity centers N_D for different values of compensation coefficient $K = N_A/N_D$ (where N_A is the number of acceptors) was determined. Only for $K = 0.2$ could the linear dependence of E_D (p) be obtained for the whole range of applied pressures and temperatures. The value $K = 0.2$ corresponds to the number of acceptors $N_A = 1.1 \times 10^{17}$ cm^{-3}. Such a value is usually reported as a typical, minimal number of acceptors in GaSb samples nominally undoped with acceptors (7,8). Considering this, it was assumed that the most likely value of the compensation ratio is $K = 0.2$. Fig. 4 presents the pressure dependence of E_D for temperatures 140K, 160K, 206K, 293K, and 350K. From the slope of the curves the pressure coefficient $\frac{\partial (E_D - E_\Gamma)}{\partial p} = -16$ meV/kbar was determined. The temperature coefficient $\frac{\partial (E_D - E_\Gamma)}{\partial T}$ was estimated to be equal to -3.1×10^{-5} eV/K.

From the temperature dependence of the Hall coefficient within the range 4.2 K – 60K, the energy of the second level E_D' was estimated to be about -10 meV at atmospheric pressure. Regarding the existence of the deeper, partly compensated, sulphur impurity level E_D, the high compensation coefficient of the shallow impurity level E_D' should be expected. Based on the measurements of the Hall coefficient as a function of pressure at temperature $T = 48.8$K, the pressure dependence of the shallow level E_D' was determined in a similar way to that used for the deep level E_D. The results of calculation (assuming considerable compensation ratio $K > 0.5$) are shown in Figure 5. The pressure coefficient $\partial (E_D' - E_\Gamma)/\partial p$ was estimated to be about -8 meV/kbar.

CONCLUSIONS

The obtained value of the pressure coefficient of the sulphur impurity level $\partial (E_D - E_\Gamma)/\partial p = -16$ meV/kbar differs from the value -10.5 meV/kbar reported by A.Ya.Vul′ et al. (2) but is in good agreement with the value determined by B.B. Kosicki (1) for pressures $p > 10$ kbar. Fig. 6 presents the pressure dependences of the value $kT \ln[\rho(p)/\rho_0]$ for different temperatures. The slope of such a dependence taken for room temperature was accepted by A.Ya. Vul′ et al. to be independent of temperature and considered as the pressure coefficient of the sulphur impurity level. It can be seen that the

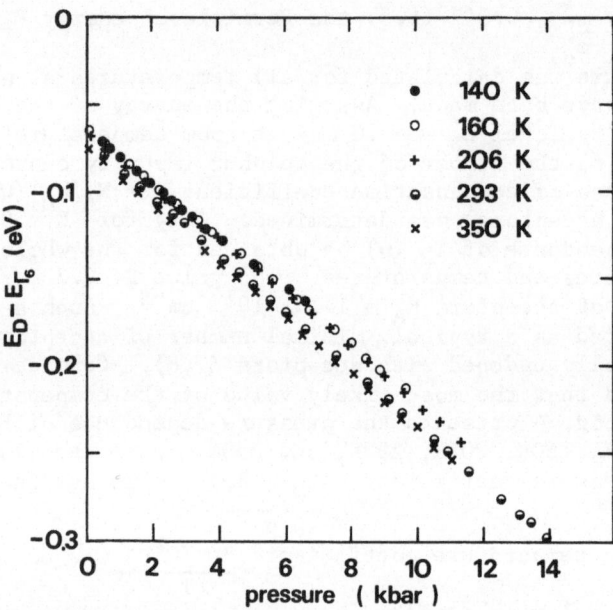

Figure 4. Pressure dependence of the sulphur impurity level at
           different temperatures.

slopes of the curves $kT\ell n[\rho(p)/\rho_o]$ strongly depend on temperature.
It gives evidence that in GaSb even at fairly low temperatures, especially in the presence of pressure, both the Γ and L minima take
place in the electronic transport. In the light of that assumption
accepted by A.Ya.Vul´ et al. that $\gamma_D = \partial(E_D - E_\Gamma)/\partial p = \ell n[\rho(p)/\rho_o]$
x kT/p is not justifiable. Considering the strong dependence of
$kT\ell n[\rho(p)/\rho_o]$ on temperature it seems that the piezoresistance tensor
symmetry in paper (2) could be also calculated incorrectly. We conclude that for pressure p<10 kbar as well as for p⩾10 kbar the same
sulphur impurity level with the pressure coefficient $\partial(E_D - E_\Gamma)/\partial p$
= - 16 meV/kbar is observed. Temperature as well as pressure
dependence of the Hall coefficient showed that, beside the sulphur
donor level active at temperatures T>120K, there is a second impurity level prominent at temperatures T<60K. It is possible that
this level may also originate from sulphur. The existence of two
energy levels with different pressure coefficients as well as slow
relaxation of the free electron concentration and metastable occupation of sulphur donor level at low temperatures, suggest that

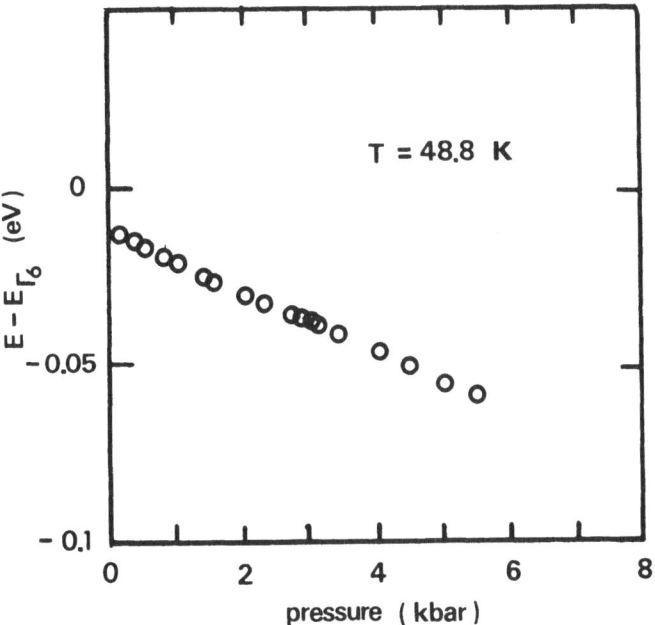

Figure 5.   Pressure dependence of the shallow impurity level ob-
            served at temperatures T<60K.

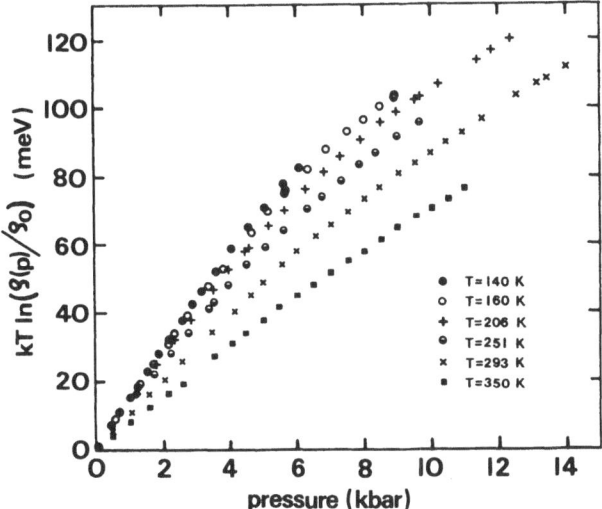

Figure 6.   Pressure dependences of kT $\ln[\rho(p)/\rho_o]$ for different tem-
            peratures. The slope of the curve for room temperature
            and low pressures was considered in paper (2) as the pres-
            sure coefficient of sulphur impurity level.

in GaSb, sulphur forms donor centers with two possible configurations and the potential barrier to the transfer between them (8). Two different donor levels observed in the transport measurements may be connected with two different configurations of the center.

REFERENCES

1. B.B. Kosicki, Harvard University Technical Report No. HP-19 1967.
2. A.Ya.Vul', G.L. Bir, Yu.V. Smartsev, Fiz. Tekh. Poluprovodn. 4, 2331 1970.
3. L. Dmowski, M. Baj, A. Iller, S. Porowski, Conf. on High Pressure and Low Temperature Physics 20-22.07.1977, Cleveland.
4. S. Zwerdling, B. Lax, K.J. Button, L.M. Roth, J. Phys. Chem. Solids 9, 320 1959.
5. G. Bordure, F. Gaustavino, C.R. ACad. Sci., Paris 267, 860 1968.
6. W.F. Boyle, R.J. Sladek, Phys. Rev. B 12, 673 1975.
7. D. Effer, P.J. Etter, J. Phys. Chem. Solids, 25, 451 1964.
8. K. Osamura, K. Nakajima, Y. Murakami, Memoirs of the Faculty of Engineering Kyoto University, Vol. XXXVII, Part 2 1975.
9. S. Porowski, M. Kończykowski, J. Chroboczek, Phys. stat. sol. (b) 63, 291 1974.

QUESTIONS AND COMMENTS

Discussion of this paper appears after the following paper.

LOW TEMPERATURE LIGHT INDUCED CHANGE OF THE CHLORINE CENTER

CONFIGURATION IN CdTe

L. Dmowski

High Pressure Research Center
Polish Academy of Sciences
Warsaw, Poland

M. Baj

Institute of Experimental Physics
University of Warsaw
Warsaw, Poland

A. Iller and S. Porowski

High Pressure Research Center
Polish Academy of Sciences
Warsaw, Poland

INTRODUCTION

The Cl atoms introduced to CdTe give two kinds of electronic levels. One of them is the simple shallow hydrogenic level observed in the temperature measurements of the Hall coefficient at low temperatures (1). The ionization energy of this level determined from experiment is close to the value predicted by the hydrogenic model. The second level was observed in the pressure measurements of the resistivity and Hall coefficient (2,3). Its energy strongly depends on pressure. Figure 1 shows the effect of the pressure induced deionization of this level on the free electron concentration at temperatures T = 297K, 240K and 186K. From the pressure dependences of the free electron concentration the energy of this level as a function of pressure was calculated. This level is situated above the bottom of the conduction band at ambient pressure (i.e.

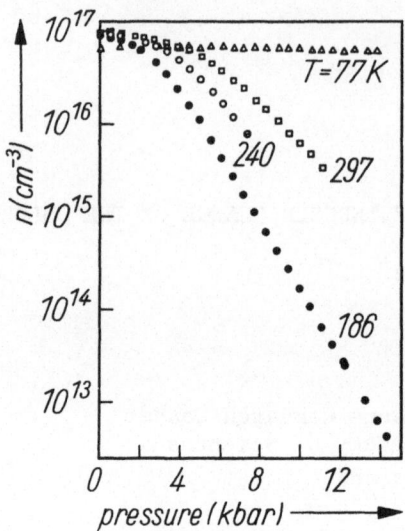

Figure 1. The free electron concentration for CdTe doped with Cl
          as a function of pressure at different temperatures.

it is the resonant state) while at 10 kbar it lies over a hundred
meV below the bottom of the conduction band. Morever it was observed (3,4,5) that under hydrostatic pressure at sufficiently low
temperatures the change of the occupation of this state takes place
with the relaxation time which at T = 160K is several minutes, at
T = 140K exceeds ten hours, whereas at even lower temperatures the
relaxation time is so long that the metastable occupation of this
level is observed. This is the reason why the free electron concentration measured at T = 77K does not depend on pressure as shown
in Fig. 1, although it could be expected that in this case the
electron concentration would decrease more steeply than at T = 186K.
However, if the sample is compressed at high temperature (e.g. 186K)
and subsequently cooled down to 77K, the free electron concentration
decreases. It becomes lower than for T = 186K shown in Fig. 1 and
remains so low even if the pressure is reduced down to one atmosphere.

In order to explain the observed effects, in papers (3.4) the
model of electron-ion excitations was used which assumes that the

change of the electron occupation of this state is followed by the change of the configuration of the system: Cl impurity center and its neighbourhood. This type of model was for the first time proposed for n-InSb (5) and recently also for CdF_2/In/ (6) and ZnTe /Cr/ (7). The model is presented in Fig. 2 where the energy E_d is the ionization energy of the Cl hydrogenic level in CdTe observed in paper (1) and the energy $E_{therm.}$ is the ionization energy determined from the electronic transport measurements under hydrostatic pressure. The states 1,2 and 3 represent the possible states of the whole system: Cl impurity center and its neighbourhood. Because of very low energy E_d, state 2 can be occupied only at very low temperatures T<<77K. At T$\geq$ 77K only states 1 and 3 can be occupied. It is evident from the above model that both states 1 and 3 must be separated by the potential barrier. At sufficiently high temperatures this barrier is of no importance and the transition between states 1 and 3 can be observed. The pressure increases the energy $E_{therm.}$ and thus it increases the occupation of state 3. However, if the temperature is sufficiently low, the transitions through the barrier are practically impossible and the metastable occupations can be observed.

In order to verify the model and to obtain further information about its parameters the optical investigations were undertaken. The above model (according to the Franck-Condon rule) predicts the possibility of optical transitions from state 3 without changing the system configuration. Such a transition is followed by the lattice relaxation and the system achieves the energy minimum in state 1. The energy of this transition $E_{opt.}$ should be much higher than the energy $E_{therm.}$ (see Fig. 2). To enable the observation of this transition, the occupation of state 3 should be large. This can be obtained by the application of high pressure and then by lowering the temperature. When the temperature is so low that thermal transitions between states 1 and 3 do not take place (i.e. they take place with the relaxation time much longer than duration of the experiment) then photons of energy higher than $E_{opt.}$ would cause the persistent increase of the free electron concentration resulting from the irreversible transitions from state 3 to state 1.

EXPERIMENTAL RESULTS

The optical transitions leading to the persistent change of the free electron concentration were investigated in CdTe /Cl/. The persistent change of the conductivity as a function of the photon energy was measured at 77K. The measurements were carried out using the high pressure Be-Cu cell with the Al_2O_3 window (9). Helium was used as a pressure transmitting medium. The conductivity measurements were carried out by the standard DC method. In order

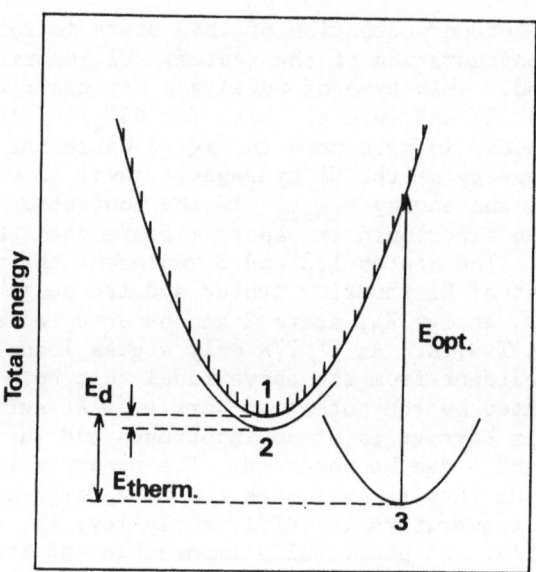

Figure 2. The energy of the system: Cl center and its nearest neighbourhood versus configuration coordinate. Curves 2 and 3 correspond to the neutral Cl center while curve 1 represents the energy of the ionized Cl center and the electron at the bottom of the conduction band. The figure presents the position of the curves at high pressure.

to provide large metastable occupation of state 3 the samples were cooled down under a pressure of about 10 kbar. The measurements were carried out within the photon energy range from 0.75 eV to 1.15 eV for three values of pressure: 1 bar, 5 kbar, 8.6 kbar. The obtained results are given in Fig. 3. It shows the change of conductivity divided by the number of photons which have fallen on the sample, as a function of the photon energy. For low absorption coefficients this value is proportional to the probability of the photoionization transition from state 3. The curves presented in Fig. 3 have the form of sharp edges which approximately determine the values of energy E_{opt}.

CONCLUSIONS

As was expected from the analysis of the model of electron-ion excitations, at sufficiently low temperature, optical excitations of the Cl centers in CdTe cause irreversible transitions from

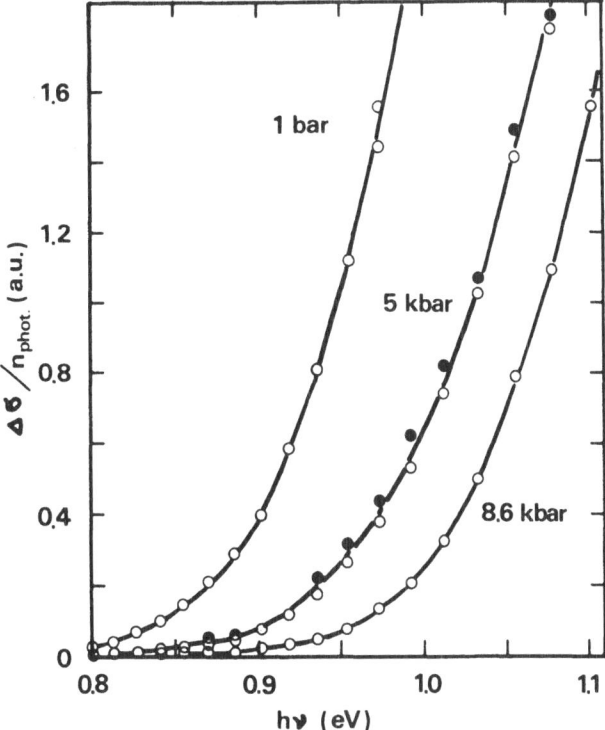

Figure 3. The change of conductivity divided by the number of
photons which have fallen on the sample, measured as a
function of photon energy at T = 77K for pressures
p = 1 bar, 5 kbar, and 8.6 kbar.

state 3 to state 1. The energy $E_{opt.}$ characteristic for these
transitions is much higher than the energy of thermal transition
$E_{therm.}$ as well as the energy barrier which separates both configurations of the center in states 3 and 1. For example, for
ambient pressure $E_{therm.}$ = - 0.05 eV (2,3), the energy barrier
is about 0.5 eV (2,3,4,8) while the obtained value of $E_{opt.}$ is
about 0.8 eV. $E_{opt.}$ increases with increase of pressure, which
agrees with the direction of the change of the energy $E_{therm.}$ with
pressure. These facts have confirmed the model of electron-ion
excitations which explains all the transport and optical effects
observed in CdTe /Cl/.

It is worth pointing out that at ambient pressure there are
observed optical transitions of electrons from the state which is
degenerated with the conduction band. However, in the light of
the applied model this is not purely electronic degeneracy, but
it should be regarded as the degeneracy of the whole system, consisting of the electron and the Cl ion.

As it has been observed for CdTe and also for other semiconduct-ors, donor centers, besides the simple hydrogenic levels, give an-other type of level. Its ionization energy strongly depends on pressure as well as on the kind of impurity. For instance, for CdTe the difference between the energy levels given by Cl and Br is about 200 meV (2). It means that the ionization energy of such a level is mostly due to the core potential and the electron is very strongly localized at the impurity center. One can expect that the ionization or deionization of such a center must be followed by the change of its configuration relative to the neighbourhood. This seems to be the origin of the observed changes of the impurity cen-ter configuration in CdTe and also in other semiconductors (10).

REFERENCES

1. N.V. Agrinskaya, M.V. Alekseenko, E.N. Arkad'eva, O.A. Matveev, S.V. Prokof'ev, Fiz.Tekh.Poluprovodn. $\underline{9}$, 320 (1975).
2. G.W. Iseler, J.A. Kafalas, A.J. Strauss, H.F. MacMilan, R.H. Bube, Solid State Commun. $\underline{10}$, 619 (1972).
3. M. Baj, L. Dmowski, M. Kończykowski, S. Porowski, phys. stat.sol. /a/ $\underline{33}$, 421 (1976)
4. L. Dmowski, M. Baj. M. Kończykowski, S. Porowski, V Conference on $A^{II}B^{VI}$ Semiconductor Compounds, 21-28.04.1975, Jaszowiec.
5. S. Porowski, M. Kończykowski and J. Chroboczek, phys. stat.sol. /b/ $\underline{63}$, 291 (1974).
6. E. Litwin-Staszewska, L. Dmowski, U. Piekara, J.M. Langer, B. Krukowska-Fulde, VII Conference on $A^{II}B^{VI}$ Semiconductor Compounds, 25.04. - 2.05.1976, Jaszowiec.
7. M. Kamińska, M. Godlewski, J. Langer, J. Baranowski, VIII Conference on Semiconductor Compounds, 24.04.- 1.05.1977, Jaszowiec.
8. D.L. Losee, R.P. Khosla, D.K. Ranadive, F.T.J. Smith, Solid State Commun. $\underline{13}$, 819 (1973).
9. W. Bujnowski, S. Porowski and A. Laissar, Prib.Tekh. Eksp. $\underline{16}$, 224 (1973).
10. L. Dmowski, M. Baj, S. Porowski, Conf. on High Pressure and Low Temperature Physics July 20-22, 1977, Cleveland.

QUESTIONS AND COMMENTS

T.A. Kaplan: Do you have any idea why this localized energy state lies above the bottom of the conduction band?

S. Porowski: No. It actually can lie above or below. There is no restriction for it. For example, for CdTe it lies above, and for GaSb it lies below it. I don't think it's important.

T.A. Kaplan: Do you think the impurity state is made out of a
                 linear combination of conduction electron states
                 near a minimum, or a special point?

S. Porowski: Yes.

T.A. Kaplan: Could this be coming from higher up in the conduc-
                 tion bands?

S. Porowski: You can think about these levels as a linear com-
                 bination of the whole conduction band. But, the
                 potential is so strongly localized that I don't
                 believe that it can be really associated with one
                 minimum.

T.A. Kaplan: It is certainly interesting to mention, as you
                 said, the possible relation between this and the
                 exciton in semiconducting samarium sulfide.

S. Porowski: Yes. If you convert one state where you have an
                 electron very far from the impurity, and other
                 which is strongly localized, you can think about
                 it as if it were a change of valency.

G.R. Barsch: You find a large change in mobility. The question
                 is, is this caused by the effect of pressure on
                 relaxation time?

S. Porowski: The conductivity takes place in different minima
                 depending on the pressure, according to the band
                 picture I presented.

G.R. Barsch: Except that you would expect three more or less
                 straight segments of straight lines.

S. Porowski: Yes, and you can see it, but the pressure has to be
                 changed between 0 and 40 kbar.

NARROW GAP SEMICONDUCTORS AS LOW TEMPERATURE PRESSURE GAUGES

M. Kończykowski, M. Baj, E. Szafarkiewicz, L. Kończewicz

and S. Porowski

High Pressure Research Center, Polish Academy of Sciences

Warsaw

INTRODUCTION

The most commonly used pressure sensor is the manganin gauge. It is particularly useful at room temperatures and for truly hydrostatic conditions. However, at low temperatures it has two main drawbacks: strong Manganin resistivity dependence on temperature and high sensitivity to nonhydrostatic stresses (1) (2). Especially, the second drawback is very serious because, in the pressure range up to 15 kbar and at temperatures below 77K, the non-hydrostatic stresses can not be completely eliminated.

In spite of the above this type of pressure gauge is also commonly used in low temperature experiments because of the lack of a better one (3)(4).

n-InSb pressure gauges presented in this paper are characterized by high sensitivity, temperature independence of readings, low sensitivity to nonhydrostatic stresses, and the reasonable linear resistance variation.

PRINCIPLE OF OPERATION

The pressure changes of resistivity in semiconductors are usually caused by changes of mobility and concentration of carriers. If the changes of concentration of carrier can be neglected in com-

parison with the changes of mobility good performance of the pressure gauge can be obtained. This condition can be easily fulfilled
for several narrow gap semiconductors. These materials are characterized by very small effective mass of electrons in the conduction band. That is why in heavily doped materials the Coulomb's
potential of donors does not create any localized states. As a
result of the above, at low temperatures the concentration of
electrons in the conduction band is independent of pressure and
temperature.

In narrow gap semiconductors the pressure changes of mobility
are caused by strong influence of pressure on the shape of the
conduction band. For a majority of n-type materials with the Kane's
type conduction band the mobility is linearly dependent on pressure,
in the amount of tens percent per 10 kbars, (5) (6) (7). For
lightly doped samples the pressure dependence of mobility is stronger than for heavily doped. Unfortunately, temperature sensitivity
of these samples is also higher.

To assure the low temperature sensitivity of the gauges the
material should be doped strongly enough to neglect scattering by
phonons in comparison with scattering by ionized impurity potentials
(iip) and short range potentials (srp) (5) (8) (9).

On the other hand doping should be heavy enough to achieve the
Fermi energy $E_F \gg kT$.

This is the condition for temperature independence of the
scattering on (iip) and (srp). For narrow gap semiconductors
both these conditions can be easily satisfied.

EXPERIMENT

The pressure sensors have been cut from the bulk InSb single
crystals heavily doped with S, Se or Te. The sizes of the samples
varied from 0.2 mm x 0.2 mm x 3.0 mm up to 1 mm x 1 mm x 8 mm.
Every gauge has one pair of current contacts and one or two pairs
of potential contacts. All contacts were soldered with Indium.
The resistance of such sensors is about 0.01 to 0.05Ω. Direct
current voltage measurements were carried out with accuracy usually
better than 10^{-4}. The typical pressure dependence of resistivity
of such a sensor is given in Fig. 1. At room temperature measurements the pure petrol was used as a pressure transmitting medium.
Pressure was measured by a Manganin gauge. At 77K the measurements were carried out using helium-gas compressor IF-012 and a
low temperature BeCu pressure cell IF-022 connected to the compressor by BeCu capillary. Pressure in the system was also measured by a Manganin gauge which was placed at room temperature in

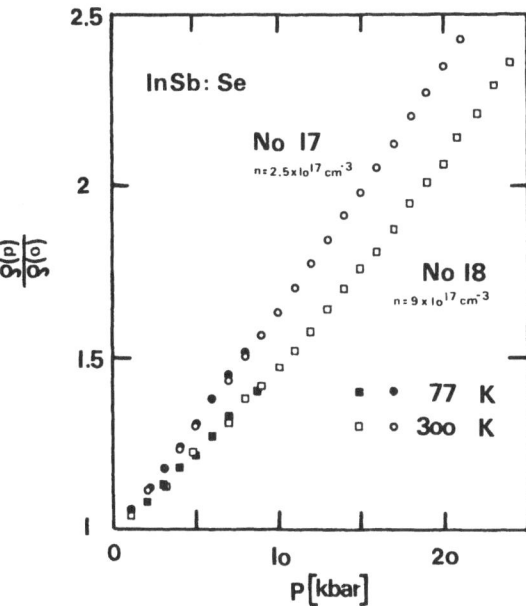

Figure 1.   Typical pressure dependence of resistivity for InSb gauges doped with Se at 300K and 77K.

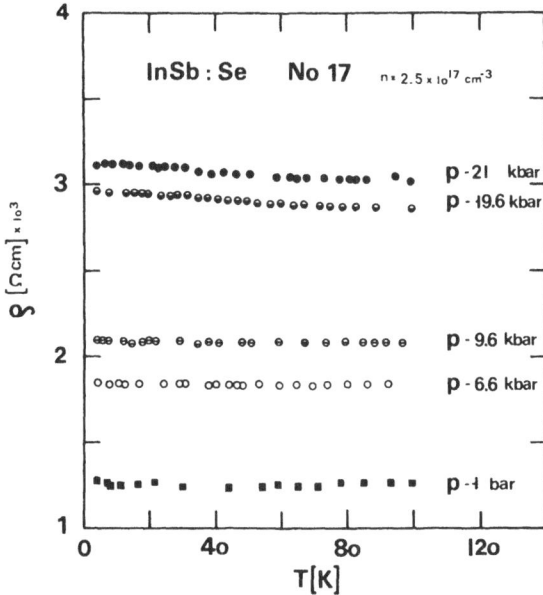

Figure 2.   Typical temperature dependences of resistivity for InSb gauge at various pressures.

the third stage of the compressor. As can be seen from Fig. 1 the
ρ(p) is nearly linear for pressures up to 10 kbar. In Table I the
pressure coefficients of resistivity α are given for this pressure
range.

Fig. 2 gives the typical temperature dependence of resistivity
of InSb gauges for several pressures. The measurements were carried
out using the clamp type pressure cell with pure petrol as a
pressure transmitting medium (Manostat If-04). In heavily doped
InSb it was shown that the pressure coefficient of resistivity in
the temperature range 77 - 300K does not change (5). In this paper
it was assumed that the pressure coefficient is also constant in
the range of 4.2 - 77K. The pressure in the manostat was measured
using the calibration curve from Fig. 1. As it can be seen from
Fig. 2 the temperature dependence of the resistivity of the gauge
is very small.

Table I gives the maximum temperature sensitivity for several
gauges in the temperature range 4.2 - 77K. In Table I the values
of α and β for Manganin sensors are also given for the same temperature range. The ratio α/β for all InSb gauges is about two orders
of magnitude better than for Manganin gauges.

The characteristic of the InSb sensors is expecially good at
the temperature range below 77K but they can be also successfully
used in the higher temperature range. The InSb gauges have been
used in various pressure cells, such as: manostat IF-04, the low
temperature helium pressure cell If-022, which is pressurized with
helium gas at high temperature and then cooled down to helium temperature, and the piston-cylinder cell which compresses helium
directly at 4.2K (10). They were found to be insensitive to small
nonhydrostatic stresses which always exist in such pressure equip-
ment.

                                DISCUSSION

For all measured samples the electric conductivity over whole
pressure and temperature ranges was determined by electrons from
Γ minimum of the conduction band. The pressure changes of the
carrier concentration n can be neglected as compared with the pressure changes of mobility (2) (5) (11). Independence of n on pressure and temperature was due to the lack of electron excitation
through the gap, because $E_g + E_f \gg kT$, and also due to the lack
of localized impurity levels in the vicinity of Fermi level E_f.
As has been shown (11) (12) the closest states which can influence
resistivity at high pressures are the resonance states lying fairly
high above E_F i.e. $E_S = 0.45$ eV, $E_{Se} = 0.52$ eV, $E_{Te} = 0.62$ eV.
Actually, at low temperatures only for gauges heavily doped with
S, the level E_S limits the maximum pressure to 15 kbars. For two

Table I

Sample No.	Dopant	Concentration at 77K $[cm^{-3}]$	$\alpha = \dfrac{\Delta\rho}{\rho_o P}10^5$ $[bar^{-1}]$	$\beta = \dfrac{\Delta\rho}{\rho\Delta T}10^4$ $[K^{-1}]$	$\dfrac{\alpha}{\beta}10^2$ $[bar^{-1}K]$
No. 17	Se	2.5×10^{17}	6.8	4	17
No. 18	Se	9×10^{17}	4.6	4	11
No. 19	Te	1.6×10^{17}	7.4	3	25
No. 32	S	3.3×10^{18}	5	3	17
No. 64	S	4.7×10^{18}	4.5	4	11
Manganin	–	–	0.23	12	0.2

other dopants pressure is limited only by phase transition which at room temperature takes place under 25 kbars (13).

The increase of resistivity with pressure is mainly caused by the increase of effective mass of conduction electrons.

$$m_n (p) = \frac{Eg (p)}{Eg (o)} m_n (o)$$

$$\frac{dEg}{dp} = 15 \times 10^{-6} \text{ eV/bar}$$

where, m_n = effective mass at the bottom of the conduction band.

Eg = energy gap.

The dominative scattering mechanism of electrons are iip and srp. The phonon scattering can be disregarded even for the most lightly doped samples.

It was shown (2) that at 77K phonon scattering for sample No. 17 is less than 2%. Due to the very small density of states in Γ minimum (m_n = 0.0145 m_o), electron gas is strongly degenerated at low temperatures. Therefore the condition for temperature independence of scattering by iip and srp is easy to satisfy. $E_F \gg kT$ over the whole pressure range, and this is the origin of described temperature independence of resistivity. The small changes which are observed, are mostly due to the decrease of m_n with temperature. However, the temperature changes of m_n are much smaller than the pressure changes. But even these undesirable temperature changes of m_n are partially compensated by the phonon scattering.

REFERENCES

1. J. Lees, High Temperature; High Pressures $\underline{1}$, 477, (1969).
2. E. Szafarkiewicz, Master Thesis, Department of Physics, University of Warsaw - 1973.
3. J. Stankiewicz, W. Giriat, A. Bienenstock, Phys. Rev. B $\underline{4}$ 4465 (1972).
4. G. Fujii, N. Mori, H. Nagano and S. Minomura, Low Temp. Technol. $\underline{4}$, 226, (1969).
5. E. Litwin-Staszewska, S. Porowski and A.A. Filipchenko, Phys. stat.sol. (b) $\underline{48}$, 519 (1971).
6. M. Baj, S. Porowski, High Temp.-High Pressures, $\underline{6}$, 95 (1974).
7. L. Kończewicz, I.K. Polushina, S. Porowski, V International Conference on High Pressure Physics and Technics, 26.05.-1.06.1975, Moscow.
8. E. Litwin-Staszewska, S. Porowski and A.A. Filipchenko, phys. stat.sol. (b) $\underline{48}$, 525 (1971).
9. W. Zawadzki and W. Szymanska, J. Phys. Chem.Solids, $\underline{32}$, 1151 (1971).
10. S. Porowski, M. Kończykowski, W. Bujnowski, Conf. on High Pressure and Low Temperature Physics, 20 - 22.07.1977, Cleveland.
11. L. Kończewicz, E. Litwin-Staszewska, S. Porowski, III Conf. of Physics of Narrow-Gap Semiconductors 12 - 15.09.1977, Warsaw, to be published.
12. S. Porowski, E. Litwin-Staszewska, A.A. Filipchenko, II Conf. on $A^{II}B^{VI}$ Semiconductor Compounds 8 - 24.04.1971, Jaszowiec.
13. M.D. Banus, M.C. Lavine, J. Appl. Phys. $\underline{40}$, 1, 409, (1969).

QUESTIONS AND COMMENTS

C.W. Chu: How reproducible are these gauges?

S. Porowski: Within the experimental error, we measured resistivity with an accuracy of 10^{-4}. We didn't see anything irreproducible.

C.W. Chu: Can you cycle it?

S. Porowski: Yes, you can cycle many times.

C.W. Chu: What is the value of the resistance of your gauge?

S. Porowski: It's quite small. It's usually about a few hundredths of an ohm or a little more.

Bi PHASE TRANSITIONS BELOW 80°K TO 150 KBARS IN AN Ar PRESSURE MEDIUM

Clarke G. Homan, Julius Frankel and David P. Kendall

Watervliet Arsenal

Watervliet, New York

ABSTRACT

Resistometric measurements on thin film Bi specimens embedded in a solid Ar pressure medium were made to 150 kbars in the temperature range of 40° to 80°K. The ratio of longitudinal to transverse acoustic velocity was simultaneously measured on the Ar medium and the EOS of Ar deduced. Correlation is made between the changes in the resistance due to polymorphic phase transitions in Bi and the pressure deduced from the Ar EOS. The present data will be compared with earlier results on Bi embedded in a pyrophyllite medium which suggests large non-hydrostatic medium effects in pyrophyllite at low temperatures. Experimental details will be discussed.

INTRODUCTION

Since Bridgeman found that several polymorphic phase transitions occurred in Bi below 100 kbars at room temperature, these transitions have been used as a calibration material for solid high pressure systems. This early work is not without controversy, as there is considerable debate as to the existence of the transitions, III-IV, at 40 kbar and IV-V near 52 kbars[2] [Bridgeman-Bundy phase notation will be used throughout this paper].

Recent investigations of the low temperature portion of the Bi phase diagram have been mainly limited to pressures below 40 kbars, however, the determination of the pressure effect on the superconducting transition temperature T_c of the high pressure phases by Il'ina and Itshevich to 90 kbars[3] and the phase diagram based on the resistometric measurements by Homan[4] to 140 kbars have recently

exceeded these pressures, again controversy exists in the interpretation of these results[2,4]. It is interesting to note that nearly
identical resistometric traces at LN_2 temperatures have led to the
widely divergent interpretation that 1) no triple point I-II-III
exists by one author[5], and 2) that not only does the triple point
I-II-III exist but there is a clear evidence for the transition
III-IV by another[4]. Recent X-ray evidence by Skelton et al[6] support the latter interpretation by finding no evidence of Bi II at
33 kbars and 8°K.

In order to make cryogenic high pressure measurements above
25 kbars, a solid high pressure system must be used in which the
sample is embedded in a combined gasket and pressure medium material
such as pyrophylite or in a soft solid pressure medium such as
NaCl contained within gaskets. At low temperatures, such systems
may lead to large frictional corrections to the pressure calibration
which are difficult to assess. The earlier work in Bi used the
former method[4] in which the arbitrary assumption was made that the
frictional component of load was temperature independent.

It is the purpose of this paper, to report on a resistometric
determination of the polymorphic phases in Bi below 80°K for a
sample embedded in solid Ar. Simultaneous measurements of the
acoustic velocity ratio of the Ar pressure medium allows an estimate of the pressure to be made from the Ar equation of state (EOS).
The Bi phase transitions are also related to a pressure scale based
on the Ar velocity ratios.

EXPERIMENTAL

Details of the cryogenic pressure system can be found in the
previous paper[4], however, Figure 1 illustrates the changes in the
sample region configuration used in this work. Three teflon spacers
have been introduced to allow the entire sample area to be filled
with liquid Ar at 95°K and 30 bar before entrapment. A slight
motion of the upper anvil entraps a solid Ar sample within the pyrophylite gaskets at a pressure of several kbars and makes electrical
contact with the Bi specimen.

The thin Bi specimen has been vapor deposited on a boron nitride ring previously coated with parlodium glue to prevent metallic
adsorption. The ring fits loosely in the sample area so that it is
surrounded by the Ar pressure medium on all cylindrical surfaces and
is in contact with the anvils on the end faces only. It is interesting to note that after several runs to pressures over 150 kbars on
the sample, the boron nitride could be removed intact. Subsequent
measurements indicated a permanent shape conserving deformation of
about 10% had occurred. This deformation cannot be used as a measure
of the final high pressure volume, since large recoverable deforma-

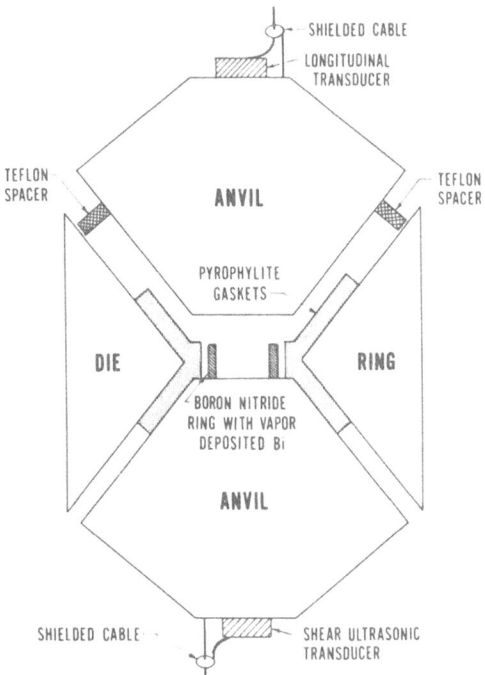

Figure 1.   Sample area configuration.

tions have also occurred during pressurization. However, this
ring has experienced the complete pressure cycle of the Ar pressure medium and suggests that a hydrostatic deformation occurred
for each run, a fact used in obtaining the Ar EOS from our ultrasonic measurements.[7]   In a NaCl experiment this ring is completely
pulverized suggesting that the deformations in that medium are not
hydrostatic as has been noted earlier.[8]

The ultrasonic transducers on each anvil are exposed to approximately the same temperatures as the samples, however, they are
at ambient pressures. These transducers allow acoustic interfero-
metric measurements to be made on the Ar pressure medium by a technique described by Frankel et al[8]. This technique measures the
resonance interference of both the longitudinal and shear acoustic
modes in a polycrystalline sample from which the respective velocities can be determined if the spacing between the anvils can be
measured or inferred. The ratio of acoustic velocities is independent of this spacing and is equal to the ratio of fundamental
frequencies. This ratio can be used as a pressure calibration
parameter itself.[9]

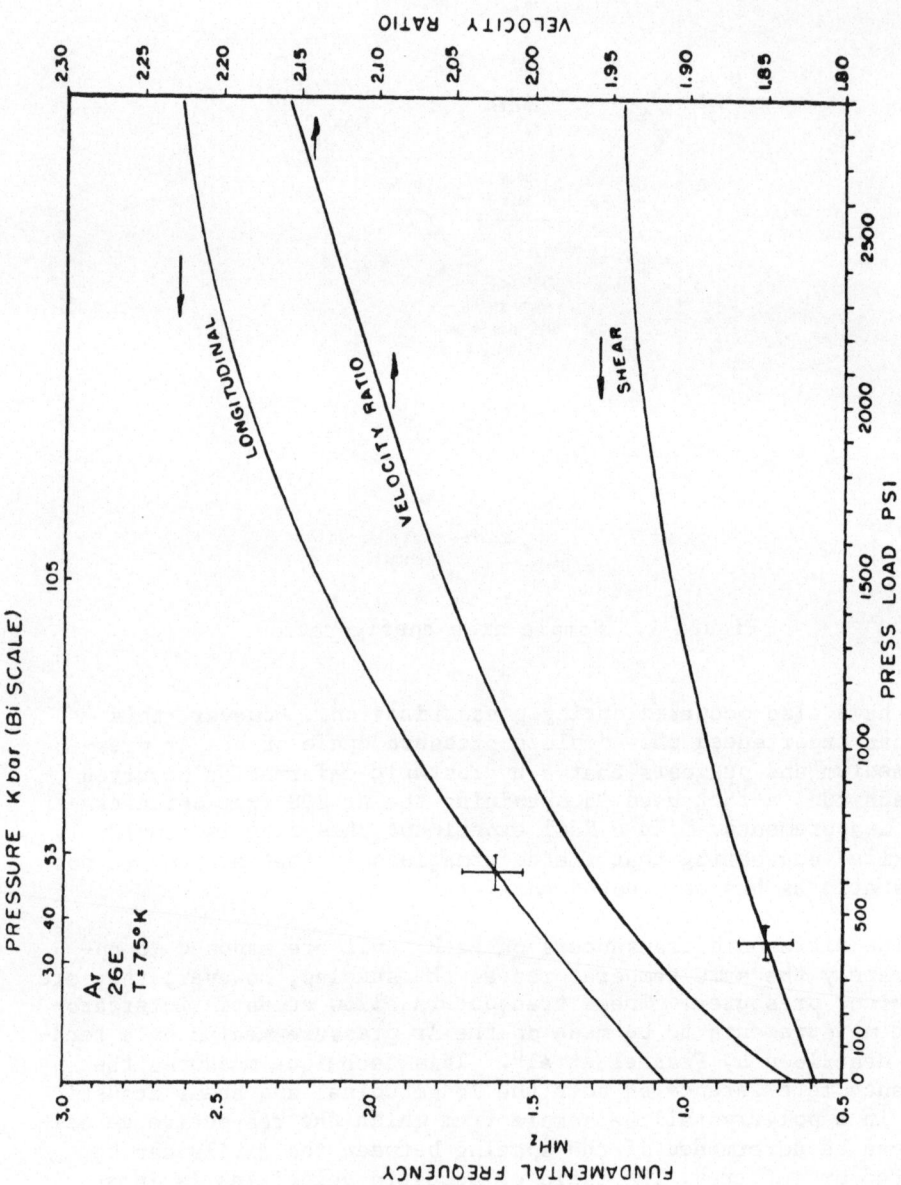

Figure 2.   Ultrasonic data reduced to fundamental frequencies and smoothed for Ar. Sample 26 run E.

Figure 2 is a plot of our raw ultrasonic data for Ar smoothed and reduced to fundamental frequencies as a function of load in a sample at 75°K. Also shown is the measured velocity ratios, which can be shown[7] to extrapolate in good agreement with the room pressure measurements of Barker and Dobbs[10]. Zero press load is <u>not</u> room pressure, due to hysteristic entrapment in the pressure cell, in fact the lower pressure level of each may vary from 4 to 10 kbars and in the run shown was determined to be near 7 kbars.

Figure 3 is a plot of the simultaneous resistometric trace of Bi in the same run. As previously mentioned the general shape of this curve up to the Bi IV-VIII transition is almost identical with the trace of Mori et al[5]. at this temperature except for the absolute resistance changes at the 30 and 40 kbar transitions. Since Bi I becomes a semiconductor at cryogenic temperatures, the shape of resistometric traces should be quite impurity dependent in the ppm range through band gap effects. In addition the pressure entrapment between successive runs on each apparatus and the differences in pressure profile between apparati significantly alter the absolute resistance traces obtained. Thus we can only assert that the general shape of our resistometric trace is in agreement with the previous data[4,5]. Above that load, we observed the large drop in resistance previously reported as Bi VII-IX[4] and although the present sample had a higher initial resistance no measurable resistance drop could be determined for Bi IX. The upper scale indicates the measured velocity ratios in Ar at which each of these corresponding points occurred in the Bi trace.

BISMUTH PRESSURE SCALE

Experimentalists use specimens of low shear strength (as in NaCl) to obtain nearly hydrostatic conditions even though the specimen boundaries don't contract according to hydrostatic considerations. Under such conditions, it would be necessary to measure the spacing between the anvils in order to obtain the absolute velocities and subsequently the Ar EOS from our ultrasonic measurements.

However, if hydrostatic deformations can be justified, then it is possible to calculate an EOS directly from the measured frequencies, if a reasonable pressure scale can be ascertained.[8] Using the previously reported transition pressures[4] yielded an Ar EOS in reasonable agreement with the precise piston cylinder data of Stewart[11] and Anderson and Swenson[12] to 20 kbars. The fit was poorest at the highest pressure of their data and in very poor agreement with the extrapolated 4.2°K isotherm of Anderson and Swenson.

Our smoothed data was then fit to the Anderson-Swenson data allowing only the linear load to pressure conversion to vary. A

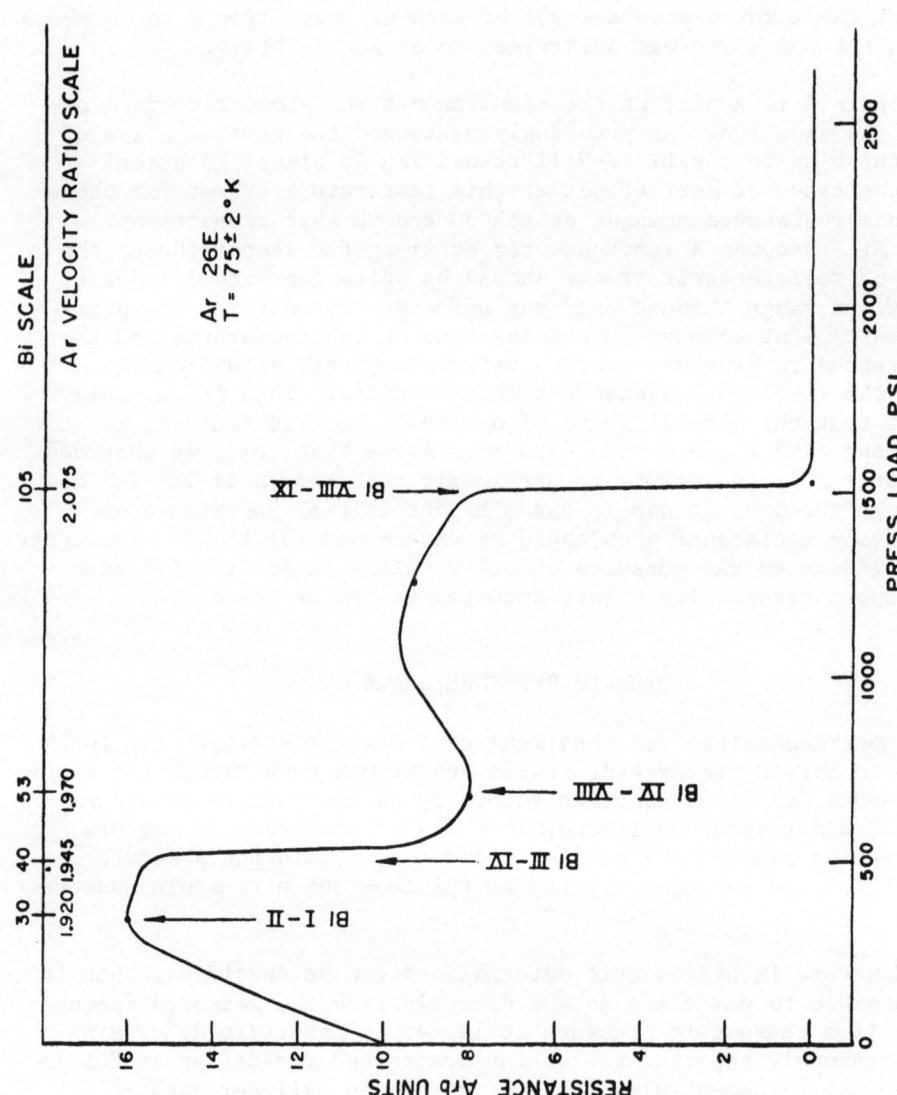

Figure 3. Resistometric trace of Bi embedded in Ar sample 26E.

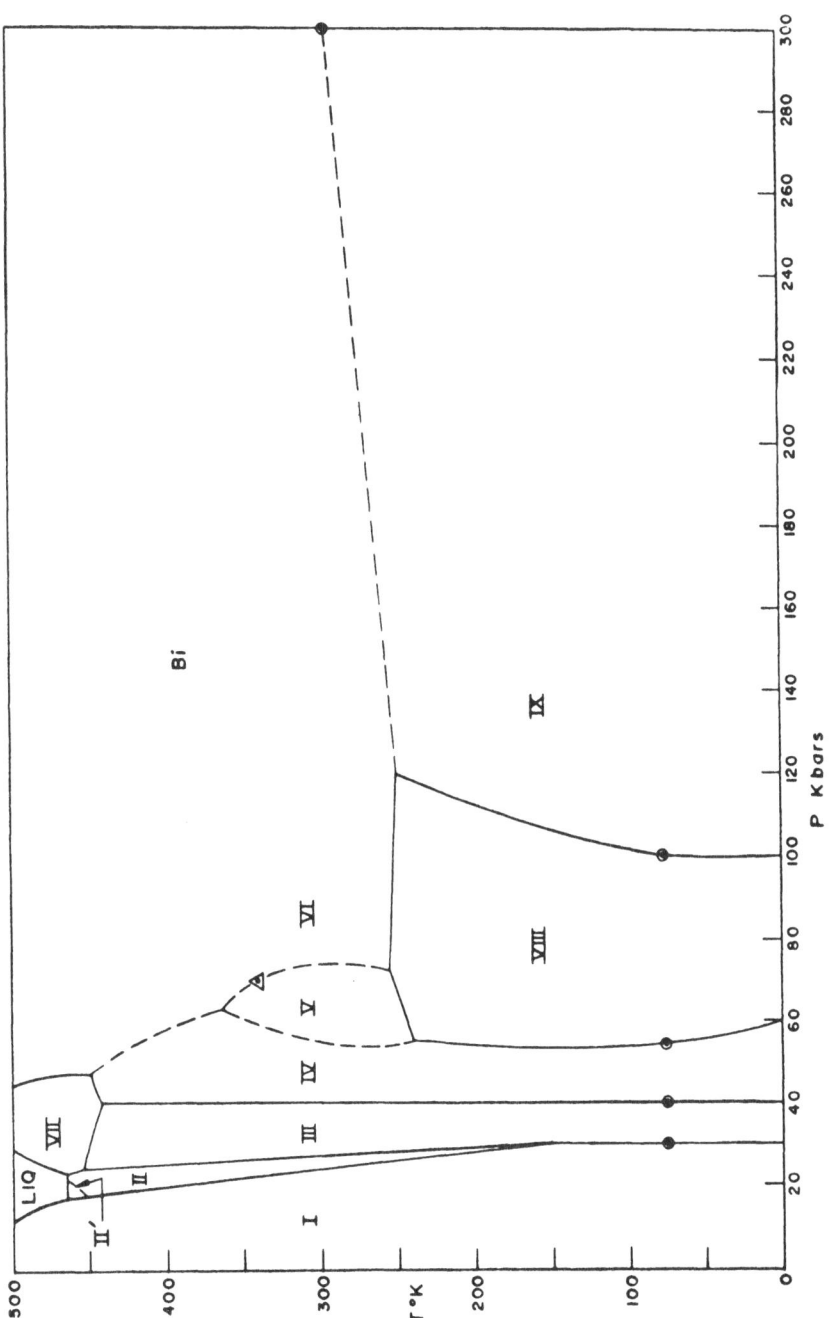

Figure 4. Bi phase diagram. Data obtained in this experiment O. Data from shock experiments, reference 13 Δ. Dotted lines indicate uncertain phase lines although Bi V-VI transition
at 72-74 kbars at ambient temperature is well established.

good fit was obtained for the two sets of data for the transition pressures shown on the upper scale of Figures 2 and 3. The maximum change in pressure was the 12% decrease in the VIII-IX transition pressure. The fit to the 4.2°K isotherm remains poor and suggests that Ar is a more compliant material than originally thought.[7]

DISCUSSION

Figure 4 shows our proposed Bi phase diagram incorporating our new data. The data at 75°K is shown since it was only at this temperature that sufficient data allowed both a Bi trace with an analysis of Ar EOS. Other isotherms to 40°K were obtained, however, transducer bonding difficulties prevented Ar EOS determination, although press load differences between transitions were consistent. Also shown are recent determination of the V-VI transition by Romain[13] using a shock technique and a room temperature redetermination of the Bi VI-IX by this laboratory using diamond tipped anvils to avoid possible extraneous short circuits.[14]

We note that the Bi III-IV and Bi VIII-IX resistometric trace changes occur over a smaller pressure range in the Ar medium than in the pyrophylite medium used earlier.[4] We interpret this to mean that Ar is a much more hydrostatic medium, however, the Bi IV-VIII is still quite broad.

One of the advantages of the technique where the pressure medium is the pressure sensor is that one can eliminate cell hysteresis effects that plague solid high pressure measurements. Measurements of a transition on the upstroke and downstroke can thus be correlated via the velocity ratio to study the "true" hysteritic-metastability of the transition.

CONCLUSION

The phase diagram previously reported was confirmed by our results below 80°K, although some downward revision in pressure scale is required to correlate our results with the precise Ar EOS data.

The velocity ratio pressure scale was shown to be a valid pressure calibration scheme at cryogenic temperatures in the large sample volume, optically opaque pressure systems.

ACKNOWLEDGEMENTS

The authors would like to thank Profs. M.S. Anderson and

C.A. Swenson for sending their data on Ar EOS and the valuable discussion with Dr. D.B. McWhan. Also we would like to acknowledge the aid and assistance of our colleagues, Dr. T.D. Davidson and J.A. Barrett and the technical assistance of W. Korman and J. Hart.

REFERENCES

1. P.W. Bridgman, Phys. Rev. 48, 893 (1935); Phys. Rev. 60, 351 (1941); see also "The Physics of High Pressure" Bell, London 1952.
2. J.F. Cannon, J. Phys. and Chem. Ref. Data 3, 781 (1974).
3. M.A. Il'ina and E.S. Itskevich, Sov. Phys. Sol. State 8, 1873 (1967).
4. C.G. Homan, J. Phys. Chem. Sol. 36, 1249 (1975).
5. N. Mori, S. Yomo and T. Mitsui, Phys. Lett. 34A, 190 (1971).
6. E.F. Skelton, I.L. Spain, S.C. Yu, C.Y. Liu and E.R. Carpenter, Rev. Sci. Inst. 48, July 1977 (in press).
7. C.G. Homan, J. Frankel, D.P. Kendall, J.A. Barrett and T.E. Davidson, "Acoustic Velocity Ratio is Solid Ar at 75°K up to Static Pressures of 150 Kbar, "Submitted to 6th AIRAPT International High Pressure Conference, Boulder, CO, July 1977.
8. J. Frankel, F.J. Rich, and C.G. Homan, J. Geophys. Res. 81, 6357 (1976).
9. J. Frankel, F.J. Rich, C.G. Homan, M.A. Hussain and R.D. Scanlon, "The Use of the Isotropic NaCl Acoustic Velocity Ratio in Ultra High Pressure Physics," Submitted to the 6th AIRAPT Meeting (ref. 7).
10. J.R. Barker and E.R. Dobbs, Phil. Mag. 46, 1069 (1955).
11. J.W. Stewart, J. Phys. Chem. Solids 29, 641 (1968).
12. M.S. Anderson and C.A. Swenson, J. Phys. Chem. Solids 36, 143 (1975).
13. J.P. Romain, J. App. Phys. 45, 135 (1974).
14. D. Kendall private communication.

QUESTIONS AND COMMENTS

J.B. Clark: You've explained just about every single feature on the curve in terms of a particular transition, except for the maximum on the resistance curve. This takes place after a minimum in the curve which has been ascribed to the IV-VIII transition. Is the maximum actually the start of the VIII-IX transition?

C.G. Homan: I don't believe so. In order to analyze the resistometric data, you have to have data at many different temperatures and relate the features of

the trace which are reproducible and have a physically reasonable PT variation to some phase diagram established by other techniques, which we
have done in our earlier paper. This worked fine
for all the transitions except the 74 kbar transition, which forms a triple point near 250 K.
The two new phases we found, VIII and IX, were
quite unexpected. In our earlier paper, our samples were embedded in a pyrophylite medium and
all the transitions were quite smeared out. In
fact, one measure of the hydrostaticity of a pressure medium is the amount of spread in the transitions. In the present experiment, all the transitions were very sharp except for the IV-VIII
transitions. Therefore, there is something about
that transition we just don't understand. We believe that the IX transition occurs at the sharp
discontinuity at 1500 psi load as shown in fig. 3.

C.W. Chu: You mention the resistivity of phase IX as being
                    lower than the resistivity of copper. Did you
                    use the four lead method?

C.G. Homan: No. We estimated the resistivity from the resis-
                    tance drop at pressure and the dimensions of the
                    sample at STP. We could not detect any drop for
                    Bi IX at these temperatures, being much less than
                    the resistivity of Bi VIII, which was of the order
                    of Cu.

PISTON-CYLINDER APPARATUS OPERATING AT LIQUID HELIUM TEMPERATURE

S. Porowski, M. Kończykowski, W. Bujnowski

High Pressure Research Center
Polish Academy of Sciences
Warsaw, Poland

INTRODUCTION

Pressure experiments which demand truly hydrostatic conditions become very difficult in the range of temperatures below 77K and pressures up to 15 kbar. This is mostly due to the fact that all fluids become solid in this range, and therefore small nonhydrostatic stresses cannot be completely avoided. Several techniques were developed for experiments in this range. Low temperature clamped cell technique uses a liquid pressure transmitting medium which is pressurized at room temperature and then cooled down to helium temperature (1) (2). The main disadvantages of this technique are: fairly high nonhydrostatic stresses which are generated during freezing of the pressure transmitting medium and the fact that the pressure can be changed only at room temperature. Much better results can be obtained by a similar technique which uses helium as the pressure transmitting medium. The pressure cell is pressurized with a helium gas compressor and then is cooled down to helium temperature (3). Nonhydrostatic stresses can be very nicely eliminated in this technique. The disadvantages of this technique are: The pressure can be changed only at temperatures above freezing of helium, and the system which consists of helium gas compressor, capillary, and the low temperature cell is usually difficult to operate. The technique developed by Stewart (4) is free from the drawbacks of the two previous techniques. In Stewart's method liquid and solid helium is directly compressed in the piston cylinder apparatus at 4.2K. But unfortunately the sealing of the piston is not very positive. Especially when temperature has to be changed over a wide range or pressure has to be cycled, this technique is difficult to apply.

The apparatus described in this paper has all the good features
of the Stewart apparatus but much better sealing of the piston enables changes of temperature over a wide range and cycling of pressure. The apparatus is very useful for experiments when pressure
has to be generated at low temperature. There is a group of materials for which the sequence of cooling down and exerting pressure
is important. For example these are the materials with defects
generated by irradiation at low temperatures or semiconductors for
which metastability at low temperatures is observed (5).

CONSTRUCTION AND PRINCIPLE OF OPERATION

Fig. 1 shows the scheme of the apparatus. It consists of a
hydraulic cylinder which is joined to compression and tension components. The high pressure cell is fixed at the end of the tension component. The piston is placed at the end of the compression
component. To assure good thermal insulation between high pressure
cell and hydraulic cylinder the components are stainless steel
about 1 m long. The hydraulic cylinder generates forces up to 30
tons. The high pressure cell is designed for a maximum pressure
of 20 kbar. The lower part of the apparatus is placed in a helium cryostat. The superconducting coil placed at the bottom of the
cryostat was used for galvanomagnetic measurements. About 15 liters of helium is needed for cooling down the apparatus, and current helium consumption during the experiment is about 1 1/h.
Fig. 2 shows the scheme of the BeCu high pressure cell. The bottom end is closed by the typical plug with electrical leads. It
is sealed by the brass washer covered with indium. Multiwire
electrical feedthrough is used. The wires are pressed in the pyro-
phillite powder layer between the conical surfaces. See Fig. 3.
In some experiments the same type of electrical feedthrough was
also fitted in the piston. For example this construction allows
for placing the sample at the end of the piston, that is essential
in the low temperature radiation damage experiments. Fig. 4 shows
the sealing principle of the piston. The beginning of the piston
forms a flat cone which slowly decreases to zero. The main part
of the piston is cylindrical and enables large displacement of the
piston. It is important because liquid helium is very compressible:
to increase pressure up to 5 kbar, its volume has to be reduced
more than 5 times. The upper end of the high pressure cell has
a conical opening of the same angle as the piston. ID of the cell
slowly increases about 0.4 mm below the conical part. As can be
seen from Fig. 4, the diameter of the cylindrical part of the piston is bigger than the ID of the cell at level A (the difference
of diameters will be called b). When the piston is pressed into
the cell it seals at this level. Using pistons giving different
values of b, the maximum sealing pressure can be selected. Pistons
of two typical dimensions were used. In the experiments in which
maximum pressure did not exceed 5 kbar b was 0.10 mm, and for higher pressure range b was 0.30 mm. To reduce friction, the piston

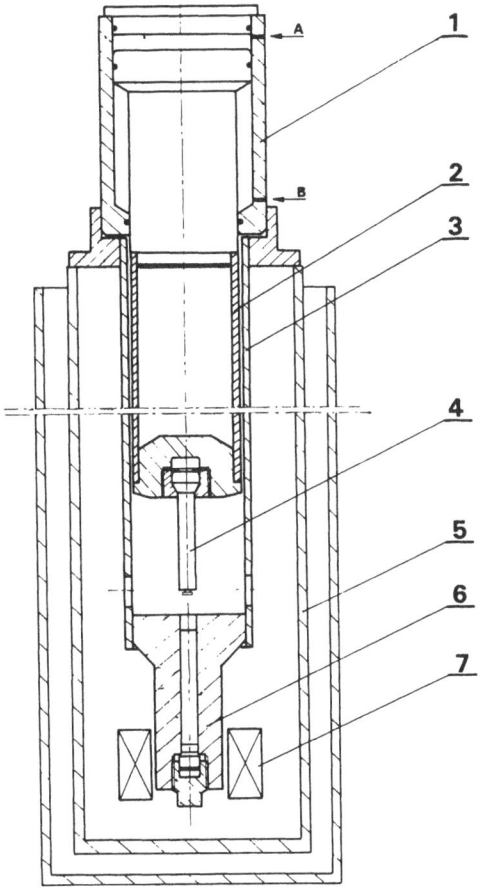

Figure 1. Construction of low temperature piston—cylinder apparatus.
1 - hydraulic cylinder, 2 - compression component,
3 - tension component, 4 - piston, 5 - He cryostat,
6 - high pressure cell, 7 - superconducting coil.

was covered with a MoS_2 layer a few microns thick. Usually the
force of friction was below 0.5T for b = 0.10 mm and 1T for b =
0.30 mm. Initial position of the piston is a few millimeters above
the cell. After the apparatus is cooled down to 4.2K and the level
of liquid helium is raised about 200 mm above the cell, oil pressure in the hydraulic cylinder is increased and the piston is forced
into the cell. The cell is sealed off almost immediately after the
end of the piston passes level A. The full pressure sealing occurs
when the beginning of the cylindrical part of the piston passes
level A. After that, the pressure cell is completely tight even
when the temperature increases up to 100K. It is also possible

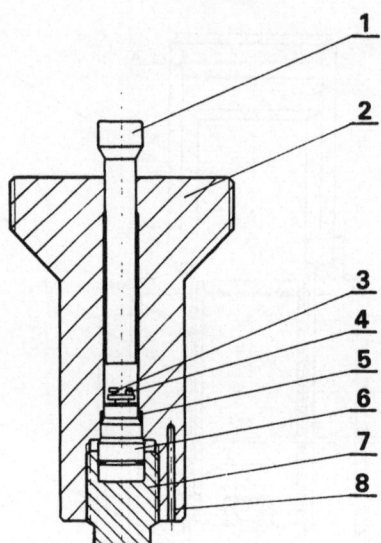

Figure 2.   Cross-section of the BeCu high pressure cell.
             1 - piston, 2 - cylinder, 3 - sample, 4 - InSb
             gauge, 5 - brass washer covered with indium,
             6 - plug, 7 - nut, 8 - cavity for the thermometer.

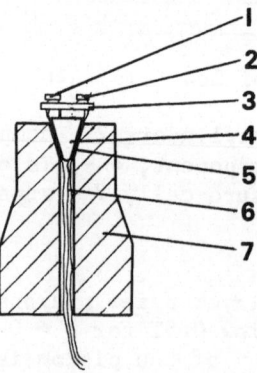

Figure 3.   The bottom plug with multiwire electrical feedthrough.
             1 - sample, 2 - InSb gauge, 3 - mounting plate,
             4 - BeCu cone, 5 - pyrophillite powder layer,
             6 - electrical leads, 7 - Be-Cu plug.

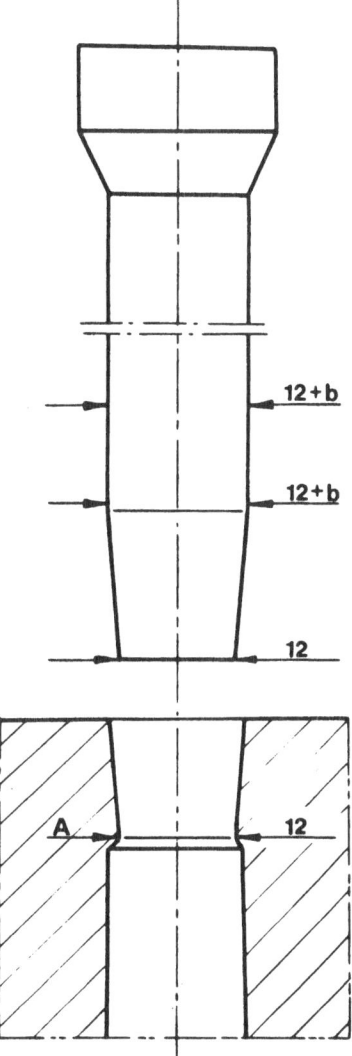

Figure 4.   The principle of piston sealing.
Angle of cone about 1°.

to increase and reduce the pressure a few times during an experiment.   Until now, the apparatus was tested up to 9 kbar.   The pressure was measured by the InSb gauge mounted flat at the mounting
plate.   If it was properly mounted, we did not observe any influence
of the nonhydrostatic stresses on the gauge up to maximum pressure.

EXPERIMENTAL RESULTS - HgTe

The influence of pressure on the Hall coefficient R_H and conductivity σ for pure HgTe samples at low temperatures, was previously measured using the clamp cell technique (6). Very strong increase of R_H and decrease of mobility μ have been observed. It
was also noticed that the minimum at 25K which is characteristic
of the temperature dependence of conductivity at normal pressure
disappeared at high pressures. These effects can not be explained
based on the current theory. Because HgTe is a zero gap semiconductor, it was possible to suspect that it is very sensitive for
nonhydrostatic stresses. Therefore similar measurements have been
carried out using the above described apparatus, in which nonhydrostatic stresses are much smaller than in the clamp cell technique.
Samples with electron concentration of about 1.7×10^{15} cm^{-3} and
mobility of 5×10^5 cm^2 V/s at 4.2K have been measured. They were
bar shaped 0.8 x 0.8 x 5 mm. Silver paste contacts were used. A
small superconducting coil was used for R_H measurements. The magnetic field did not exceed 100 Gs. Pressure was measured with
InSb gauge (7), temperature with GaAs thermometer and Cu-Constantan
thermocouple. The HgTe sample and InSb pressure gauge were lying
freely on the mounting plate. For pressures exceeding 1 kbar the
influence of nonhydrostatic stresses on HgTe has been observed.
After each pressure change very strong increase of resistivity was
observed. Resistivity was also time dependent due to the relaxation of nonhydrostatic stresses. It was possible to eliminate
these effects by passing strong current pulses heating the sample
after each change of pressure. During the pulses the temperature
of the sample increased above the melting point of He. Necessity
of elimination of nonhydrostatic stresses limited the pressure range
in this experiment to 3.5 kbar. Above this pressure we could not
melt the helium surrounding the sample without breaking down electrical contacts. We did not observe any effects of nonhydrostatic
stresses on n-InSb. Fig. 5 shows the pressure dependence of R_H
at 4.2K. One can see that the change for 1 kbar is about 6% while
in paper (6) changes above 100% were observed. In this experiment
the mobility μ decreased about 10% per 1 kbar while in paper
(6) the increase of more than two times per 1 kbar was observed.
Fig. 6 shows the temperature dependence of conductivity σ for two
values of pressure. The increase of conductivity due to the decrease of electron effective mass in the conduction band is observed. One can also notice that the characteristic minimum does
not disappear at high pressure as was observed in (6). We believe
that the experimental discrepancies presented above, are caused
by the nonhydrostatic stresses which always occur in the clamped
cell technique and interfered with the measurements presented in
paper (6).

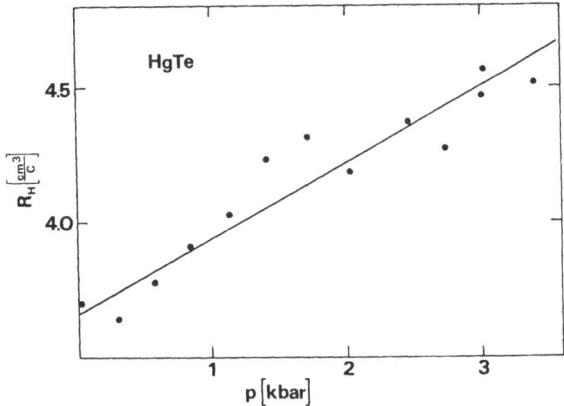

Figure 5.   The pressure dependence of the Hall coefficient for
            pure HgTe at 4.2K.

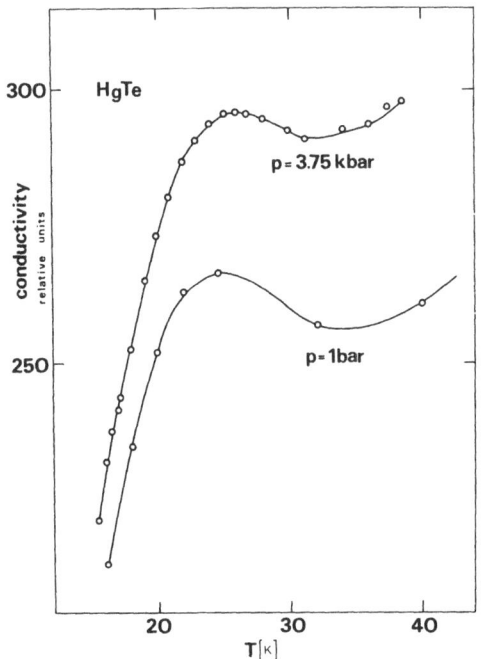

Figure 6.   The temperature dependence of conductivity σ for
            pure HgTe.

ANALYSIS

Assuming the parameters of the conduction band like in (6) and $R_H = \frac{1}{ne}$ the Fermi energy calculated for normal pressure and temperature of 4.2K equals 1.9 eV. Knowing that, the pressure dependence of the effective mass is given by:

$$m_n\,(p) = m_n\,(o)\,\frac{E\,(p)}{E\,(o)}$$

we can calculate the Fermi level for the whole pressure range. It follows from the above, that E_F is constant in this pressure range within 0.1 meV. This result is consistent with the existence of the pressure independent acceptor level with energy of 2.2 meV. This level was observed for HgTe in magneto-absorption measurements (8). The pressure coefficient of this level should be smaller $\gamma < 10^{-7}$ eV/bar. The observed pressure increase of σ, see Fig. 2, at the temperature range of 20 - 40K is mainly caused by the decrease of m_n with pressure. The opposite direction of the mobility changes at 4.2K is probably due to the increase of resonant scattering of electrons by the acceptor level.

REFERENCES

1. P.F. Chester, G.O. Jones, Philos. Mag. <u>44</u>, 1281 (1953).
2. E.S. Itskevich, Cryogenics <u>4</u>, 365 (1964).
3. J.E. Schirber, Cryogenics <u>10</u>, 418 (1970).
4. J.W. Stewart, Modern Very High Pressure Techniques ed. R.H. Wentorf (Butterworths 1962) p. 181.
5. L. Dmowski, M. Baj, A. Iller, S. Porowski, Conf. Cleveland 1977.
6. J. Stankiewicz, W. Giriat, Phys. Rev. B, <u>13</u>, 665 (1976).
7. M. Kończykowski, M. Baj, E. Szafarkiewicz, L. Kończewicz, S. Porowski, Conf. 20 - 22.07.1977, Cleveland.
8. G. Bastard, Y. Guldner, C. Rigaux, N'guyen Hy Hau, J.P. Vieren, M. Menant, A. Mycielski, Phys. Lett. A, <u>46A</u> 99 1973.

QUESTIONS AND COMMENTS

I.L. Spain: I am surprised that HgTe shows such a strong dependence on uniaxial stress, since the conduction and valence band extrema occur at the center of the Brillouin zone. Would not a semiconductor with conduction band minima located between the center of the zone and the zone boundaries (e.g. ΓL or ΓX) show more pronounced effects? In

this case application of uniaxial stress would
cause a lowering of symmetry, thus raising and
lowering energies.

S. Porowski: HgTe is very sensitive because the uniaxial stress
                lifts the degeneracy of conduction and valence
                bands at k = 0.

I.L. Spain: The other thing was that you talked about anneal-
                ing helium. Did you do this with a wire inside?

S. Porowski: No, we just passed the current through the sample.
                Or, sometimes we just heated the cell. If you
                use this kind of sealing, you can heat up the cell
                and it stays sealed up to about 100K. That's as
                far as we went.

L.N. Mulay: I wonder if you can, perhaps, modify your appara-
                tus for making magnetization or magnetic suscep-
                tibility type measurements of the kind reported
                yesterday by Dr. Guertin. Can you make routine
                magnetization measurements under pressure by an
                inductance type technique?

S. Porowski: I don't know. The weight of this apparatus is
                about 10 kg.

L.N. Mulay: What is the size of your sample, for example,
                for the Hall effect or susceptibility measure-
                ments? How small or how large is your sample
                for such measurements?

S. Porowski: We usually try to make them rather small,
                0.2 x 0.2 x 1 mm up to about 1 x 1 x 8 mm.

He^3 CRYOSTAT WITH HIGH PRESSURE HYDROGEN CHAMBER FOR SUPERCONDUCTIVITY MEASUREMENTS IN METAL-HYDROGEN ALLOYS

L. Sniadower and J. Igalson

Institute of Physics, Polish Academy of Sciences

Warsaw, Poland

ABSTRACT

A sorption pumping He^3 cryostat has been constructed, which enables one to obtain Pd-H_x alloys in the concentration range of hydrogen corresponding to the occurrence of superconductivity, using high pressure techniques. After hydrogenation, the sample is cooled to a temperature of 0.4 K, in such a way as to avoid hydrogen desorption. The apparatus is used for determing $T_c(x)$ dependence and for tunneling measurements of Pd-H_x below 1 K.

The discovery of superconductivity in palladium hydride (Pd-H_x) by Skoskiewicz[1] in 1972 aroused new interest in the hydrides of transition metals. Until recently the low temperature measurements on Pd-H_x have been done only to about 1.5 K. There are some technical difficulties in cooling the high hydrogen concentration Pd-H_x system down to temperatures below 1 K and so no such measurements have yet been reported. However, there are two reasons why measurements at the lower temperatures are of interest. Firstly there is the question of the lowest concentration of hydrogen at which superconductivity can occur. Secondly at lower temperatures it would be possible to observe sharper tunneling curves and to determine microscopic parameters like the superconducting energy gap, electron-phonon interaction etc.

As has been shown[2,3], the high pressure technique (pressure of a few kilobars) is an efficient method of obtaining the required concentration of hydrogen in palladium for superconductivity to occur. The samples are more homogeneous than those prepared by other techniques such as electrolysis and ion implantation. Also it is possible to measure the hydrogen concentration precisely.

Because of the need to avoid the desorption of hydrogen from palladium, a method has to be used in which high pressure and cooling occur simultaneously. For this reason in the present He^3 refrigerator, the high pressure hydrogen chamber is fixed in mechanical thermal contact with the He^3 chamber. After charging with high pressure hydrogen at room temperature the palladium hydride sample is cooled down below 1 K without demounting.

Figure 1 shows the layout of the He^3 refrigerator. The system consists of a liquid He^3 chamber, a liquefaction part immersed in He^4 bath, a sorption pump for lowering the liquid He^3 temperature and a He^3 handling system. The He^3 chamber is joined to the high pressure chamber which is connected to the high pressure compressor by a stainless steel capillary tube. The capillary goes through the liquid He^3 and the liquid He^4 baths to minimize the heat transport to the sample.

Figure 2 shows the details of the He^3 and high pressure chambers. The volume of the He^3 chamber is about 7 cm^3 thus retaining liquid He^3 for many hours. The whole He^3 chamber is made from copper except for a plug of copper-beryllium for sealing the high pressure chamber. The high pressure chamber is made from pure copper in order to avoid a temperature gradient. The diameter of the copper chamber is 22 mm, the diameter of the sample space is 8 mm and the thickness of the chamber wall is 7 mm. The length of the chamber is 5 cm and its mass is 150 g. The sample is supported on the feed-through which is sealed with soft copper into the bottom of the chamber. The temperature is measured by a thermometer attached to the outside of the chamber. There is also a heater for temperature control.

The hydrogenation-cooling schedule is given in Figure 3 which shows the temperature of the sample plotted against the time in hours. The initial room temperature charging at high pressure takes about half an hour. The high pressure chamber is then cooled at constant pressure. Some hydrogenation occurs (close to room temperature) during this period. At liquid nitrogen temperature (LN_2) the pressure is released. Subsequently there is no desorption of hydrogen from the sample during the experimental measurements. The next step in the schedule is the filling of the cryostat with liquid helium which causes the pressure chamber to cool to 4.2 K. This is followed by pumping of the He^4 and liquefaction of He^3 occurs at about 2 K. The last step is He^3 sorption pumping. The final temperature reached is about 0.4 K with a cooling power of about 1 mW. Temperature is stable to within $\pm$ 1 mK.

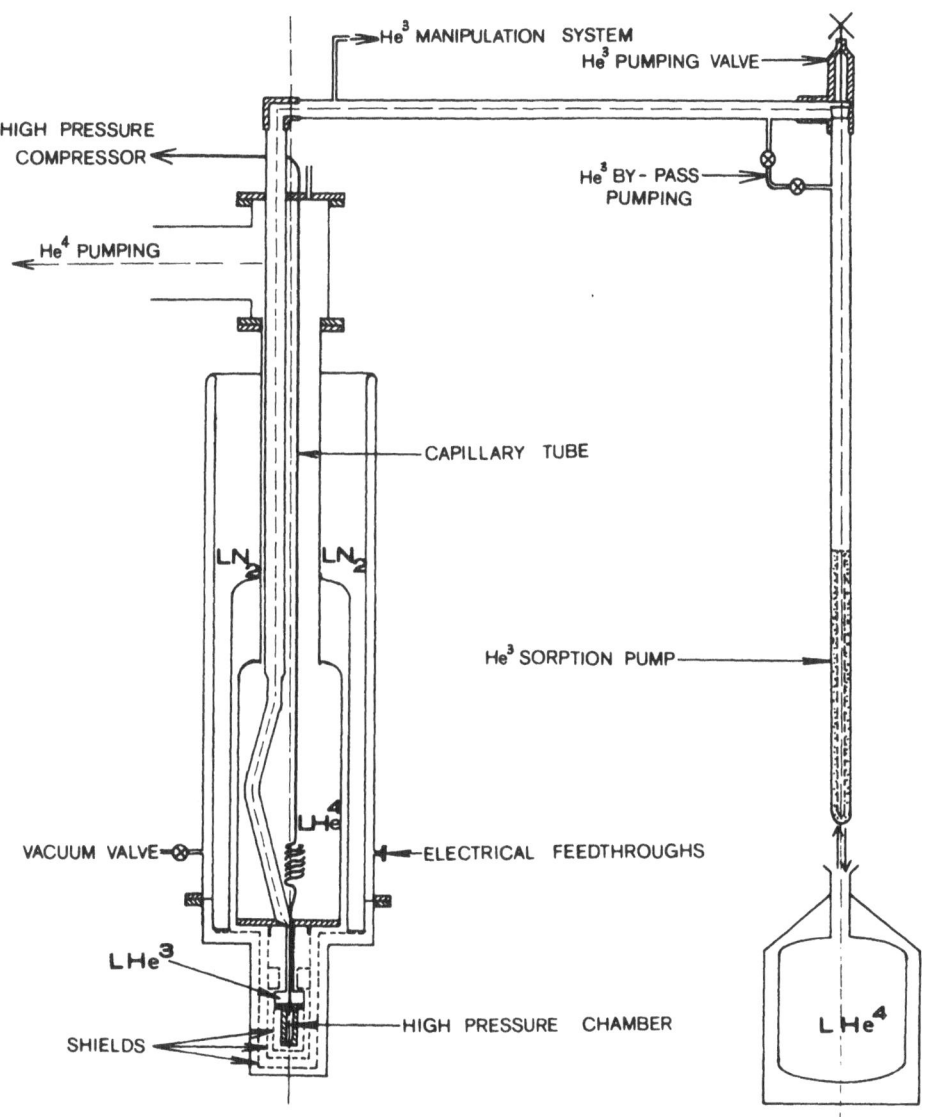

Figure 1. The layout of the He³ refrigerator.

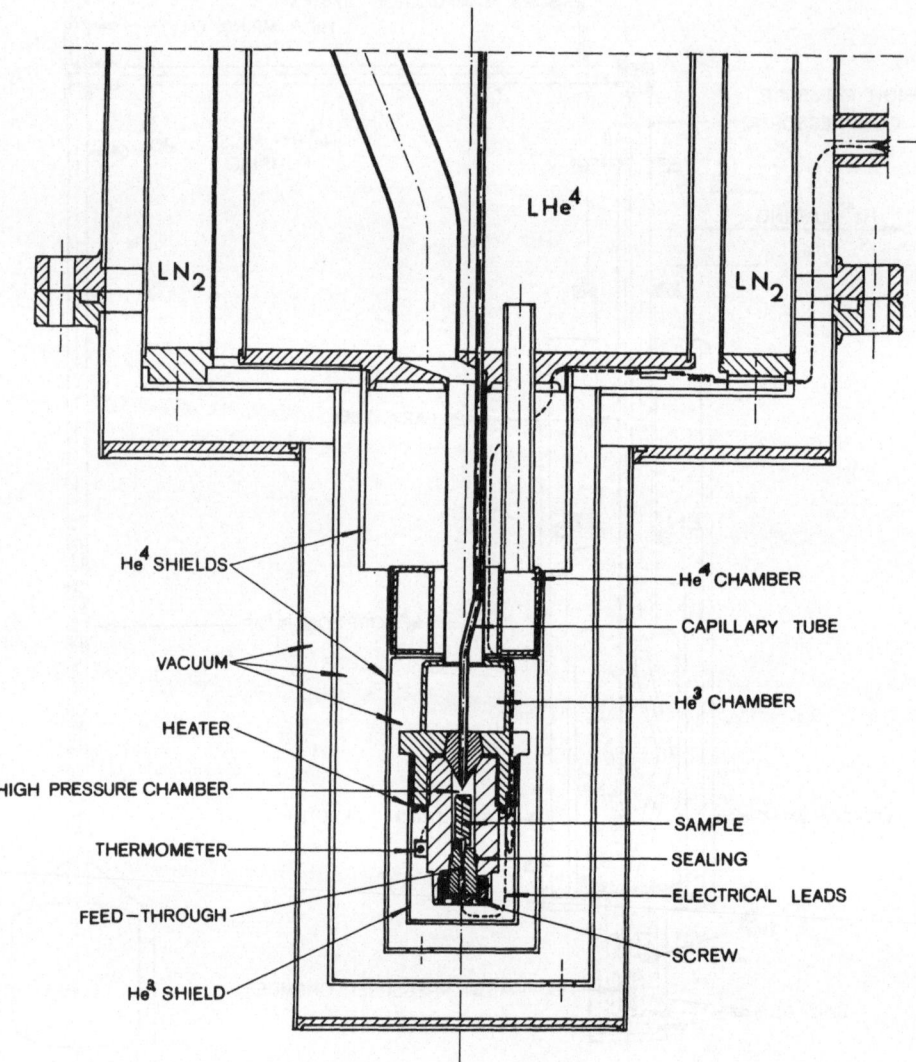

Figure 2.   He^3 and high pressure chambers.

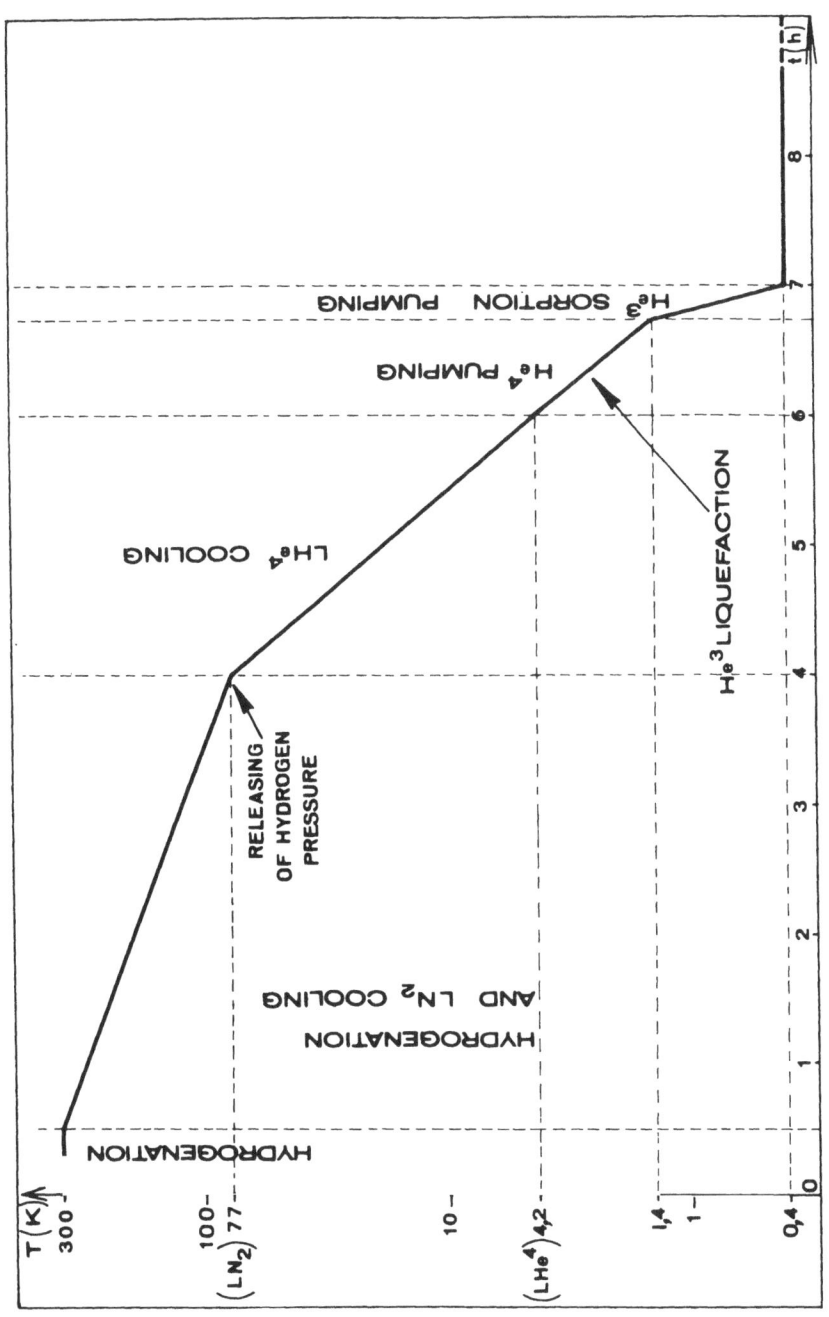

Figure 3.   Hydrogenation-cooling schedule.

Although the hydrogen chamber was designed for pressures up to about 3 kbar, up to now it has been tested to 1600 bar which is sufficient for the studies of the onset of superconductivity. The higher pressures are important only for the tunneling measurements of Pd-H_x samples with high T_c where the highest hydrogen concentrations are required.

REFERENCES

1. T. Skośkiewicz, Phys. Stat. Sol. (a) 11, K123 (1972).
2. J.E. Schirber, Phys. Lett. 45A, 141 (1973).
3. T. Skośkiewicz et al., J. Phys. C 7, 2670 (1974).

EFFECTS OF PRESSURE, TEMPERATURE, AND DEFECT CONCENTRATION ON

POSITRON BEHAVIOR IN METALS

John D. McGervey

Department of Physics, Case Western Reserve University

Cleveland, Ohio 44106

ABSTRACT

Positron lifetime distributions and the angular correlation
of annihilation radiation are being used by numerous investigators
to determine vacancy formation enthalpies and to reveal the presence of defects which may be precursors of voids in irradiated
materials. However, the analysis that is needed to find vacancy
formation energies is complicated by the presence of vacancy-
independent temperature effects. These effects may be isolated
by the application of sufficient pressure to remove the volume
changes that result from increasing the temperature. For example,
we find that the positron mean lifetime in cadmium at 296 K and
9.8 kb is 178 $\pm$ 2 ps, significantly larger than the lifetime of
170 $\pm$ 2 ps observed at 77 K and 0 kb, even though the volume of
the sample is smaller in the former case, demonstrating that the
effect of temperature is more than that of a simple volume change.

The simple trapping model[1,2,3] has been quite successful in
explaining the temperature dependence of positron lifetime distributions and of the angular correlation of annihilation radiation
from metals. From this understanding has developed a method for
determining the vacancy-formation enthalpy E_v in metals.

According to the simplest form of this model, a thermalized
positron may be trapped by a vacancy in a metal, with a trapping
probability per unit time, α, that is directly proportional to the
vacancy concentration. It is assumed that positrons, once trapped,
remain trapped until they are annihilated; it follows that the
fraction of the total positron population that is annihilated in

the trapped state is simply

$$f = \frac{\alpha}{\alpha + \lambda_b} \quad , \tag{1}$$

where λ_b is the annihilation probability per unit time (the reciprocal of the mean lifetime) in the "bulk" or untrapped state.

In an angular correlation measurement, the rate of emission of
gamma-ray pairs per unit solid angle centered on a given relative
angle is then the sum of two contributions, one from trapped positrons, the other from positrons in the bulk. In particular, the
peak rate P, occurring when the angle between gamma rays is 180°,
is a superposition of P_b and P_t, where P_b is the peak rate that
would result if all annihilations were in the bulk, and P_t the
peak rate that would result if all annihilations were in the trapped state. Thus $P = fP_t + (1 - f)P_b$. Substitution of the value
of f from Eq. 1 yields

$$\alpha = \lambda_b \left(\frac{P - P_b}{P_t - P} \right)$$

and since α is proportional to the vacancy concentration, which in
turn is proportional to exp $(-E_v/kT)$, we arrive at

$$\frac{P - P_b}{P_t - P} = (A/\lambda_b)\, e^{-E_v/kT} \tag{2}$$

where A is a constant.

Eq. 2 suggests that the values of P obtained at various temperatures should fit an Arrhenius plot; that is, a graph of $\ln\{ (P-P_b)/
(P_t - P) \}$ vs. $1/T$ should be a straight line, whose slope is pro-
portional to E_v. Data on a number of metals have borne out this
expectation,[4-6] and have yielded values of E_v that are significantly more precise than those obtained by other methods.

In making such a graph, the difficulty is that there are four
unknown parameters (P_b, P_t, E_v, and A/λ_b) that cannot be determined
separately and must simply be chosen to give the best fit to the
data. Fortunately, P_b and P_t can be closely approximated at the
outset, because P approaches P_b at low T and P_t at high T.
(Near the melting point, almost all of the positrons are annihilated
in vacancies.) But the fitting procedure is nevertheless complicated
by the fact that P_b as well as λ_b are not constants. (Nor is A in
some versions of the model.)[5]

At temperatures below the onset of vacancy formation, one can
use angular correlation and positron lifetime measurements to
determine directly the temperature dependence of P_b and of λ_b, respectively. Such measurements have shown a linear temperature dependence for P_b in many metals, which has been explained as being

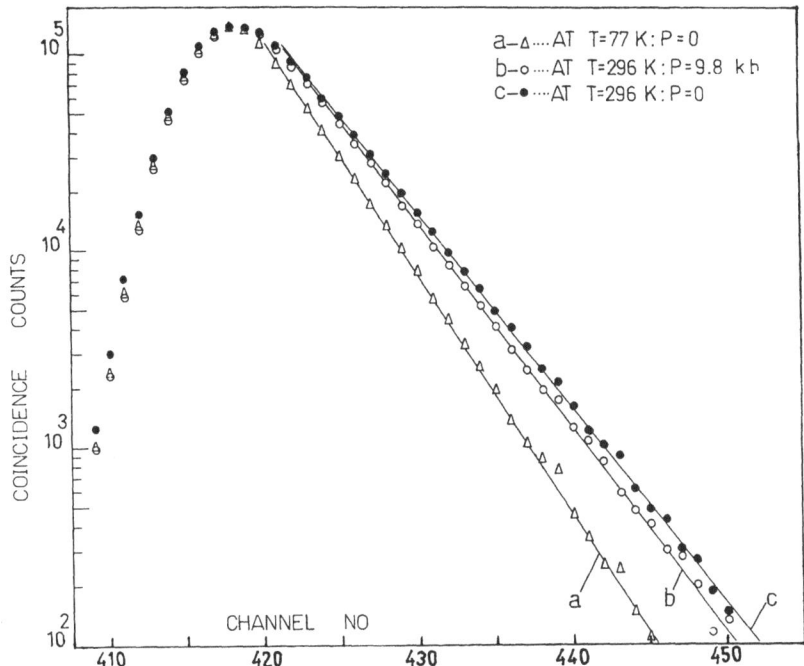

Figure 1.   Positron lifetime spectra in cadmium at (a) T = 77 K,
P = 0;   (b) T = 296 K, P = 9.8 kb; and (c) T = 296 K,
P = 0.   The source contribution has been subtracted
from all these spectra.   Time calibration is 46 ps/chan-
nel.

a result of thermal expansion of the lattice.[6]   This linear temperature dependence of P_b has been extrapolated into the high-T region
(where P_b can no longer be directly observed) to improve the above-
mentioned fitting procedure.

But there is evidence for some metals that the temperature dependence of P_b and λ_b is more complicated than one would expect from
a simple "static lattice" model of thermal expansion.[7-14]   It has
been suggested that this temperature dependence results from a combination of thermal expansion and attraction of the positron into
momentarily large interstitial sites produced by lattice vibrations.[14]
Both expansion and vibration would tend to increase P_b and to decrease λ_b, by permitting the positron to avoid the high-momentum
electrons in the ionic cores.

We have now begun a series of experiments that will permit us to
separate the contributions of these two physically distinct processes.

In these experiments we measure the positron lifetimes as a function
of temperature and pressure to obtain $\lambda_b(T,p)$. This is an extension
of the usual experiment, in which one measures the lifetime, or λ,
at various temperatures but at constant pressure, to obtain $(\partial\lambda/\partial T)_p$.
We may separate the effect of change in volume by means of the ex-
pansion

$$(\partial\lambda/\partial T)_p = \beta V(\partial\lambda/\partial V)_T + (\partial\lambda/\partial T)_V \tag{3}$$

where $\beta = V^{-1}(\partial V/\partial T)_p$ is the thermal expansion coefficient. If
the change in λ with temperature were a result of lattice expansion only, then $(\partial\lambda/\partial T)_V$ would be zero. To test this, it is only
necessary to apply pressure as T is increased, in such a way as
to keep V constant, and to measure λ at various values of T. A
minimal test requires measurement of only three values of $\lambda(T,p)$:
$\lambda(T,0)$, $\lambda(T + \Delta T,0)$, and $\lambda(T + \Delta T,p)$, where p is the pressure required to give the crystal the same volume that it had at tempera-
ture T and zero pressure. Both T and T + ΔT of course must be below the vacancy-formation region of temperature, so that one is
truly measuring bulk effects. Then, if $(\partial\lambda_b/\partial T)_V$ is zero throughout the range of temperatures between T and T + ΔT, $\lambda(T + \Delta T,p)$
must be equal to $\lambda(T,0)$.

Our first test was done on cadmium, with T = 77 K, ΔT = 219 K,
and values of p ranging from 4.9 to 9.8 kb. Cd was chosen because the complications in the temperature dependence of P_b and
λ_b are especially pronounced in this metal. Unfortunately, in Cd
it is not possible to use pressure to reproduce, at 296 K, the same
values of both lattice constants (c and a) that are obtained at
77 K, even though the volume can be held constant. Therefore one
of the runs at 296 K was done at 9.8 kb, a pressure great enough
to reduce both c and a to less than their values at 77 K.

Details of our lifetime measurement apparatus have been described elsewhere.[15] For this application we used 1" x 1" cylinders of NE 111 coupled to RCA C31024 photubes. Our prompt time
resolution (fwhm) was 240 ps taken with a ^{60}Co source at the experimental positron window settings with 40% energy windows. The
source was ∿ 15 µCi of [22]Na, sandwiched between two pieces of
cadmium (99.9999%), 3 mm x 4 mm rectangles and 2 mm thick. The
source-sample sandwich was sealed with M-Bond adhesive Type AE
and curing agent Type 10 from Micro-Measurement Inc. around the
edges, and then placed in a hydrostatic pressure apparatus designed by C.W. Chu.[16] A mixture of kerosene and silicone fluid
contained in a teflon cup inside a high pressure clamp was used
as the pressure-transmitting medium. The 3/8" thick copper walls
of the pressure clamp caused considerable attenuation of the gamma
rays; the coincidence counting rate was 4 c.p.s. and each run
accumulated about 1.4 x 10[6] counts. Fig. 1 shows the positron
lifetime distribution spectra in cadmium at (a) T = 77 K, p = 0;

Table 1. Positron Lifetime Results at Various Temperatures and
Pressures

Run No.	Temperature K	Pressure kb[+]	Mean Lifetimes ps
1	77	0	170 ± 2
2	296	0	190 ± 2
3	296	4.9	188 ± 2
4	296	5.6	182 ± 2
5	296	9.8	178 ± 2

[+]Uncertainty in pressures is $\sim$ 5%.

(b) T = 296 K, p = 9.8 kb; and (c) T = 296 K, p = 0.

The lifetime data were fitted by a routine similar to that described by Kirkegaard and Eldrup.[17] The results are shown in Table
1. A source component of about 7% intensity and about 540 ps mean
lifetime was present in all the runs. The constancy of this component showed that the source remained sealed and isolated from
the surrounding fluid medium during all runs. Runs 1 and 2 in
Table 1 were also taken with a much better time resolution (fwhm)
of 200 ps using small conical scintillators.[15] The results obtained were very similar, within errors, to those obtained by 1" x
1" cylindrical scintillators.

The change in lifetime produced by ΔT is approximately twice
the change produced by a p of 5.6 kb, indicating that the two terms
on the right in (3) contribute about equally to the lifetime change
in Cd in this temperature range. There may be some ambiguity about
this result, because the c/a ratio for run No. 4 differs from that
of run No. 1 even though the volume is the same. However, even at
9.8 kb the pressure effect remains substantially smaller than the
temperature effect. At 9.8 kb and 296 K the values of c and a
are 5.498 Å and 2.971 Å respectively, compared with c = 5.546 Å
and a = 2.969 Å at P = 0 and liquid N_2 temperature,[18,19] yet the
positron lifetime is longer at 9.8 kb. This certainly indicates
that the term $(\partial\lambda/\partial T)_V$ in Eq. (3) is not small in Cd, and that
temperature effects other than volume expansion are important.
One possible source of a non-zero $(\partial\lambda/\partial T)_V$ contribution is the
vibrational effect referred to above.

We are currently extending these investigations to other metals, and hope eventually to observe the effect at various temperatures with smaller values of the temperature interval ΔT.

I would like to thank my collaborators in this work, Dr. P. Sen, I.K. MacKenzie, and T. McMullen, in particular Dr. MacKenzie for suggesting the experiment and Dr. McMullen for elucidating the theory. I would also like to thank Dr. C.W. Chu for making available the pressure vessel used in the present studies and Charles Knox for valuable help in making the measurements.

REFERENCES

1. W. Brandt: In Positron Annihilation, Ed. by A.T. Steward and L.O. Roellig, (Academic Press, New York 1967) p. 180.
2. B. Bergersen, M.J. Stott: Solid State Commun. 7, 1203 (1969).
3. D.C. Connors, R.N. West: Phys. Letters 30 A, 24 (1969).
4. B.T.A. McKee, W. Triftshauser, A.T. Stewart: Phys. Rev. Letters 28, 358 (1972).
5. A. Seeger: J. Phys. F (Metal Phys.) 3, 248 (1973).
6. W. Triftshauser and J.D. McGervey: Appl. Phys. 6, 177 (1975).
7. Bhattacharyya, P. and Singwi, K.S., Phys. Rev. Lett. 29, 22-5 (1972).
8. West, R.N., Adv. Phys. 22, 263-383 (1973).
9. Jamieson, H.C., McKee, B.T.A. and Stewart, A.T., Appl. Phys. 4, 79-82 (1974).
10. Lichtenberger, P.C., Schulte, C.W. and MacKenzie, I.K., Appl. Phys. 6, 305-7 (1975).
11. Seeger, A., Appl. Phys. 7, 85-92 (1975).
12. Singh, K.P. and West, R.N., J. Phys. F: Metal Phys. 6, L267-70 (1976).
13. Smedskjaer, L.C., Fluss, M.J. Chason, M.K., Legnini, D.G. and Siegel, R.W., Proc. of the 4th Inter. Conf. on Positron Annihilation, Helsingor, C4, 10-1 (1976).
14. Stoll, H., Trost, W., Herlach, D., Maier, K., Metz, H. and Schaefer, H.E., Proc. of the 4th Inter. Conf. on Positron Annihilation, Helsingor, C6, 18-23 (1976).
15. McGervey, J.D., Vogel, J., Sen, P. and Knox, C., Nucl. Instr. Meth. 143, 435 (1977).
16. Chu, C.W. and Testardi, L.R., Phys. Rev. Lett. 32, 766-9 (1974).
17. Kirkegaard, P. and Eldrup, M., Comput. Phys. Comm. 3, 240-355 (1972).
18. Perez-Albuerne, E.A. Clendenen, R.L., Lynch, R.W. and Drickamer, H.G., Phys. Rev. 142, 392-9 (1966).
19. Kirby, R.K., Hahn, T.A. and Rotchrock, B.D., Amer. Inst. Phys. Handbook, 3rd Ed. McGraw-Hill Book Co. 4, 119-42 (1972).

QUESTIONS AND COMMENTS

D. Debray: You mentioned defects. Did you mean vacancies?

J. McGervey: Well, vacancies are the one kind of defects that
                  positrons sense, there may be other kinds of de-
                  fects in there.

D. Debray: Defects like interstitials?

J. McGervey: Yes, right.

B. Hamermesh: What is the source, and what is its strength, and
                  what is your source counting rate?

J. McGervey: In the angular correlation, we have a 12 mCi source
                  of sodium 22, and the counting rate for coincidences
                  is something on the order of 2 or 3 per second, and
                  we've run it for several days. In a positron life-
                  time experiment, we put it in the pressure bomb and
                  we have attenuation by the walls of the bomb. We
                  use a source of something like 10 μ C, and have a
                  counting rate which allows us to get half a million
                  to a million counts per day.

DIMENSIONALITY EFFECTS AND PHYSICAL PROPERTIES IN 1D COMPOUNDS

D. Jérome

Laboratoire de Physique des Solides
Université Paris - Sud
91405 Orsay - France

ABSTRACT

Large fluctuation effects in one-dimensional conductors prevent the occurrence of metal-insulator transitions at finite temperature.

However, in the charge transfer salts of the TTF-TCNQ family, the existence of well defined M-I transitions at $T = T_p$ is attributed to a non zero 3 dimensional coupling.

Effects of high pressure on conducting charge transfer salts will be reviewed[1]:

1) on the low temperature insulating phase, with emphasis on the stabilisation of the metallic phase in the HMTTF-(HMTSF)-TCNQ structure[2].
ii) on the metallic phase: conductivity, anisotropy, magnetic properties.
iii) on the hopping rate $\tau_\perp^{-1}$ between conducting chains measured by the field dependence of the nuclear relaxation rate[3].

We present a unified picture of quasi 1-D conductors based on two parameters; the tunneling interchain coupling $t_\perp$ and the intrachain electron scattering time τ_v.

Within this model the electronic properties exhibit a crossover at $T = T*$ between a high temperature 1-D metallic Fermi surface and a low temperature 3D semi-metallic surface[4]. The validity of this model is corroborated by the temperature dependence of the Hall effect,[5] magnetoresistance and susceptibility in HMTSF-TCNQ[6] (for which $T_p < T^*$) and by pretransitional effects in TTF-TCNQ (for which $T_p > T^*$)[7].

BIBLIOGRAPHY

1. An extensive review of the electronic properties of the charge transfer (CT) salts: band structure and pressure effects has been published in:
    D. Jérome and M. Weger, Chemistry and Physics of One-Dimensional Metals, H.J. Keller editor, Plenum Press NY 1977.
2. For HMTSF-TCNQ see: J.R. Cooper, M. Weger, D. Jérome, D. LeFur, K. Bechgaard, A.N. Bloch and D.O. Cowan, Solid State Comm. 19, 749, (1976).
3. For a detailed interpretation of NMR in one dimensional conductors, see: G.Soda, D. Jérome, M. Weger, J. Alizon, J. Gallice, H. Robert, J.M. Fabre and L. Giral, J. Physique 38, 931 (1977).
4. A classification of the various CT salts, with respect to the stabilization of the metallic state at low temperature under pressure is given in:
    R.H. Friend, D. Jérome, J.M. Fabre, L. Giral and K. Bechgaard, J. Phys. C. to be published (1977).
5. J.R. Cooper, M. Weger, G. Delplanque, D. Jérome and K. Bechgaard: J. Physique Lett 37, L-349 (1976).
6. G. Soda, D. Jérome, M. Weger, K. Bechgaard and E. Pedersen: Solid State Comm 20, 107 (1976).
7. For the Hall effect in TTF-TCNQ see J.R. Cooper, M. Miljak, G. Delplanque, D. Jérome, M. Weger, J.M. Fabre and L. Giral: J. Physique 38, 1097, (1977).

QUESTIONS AND COMMENTS

J.A. Woollam: You mentioned Hall effect measurements. Aren't they terribly hard to make on such tiny samples?

D. Jérome: I don't know any other measure of Hall effect at low pressure. I want to tell you that working with this is not as easy as working with other bulky samples. It is an extremely small, brittle sample, and even when you look at them they tend to break!!

J.A. Woollam: How do you put leads on?

D. Jérome: For Hall effect measurements you put 5 contacts on, usually by silver paste.

NEUTRON DIFFRACTION STUDY OF THE COMPRESSIBILITY OF TTF-TCNQ and KCP UNDER HYDROSTATIC PRESSURES UP TO ABOUT 20 kbar

D. Debray
Laboratoire Léon Brillouin and DPh-G/PSRM
CEN-Saclay, BP n°2, 91190 Gif-sur-Yvette, France

D. Jérome
Laboratoire de Physique des Solides
Université Paris-Sud, 91405 Orsay, France

S. Barisic
Institute of Physics of the University
41001 Zagreb, Yugoslavia

The organic charge-transfer salt TTF-TCNQ (tetrathiafulvalene-tetracyanoquinodimethane) and the Krogmann salt KCP [$K_2Pt(CN)_4Br_{0.3}$ $.3H_2O$] have been the focus of considerable recent work because of their unusual quasi-one dimensional electronic transport properties and, in case of TTF-TCNQ, its sequence of Peierls and collective phase transitions at low temperatures. We have measured the linear axial compressibilities of deuterated polycrystalline samples of these two compounds under purely hydrostatic pressures up to about 20 kbar by neutron diffraction measurements at room temperature. The high-pressure cell is a piston-cylinder device and a liquid is used as the pressure transmitting medium. The pressure is monitored by a manganin pressure gauge placed directly in the transmitting liquid and calibrated previously by reference to the NaCl pressure scale.

The observed compressibilities for TTF-TCNQ along the three crystallographic axes are, respectively, $k_a = 2.7$, $k_b = 4.7$ and $k_c = 3.2 \times 10^{-12}$ cm^2/dyne. The rather surprising feature of our results is that the compressibilities along the $\vec{a}$ and $\vec{c}$ axes (i.e. perpendicular to the chain axis) are smaller than that along the $\vec{b}$ axis (chain axis). These results thus indicate a strong interchain binding, stronger than that along the chain. The pressure dependence of the compressibilities yields the following values for the Grüneisen's

constants along the three crystallographic axes : γ_a=9.57 ,
γ_b = 3.57 and γ_c = 7.92 , respectively. These results indicate a
strong anharmonicity of the interchain binding focus. We suggest
that the dominant part of the crystal cohesion in TTF-TCNQ results
from a stacking of dynamically polarized TTF and TCNQ molecules
acting as effective molecular dipoles. This is a consequence of the
high polarizability of the molecules combined with a small intermolecular separation which is of the same order as the length of the
molecules themselves. The bonding then results from the dominant
attractive interaction between adjacent dipoles. In such a case,
one also expects a strong anharmonic effect. This is evidenced by
the high values of the Grüneisen's constant along the $\vec{a}$ and $\vec{c}$ crystallographic axes.

The observed compressibilities for KCP along the $\vec{a}$ and $\vec{c}$ (chain
axis) crystallographic axes are, respectively, k_a = 1.3,
k_c = 1.4×10^{-12} cm^2/dyne. KCP is thus less compressible than TTF-TCNQ.
These results indicate that the interchain binding is as strong as the
binding along the chain. Also note that the anharmonicity of the
interchain binding in KCP is much less pronounced than that in case
of TTF-TCNQ, indicating the absence of any dipolar-type contribution
to crystal cohesion.

We conclude that though quasi-one dimensional with respect to
electronic transport properties, both TTF-TCNQ and KCP are fully
three-dimensional materials as far as crystal binding is concerned.

QUESTIONS AND COMMENTS

E.F. Skelton: I recall 3 or 4 years ago, Tom Kistenmacher at
                  Johns Hopkins published an X-ray structure of
                  TTF-TCNQ. Have you compared your neutron data
                  with that structure?

D. Debray: No. We didn't determine the crystal structure.
                  The crystal structure was already known from X-ray
                  work. All that we did by neutron diffraction meas-
                  urements was to determine the compressibility, in
                  other words, how the lattice parameters of the unit
                  cell vary as a function of pressure.

E.F. Skelton: Couldn't you do that with X-ray?

D. Debray: Well, I just wanted to point out that it is one
                  of the techniques. This is a monoclinic structure,
                  and you have very large lattice parameters. If
                  you take an X-ray picture of this compound you can
                  see very clearly only a few lines. The number of

lines you see clearly, I think, is insufficient to do a least squares fitting, to find the lattice parameters from the positions of the Bragg peaks.

E.F. Skelton: Well, I think it can be done. The point is, the more lines there are the better you can identify them.

D. Debray: Sure, it can be done. For a least squares you need at least 7 or 8 lines. Of course, this can be done, but the question is how well you can rely on them.

E.F. Skelton: Ideally you can turn to a single crystal.

D. Debray: Sure, but with TTF-TCNQ you can't grow very large single crystals--big enough for neutron work at a medium flux reactor.

I.L. Spain: I'd like to perhaps draw an analogy between the layer structure system to these quasi-one-dimensional systems. Very often when one has a weak bond, one calls it a Van der Waals bond and attributes it to a kind of fluctuating dipole-dipole interaction. In layer materials I think it is true to say that if one calculates elastic constants, etc., based on these forces, one always fails to explain the experimental data. The term Van der Waals is, perhaps, not appropriate to the type of bonding in one-directional systems. I would like to ask a question of both you and Jérome. If these systems really do have very weak bonding then one expects defects of various kinds to be very important and might control many properties. What kind of information is there available for these crystals about defects and their importance in measurements, particularly, electrical measurements?

D. Debray: Well, I don't know how much I should agree with you. I agree with you to the extent that defects would influence electrical properties. But I don't think the compressibility or the elastic constants should be much affected by defects.

I.L. Spain: How do you know this? I don't think the X-ray pattern will tell you.

T.M. Rice: I would just like to make a comment. It is not possible to describe these bonds as Van der Waals

bonds. In fact, these crystals are not
Van der Waals type at all. They are all charge
transfer salts; for example, KCP and TTF-TCNQ
are charge transfer salts. The binding is due
to a Madelung interaction between oppositely
charged molecules.

D. Jérome: I don't like Van der Waals because Van der Waals
is for neutral molecules usually. This is why in
the paper we called them polarization forces.

D. Debray: I did mention in the very beginning that the binding is an order of magnitude larger than
Van der Waals. I just wanted to compare the values.
What I think is this. Call this binding whatever
you like, Van der Waals, polarization, or whatever.
It is not the usual kind of attraction you find in
neutral molecules like organic molecular crystals.
It is a different type. It is an induced dipole-
dipole moment interaction, but the strength is very
much different because you have a high polarizability of the molecules, and these molecules are
charged. You are right, because the molecules
are charged, the attraction is much higher. So
there are two points actually, the high polarizability and the charge of the molecules. And these
two things combined can induce very high fluctuating dipole moments which can lead to a strong
interaction.

D. Jérome: Well, there could be a small isotope effect probably.

THE ROLE OF LIBRATIONS IN TTF-TCNQ

M. Weger

The Hebrew University of Jerusalem
and
Nuclear Research Center
Negev

The temperature dependence of the resistivity of TTF-TCNQ has been a mystery for a long time, and received considerable attention since Heeger's talk in SanDiego early in 1973. The resistivity follows approximately a T^2 law $\rho_\parallel(T,P) = A + B(P) T^2$ in the temperature region 60 - 300 K. Above 200 K, at ambient pressure, $\rho_\parallel$ is slightly larger, and a fit to $T^{2.3}$ would be better; also, close to the transition at 52.5 K, up to about 60 K, there are slight upward deviations. However, by and large, this T^2 law is followed, and the pressure variation of B is very large. B(P) falls by about a factor of 2 at 4 kbar.[1]

This experimental result is known for quite some time, and has been observed in many laboratories; the strong pressure variation of the resistivity was first reported by Chu et al[2]. In contrast the theoretical situation is confused. About a dozen theories, mutually inconsistent, have been proposed to account for this resistivity behaviour.

TTF-TCNQ is a one-dimensional metal, in the sense that the electronic motion along the chains is wavelike and coherent: $\hbar/\tau_\parallel < \varepsilon_F$, where $\tau_\parallel$ is the collision time for motion along the chains, derived from the dc resistance $\rho_\parallel = m^*/ne^2\tau_\parallel$; while the electronic motion perpendicular to the chains is diffusive: $\hbar/\tau_\perp > t_\perp$, where $t_\perp$ is the inter-chain integral, obtained for example from the frequency variation of the NMR relaxation time T_1. The relationship $t_\perp < \hbar/\tau_\parallel < \varepsilon_F$ holds for TTF-TCNQ at all temperatures (in the metallic state) and pressures. (In HMTSF-TCNQ, the

transverse motion is coherent at low temperature and there the resistive behaviour is entirely different, namely, $\rho_{\parallel}$ is virtually
temperature and pressure independent, and probably due to defects).

We can understand the T^2 law in a very simple way, when we re-
alise that the TCNQ and TTF molecules librate strongly, the libration frequency being very low (about 30-50 cm^{-1} for librations a-
round an axis perpendicular to the molecular plane). These librations were first proposed by Morawitz; Merrifield and Suna pointed
out that the interaction between them and the electronic motion
along the chain is a second-order process; Weger and Friedel[3] investigated them in detail and showed that they may be responsible
for the $4k_F$ reflections and the cascade of phase transitions at
52.5, 49, and 38 K.

When a TCNQ, or TTF, chain undergoes a libration, it possesses
a glide-plane symmetry element; namely, a translation along the
chain by $\Lambda/2$ (where Λ is the wavelength of the librational mode),
followed by a reflection $x \to -x$, transforms the libration into
itself. This glide-plane symmetry has the property that the fundamental (and odd-harmonic) reflections along the chain (and in the
x=0 plane) are strictly forbidden by symmetry. This basic selection
rule applies to all waves - X-rays, neutrons, electrons. In our
case, it implies that a libration with wavevector $q = 2\pi/\Lambda$ cannot
reflect an electron moving along the chain, with a wavevector-change
$\Delta k = q$. However, since even-harmonic reflections are allowed, reflections with $\Delta k = 2q$, and reflections caused by two librons with
wavevectors q, q' with $\Delta k = q + q'$, are allowed. Phrased alternatively, the electron-libron interaction is a second-order one, unlike the usual Frohlich-type electron-phonon interaction which is
first order, because the electron cannot distinguish between a
libration of amplitude A and one of amplitude -A. Thus, the
Hamiltonian is quadratic in the libron operators:

$$\mathcal{H} = \Sigma \ g_{k,q,q'} \ c^{\dagger}_{k+q+q'} \ c_k \ (a_q + a^{\dagger}_{-q}) \ (a_{q'} + a^{\dagger}_{-q'})$$

This interaction accounts for the observed resistivity in a
very simple way. The collision time is given by the Golden Rule

$$\tau_{\parallel}^{-1} = \frac{2\pi}{\hbar} \ (\frac{\partial^2 t_{\parallel}}{\partial\theta^2}<\theta^2>)^2 \ n(\varepsilon_F)$$

where $t_{\parallel}$ is the transfer integral along the chain, and θ the angle
between two librating neighbouring molecules. (The absence of a
first-order coupling implies the vanishing of the first derivative
$\partial t_{\parallel}/\partial\theta$). The amplitude of the libration[4] is given by

$$<\theta^2> = \frac{k_B T}{I\omega_{Lib}^2(P)} \qquad\qquad \text{Thus,}$$

$$\rho_{\parallel} = (m*/ne^2) \ \tau_{\parallel}^{-1} = (m*/ne^2) \cdot (2\pi/\hbar) \cdot \partial^2 t_{\parallel}/\partial\theta^2 \cdot n(\varepsilon_F) \cdot$$

$$\cdot [k_B T/I\omega_{Lib}^2(P)]^2 \quad \text{accounting}$$

immediately for the T^2 law, and for the pressure-dependence (due to
the pressure dependence of ω_{Lib} , which occurs in the fourth power)[4]

In order for this simple expression for the resistivity to apply,
a number of requirements have to be met:
1) The interaction has to be a second-order one, since a linear
   electron-phonon coupling will cause a linear-T dependence, as
   is the case for ordinary metals.
   This follows from the one-dimensional motion of the electrons
   along the chains. Electrons moving diagonally (with a component of k in the x-direction) do interact with the librons in
   the first order.
2) The libration frequency has to be low, $\hbar \omega_{Lib} < k_B T$, in order
   for Boltzmann statistics (rather than Bose-Einstein statistics)
   to apply.
   This condition is satisfied above 60 K.
3) The matrix element $\partial^2 t/\partial\theta^2$ has to be very large, to account for
   the large observed resistance. Since the molecules are long,
   a relative rotation of $3°$ or so is sufficient to reduce the
   overlap by a factor of 2 or so, thus this condition is met.
4) The libron frequency must be strongly pressure dependent, increasing by about 20-25% at a pressure of 4 kbar.
   Such a strong pressure dependence is common in organic solids.
   Thus the explanation of the mystery of the resistance of TTF-
TCNQ is simple and straightforward, if not trivial.

When something works so simply and nicely, we have to start to
worry whether the situation is not too good. Didn't we ignore something?
In order to be candid, let us point out here also the weakest
point of this derivation. In this derivation, we assume that the
libron frequency is temperature independent, and that only one
libron mode is present. We have as yet no justification for these
assumptions.

The amplitude of libration is very large (about $3°$ at ambient).
For such a large amplitude, anharmonic forces undoubtedly come into
play, causing changes in the libron frequency with temperature.
The large thermal expansion is in some sense a manifestation of
these anharmonic forces; another manifestation is the non-linear
compressibility found by Debray et al.[5]

In addition to librations in the molecular plane, librations
perpendicular to it should be excited thermally at ambient, since
their frequency does not exceed 100-150 cm^{-1} or so. It might be
tempting to ascribe the deviation from the T^2 law above 200 K to

excitation of such modes; however, the T^2 law holds under pressure up to ambient temperature. (The deviation from the T^2 law occurs when $\hbar/\varepsilon_F \tau_\parallel$ becomes of order unity, and deviations of this sort must occur. According to the general idea of Mott, once localisation starts, it proceeds very rapidly, thus $\rho_\parallel$ should increase rapidly once $\hbar/\tau_\parallel$ is no longer small compared with ε_F).

Thus, the librations are at the heart of the unusual properties of TTF-TCNQ. $\rho_\parallel$ (T,P) is due to them, as shown here. $\rho_\perp$ (T,P) is given by $\mu_\perp \simeq e\, D_\perp / \varepsilon_F$ (Einstein relation) with $D_\perp \simeq \ell^2/\tau_\perp$, where ℓ is the hopping distance (a/2 or c/2), and $\tau_\perp^{-1} \simeq (2\pi/\hbar)$ $t_\perp^2 (\tau_\parallel/h)$ (a slight modification of the Golden Rule for 1-D metals). Thus, the temperature and pressure variation of $\rho_\perp$ follow from that of $\tau_\parallel$. This behaviour has been verified experimentally by NMR[6]. The diffusion constant of the electrons along the chains, $D_\parallel$ (T,P), which dominates the NMR properties, the Thermoelectric power, and the thermal conductivity, is given by $D_\parallel (T,P) \simeq v_F^2\, \tau_\parallel$ (T,P), and is thus dominated by the librations as well. The susceptibility is proportional to the average of the density of states over a region of width $\hbar/\tau_\parallel$ around ε_F; since $n(\varepsilon)$ is concave upwards $(d^2 n(\varepsilon)/d\varepsilon^2 > 0)$ for 1-D systems, this average is larger than $n(\varepsilon_F)$, and χ is larger than the Pauli value, and its temperature and pressure variation follows (in part) from that of $\tau_\parallel$. The high conductivity of HMTSF-TCNQ follows from the rigidity of the HMTSF stack, increasing ω_{Lib} and thus decreasing $\rho_\parallel$.

The more complicated features of TTF-TCNQ, namely the $4k_F$ reflections, the cascade of phase transitions and the behaviour of the transverse period of the CDW, follow directly from the presence of these librations[3].

We might question, why have not these all-important librations been seen directly? It turns out, that the librations are very elusive. Since they do not possess an electric dipole moment, it is hard to see them by IR spectroscopy. Because of their low frequency, Raman and neutron work is diffiuclt. Obviously they do not couple to a magnetic field. Pressure seems to be the physical variable which couples to the librations most strongly, therefore it plays a key role in the understanding of the properties of organic metals. Continuous, intense awareness of the strong pressure dependence of the properties of organic metals seems to be a key factor in the theoretical understanding of these exciting systems.

REFERENCES

1. J.R. Cooper, M. Weger, D. Jerome and S. Etemad, J. Physique Lett. 36, L219(1975).
2. C.W. Chu, J.M.E. Harper, T.H. Geballe and R.L. Greene, Phys. Rev. Lett. 31, 1431(1973).

3. M. Weger and J. Friedel, J. Physique 38, 241 (1977); ibid 38, 881 (1977).
4. H. Gutfreund and M. Weger, Phys. Rev. B16, 1753 (1977).
5. D. Debray et al, J. Physique Lett. 38, 224, (1977).
6. G. Soda et al, J. Physique 38, 931 (1977).

QUESTIONS AND COMMENTS

T.M. Rice: Your theory focuses on the role of the librons which couple to the electrons only in second order and have no coupling in first order. I would like to point out that there are other theories, which are based on effects which couple in first order, i.e., acoustic phonons and Coulomb electron-electron interactions, and which can explain these materials at least semiquantitatively.

B.T. Matthias: Whatever happened to ATTF-TCNQ which 4 years ago was announced as a fine-tuned TTF-TCNQ?

M. Weger: Not by us, not by the Orsay group. You're not going to make us responsible for the statement.

B.T. Matthias: I just asked the question.

M. Weger: Some people went to the New York Times. We did not-- I don't claim that that material is important to industry. I don't make that statement. I say we do things under pressure and we understand it.

D.B. McWhan: Isn't it common practice in the molecular organic crystals, when you do a crystal structure, to define the structure in terms of rigid molecules and then put in your temperature factor, which in fact involve the librations and so forth? So, if there's evidence for your librational motion, it must be there in the X-ray data, and you should be able to see it. It's an enormously large rotation.

M. Weger: Well, in some work by Stukey and Schultz: JACS 98, 3194 (1976) they do X-ray form factor measurements at low temperature. It is hard to get machine time, and the reason is simply that people do the X-ray work with on-line computers. When they take a new material they can get a paper out in one day. If somebody wanted to do an analysis of the thermal vibration effects it would take 1 or 2 months, and we can't get so much machine time.

D. Bloch: A very simple question; maybe the resistivities
                         are largely dependent on temperature, since they
                         are largely dependent on pressure. When you cool
                         a sample from room temperature to low temperature
                         you decrease the dimension of the sample as well
                         as on applying pressure on the sample. What is
                         the volume contribution to the thermal dependence
                         of the resistivity of TTF-TCNQ?

M. Weger: If you take copper, and change the temperature
                         from room temperature to the temperature of nit-
                         rogen, the resistance will change by more than
                         10. If you apply a pressure of 2 kbar, the re-
                         sistance will change by 1 percent. The effect
                         of temperature in copper and TTF-TCNQ is about
                         the same. Now in metals the temperature has a
                         huge effect on resistance but pressure has a
                         small effect. But here, pressure has pretty much
                         the same effect as temperature change on electric-
                         al resistance, and that would be very unusual for
                         metals.

THE COMMENSURATE-INCOMMENSURATE PHASE TRANSITION

T.M. Rice, S.A. Jackson and P.A. Lee

Bell Laboratories
Murray Hill, New Jersey 07974

ABSTRACT

Recent theoretical work has led to the prediction of large
spatial variation in the order parameter in an incommensurate
phase when a commensurate phase is nearby. The application of
pressure is an ideal variable to map out the phase diagram for the
commensurate-incommensurate transition.

I. INTRODUCTION

In solids there is a variety of possible phases with a periodic order parameter charge density waves (CDW), spin density
wave (SDW), orbital waves, and periodic lattice distortions where
the period of the order parameter can be incommensurate with that
of the underlying lattice. The period can be determined by a dimension of the electron Fermi surface, or a balance of long range and
short range forces. There are, however, forces in a crystal which
favor a locking-in of the period to that of the underlying lattice
or a phase in which both periods are commensurate. Hydrostatic
pressure is a unique variable by which one can change the balance
between the forces while maintaining the symmetry and long range
order of the crystal. Other variables such as alloying destroy
one or both of these properties. Pressure therefore can cause transitions between commensurate (C) and incommensurate (I) phases and
enable one to map out a phase diagram for this transition. The
discovery in recent years of charge density waves in a number of
layered compound and organic compounds and which also undergo
transitions C-I transitions has stimulated interest anew in this

topic. In this brief review we will discuss some recent theoretical developments specifically as they relate to charge density waves[1-5] Of particular interest are the possibility of finding large spatial variation in the order parameter in I-phase and a change in character from discontinuous to continuous C-I transitions. Also, as we shall see there is the possibility of a change in the order of the phase transition between the normal (N) and incommensurate (I) phases.

II. SIMPLE THEORY

We shall begin by writing down a phenomenological or Landau free energy which describes the free energy of the CDW phase. The order parameter can be written as

$$\rho(X) = \overline{\rho} + \text{Re}\{\psi(X) \exp(i\vec{G}\cdot\vec{X}/p)\} \tag{1}$$

where we have specialized to single $\vec{Q}$ CDW. $\overline{\rho}$ is the average density, $\vec{G}$ is a reciprocal lattice vector of the underlying lattice, and $\psi(X)$ is a slowly varying complex order parameter. The magnitude of ψ measures the amplitude of the CDW. If $\psi(X)$ is chosen as a real constant (ψ_c) then the period of the CDW is $2\pi p/G$ and is commensurate and the integer p is the order of commensurability. Alternatively if $\psi(\vec{X}) \equiv \psi_I(\vec{X}) = \psi_0 \exp(i(\vec{q}\cdot\vec{X} + \phi))$ then the period is $2\pi/|p^{-1}\vec{G}+\vec{q}|$ and is incommensurate with the underlying lattice. The phase ϕ in this case measures the displacement of the CDW. Since an incommensurate CDW may be displaced arbitrarily with respect to the underlying lattice at no cost in energy, the free energy must be independent of the choice of the constant ϕ.

The Landau expansion for the free energy difference to the normal or undistorted state is

$$F - F_N = a \int |\psi(X)|^2 dX + c \int |\psi(X)|^4 dX - b\text{Re} \int \psi^p(X) dX \tag{2}$$

$$+ e \int |d/dX-iq)\psi(X)|^2 dX$$

where we assume that $\psi(X)$ varies only along X - the direction of the $\vec{Q}$-vector. The constants b, c, and e are taken to be independent of temperature. The free energy of the I-phase $(\psi(X)=\psi_I(X))$ is given by

$$F_I - F_N = -a^2/4c; \quad \psi_0^2 = -a/2c , \tag{3}$$

and the N-I transition temperature is given by the condition $a(T_{N-I}) = 0$. The free energy of C-phase is found by minimizing

$$F_C - F_N = \psi_c^2 + c\psi_c^4 - b\psi_c^3 + eq^2\psi_c^2 , \tag{4}$$

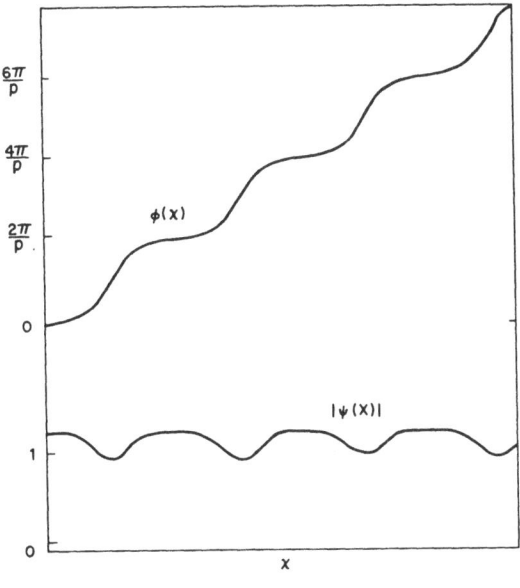

Figure 1

w.r.t. ψ_c. If we simply compare these energies we obtain a first order C-I transition when $F_C = F_I$.

III. AN EXACTLY SOLUBLE MODEL

The two cases discussed above are special solutions of the more general problem of minimizing Eq. (2) w.r.t. to $\psi(X)$. If one specializes to the form $\psi = \psi_0 \exp(i\phi(X)$, i.e., allows only a spatially varying phase, then the minimization problem is exactly soluble. The problem has a long history and to our knowledge was first discussed many years ago by Frenkel and Kontorova,[6] and Frank and Van der Merwe[7] in the context of dislocations in an epitaxial monolayer on a substrate. McMillan[1] independently introduced the model for the CDW state and the exact solution for this case was given by Bak and Emery.[3] In the present CDW case the approximation of neglecting variations in the amplitude ψ_0 can be expected to hold only at temperatures $T \ll T_{N-I}$. We will return to this point in our next section.

The free energy in the phase only approximation can be written as

$$F[\phi(X)] = e\psi_0^2 \left[\int (d\phi/dX - q)^2 \, dX + Y \int (1 - \cos p\phi(X)) dX - Y \right] + F_I \, ,$$

(5)

where $Y = b\psi_0 p^{-2}/e$. Taking the derivative w.r.t. $\phi(X)$ leads to the equation

$$\frac{2d^2\phi}{dX^2} - pY\sin p\phi = 0 \quad, \tag{6}$$

which can be cast into the standard Sine-Gordon form by the substitutions $\phi(X) = p\phi(X)$ and $S^2 = p^2 YX^2/2$ giving

$$\frac{d^2\phi}{dS^2} = \sin\phi \quad. \tag{7}$$

This equation has been extensively studied. It can be integrated at once to give

$$\left(\frac{d\theta}{dX}\right)^2 = -\frac{2}{p}\cos p\phi + \varepsilon \quad, \tag{8}$$

where ε depends on the boundary condition. This equation can in turn be integrated again and the answer expressed in terms of incomplete elliptic functions. Details of the mathematical solutions can be found in the paper by Frank and Van der Merwe.[7] We will summarize the physical results here.

For $Y \ll q^2$, the effect commensurability energy is to induce weak harmonics of the original period $2\pi/q$. However, as Y increases the magnitude of the harmonics continues to increase until at a critical value of $Y(=\pi^2 q^2/8)$ a continuous transition occurs to the C-phase. In the I phase the C-I phase boundary the free energy can be written as [3]

$$F = e\psi_0^2 \left[\frac{8}{\pi} \mid \delta \mid \sqrt{\frac{Y}{2}} \left\{1+4 \exp\left(-\frac{2\pi}{\delta}\sqrt{\frac{Y}{2}}\right)\right\} - 2\delta q + q^2 + F_I \right. \quad. \tag{9}$$

The form of the phase function $\phi(X)$ in this regime is shown in Fig. 1. It takes the form of a periodic array of equally spaced kinks of magnitude $2\pi/p$. These phase kinks are variously referred to in the literature as discommensurations, solitons, etc. The first term on the l.h.s. of Eq. (9) is the creation energy for a single discommensuration. The second term is a repulsive overlap term which depends exponentially on the separation. The third term arose from the separation of q from $p^{-1}\vec{G}$. This term acts as a chemical potential in Eq. (9) to create discommensurations. The actual separation between discommensuration is determined by minimizing Eq. (9) w.r.t. δ and is $2\pi/p\delta$. The discommensurations can be viewed as particles with a repulsive interaction and because of this repulsive interaction their onset is continuous as the chemical potential moves through a critical value. The repulsive interaction is the key to the continuous C-I transition.

The presence of the discommensurations will show up in scattering experiments through large values of the δ-harmonics of the order

parameter. Indeed McMillan[1] solved the problem numerically by a
harmonic expansion and his results give the intensity of the harmonic amplitudes.

IV. AMPLITUDE MODULATIONS

In the previous section we discussed a model in which only the
phase was spatially modulated. Since the phase is proportional to
the displacement of the commensurate wave, this analysis can be
applied to any incommensurate system. However, for a CDW at temperatures not far below the N–I transition, a modulation of the
amplitude of the CDW will also be important. Note that the amplitude fluctuations at very low temperatures which are costly
in energy will be frozen out and the phase only model will apply.

The free energy (Eq.2) can be readily expressed in terms of
the amplitude $\Psi(X)$ and phase $\phi(X)$ of the order parameter. Taking
the derivative to obtain the conditions for a minimum yields

$$a\Psi(X) + 2c\Psi(X)^3 - \frac{1}{2} pb\Psi^{p-1}(X) \cos p\psi + e\Psi(X) (d\phi/dX-q)^2 \tag{10}$$

$$= e d^2\Psi/dX^2 \quad ,$$

$$\frac{d^2\phi}{dX^2} - \frac{p}{2e} \Psi(X)^{p-2} \sin p\phi = 0 \quad . \tag{11}$$

In general these equations lead to a coupling between spatial variations the phase and amplitude. For the case $p = 2$, however, the
phase always satisfies a Sine-Gordon equation. At low temperatures
the coefficient a is large and negative and variations in $\Psi(X)$ can
be linearized about the constant Ψ solution. In this case it is
clear that the phase only solution, discussed in the previous
section, is modified by making small dips in the amplitude in the
region of rapidly varying phase. The amplitude modulation falls
off exponentially in a manner similar to the deviation in phase
from the commensurate value.

The present authors[4] considered the case $p = 3$ and a single
$\vec{Q}$-vector CDW. A general phase diagram for the three N,I, and C
phases can be constructed. Consider first the extremities of the
phase diagram (a) the N–C transition is first order because of
the cubic in the order parameter, (b) the I–C transition at very
low temperature can be described by the phase only model and is
continuous, and (c) the N–I transition away from the C boundary
is also continuous and second order. The outlines of the phase
diagram are shown in Fig. 2. The axes are labelled T which varies
linearly with $-a$, and P which is proportional to q^2. In practice
to construct a temperature and pressure phase diagram the variation of all the Landau coefficients must be taken into account.

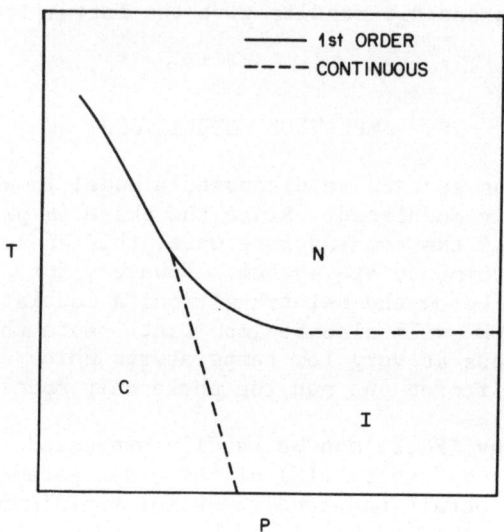

Figure 2

It is clear from the above considerations that in this model there is
at least one tricritical point, where the transition changes from
first to second order, must occur. The calculations show that the
inclusion of amplitude modulations leads to an attractive contribution to the overlap energy between two discommensurations. As one
moves up the I-C phase boundary from low temperatures, the importance
of the attractive overlap energy between the discommensurations grows.
However, at a tricritical point the total interaction adding both
phase and amplitude modulations must become attractive. Nakanishi[8]
has shown by a detailed numerical solution that the tricritical point
of the I-C phase transition for the free energy (2) occurs exactly
at the triple point. Approaching the triple point along the N-I phase
boundary leads to a similar conclusion. In this region the amplitude
of the order parameter is small and the restoring force against amplitude fluctuations is weak. Therefore in this region there is no
need to distinguish between amplitude and phase modulations. A direct
expansion of the total order parameter in harmonics is more appropriate[9]

$$\psi(X) = \Phi_0 \, e^{-i\delta X} + \Phi_1 \, e^{2i\delta X} + \cdots \tag{12}$$

Upon minimizing w.r.t. Φ_1 and δ, one finds that the coefficient of
the Φ_0^4 term in the free energy vanishes at the triple point where
the C, I and N phases meet. The triple point in this model is an
unusual form of multicritical point. In fact Nakanishi[8] has shown

that at this point the free energy is <u>independent</u> of the average periodicity δ. This highly unusual behavior, however, is an artifact of the model. If one keeps a term $\sim \int |\psi(X)|^6 dx$ in (2), as one must to be correct physically, then the tricritical point occurs on the N-I boundary at temperatures below the triple point. The C-I transition remains continuous up to the triple point. This phase diagram is shown schematically in Fig. 2.

At present we know of no material to which the single $\vec{Q}$-vector, $p = 3$ model can be applied. The layered compound 2H–TaSe$_2$ has an I-C transition to a $p = 3$ phase[9] but in this case there are 3 co-existing CDW. A $3\vec{Q}$-vector model has been studied numerically by Nakanishi and Shiba[5] for 1T – TaS$_2$ using an expansion of the form (12). This material has an intermediate phase between the commensurate phase at low temperatures and the simple incommensurate phase at high temperature. The intermediate phase is bounded above and below by first order transitions. Nakanishi and Shiba's theory successfully accounts for the existence of the intermediate phase and further they obtain agreement with the experiments on the intensity of the harmonics by Nakanishi et al.[10] These experiments show that discommensurations are present in the intermediate phase.

V.   CONCLUSIONS

Recent theories have predicted existence of discommensurations near the C-I phase boundary. This phase transition can be of first or second order and model calculations show that a complex phase diagram is possible. Pressure is an ideal variable to explore these phase diagrams and the study of pressure effects on this phase transition is a promising area for future research.

ACKNOWLEDGMENT

We are grateful to Dr. K. Nakanishi for pointing out errors in our original numerical work and for informing us of his results prior to publication.

REFERENCES

1. W.L. McMillan, Phys. Rev. <u>B14</u>, 1492 (1976).
2. A. Kotani, J. Phys. Soc. Japan <u>42</u>, 408 (1977).
3. P. Bak and V.J. Emery, Phys. Rev. Lett. <u>36</u>, 978 (1976).
4. S.A. Jackson, P.A. Lee and T.M. Rice (to be published).
5. K. Nakanishi and H. Shiba (preprint).
6. J. Frenkel and T. Kontorova, Phys. Z. Sowjet. <u>13</u>, 1 (1938).
7. F.C. Frank and J.H. Van Der Merwe, Proc. Roy. Soc. <u>198</u>, 205 (1949).

8. K. Nakanishi (private communication).
9. D.E. Moncton, J.D. Axe and F.J. DiSalvo, Phys. Rev. Lett. 34,
    734 (1975).
10. K. Nakanishi, H. Takatera, Y. Yamada and H. Shiba (preprint).

QUESTIONS AND COMMENTS

D. Jérome: Is P the high temperature value?

T.M. Rice: Yes, the high temperature value is P. We simply
                 assume that with pressure you can vary this value.

C.W. Chu: Could you explain why you expect that the pressure
                 coefficient for the T_{NI} is much smaller than T_{CI} or
                 do you just assume that?

T.M. Rice: I just said that we label that axis P, but the
                 phase diagram is in terms of two parameters in the
                 free energy model. It was not meant to be a phase
                 diagram of actual pressure because the pressure will
                 change other variables as well. This was meant to
                 be an illustration of how by varying this parameter
                 with pressure you can move these phase boundaries
                 around. To determine in an actual crystal, the
                 actual pressure derivatives of the two phase bound-
                 aries, one will have to include the pressure de-
                 pendence of other terms which I've ignored.

C.W. Chu: Yes, because the incommensurate-normal transition
                 depends on pressure just as sensitively as the
                 others, then it seems to me the amplitude fluctu-
                 ates only slightly. At low temperature the fluc-
                 tuation terms become unimportant.

T.M. Rice: But there are cases where you could really sep-
                 arate the two. I mean, for example chromium, you
                 know that the commensurate-incommensurate boundary
                 varies much more rapidly with pressure than the
                 other boundaries.

THE CONFERENCE: ONE MAN'S VIEW

Bernd T. Matthias

Institute for Pure and Applied Physical Sciences*

University of California, San Diego,
La Jolla, California 92093

and

Bell Laboratories
Murray Hill, New Jersey 07974

Well, I have nothing to project, only to reflect. Obviously, I'm not going to rehash the whole Conference, I'll just give you my opinion. A conference reflects the mood of its scientific societies and the establishment. They act as a kind of gauge of the atmosphere in the field. From this point of view, this Conference was quite a revelation. First of all, it was an extremely good Conference, I thought, probably also because it was small. Secondly, while I hate to bring politics into this, yet in my opinion the Conference was distinguished by a certain post-Watergate candor. People were really very candid and many were quite willing to admit their ignorance or inability to explain their data. They simply cited their results without any implications for theories. So, of course, I found that very refreshing.

Had I organized this Conference (I mean, all by myself) I would have made one change. I would have prevented the use of the word one-dimensional (I hate the word one-dimensional because there is nothing one-dimensional to begin with). I would have put the

*Research in La Jolla sponsored by the Air Force Office of Scientific Research, Air Force Systems Command, USAF, under AFOSR contract #F49620-77-C-0009.

whole TTF-TCNQ affair at the beginning, and at the end I would have
put the metallic hydrogen, in order not to have the Conference end
with non-superconducting TTF-TCNQ, but with an upbeat note for
metallic hydrogen. Both of them have something in common as you
probably know; they are both promises. Only, TTF-TCNQ is a broken
promise, and metallic hydrogen is still a possibility. All you
have to do is look inside of Jupiter and you know that it exists.
However, I wondered what one could compare TTF-TCNQ to. Then I
decided that the best comparison I could think of (I'm sure you can
think of a better one) is if somebody from a different star came on
Earth and decided he wanted to buy something (he had a lot of money).
So, he looked through advertisements and saw a Ford advertised that
was parked in front of a huge castle. It was on enormous grounds,
an estate with everything behind it, nice people and all that. So
the visitor buys it. Soon he realizes, however, that all he gets
is the Ford. There is nothing else. No castle-- He is stuck with
what he has, the Ford to drive around. TTF-TCNQ is pretty much
that.

Four and a half years ago, I was in San Diego when TTF-TCNQ
was still superconducting, and here at this Conference many uses
or theories or applications of TTF-TCNQ have been shown. It would
be a good polarizer if it were transparent, it would be a good
dielectric if it wouldn't conduct, it would be a good metal if it
had some conductivity--all of these. One thing where TTF-TCNQ is
better than anything else, is as a bait. To collect money, TTF-
TCNQ for the last four years has been absolutely infallible.

Now, where are we today? The only thing TTF-TCNQ has in common with metallic hydrogen (which in my opinion is still sort of
a certain future though not a very bright one, but still it's a
future you know--at least it will be a metal) is that both appeal
to a very human instinct which must underline all of this. Even
though neither at the beginning of the Conference nor at the end
has anybody hardly mentioned it, but what is in everyone's mind
is superconductivity at high temperatures, superconductivity at
room temperature, superconductivity at temperatures where you
would get contracts and fortunes. And this was the motivation
for TTF-TCNQ, and actually even much earlier, metallic hydrogen.

Why have people dreamed so long of metallic hydrogen? Well,
there are various reasons, but, the official explanation, at least
for metallic hydrogen, is that it is very interesting. That's true.
It may be metastable. Well, I heard the man's talk, and I was very
much intrigued by coating the surface in order to remove the instability. Only, I hope I will be far away when people start coating! However, it is a very interesting thing. Yet, nobody in his
right mind could ever believe that this would be a superconductor
at room temperature. That is completely out of the question. It

will be superconducting if you believe the theory or if you apply
extrapolations. And, it probably will be superconducting around
20 degrees. No better--no worse--than all the rest that we have.

As far as TTF-TCNQ is concerned, one might say it is interesting. There, I can sincerely say that this is a hypocritical statement. Anything that doesn't conduct at one degree just isn't a
metal. All these organic conducting materials (I don't even want
to call them conducting) have been with us since 1968. And, today,
they are ferroelectric as I had predicted in 1968 and which was
verified by Buravov in a very beautiful work shortly afterwards.
You also always hear about the Peierls transition. You should read
Physica Status Solidi (1972), page 1. There Shegolev published
everything that was to be known on TCNQ addition compounds and it
is a very beautiful work. He includes Peierls instabilities, everything is right there including ferroelectricity. But, of course,
nobody paid the slightest bit of attention simply because Shegolev
never mentioned the word superconductivity. So, no one cared. What
I am trying to say to you is that this Conference began and ended
with the underlying thought-- superconductivity at high temperature.
It was such an open and honest Conference because nobody ever mentioned these underlying motives. But mind you, they stay with us.

Now, as I'm sure you know, neither of these approaches to superconductivity has any chance whatsoever. One might say, "What about
Vereshagin's data, which were not presented at this Conference, on
non-metal metal transitions?" I don't know if you have read his
latest papers which were published where sodium chloride, ice,
silicon dioxide, magnesium oxide, just about everything became metallic, not superconducting, metallic and Kuwai's data seem to support it. Well, some of my best friends are theorists and some of
them I believe, and while I don't think they could get you a very
accurate statement, they are certainly correct within two orders
of magnitude. David Liberman at Los Alamos has calculated where
these non-metal metal transitions take place. Recorded values are
between 1 and 2 megabars. Practically nothing will become metallic at these pressures, perhaps iron and xenon, but nothing else.
But, magnesium oxide and sodium chloride as recorded haven't got
a chance of becoming metallic at these pressures because even very
primitive theoretical calculations indicate that you cannot rely on
these published experiments. They have shown that there was no
chance for a metal transition anywhere below 50 megabars for most
of these ionic crystals. So, I don't think that any of the data
that we have today on metallic hydrogen gives you any great hopes
for the future as far as superconductivity is concerned. However,
this is now the beginning of the end of the Conference as I say,
and I think that it should have been reversed.

Now let me come to the Conference itself which I thought was
really an absolutely excellent Conference, particularly with its

openness and candor. Now, as I said before, I'm giving you my personal opinion which I'm sure will not agree with all of you, maybe with none of you. But, that doesn't matter, because I'll tell you what really impressed me most with this Conference was its frankness and honesty. For instance, yesterday, the people who are working on materials, A-15's which are very crucial in our technology made a show of data, in particular, Fred Smith concerning the expansion of V_3Si. I found it stunning. What was stunning also was his honesty. He said, "Nothing fits." How often have you gone to a Conference where someone presents beautiful data and says, "Nothing fits"!

I must say there were a few standard talks at this Conference where people were rehashing old chestnuts which we have known for many years. However, the real tenor of the Conference was, new things which we had not known and new data which were uncluttered by unfounded speculations.

Another thing which impressed me very much, because nobody tried to draw any deep and profound conclusions from it, were the data by Baranowski and Skoskiewicz showing a second maximum in nickel. Usually magnetic materials aren't exactly the right thing for superconductors and yet these palladium hydrides, which in my opinion are amongst the most interesting things right now, are also one of the newest things in superconductivity. That the nickel should give a second and such extreme maximum, I found to be a really very intriguing thing.

Schirber had some really beautiful data. However, he stated one phrase which baffled me because I had never really thought about it. He said, ..."the conscience of the theorists..." He had thought of it! So, let me quickly come back to TTF-TCNQ.

Four and a half years ago at the APS March Meeting in San Diego there were quite a number of very good theorists. Everyone said to me, "My God, the superconductivity of TTF-TCNQ is the greatest development in Solid State Physics. It's true, you may not like it, but I assure you it is true. It has been predicted by the theory. You realize that that is perfectly correct; that the theory had predicted a linear organic superconductor for almost two decades". It is the theory, therefore, that bears a certain amount of responsibility. However, now the experimentalists really have caught up with them. If you have to look at bad experimentalists today, you'll find worse ones than you'll find among bad theorists. Except perhaps for Anderson's statement about band calculations (which is a contradiction in itself--either you calculate something legitimately or you make a band calculation, never the two shall meet).

Now and then, old data, old conclusions were dressed up as

new ones, but this didn't really bother the Conference because so
many things were radically new. For instance, one thing that I was
intrigued by, and it had nothing to do with local patriotism, were
the papers by Samara and Peercy. I had not realized the radical
difference between normal potassium dihydrogen phosphate and the
deuterated salt. It all fits so beautifully that even though perhaps some had been known before, it is a good thing that we have
been reminded of it. That is what intrigued me very much in Paul's
(Chu) talk. Very often, the history of a scientific development
is omitted. But Paul made it very clear, though he didn't go back
quite far enough. Instabilities of high temperature superconductors
are a long story. In fact, it dates back to 1950 except it wasn't
published. At Purdue, Frölich presented his first idea about the
mechanism of lattice electron interaction which was not quite like
the ones today, but it was a lattice electron interaction. He then
gave me the manuscript. I took it to Gregor Wentzel who is one of
the great theorists. He looked at it and said, "I am afraid that
is an unstable lattice. That means superconductivity should never
have happened at all, certainly no high temperature conductivity.
This is very depressing." Maria Mayer and Walter Kohn soon came
to the same conclusion. That is why it was so good that Fong reminded us here that very few things are radically new. They have
been with us, only we seem to forget them. I think it would be
much better if now and then one would refer to things done previously in one's presentation and discuss the development. Because,
results don't really come out of thin air! There is usually a
tradition on which things are built.

Now, I'm not going to bore you any longer. All I wanted to
say is that I thought it was an extremely refreshing Conference.
I was so intrigued yesterday by the last talk (McGervey) that one
can really deal with <u>real</u> <u>metals</u>. One can see what's so wrong about
a metal with his positrons. I thought it was an extremely elegant
technique.

So, to sum it up, I think it was a very, very good Conference
in all ways.

QUESTIONS AND COMMENTS

M. Weger: Did you say HMTSF-TCNQ is not a metal?

B.T. Matthias: No, TTF-TCNQ is not a metal at 1 K. HMTSF is the
                   only thing. Now, if you really want to go into
                   details, I'll tell you. I talked to Calvin and
                   I said, "Tell me, if you grow these things (they're
                   all black and sort of weird), what happens if some
                   of the solvent is incorporated into the crystal?"

He said, "It will conduct." Now, with the exception of HMTSF all of these are not even conducting at 1 K. And I think HMTSF is conducting because it's dirty. If you have less than one percent of the solvent in it, it will conduct. You mean you can guarantee there is not one percent of the solvent?

M. Weger: First of all there _is_ a guarantee; secondly a poor quality sample is not conducting.

B.T. Matthias: Anyhow, what I will say now is that all these things (maybe with the exception of HMTSF, which of course doesn't become superconducting), fine tuned or otherwise, when you cool them down, certainly are insulators again. Now, I have a very old-fashioned approach, the metal is something that conducts at low temperatures, and so, to call them organic metals is, of course, not the right way to call it. To call them one-dimensional conductors is not the right way to call them because the one-dimension argument would never give an X-ray diffraction, by definition. Jérome doesn't agree with me.

D. Jérome: What about diffuse streaks observed in quasi-one-dimensional conductors?

B.T. Matthias: I saw the elastic constants. How can you have three different elastic constants for a one-dimensional affair? That would require quite some mental gymnastics, wouldn't it?

International Conference on

HIGH PRESSURE AND LOW TEMPERATURE PHYSICS

July 20 - 22, 1977

ATTENDEES

Achar, B.N.N.
105 Materials Research Lab
Pennsylvania State University
University Park, PA 16802

Anderson, J.R.
Department of Physics
University of Maryland
College Park, Maryland 20742

Aron, P.R.
NASA-Lewis Research Center
21000 Brookpark Rd.
Cleveland, Ohio 44135

Baranowski, B.
Institute of Physical Chemistry
Polish Academy of Sciences
Warsaw, Kasprzaka 44/52
Poland

Barsch, G.R.
107 Materials Research Lab.
Pennsylvania State University
University Park, PA 16802

Bauer, H.J.
Physics Dept., Univ. of Munich
D-8000 Munich 40
Schellingsstrasse 4/III
W. Germany

Bloch, D.
Lab. de Magnetisme, CNRS
BP 166, Grenoble
Cedex 38062
Grenoble, France

Brown, G.V.
NASA-Lewis Research Center
21000 Brookpark Rd.
Cleveland, Ohio 44135

Chang, Z.P.
104 Materials Research Lab.
Pennsylvania State University
University Park, PA 16802

Chu, C.W.
Dept. of Physics
Cleveland State University
Cleveland, Ohio 44115

Clark, J.B.
National Physical Research Lab.
P.O. Box 395, Pretoria
South Africa 0001

Cordero, C.
Depto. de Fisicay Matematicas
Colegio Universitario de Cayey (UPR)
Cayey, Puerto Rico
(presently at PSU)

Crummett, Wm. P.
Solid State Division
Oak Ridge National Lab.
Oak Ridge, Tenn. 37830

Debray, D.
DPH-SRM, C.E.N. de Saclay, B.P. no.2
Gif-sur-Yvette
France 91190

Delong, L.
University of California, San Diego
Mail Code B-019
La Jolla, California 92093

589

Dietrich, M.
Kernforschungszentrum
Karlsruhe
W. Germany

Eichler, A.
Inst. f. Techn. Physik
Mendelssohnstr. 1 B
3300 Braunschweig
W. Germany

Finnemore, D.K.
Department of Physics
Iowa State University
Ames, Iowa   50011

Fontanella, J.
Physics Dept.
U.S. Naval Academy
Annapolis, Maryland   21402

Frankel, J.
Bldg. #115
Watervliet Arsenal
Watervliet, New York   12189

Gordon, Wm. L.
Dept. of Physics
Case Western Reserve University
Cleveland, Ohio   44106

Guertin, R.P.
Dept. of Physics
Tufts University
Medford, Massachusetts   02155

Hambourger, P.D.
Physics Department
Cleveland State University
Cleveland, Ohio   44115

Hammerberg, J.E.
Dept. of Physics
Simon Fraser University
Burnaby, B.C.
Canada   V5A 1S6

Hein, R.A.
Dept. of the Navy
Naval Research Lab.
Washington, D.C.   20375

Hochheimer, H.D.
Fachbereich Physik der Universität
D-8400 Regensburg
Universitaetsstr.   31
W. Germany

Holzapfel, W.B.
Max-Planck-Institut
f. Festkoerperforschung
7000 Stuttgart 80
Busnauer Str. 171
W. Germany

Homan, C.
Watervliet Arsenal
Watervliet, New York   12189

Huang, C.Y.
Mail Stop 764
Los Alamos Scientific Lab.
Los Alamos, New Mexico   87545

Huang, S.
Dept. of Physics
Cleveland State University
Cleveland, Ohio   44115

Jérome, D.
Université de Paris - SUD
Physique des Solides
Batiment 510, 91 Orsay
France

Johnston, D.C.
Physics Dept.
University of California, San Diego
La Jolla, California   92093

Kaplan, T.A.
Dept. of Physics
Michigan State University
East Lansing, Michigan   48824

Lowrey, W.H.
Code 6434- Naval Research Lab.
Washington, D.C.   20375

Mahanti, S.D.
Dept. of Physics
Michigan State University
East Lansing, Michigan   48824

Maple, M.B.
Dept. of Physics
Univ. of California, San Diego
La Jolla, California   92093

Matthias, B.
Dept. of Physics
Univ. of California, San Diego
La Jolla, California   92093

McGervey, J.
Case Western Reserve University
Cleveland, Ohio   44106

McMahan, A.K.
Lawrence Livermore Lab., L-450
University of California, Box 808
Livermore, California   94550

McWhan, D.B.
Bell Laboratories
Murray Hill, New Jersey   07974

Minomura, S.
University of Tokyo
Solid State Physics
Roppongi, Minato-ku
Tokyo, Japan   106

Moodenbaugh, A.R.
Dept. of Physics
Illinois Inst. of Technology
Chicago, Illinois   60616

Mulay, L.N.
139 Materials Research Lab.
Pennsylvania State University
University Park, PA   16802

Murray, W.K.
3330 W. 100 St.
Cleveland, Ohio   44111

Okai, B.
National Institute for Researches
in Inorganic Materials
Sakura-mura, Niiharigun, Ibaraki
Japan   300-31

Onodera, A.
Faculty of Engineering Science
Osaka University
Toyonaka, Osaka
560, Japan

Ott, H.R.
Lab. f. Festkoerperphysik, ETH
8093 Zuerich, Switzerland

Oyanagi, H.
Institute for Solid State Physics
Tokyo University
Roppongi, Minato-ku
Tokyo 106, Japan

Peercy, P.S.
Division 5112, Sandia Labs.
Albuquerque, Mexico   87115

Phillips, N.E.
Dept. of Chemistry
University of California
Berkeley, California   94720

Porowski, S.
High Pressure Research Center
Polish Academy of Sciences
Warsaw, ul. Sokolowska 29
Poland

Probst, Chr.
Zentralinstitut f. Tieftemperatur-
forschung
D-8046 Garching, Hochschulgelaende
W. Germany

Rachford, F.J.
Dept. of Physics
University of Maryland
College Park, Maryland 20742

Rice, T.M.
Bell Laboratories
Murray Hill, New Jersey 07974

Ruoff, A.L.
Material Science and Engineering
Cornell University
Ithaca, New York 14853

Samara, G.A.
Sandia Laboratories
Dept. 5130
Albuquerque, New Mexico 87115

Satterthwaite, C.B.
Dept. of Physics
University of Illinois
Urbana, Illinois 61801

Saur, E.J.
Inst. Angewandte Physik
Universitaet Giessen
Leihgesterner Weg 106
W. Germany

Schilling, J.
Exp. Phys. IV
U. Bochum
4630 Bochum
Postfach 2148, W. Germany

Schirber, J.E.
Sandia Laboratories
Department 5150
Albuquerque, New Mexico 87115

Schmidt, V.H.
Dept. of Physics
Montana State University
Bozeman, Montana 59715

Shelton, R.N.
Dept. of Physics, B-019
Univ. of California, San Diego
La Jolla, California 92093

Skelton, E.F.
Naval Research Lab., Code 6434
Washington, D.C. 20375

Skoskiewicz, T.
Institute of Physical Chemistry
Polish Academy of Sciences
01-224 Warsaw, Kasprzaka 44/52
Poland

Skove, M.J.
Dept. of Physics
Clemson University
Clemson, SC 29631

Smith, T.F.
Dept. of Physics
Monash University
Clayton, Victoria 3168 Australia

Sniadower, I.
Institute of Physics
Polish Academy of Sciences
02-668 Warsaw, al. Lotnikow 32/46
Poland

Spain, I.L.
Dept. of Chemical Engineering
University of Maryland
College Park, Maryland 20742

Suski, T.
High Pressure Research Center
Polish Academy of Sciences
Warsaw, ul. Sokolowska 29
Poland

Ting, C.S.
Dept. of Physics
University of Houston
Houston, Texas 77004

Tomasch, W.J.
Dept. of Physics
University of Notre Dame
Notre Dame, Indiana 46556

Vettier, Chr.
Laboratoire Louis NeeP - CNRS
Grenoble, France 38042

Viswanathan, R.
Physics Dept.
Brookhaven National Lab.
Upton, N.Y. 11973

Weger, M.
Racah Institute of Physics
Hebrew University of Jerusalem
Israel

Wittig, J.
Institut f. Festkoerperforschung
Kernforschungsanlage
D-5170 Juelich
W. Germany

Woollam, J.
NASA-Lewis Research Center
21000 Brookpark Rd.
Cleveland, Ohio 44135

Wortmann, G.H.
Argonne National Lab.
Argonne, Illinois 60439

1. H.R. Ott; 2. S. Huang; 3. P.D. Hambourger; 4. C.Y. Huang; 5. J. Wittig; 6. D.K. Finnemore;
7. W.J. Tomasch; 8. B.N.N. Achar; 9. D. Debray; 10. T. Skoskiewicz; 11. C.B. Satterthwaite;
12. D.B. McWhan; 13. B. Matthias; 14. J.R. Anderson; 15. D. Bloch; 16. T.F. Smith;
17. W.B. Holzapfel; 18. B. Baranowski; 19. A.K. McMahan; 20. S. Minomura; 21. Z.P. Chang;
22. J.E. Schirber; 23. R.N. Shelton; 24. E.F. Skelton; 25. J. Schilling; 26. Wm.P. Crummett;
27. G.H. Wortmann; 28. P.S. Peercy; 29. M. Weger; 30. G.V. Brown; 31. D.C. Johnston;
32. A.R. Moodenbaugh; 33. R. Viswanathan; 34. M. Dietrich; 35. A. Onodera; 36. H. Oyanagi;
37. S.D. Mahanti; 38. T.A. Kaplan; 39. C. Cordero; 40. C.W. Chu; 41. J. Woollam;
42. M.B. Maple; 43. G.A. Samara; 44. C.S. Ting; 45. Chr. Probst; 46. J. Fontanella;
47. B. Okai; 48. T. Suski; 49. C. Homan; 50. V.H. Schmidt; 51. J. Frankel; 52. S. Porowski;
53. E.J. Saur; 54. M.J. Skove; 55. J.E. Hammerberg; 56. T.M. Rice; 57. N.E. Phillips;
58. A. Eichler; 59. J.B. Clark; 60. R.A. Hein; 61. R.P. Guertin; 62. W.H. Lowrey;
63. I.L. Spain; 64. F.J. Rachford; 65. P.R. Aron; 66. G.R. Barsch.

INDEX